INTRODUCTION TO
BASIC ELECTRICITY AND
ELECTRONICS TECHNOLOGY

INTRODUCTION TO
BASIC ELECTRICITY AND
ELECTRONICS TECHNOLOGY

EARL GATES

DELMAR
CENGAGE Learning·

Australia • Brazil • Japan • Korea • Mexico • Singapore • Spain • United Kingdom • United States

Introduction to Basic Electricity and Electronics Technology
Earl Gates

Vice President, Careers and Computing:
Dave Garza

Director of Learning Solutions: Sandy Clark

Acquisitions Editor: Stacy Masucci

Director, Development - Careers and Computing:
Marah Bellegarde

Managing Editor: Larry Main

Product Manager: Mary Clyne

Editorial Assistant: Kaitlin Schlicht

Brand Manager: Kristin McNary

Market Development Manager: Erin Brennan

Senior Production Director: Wendy Troeger

Production Manager: Mark Bernard

Content Project Manager: Barbara LeFleur

Senior Art Director: David Arsenault

Media Editor: Deborah Bordeaux

Cover image(s): © Dariush M./
www.Shutterstock.com

For product information and technology assistance, contact us at
Cengage Learning Customer & Sales Support, 1-800-354-9706
For permission to use material from this text or product,
submit all requests online at **www.cengage.com/permissions**.
Further permissions questions can be e-mailed to
permissionrequest@cengage.com

Library of Congress Control Number: 2012949451

ISBN-13: 978-1-133-94851-3

ISBN-10: 1-133-94851-0

Delmar
5 Maxwell Drive
Clifton Park, NY 12065-2919
USA

Cengage Learning is a leading provider of customized learning solutions with office locations around the globe, including Singapore, the United Kingdom, Australia, Mexico, Brazil, and Japan. Locate your local office at:
international.cengage.com/region

Cengage Learning products are represented in Canada by Nelson Education, Ltd.

To learn more about Delmar, visit **www.cengage.com/delmar**

Purchase any of our products at your local college store or at our preferred on-line store **www.cengagebrain.com**

Notice to the Reader
Publisher does not warrant or guarantee any of the products described herein or perform any independent analysis in connection with any of the product information contained herein. Publisher does not assume, and expressly disclaims, any obligation to obtain and include information other than that provided to it by the manufacturer. The reader is expressly warned to consider and adopt all safety precautions that might be indicated by the activities described herein and to avoid all potential hazards. By following the instructions contained herein, the reader willingly assumes all risks in connection with such instructions. The publisher makes no representations or warranties of any kind, including but not limited to, the warranties of fitness for particular purpose or merchantability, nor are any such representations implied with respect to the material set forth herein, and the publisher takes no responsibility with respect to such material. The publisher shall not be liable for any special, consequential, or exemplary damages resulting, in whole or part, from the readers' use of, or reliance upon, this material.

Printed in the United States of America
1 2 3 4 5 6 7 17 16 15 14 13

CONTENTS

PREFACE

INTENDED AUDIENCE

Introduction to Basic Electricity and Electronics Technology is written to meet the needs of a one-year program for electronics in high schools, vocational schools, and technical colleges, and as an introduction to electricity and electronic in four-year colleges. This textbook may also be used in a survey course in electricity and electronics for electronics technology, computer technology, and telecommunications. This first edition is designed to give students the basic background that more closely relates to the needs of industry. With guided instruction it can provide the hands-on skills required by industry.

BACKGROUND OF THIS BOOK

This first edition objective is to provide a text and reference book that summarizes in understandable terms those principles and techniques that are the basic tools of electronics. In keeping with current trends, increased emphasis is placed on the general techniques of electronics.

During my teaching in public school I completed a study on what industry wanted from students graduating with a background in electronics. I found that industry valued students' ability to do more than their ability to know. I found that industry wanted less time spent on teaching theory and more time spent on instructing hands-on applications.

After I had rewritten my curriculum, I found I had to use several textbooks to teach it. This textbook is intended to provide the students with all the information required by the curriculum in one easy-to-use textbook. The curriculum I used is included in the Instructor's Guide.

TEXTBOOK ORGANIZATION

Due to the rapid growth of electronics, it becomes impossible to cover all of the important topics in a one-year course. *Introduction to Basic Electricity and Electronics Technology* provides instructors with an opportunity to select those topics they wish to emphasize and at the same time provides the student with a reference book of basic electricity and electronics coverage and continuing value.

Teachers can guide students to concentrate on the material related to a particular course syllabus, leaving the remaining subject matter as enrichment should students wish to extend their knowledge. Alternatively, instructors can choose to cover a series of selected topics, such as DC and AC circuits.

Another possibility is to concentrate on the material related primarily to linear electronics circuits or another topic of choice. Many other combinations are possible.

The emphasis is the coverage of electronics combined with a presentation that allows the student to study a particular topic without having to read the entire text.

The textbook is divided into seven separate sections.

SECTION 1—Introduction to Electricity and Electronics discusses careers in electricity and electronics, certification in electricity and electronics, work habits and issues, calculators for electricity and electronics, electronic circuit design, software for electronics, safety, tools and equipment, and hazardous materials.

SECTION 2—DC Circuits discusses fundamentals of electricity, current, voltage, resistance, Ohm's law, electrical measurements meters, power, DC circuits, magnetism, inductance, and capacitance.

SECTION 3—AC Circuits discusses alternating current, AC measurement, resistive AC circuits, capacitive AC circuits, inductive AC circuits, resonance circuits, and transformers.

SECTION 4—Semiconductor Devices discusses semiconductor fundamentals, P–N junction diodes, zener diodes, bipolar transistors, field effect transistors (FET), thyristors, integrated circuits, and optoelectric devices.

SECTION 5—Linear Electronic Circuits discusses power supplies, amplifier basics, amplifier applications, oscillators, and waveshaping circuits.

SECTION 6—Digital Electronic Circuits discusses binary number systems, basic logic gates, simplifying logic circuits, sequential logic circuits, combinational logic circuits, and microcomputer basics.

SECTION 7—Practical Applications discusses project design, printed circuit board fabrication, printed circuit board assembly and repair, and basic troubleshooting.

GLOSSARY—This valuable resource contains key terms and definitions.

Appendices—Included are the following:

APPENDIX 1—Periodic Table of Elements
APPENDIX 2—The Greek Alphabet
APPENDIX 3—Metric Prefixes Used in Electronics
APPENDIX 4—Electronics Abbreviations
APPENDIX 5—Common Reference Designators
APPENDIX 6—DC and AC Circuit Formulas
APPENDIX 7—Formula Shortcuts
APPENDIX 8—Resistor Color Codes
APPENDIX 9—Common Resistor Values
APPENDIX 10—Capacitor Color Code
APPENDIX 11—Electronics Symbols
APPENDIX 12—Semiconductor Schematic Symbols
APPENDIX 13—Digital Logic Symbols
APPENDIX 14—Activity Assessment Rubric
APPENDIX 15—Safety Test
Self-Test Answers are included for students.

FEATURES

The following list provides some of the significant features of the textbook:
- Chapters are brief and focused.
- Objectives are clearly stated with the learning goals at the beginning of each chapter.
- Colorful illustrations are generously used throughout the text to strengthen concepts learned.
- Four-color photographs are used to show the learner exactly what is addressed in the text.
- Four-color layouts focus attention to important features in the text.

- Cautions and notes are color coded for easy identification throughout the text.
- Many examples have been developed into Multisim version 12 to learn firsthand what is happening in a circuit.
- Review questions appear at the end of every chapter subdivision to allow a comprehension check.
- All formulas are written using fundamental formulas only.
- Many examples show math and formulas in use throughout the text.
- Summaries are included at the end of each chapter for reviewing important concepts.
- Self-tests are included at the end of each chapter as a learning tool.
- Numerous examples incorporate the chapters' material with real-life applications.
- Section activities provide an opportunity to reinforce concepts with hands-on problem-solving projects.

THE LEARNING PACKAGE

A robust complement of supplemental learning solutions has been developed to achieve two goals:
1. To assist students in learning the essential information needed to prepare for the exciting field of electronics
2. To assist instructors in planning and implementing their instructional programs for the most efficient use of time and other resources

The following components of this package are offered to help you achieve these goals:

Lab Manual

Labs provide students with the opportunity to learn the terminology and transfer theory provided in class to hands-on practical applications. Projects serve to reinforce the students' learning, providing opportunities to see theory become practice. (ISBN: 978-1-1339-4852-0)

CourseMate for Basic Electricity and Electronics

CourseMate for Basic Electricity and Electronics is an integrated, Web-based learning solution for building the knowledge and skills needed to succeed in electricity and electronics. The CourseMate Website includes the following suite of resources and study tools:

- An interactive eBook with highlighting, note-taking, and search capabilities.
- Interactive quizzes
- A set of student-focused PowerPoint slides
- Flashcards, crosswords, and other skill-building games
- The CourseMate website also includes Multisim™ circuit files. Students can use these precreated files for troubleshooting and simulation. Textbook figures created as Multisim files are identified by a Multisim icon throughout the text.

HOW TO ACCESS THE *COURSEMATE FOR BASIC ELECTRICITY AND ELECTRONICS* SITE

To access these supplemental materials or to see a CourseMate demo, please visit **http://www.cengagebrain.com**. At the CengageBrain.com homepage, enter the ISBN of your title (from the back cover of this book), using the search box at the top of the page. This will take you to the product page where these resources can be found.

INSTRUCTOR'S COMPANION WEBSITE. This educational resource creates a truly electronic classroom with array of tools and instructional resources that will enrich the learning experience and expedite your preparation time. The instructor resources directly correlate with the text, so the text and website combine to provide a unified instructional system. With the following instructor resources, you can spend more time teaching and less time preparing to teach:

INSTRUCTOR'S GUIDE. The Instructor's Guide contains solutions to textbook section questions and to the lab manual experiments. To assist the instructor or teacher in preparing the program, a curriculum guide is provided in the Instructor's Guide. It helps instructors to provide a program that will develop a student's interest in the field of electronics.

POWERPOINT PRESENTATION. Slides cover every chapter in the text, providing the basis for a lecture outline that helps present concepts and material. Key points and concepts can be graphically highlighted for student retention.

COMPUTERIZED TESTBANK. The testbank includes questions in multiple-choice format, so students' comprehension can be assessed.

IMAGE LIBRARY. More than 200 images from the textbook are included to create transparency masters or to customize PowerPoint slides. The Image Library comes with the ability to browse and search images with key words and allows quick and easy use.

ABOUT THE AUTHOR

- Retired as an Associate Professor from the State University of New York at Oswego, where he taught electronics technology
- Has 23 years experience in public education as a teacher and administrator
- Retired from the U.S. Navy as an Electronics Technician Senior Chief
- Recently, as an educator he taught VoTech at Choices Charter School of Florence and Darlington in Florence, South Carolina
- President of TEK Prep, a small business that does education consulting, training, and evaluation
- As an education consultant, provides training for teachers and adults in Florida, New York, and South Carolina

ACKNOWLEDGMENTS

I would like to thank John Millhouse, a retired Navy Chief Electronics Technician who served with me in the U.S. Navy. He has retired and now works as a consultant electronics engineer in Florida. He helped with the Multisim examples and sample problems used throughout the text. Thanks also to Lois Dodge for her help in proofing the manuscript.

I would also like to thank Avi Hadar, owner of Kelvin, the exclusive dealer in the United States for New Wave Concepts' Circuit Wizard, for his support; Jim Good, who helped shape the concept of the book when we were at Greece Central School; and Gerald Buss, retired president of EIC Electronics, who provided help by reading the book, answering questions, and lending his support from the industrial sector.

Thanks are also due to the numerous teachers who identified areas to include, expand, or improve in the textbook. Thanks to the staff at Delmar, Cengage Learning for their faith that the book could be done.

The author and Delmar, Cengage Learning wish to thank the reviewers for their suggestions and comments during development of this edition.

Finally, I would like to thank my wife, Shirley, who has supported me in the preparation and development of this textbook.

Earl D. Gates
Cincinnatus, New York
2013

SECTION 1

INTRODUCTION TO ELECTRICITY AND ELECTRONICS

PEOPLE AND EVENTS IN ELECTRONICS

Age Discrimination in Employment Act (ADEA) of 1967 (December 15, 1967)

United States Public Law 90-202, 81 Statue 602 prohibits employment discrimination against persons 40 years or older. It also applied to standards for pensions and benefits provided by employers to be provided to the general public. The ADEA was later amended in 1986 and again in 1991.

American Disabilities Act (ADA) of 1990 (July 26, 1990)

Title I prohibits private employers, state and local governments, employment agencies, and labor unions from discriminating against qualified individuals with disabilities.

Earle C. Anthony (1880–1961)

Anthony is credited with the invention of the electronics breadboard; in 1922, he created a circuit using an actual breadboard on his kitchen table.

William Seward Burroughs (1857–1898)

Burroughs invented the first practical adding and listing machine and submitted a patent application for the machine in 1885.

Canadian Standards Association (CSA) (1944–)

CSA is a global organization that provides product testing and certification for products that conform to CSA standards.

Certified Electronics Technician (CET) Program (1965–)

The CET program was designed by the National Electronics Association to measure the theoretical knowledge and technical proficiency of practicing electronics technicians.

Civil Rights Act of 1964 (Enacted July 2, 1964)

United States Public Law 88-352, (78 Statue 241) outlawed all major forms of discrimination against racial, ethnic, national, and religious minorities and women.

Electronics Technicians Association, International (ETA-I) (1978–)

ETA-I is a professional association that promotes excellence through testing and certifying electronics technicians.

James Hodgson (1915–)

In 1970, Hodgson helped to craft the Occupational Safety and Health Act, which paved the way for the OSHA agency, established in 1971.

International Electrotechnical Commission (IEC) (1906–)

IEC is an international standards organization that prepares and publishes international standards for all electrical, electronic, and related technologies.

International Society of Certified Electronics Technicians (ISCET) (1975–)

ISCET is a professional organization whose main function is the direction and administration of the CET program; in addition, it promotes obtaining technical certification worldwide and provides a location for certified technicians to band together for professional advancement.

Jan Lukasiewicz (1878–1956)

Lukasiewicz invented reverse Polish notation around 1920.

National Electrical Code (NEC) (1897–)

The *NEC* is a U.S. standards guide that is regionally adopted for the safe installation of electrical wiring and equipment and is published by the National Fire Protection Association (NFPA).

OSHA (1970–)

In 1986, OSHA began requiring material safety data sheets (MSDSs) for all hazardous materials.

Ronald J. Portugal

In 1971, Portugal invented the solderless breadboard while working for EI Instruments.

Résumé (1066–)

The *résumé* originated in feudal England as a document created by a Lord or the head of a local guild to serve as a letter of introduction for someone traveling long distances.

Underwriters Laboratories (UL) (1894–)

UL develops standards and test procedures for products, materials, components, assemblies, tools, and equipment, for product safety.

Careers in Electricity and Electronics

OBJECTIVES

After completing this chapter, the student will be able to:

■ Identify careers in the electrical and electronics field.
■ Describe the difference between work done by an electronics engineer and by an electronics technician.
■ Identify the traits for successful employment.
■ Describe the parts of a basic résumé.
■ Discuss the purpose of the résumé and the letter of application.
■ Identify how experience in the workforce can be obtained.
■ Describe the importance of an apprenticeship.
■ Explain how a co-op program functions.
■ Discuss the differences among an externship, an internship, and job shadowing.
■ Identify the role a mentoring program plays in selecting a chosen career field.

KEY TERMS

1-1 automation mechanic
1-1 automotive mechanic
1-1 computer engineer
1-1 computer technician
1-1 electrical engineer
1-1 electrician
1-1 electronics engineer
1-1 *National Electrical Code® (NEC®)*
1-1 *NEC* Guidebook
1-2 electronics engineering
1-2 electronics technician
1-2 Federal Communications Commission (FCC)

1-2 Institute of Electrical and Electronics Engineers (IEEE)
1-3 *Dictionary of Occupational Titles (DOT)*
1-3 job interview
1-3 letter of application
1-3 *Occupational Outlook Handbook*
1-3 résumé
1-4 apprenticeship
1-4 cooperative education
1-4 externship
1-4 internship
1-4 job shadowing
1-4 mentoring

This chapter looks at careers in the electrical and electronics industry. It also looks at some useful information for locating a job and developing skills to secure a job, as well as some pointers to keep a job. It is not realistic for an individual to make a career choice after reading this chapter; rather, this chapter is a starting point to help define the process of getting a job.

1-1 CAREERS IN THE ELECTRICAL AND ELECTRONICS FIELD

Many exciting career opportunities exist in the electrical and electronics field. The following pages provide a sample of the available opportunities. Check for other career opportunities at the career information center in your school or community.

Automation Mechanic

An **automation mechanic** maintains controllers, assembly equipment, copying machines, robots, and other automated or computerized devices (Figure 1-1). A person with this job installs, repairs, and services machinery with electrical, mechanical, hydraulic, or pneumatic components. Precision measuring instruments, test equipment, and hand tools are used. Knowledge of electronics and the ability to read wiring diagrams and schematics is required.

Becoming an automation mechanic requires formal training, which is offered by the military, two-year colleges, vocational-technical schools, and in-house

■ FIGURE 1-1

An automation mechanic needs knowledge of electricity and electronics as well as of computers, hydraulics, mechanics, and pneumatics.

© iStockphoto/baranozdemir

apprenticeship programs. Although most training is provided through formal classroom instruction, some may only be obtained through on-the-job training.

Automation mechanic is one of the fastest growing vocations in the industry. This rapid growth is expected to continue annually.

Automotive Mechanic

There are currently more computers aboard today's automobile than aboard our first spaceship. A typical automobile contains approximately 10 to 20 computers that operate everything from the engine and radio to the driver's seat. As a result, **automotive mechanics** now need a greater knowledge of electronics.

To be able to distinguish an electronic malfunction from a mechanical malfunction, automotive mechanics must be familiar with the minimum basic principles of electronics. In addition, they must be able to test and replace electronic components.

Becoming an automotive mechanic requires formal training, which is offered by the military, junior/community colleges, vocational-technical schools, and in-house apprenticeship programs. Although most training is provided through formal classroom instruction, some can only be obtained on the job. To reduce the amount of time invested in training a prospective mechanic, more employers are now looking for people who have completed a formal automotive training program.

Employment opportunities are good for automotive mechanics who have completed an automotive training program and better when certified by the National Institute for Automotive Service Excellence (ASE). People whose training includes basic electronics skills have the best opportunities. Employment growth is expected to increase at a normal rate annually, with a concentration in automobile dealerships, independent automotive repair shops, and specialty car-care chains. Employment in service stations will continue to decline, as fewer stations will offer repair services.

Computer Engineer

The rapid growth in computers has generated a demand for people trained in designing new hardware and software systems and incorporating new technologies into existing and new systems. These trained professionals are known as **computer engineers** and system analysts.

Computer engineers can be further broken down into hardware and software engineers. Computer hardware engineers design, develop, test, and supervise the manufacturing of computer hardware. Computer software engineers design and develop software systems for control and automation of manufacturing, business, and management processes. They also may design and develop software applications for consumer use at home or create custom software applications for clients.

There is no universally accepted preparation for a computer professional because the job often depends on the work that needs to be done. Most employers require that employees have at least a bachelor's degree. However, a passion for computers and proficiency in advanced computer skills will at times win out over a bachelor's degree.

This field is one of the fastest-growing fields today. Technological advances are occurring so rapidly in the computer field that employers are struggling to keep up with the demand for trained professionals. As the technology becomes more sophisticated and complex, more expertise and a higher level of skills will be required. Computer engineers must be willing to continue their learning process to keep up. College graduates with a bachelor's degree in computer science, computer engineering, information science, or information systems will enjoy favorable employment opportunities.

Computer Technician

A **computer technician** installs, maintains, and repairs computer equipment and systems (Figure 1-2). Initially, the computer technician is responsible for laying cables and making equipment connections.

■ FIGURE 1-2

Computer technicians install, maintain, and repair computer equipment and systems.

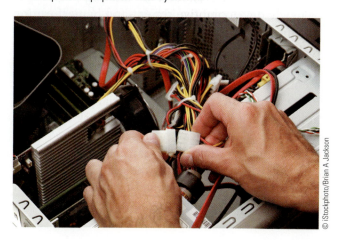

© iStockphoto/Brian A Jackson

This person must thoroughly test the new system(s), resolving all problems before the customer uses the equipment. At regular intervals, the computer technician maintains the equipment to ensure that everything is operating efficiently. Knowledge of basic and specialized test equipment and hand tools is necessary.

Computer technicians spend much of their time working with people—listening to complaints, answering questions, and sometimes offering advice on both equipment system purchases and ways to keep equipment operating efficiently. Experienced computer technicians often train new technicians and sometimes have limited supervisory roles before moving into a supervisory or service managerial position.

A computer technician is required to have one or two years of training in basic electronics or electrical engineering from a junior college, college or vocational training center, or military institution. The computer technician must be able to keep up with all the new hardware and software.

Projections indicate that employment for computer technicians will be high. The nation's economy is expanding, so the need for computer equipment will increase; therefore, more computer technicians will be required to install and maintain equipment. Many job openings for computer technicians may develop from the need to replace technicians who leave the labor force, transfer to other occupations or fields, move into management, or retire.

Electrical Engineer

Electrical engineers make up the largest branch of engineering. An electrical engineer designs new products, writes performance specifications, and develops maintenance requirements. Electrical engineers also test equipment, solve operating problems within a system, and predict how much time a project will require. Then, based on the time estimate, the electrical engineer determines how much the project will cost.

The electrical engineering field is divided into two specialty groups: electrical engineering and electronics engineering. An electrical engineer works in one or more areas of power-generating equipment, power-transmitting equipment, electric motors, machinery control, and lighting and wiring installation. An **electronics engineer** works with electronic equipment associated with radar, computers, communications, and consumer goods.

The number of engineers in demand is expected to increase annually. This projected growth is attributed to an increase in demand for computers, communication equipment, and military equipment. Additional jobs are being created through research and development of new types of industrial robot control systems and aviation electronics. Despite this rapid growth, a majority of openings will result from a need to replace electrical and electronics engineers who leave the labor force, transfer to other occupations or fields, move into management, or retire.

Electrician

An **electrician** may specialize in construction, maintenance, or both. Electricians assemble, install, and maintain heating, lighting, power, air-conditioning, and refrigeration components (Figure 1-3). The work of an electrician is active and sometimes strenuous. An electrician risks injury from electrical shock, falls, and cuts from sharp objects. To decrease the risk of these job-related hazards, an electrician is taught to use protective equipment and clothing to prevent shocks and other injuries. An electrician must adhere to the *National Electrical Code*® (*NEC*®)* specifications and procedures, as well as to the requirements of state, county, and municipal electric codes.

A large proportion of electricians are trained through apprenticeship programs. These programs are comprehensive, and people who complete them are qualified for both maintenance and construction work. Most localities require that an electrician be licensed. To obtain the license, electricians must pass an examination that tests their knowledge of electrical theory, the *National Electrical Code*®, and local electrical and building codes. After electricians are licensed, it is their responsibility to keep abreast of changes in the *NEC* guidebook, with new materials, and with methods of installation.

Employment for electricians is expected to increase annually. As population increases and the economy grows, more electricians will be needed to maintain the electrical systems used in industry and in homes. Additionally, as both new and old homes are prepared for new technologies to make them smarter, the demand requires more electricians who are trained in the new technologies.

---

*NFPA 70(R), *National Electrical Code*®, and *NEC*® are registered trademarks of the National Fire Protection Association, Quincy, MA.

■ FIGURE 1-3

These students are training to become electricians.

© iStockphoto/BartCo

Electronics Technician

Electronics technicians develop, manufacture, and service electronic equipment, and they use sophisticated measuring and diagnostic equipment to test, adjust, and repair electronic equipment. This equipment includes radio, radar, sonar, television, and computers, as well as industrial and medical measuring and controlling devices.

One of the largest areas of employment for electronics technicians is in research and development. Technicians work with engineers to set up experiments and equipment and calculate the results. They also assist engineers by making prototypes of newly developed equipment and by performing routine design work. Some electronics technicians work as sales or field representatives to give advice on installation and maintenance of complex equipment. Most electronics technicians work in laboratories, electronics shops, or industrial plants. Ninety percent of electronics technicians work in private industry.

Becoming an electronics technician requires formal training, which is offered by the military, junior and community colleges, vocational-technical schools, or in-house apprenticeship programs.

Employment of electronics technicians is expected to increase annually due to an increased demand for computers, communication equipment, military electronics, and electronic consumer goods. Increased product demand will provide job opportunities, and the need to replace technicians who leave the labor force, transfer to other occupations or fields, move into management, or retire will also increase.

1-1 QUESTIONS

1. What types of jobs do automation mechanics perform?
2. What is the difference between a computer technician and an electronics technician?
3. Computer engineers require what type of training?
4. Where does an electrical engineer work?
5. Where do most electronics technicians work?

1-2 DIFFERENCES BETWEEN A TECHNICIAN AND AN ENGINEER

Electronics engineering is an electrical engineering discipline involved with designing electronic circuits, devices, and systems. It is a broad engineering field that can be broken into *analog electronics, digital electronics, consumer electronics*, and *power electronics*. Electronics engineering is also involved with the implementation of applications and principles within the many related fields. The **Institute of Electrical and Electronics Engineers (IEEE)** is an important and influential organization for electronics engineers. Advancement for electronics engineers will often be to management.

Electronics technicians are expected to repair, maintain, assemble, and test electronics equipment. Titles for technicians are often expanded and may include the following:

- *Bench electronics technician*—works offsite to assemble, test, or repair work at a workbench.
- *Computer electronics technician*—works at a job site to install, maintain, and repair equipment and provides computer and IT (information technology) support.
- *Field electronics technician*—works at a job site to install, maintain, and repair equipment in the field.

- *Metrology electronics technician*—works at a workbench to calibrate, repair, and maintain all types of electronic standards, test, and measuring equipment. Metrology is defined as "the science of measurement."
- *Senior or Master electronics technician*—an in-house management position requiring extensive experience within an organization. A technician often has this position as a goal for advancement.

All electronics technician jobs include repair and maintenance of the employer's electronics equipment. This requires the ability to read schematics and maintenance manuals, test and troubleshoot the equipment, solder and unsolder components, and research parts availability and costs.

Specialized licenses are required when working on equipment requiring **Federal Communications Commission (FCC)** licensing or working in physically hazardous environments such as nuclear power.

Most electronics technician positions require at least an associate degree, equivalent military training, or a diploma from a vocational or technical school. This could be followed by certification, depending on the employer.

1-2 QUESTIONS

1. What two areas do electronics engineers work in?
2. What is IEEE to an electronics engineer?
3. What tasks do electronics technicians perform?
4. List the types of electronics technicians.
5. What skills does an electronics technician require?

1-3 PREPARATION FOR EMPLOYMENT

The best job for an individual is one that matches his or her interests and abilities. When first thinking about a job, think of the following individual traits:

- Abilities—skill developed through school or training.
- Aptitude—focuses on natural abilities.
- Interests—things done for personal interest or enjoyment.
- Personality—identifies how well one relates to others.

■ **FIGURE 1-4**

Two workers installing a photovoltaic (solar) panel.

© iStockphoto/LL28

Matching personal traits against available jobs is an important next step (Figure 1-4). Identifying a job title can be accomplished using the **Dictionary of Occupational Titles (DOT)** and the **Occupational Outlook Handbook** produced by the U.S. Department of Labor. Both of these documents are valuable resources when looking for a job. Electronics technicians are expected to repair, maintain, and test electronics equipment. Titles for electronics technicians are often expanded and may include the following:

- Computer electronics technician
- Field electronics technician
- Metrology electronics technician
- Bench electronics technician
- Senior electronics technician

Once you identify an appropriate job, you can begin to identify the skills you need to succeed in that job. To obtain the necessary job skills may require attending a technical school or college or an evening program, joining the military, or attending an apprentice program. Some jobs may require a four-year degree. It should be pointed out that the first job may lead to other jobs requiring different skills. Jobs could be changed four to five times over a lifetime of employment.

Once the skills are obtained, it's time to search the job market. Jobs may come from friends, teachers, instructors, classified ads, an Internet search, or a combination of all these techniques.

A basic **résumé** (Figure 1-5) should include the following information:

- Personal contact information—Include your name, address, phone number and e-mail address. A note on e-mail addresses: Keep them related to your name. Do not use complex or cute names for your résumé. Avoid names such as "2hot4u," "iluvchevy," "tinytim," and so on.
- Career objective—*This is not mandatory.* It is a personal decision whether to include it.
- Education—Electronics technician positions require at least an associate degree. If you lack the minimum education, you are unlikely to be granted an interview. Most employers perform background checks to verify any education claims.
- Work experience—List any related job held, in a reverse chronological order, with job title, employer, employer address, time period worked, and duties and responsibilities.
- Accomplishments—Itemize accomplishments and achievements.
- References—Most résumés show that references are available by adding the phrase "References supplied on request." It's best to have asked three or four people to supply references.

There are many different opinions for what to include on a résumé, but this list represents the basics. When writing your résumé, avoid complete sentences and first-person pronouns. Your résumé should focus on the highlights of your achievements—it should not read like an autobiography. Try to limit the résumé to one page, and concentrate on your achievements, not your responsibilities. Use key words that describe the position you are applying for.

When you send your résumé to an employer, you must include a **letter of application**, or cover letter (Figure 1-6). The letter of application is specific to the job you are applying for and must be brief and to the point. It includes the following parts:

- Introduction—Identify the job you are applying for and how you discovered the opportunity.
- Discussion—Discuss your work experience, education, and professional skill. Do not replicate the attached résumé.
- Conclusion—Express your desire to meet to discuss employment opportunities. Offer your contact information, and thank your reader for the opportunity to apply.

The letter of application and résumé are sent to the contact person in a classified ad or to the person who has the ability to hire. If the contact person is not known, call the employer and find out whom to send it to. Many larger employers now use an online application process, but you can still apply many of the

■ FIGURE 1-5
Sample résumé.

Charles D. Gears
456 Manhattan Ave
NY, NY 12345
Phone: 555-955-5210
Email: charles.gears@mail.com

Career Objective
To secure a challenging position as an Electronic Repair Technician in an organization where my abilities and skills are utilized to their full potential.

Education
| June 20XX | AAS in Electronics Technology, Chandler Community College, New York, NY |
| June 20XX | HS Diploma, XYZ High School, Anywhere, NY |

Certifications
August 20XX Education Systems Associate 20XX Microsoft Certified

Experience
September to August 20XX — Associate Electronic Repair Technician at Paradigm Computer Repair, New York, NY
- Assembled motherboards and installed circuit boards for new computer systems
- Tested newly assembled systems for proper operation
- Assisted senior electrical technicians in installing computer systems
- Performed basic functionality checks to avoid user error
- Attended meetings and trainings as required by the department

Professional Strengths
- Possess a wide knowledge of schematic diagrams and wiring diagram
- Ability to read and understand technical drawings
- Skilled in installation, operation, repair and maintenance of computer systems
- Ability to identify, replace and explain the proper use of electronic equipment
- Ability to quickly understand new systems and their requirements
- Possess excellent verbal and written communication skills

Technical Skills
- Familiar with Microsoft Office products

Professional
- Member of International Society of Certified Electronics Technician

■ **FIGURE 1-6**

Sample letter of application.

456 Manhattan Ave
New York, NY 12345
14 July 20XX

Mr., Samuel Wilson
AAA Repair Services
NewYork, NY

Dear Mr. Wilson

I am responding to your advertisement in the July 5, 2OXX issue of the *New York Gazette* for a Bench Electronic Technician. Because of my past year experience working as an electronic repair technician, my AAS in Electronics Technology, and my certification as Electronics Systems Associate, I believe I have the skills and knowledge you require.

I am interested in applying for your recently advertised position of Bench Electronics Technician. I have enclosed a copy of my résumé.

In addition to those skills and abilities highlighted in my résumé, 1 have always received excellent job ratings on my reliability, dependability, the quality of work produced, and my ability to meet all established production goals. Also, I have received high praise for my ability to work on my own and part of a team.

I look forward to the possibility of working with AAA Repair Services. I can be reached at 555-955-5210 and welcome the prospect of interviewing with you. I appreciate your consideration

Sincerely.

Charles D. Gears

Charles D. Gears

same résumé-building and letter-writing techniques to present your qualifications online.

The résumé serves another purpose as well, because many employers want their own job application filled out. In such cases, much of the information can be taken off the résumé. Read the full job application before filling it out, and write clearly and carefully when filling it out.

The **job interview** is an important follow-up on a successful résumé and job application (Figure 1-7). An interested employer wants to know more information. Learn everything you can about the employer before the interview. Dress appropriately for the interview, cover any tattoos, and wear clothes that are clean and pressed, with clean and neatly styled hair. Attend the interview alone. After the interview, be sure to send a follow-up letter thanking the employer for the opportunity to be interviewed.

When a job is offered, it is important to find out about the start date, working hours, salary, vacation days, and benefits. It is a personal decision to accept or reject a job, but it must be done in a timely fashion. Regardless of the situation, always be polite and considerate.

To be successful at the job will require hard work and attention to the following:

- Dress appropriately.
- Be on time and have good attendance.
- Be responsible for your actions.
- Be honest.
- Be enthusiastic.
- Display a proper attitude.

■ FIGURE 1-7

A job interview allows an employer to learn more about a prospective candidate as a future employee.

© iStockphoto/Track5

On the job, work hard, cooperate with others, and avoid gossiping, to increase productivity.

| 1-3 | **QUESTIONS** |

1. What four traits should be considered when thinking about a job?
2. List titles for electronics technicians.
3. Where can the necessary skills to become an electronics technician be obtained?
4. List the parts of a basic résumé.
5. What traits are required to be successful at a job?

| 1-4 | **GAINING EXPERIENCE IN THE WORKPLACE** |

To gain experience in a chosen career field, it is possible for a student try out a specific career in the workplace through several different options offered by the education provider and an employer. These include the following:

- Apprenticeship
- Co-op learning
- Externship
- Internship
- Job shadowing
- Mentoring

An **apprenticeship** is a program provided by an employer in which an apprentice learns a craft or trade through hands-on experience by working with a skilled worker. For those joining the workforce, an apprenticeship program combines on-the-job training with academic instruction and can last for four to five years. In an apprenticeship program, the apprentice is paid, with salary increases reflecting completion of different levels of training.

The primary purpose of an apprenticeship program is to help employers develop employees who will improve the quality of the workforce and to increase productivity.

An import concept to keep in mind is that apprentices who complete an apprenticeship program are accepted by the industry as journey-workers (journeymen). Any certification earned through an approved apprenticeship program is recognized nationwide and will often transfer.

A **cooperative education** program (co-op) allows a student to combine class time with work experience by spending part of the day in class and the balance of the day working full-time in a job related to his or her chosen career field. At the college level, the student might attend classes one semester and then spend the next semester working off-campus with pay, to gain experience in a career field that relates to the student's classes.

A co-op assignment is a supervised work experience in which the individual has identified learning goals that reflect what is to be learned throughout the co-op experience. To form a successful co-op program, a partnership is developed between the educational provider and an employer.

An **externship** is similar to an internship, but for a shorter duration; it is also similar to job shadowing. An externship may last from one day to several weeks and is intended to offer the student an idea of the work requirements in a particular career field. It also provides the student with professional contacts for future networking in the career field of choice.

The time spent with an externship can take many forms because of its short duration. Participants can ask questions, attend meetings, spend the day observing, and even get an opportunity to participate in a project. An externship provides a tour of a participating organization and an opportunity to meet with people working in the organization. Externships allow participants to learn more about and meet with people working in their chosen career fields.

An **internship** is an opportunity to mix education with a chosen career field by participating in supervised work experience (Figure 1-8). An internship is similar to an apprenticeship for trade and vocational jobs and is established between an employer and a school. Although internships are typically found at the college or university level, they are sometimes offered at the high school level.

Typically, an internship provides an exchange of the intern's services for the opportunity to gain experience in a chosen career field. It can also help students determine whether their interest in a particular career is real. Some interns find permanent employment with the companies in which they interned. Employers benefit from hiring experienced interns because they require little or no training when they start full-time employment.

Job shadowing is a work experience program that helps an individual find out what a chosen career

■ **FIGURE 1-8**

Internships are an important way to gain experience on the job.

© iStockphoto/darrenwise

field is like. Shadowing an experienced professional can help the individual choose a good career fit and identify the appropriate training or college program. The individual doing the shadowing observes a work environment and the necessary skills in practice, and gains an understanding of what further training may be needed, as well as other related potential career options.

Job shadowing increases an individual's career awareness, helps to model behavior through examples, and reinforces the connection between classroom learning and work experience. A successful job-shadowing program is formed by a partnership between an educational provider and an employer.

Mentoring describes a partnership between experienced trade professionals and less experienced workers who want to increase their knowledge and skills. Mentors share their knowledge, information, perspective, and skills, and their mentees ask for advice and seek the support of their more experienced mentors.

■ FIGURE 1-9

Mentoring programs allow trade professionals to share their knowledge, information, perspective, and skills.

© iStockphoto/philsajonesen

The power of mentoring is that it creates an opportunity for a student to be exposed to a chosen career field. The mentor provides the counseling, guidance, and wisdom to the mentee to advance his or her knowledge in the chosen career field (Figure 1-9).

1-4 QUESTIONS

1. What opportunities exist to increase knowledge in an individual's chosen career field?
2. What is the purpose of an apprenticeship?
3. How does an externship differ from an internship?
4. What function does an internship serve?
5. Where does job shadowing fit in when learning about a chosen career field?
6. What is the power behind a mentoring program?

SUMMARY

- A sampling of careers in the electrical and electronics field includes these:
  - Automation mechanic
  - Computer engineer
  - Computer technician
  - Electrical engineer
  - Electrician
  - Electronics technician
- An engineer designs electronic circuits, devices, or systems.
- Electronic technicians repair, maintain, and assemble electronics equipment.
- Most electronics technician positions require a two-year degree, equivalent military training, or a diploma from a technical or vocational school.
- Personal traits to think about when considering a career include abilities, aptitude, interests, and personality.
- A basic résumé includes personal information, objectives, education, work experience, accomplishments, and references (available upon request).
- Experience in the workforce can be achieved through programs such as apprenticeship, co-op learning, externship, internship, job shadowing, and mentoring.
- An apprenticeship provided by an employer combines on-the-job training with academic instruction over a period of time in order for an apprentice to learn a particular craft or trade.
- A co-op program allows the student to combine class time with a supervised work experience.
- An externship provides the student with an idea of the work requirements for a particular career by giving him or her the opportunity to spend a short duration of time in a chosen career field.
- An internship is a supervised work experience that mixes education with work in a chosen career field.
- Job shadowing is a work experience program to help an individual find out what a chosen career field is like by spending a short period of time following and observing an experienced professional.
- Mentoring involves a student working with a more experienced professional to gain knowledge of a chosen career field.

CHAPTER 1 SELF-TEST

1. Using the *Dictionary of Occupational Titles* from the library, or the Internet, identify three job titles that might be of interest to you.

2. Select the most interesting job title from problem 1, and research it to learn about the work, identify the pay scale, and discover other interesting facts about the position to report in an oral presentation.

3. Using the classified ads, current and past, find a job posting for the job title selected in problem 2.

4. Using a word processing program, prepare a résumé in order to identify your strengths.

5. Prepare a letter of application for problem 3.

6. Enlisting a friend or classmate, perform a mock interview for the letter of application in problem 5.

7. Find a job posting for an electronics engineer, and research the starting salary.

8. Find a job posting for an electronics technician, and research the starting salary.

9. Find a job posting for a computer engineer, and research the starting salary.

10. Find a job posting for a computer technician, and research the starting salary.

11. Make a chart for problems 7–10. Include: job description, education requirements, salary, and benefits.

12. Check with the counseling office to determine what programs exist for gaining work experience in a chosen career field.

13. Research a chosen career field to develop a list of questions regarding work requirements, job skills, and educational requirements.

14. Apply for a work experience program.

CHAPTER 2

Certification for the Electricity and Electronics Field

OBJECTIVES

After completing this chapter, the student will be able to:

- Identify who is responsible for certification of electronics technicians.
- Discuss why getting a certification is important for becoming an electronics technician.
- Discuss the different levels and certifications available for electronics technicians.
- Identify what knowledge is required for each level of the certification exams.
- Identify who is eligible to become certified as an electronics technician.
- Describe how to prepare for a career in electronics.
- Discuss the educational requirements for becoming an electronics technician.

KEY TERMS

2-1 Certified Electronics Technician (CET) Program

2-1 Electronics Technicians Association International (ETA-I)

2-1 International Certification Accreditation Council (ICAC)

2-1 International Society of Certified Electronics Technicians (ISCET)

2-2 Associate-Level CET Exam

2-2 Electronics Systems Associate (ESA) Exam

2-2 Journeyman-Level CET

2-3 bench electronics technician

2-3 computer electronics technician

2-3 meterology electronics technician

2-3 senior electronics technician

An electronic technician can satisfy specific requirements by passing an exam and earning professional certification through certification agencies. Two such agencies are Electronics Technicians Association International (ETA-I) and International Society of Certified Electronics Technicians (ISCET). These agencies offer many certifications, including Associate Electronics Technician, Senior Certified Electronics Technician, and Master Certified Electronics Technician.

Certifications can also focus on a specific area of electronics, such as fiber optics, computer software, or residential electronics.

Graduates of a diploma program may also decide to continue their education toward a Bachelor of Science in Electronic Engineering Technology or a related degree.

This chapter looks at the basics of certification required for electronics technicians.

2-1   BASIS FOR CERTIFICATION

The **Electronics Technicians Association International (ETA-I)** (Figure 2-1), a not-for-profit professional organization, was founded in 1978 by electronics technicians. These technicians had a need for meeting as a group that was independent of industry. The purpose of the group was to promote recognized professional credentials as proof of a technician's skills and knowledge in the field of electronics. ETA now offers over 50 certifications, and all

■ FIGURE 2-1

Electronics Technicians Association International logo.

Courtesy of ETA International

■ FIGURE 2-2

International Society of Certified Electronic Technicians logo.

ISCET logo copyright ISCET, the International Society of Certified Electronics Technicians.

these are accredited by the **International Certification Accreditation Council (ICAC)** and align with the ISO-17024 standards. The ICAC is an alliance of organizations dedicated to assuring competency, professional management, and service to the public through encouraging and setting standards for certification and licensing programs internationally.

The **International Society of Certified Electronics Technicians (ISCET)** (Figure 2-2) trains, prepares, and tests technicians in the electronics and appliance service industry. The **Certified Electronics Technician (CET) Program**, founded in 1965, measures the degree of knowledge and technical skills of working electronics technicians. This voluntary certification enables employers to separate knowledgeable job applicants from those with less training and fewer skills.

The CET program protects consumers by certifying, in the absence of governmental licensing, that the technician repairing their products has the knowledge, skills, and experience to do the job. The CET program has increased acceptance by technicians and employers as well as the public. Many employers encourage and some even require their technicians to be certified.

Electronics technicians who have passed any of the ISCET CET exams are encouraged to apply for membership in ISCET.

Anybody who has the knowledge and interest in electronics has the opportunity to become certified. There are more than 70 certification programs, including both stand-alone certifications and career track certifications for those interested in mastering their chosen field. The certification exams test individual proficiencies rather than product or vendor knowledge. The exams meet or exceed industry standards.

A panel of experts, comprised of educators and practitioners from each discipline, reviews each exam regularly to ensure that it meets professional and industry standards.

Certification can be a ticket to a better career, as it represents the difference between someone who claims to be a technician and someone who really is a technician.

Many government and private organizations require certification as a qualification for their positions, so passing one or more of the exams generally makes a technician more employable.

2-1 QUESTIONS

1. List the organizations responsible for certification of electronics technicians.
2. What organization accredits the certification of electronics technicians?
3. What is the purpose of getting a certification in electronics?
4. What do the certification exams test for?
5. What is the value of getting certified as an electronics technician?

2-2 IDENTIFYING CERTIFICATIONS REQUIREMENTS

There are many certification exams. Some certifications available to students and technicians starting in electronics and their requirements are listed here:

The **Associate-Level CET Exam** is available to students in electronics or technician programs. This exam focuses only on basic electronics and is part of the full credit CET exam. The exam is multiple choice and covers *basic electronics, math, DC and AC circuits, transistors*, and *troubleshooting*. It must be passed with a score of 75% or better. A successful associate CET receives a wall certificate valid for four years and is eligible to join ISCET, the International Society of Certified Electronics Technicians. *The Associate CET certification expires four years after being issued.*

The **Electronics Systems Associate (ESA) Exam** is available to anyone interested in electronics, including students and technicians. The ESA exam is a four-part exam that covers *DC, AC, semiconductors,*

and digital. Certificates are awarded following the passing of each of the four parts of the exam. The Associate CET is awarded after all the four certificates are earned. An advantage of the ESA program over the Associate level CET is that the *ESA certificates are issued for life,* but they are subject to the registration requirements.

To achieve the **Journeyman-Level CET** for becoming a fully certified technician, the technician must complete the following:

- Passing the Associate-Level Electronics test (ISCET or ETA-I) or acquiring the ISCET Associate Level Electronics Certification through the ISCET ESA program; and
- Passing one of the following Journeyman-Level CET exams:
  - Communications
  - Computer
  - Consumer
  - Industrial
  - Medical
  - Radar

2-2 QUESTIONS

1. Where are the certification practice exams available for someone getting started in electronics?
2. What does an ESA exam focus on?
3. Why would a student in electronics take the ESA exam instead of the Associate Level CET exam?
4. What is required to become certified as a Journeyman Electronics Technician?
5. What options are available at the Journeyman's level for certification?

2-3 HOW TO PREPARE FOR A CERTIFICATION

ETA-I currently provides a free practice exam at http://www.eta-i.org/free_practice_exams.html (Figure 2-3). An online store for ISCET at http://store.nesda-iscet.org/stguprte.html provides a Computerized Associate Practice Test; the study material has a database of 300 questions in 75-question tests. The tests can be taken over and over again. These organizations can change their offerings at any time, and this textbook reflects what was current at the time of printing.

■ FIGURE 2-3

Location of practice Associate CET exam on the ETA-I website.

Courtesy of ETA International

Use the online learning materials to study for the Associate examination. To use the material requires the individual to possess and understand the electronics field. The ISCET online materials help sharpen electronics knowledge in order to prepare technicians for a successful exam.

Electronics technicians repair, maintain, assemble, and test electronics equipment. Designations for electronics technicians include these:

- *Bench electronics technician*—indoor repair, test, or assembly work at a workbench.
- *Computer electronics technician*—focus on computer and IT (information technology) equipment and support for field electronics technicians; out-of-doors or off-site maintenance and repair tasks.
- *Metrology electronics technician*—equipment calibration and repair, often to NIST or other standards.
- *Senior electronics technician*—often an in-house position requiring extensive experience with the organization's products or exact field of technology.

Most electronics technician job requirements require at least an associate (two-year technical) degree, a diploma from a vocational or technical school, or equivalent military training. If the education list on a résumé is not accurate, a job interview is very unlikely to be granted. Employers perform a background check and check references to verify any education claims.

Nearly all electronics technician jobs require the technician to maintain and repair electronics equipment. This requires the technician to have the skills and ability to read schematics and maintenance

manuals, test and troubleshoot equipment, solder and unsolder components, and research availability of parts and costs. These jobs also call for the ability to determine whether further assistance is required in order to make timely repairs.

2–3   QUESTIONS

1. Where can study materials that cover what is on the certification exams be found?
2. What is required for the materials in problem 1?
3. What are three job titles for electronics technicians?
4. What is the minimum education required to work as an electronics technician?
5. What skills does an electronics technician need on the job?

SUMMARY

- Electronics technicians founded the Electronics Technicians Association International (ETA-I) for electronics technicians.
- The ETA-I promotes professional credentials as proof of a technician's skills and knowledge.
- The International Certification Accreditations Council (ICAC) accredits all certification for electronics technicians.
- The International Society of Certified Electronics Technicians (ISCET) trains, prepares, and tests electronics technicians.
- The CET program protects consumers by certifying that the technician is knowledgeable and has the necessary skills.
- The different levels of certification require the student or technician to be knowledgeable in all aspects of the electronics field.
- A technician with a certification is more employable in the electronics field.
- Many electronics technician jobs require a minimum of an associate degree, graduation from a vocational or technical school, or military training.
- An electronics technician must have the skills and ability to read schematics and maintenance manuals, test and troubleshoot equipment, solder and unsolder components, and research availability of parts and costs.

CHAPTER 2 SELF-TEST

1. What is the purpose of ETA-I and ISCET?
2. What are the benefits of getting a certification in electronics for a technician?
3. List the basic levels of certification for someone starting out in a career of electronics.
4. What is the disadvantage of taking the Associate-Level CET exam?
5. What is required to become a Journeyman Electronics Technician?
6. What are the different designations for an electronics technician once certified?
7. Make a personal career chart on the requirements for getting the education to become an electronics technician.
8. What skills and abilities must an electronics technician demonstrate once he or she is on the job?
9. Download a study exam from the Internet, using the websites listed in the text.
10. Take a practice exam for certification as an electronics technician.

Work Habits and Issues

OBJECTIVES

After completing this chapter, the student will be able to:

- Identify the purpose of having good work habits.
- Discuss techniques to reduce procrastination.
- Identify who is responsible for health and safety in the workplace.
- Discuss why young workers are injured more at the workplace.
- Define diversity.
- Identify workplace diversity issues.
- Define discrimination.
- Discuss types of employment discrimination.
- Define workplace harassment.
- Identify types of harassment in the workplace.
- Define workplace ethics.
- Discuss ethic issues in the workplace.
- Define teamwork.
- Identify techniques to be a better team member.

KEY TERMS

3-1	good work habits
3-1	procrastination
3-2	electric shock
3-2	electrocution
3-2	hazard identification
3-4	discrimination
3-5	harassment
3-6	work ethic
3-7	communications
3-7	team
3-7	teamwork

Successful people have a clear vision and mission of what they want to accomplish. They strive to become better at what they do and do not let setbacks and failures shatter their confidence. They believe they create their own success and have the courage to take calculated risks. They have good work ethics and know how to be team players.

This chapter looks at those issues that lead to success in the workplace and improve productivity. It also discusses important health and safety issues.

3-1   DEFINING GOOD WORK HABITS

People who develop good work habits tend to be more successful in their careers than those with poor work habits. **Good work habits** help to develop and improve skills that are important for improving an individual's productivity and job performance.

Procrastination is the act of replacing high-priority tasks with low-priority tasks, and thereby putting off important tasks to a later time. The leading cause of poor productivity is procrastination. People tend to procrastinate for many reasons. Some of the reasons for procrastination are listed here:

- The task is unpleasant.
- The task is overwhelming.
- The task may lead to negative consequences.

Techniques to help reduce procrastination include these:

- Determine the cost of procrastination.
- Attack the burdensome task.
- Nibble away on an overwhelming task.

Developing good work habits is a matter of developing proper attitudes and values toward work. To be successful at work:
- Develop a mission, goals, and a strong work ethic.
- Have good attendance and be punctual.
- Value time by staying on task.
- Operate with neatness, orderliness, and speed.
- Work smarter, not harder.
- Appreciate the importance of rest and relaxation after the task is complete.

Techniques to help become more productive include the following:

- Keep workspace clean and organized.
- Master and make good use of technology.
- Focus on one task at a time.

- Streamline and emphasize important tasks.
- Create quiet uninterrupted time for planning.
- Stay in control of paperwork and e-mail.

Here are some suggestions for overcoming wasting time:

- Minimize daydreaming.
- Prepare a time log to evaluate time usage.
- Avoid being a cyber-loafer.
- Set a time limit for each task.
- Schedule doing similar tasks together.
- Become decisive and finish things in a timely manner.

Work habits that help produce results on a job include these:

- Volunteer for assignments.
- Be nice to people.
- Prioritize work.
- Stay positive.
- Do not just identify a problem, but also supply a solution.

3-1   QUESTIONS

1. What are the signs of a successful person?
2. Why is procrastination not wanted in the workplace?
3. Identify three good work habits.
4. Identify three techniques to increase productivity.
5. Identify three work habits that produce results.

3-2   HEALTH AND SAFETY ISSUES

Health and safety at the workplace is everyone's responsibility, so it is important that a general understanding of health and safety be provided. While at a workplace, the employer is responsible for ensuring the workplace is safe, which includes providing the following:

- Safe work areas
- Information, instruction, training, and supervision
- Personal protective equipment

The workers are responsible for themselves, their own health and safety. The worker must not put others at risk on the job. However, the employer must ensure that a worker's health and safety are not harmed in any way. Worker responsibility includes these procedures:

- Following safety instructions
- Using equipment carefully
- Reporting hazards and injuries

> **NOTE**   Young workers usually work hard on a task, but they also have a higher risk of being injured on the job than older workers.

One reason young workers are injured is they are not aware of their rights to be properly trained, supervised, cared for, and provided with a safe and healthy environment.

Common hazards exist in the electrical and electronics industry. It is important to learn about these hazards and how to work safely in these environments. Within the industry, there are a wide range of work activities such as using measuring instruments, hand and power tools, and specialist tools; soldering; and constructing circuits (Figure 3-1).

> **NOTE**   Only licensed electricians can legally perform electrical work. This includes new electrical installations and alterations as well as repairs to existing installations.

■ FIGURE 3-1

Working with electrical test equipment can pose serious safety hazards. Young employees especially should learn the rules for using these tools.

© 2014 Cengage Learning

Hazard identification is a key step in preventing injury in the workplace. Where there's a risk, stop and think about what action is being done, and identify the potential risks to other workers.

Here is a three-step approach to hazard identification:

1. *Spot the hazard.* Be alert and notice potential hazards, and identify what is the problem.
2. *Assess the risk.* Talk to the job supervisor about the hazard, and suggest a way to resolve the problem.
3. *Make the changes.* Perform the agreed-upon solution.

It is important that every electrical worker works in a safe manner, not only for personal safety but also the safety of others. Working safely includes wearing the appropriate clothing and safety gear, using tools properly, and performing the task according to directions.

An **electric shock** occurs when a worker becomes part of an electrical circuit and current flows through his or her body. The most common cause of **electrocution** in the workplace is contact with overhead wires. This is because when carrying equipment like poles and ladders or operating equipment with height extension such as cranes, people can misjudge heights and distances between the ground and overhead wires.

Accidental death can also result from equipment becoming energized due to electrical faults, lack of maintenance, or short circuit. Electrical accidents are most often caused by a combination of factors such as these:

- Lack of supervision
- Lack of training
- Inadequate work practices
- Poorly maintained equipment or installation
- Hazardous workplace environment
- Unauthorized electrical repairs

General safety precautions include these practices:

- Keep tools and equipment in safe and proper working order through regular inspections and a preventative maintenance program.
- Repair immediately any tools and equipment that have frayed cords.
- When a tool and equipment has been disconnected from the power source, activate it to release the stored energy.
- Always switch off tools and equipment at the power point, before pulling out the plug.
- Keep electrical cords picked up off the floor to prevent damage from sharp objects.
- Know the location of the main electrical disconnect.

- Keep electrical tools and equipment away from any wet areas.
- Don't overload circuits by plugging in too many plugs on one circuit.

3-2   QUESTIONS

1. What is the employer responsibility for ensuring workplace safety?
2. What is the worker's responsibility regarding safety?
3. Why are young workers injured more than older workers?
4. What is the three-step approach to hazard identification?
5. What causes electrical accidents?

3-3   WORKPLACE DIVERSITY

Diversity refers to creating a workplace that respects the differences of each employee, recognizes the unique contributions that a diverse workforce can make, and seeks to bring out the full potential of all employees.

Our current understanding of workplace diversity began to develop in the 1960s. A series of executive orders required federal contractors to comply with equal employment opportunity objectives.

Today, diversity in the workplace means seeking to improve job performance by embracing the cultural differences among employees.

Workplace diversity is most often associated with the following:

- **Age:** Older workers have experience and know time-tested practices that may not be taught elsewhere.
- **Cultural Differences:** Differences in thinking through upbringing and education bring about a greater variety of solutions to a problem.
- **Demographics:** The most basic type of workplace diversity, it comprises age, ethnicity, and gender.
- **Disability:** Accommodations provide a new perspective to solving problems and creating design solutions.
- **Education:** Differences in educational background afford a different base of knowledge.
- **Ethics:** This field encompasses fundamental differences of values and objectives as well as the function, purpose, and importance of the work to be performed.
- **Ethnicity:** Cultural differences bring a broader perspective to problem solving and design.
- **Experience:** Life experiences can result in an employee having unique ideas and different perspectives.
- **Gender:** Men and women often bring different ways of thinking that result in different solutions.
- **Socioeconomics:** This entails the combined sociological and economic measure of an employee's work experience based on education, income, and occupation.
- **Veterans:** These people bring a work ethic and experience not provided by a purely academic education, as well as the ability to work long hours and with a diverse workforce.
- **Work History:** Different experiences from various employments can have a positive effect when looking for solutions.

Diversity can be used to bring people together to share a common experience in the workforce. Workplace diversity can also develop creative discussion and inventive solutions based on the differences in experience and knowledge that each employee brings to the team (Figure 3-2).

Creating a fully diverse workplace requires a lot of thought, effort, and action. Remember, each workplace is established through different sets of standards. Employees tend to align themselves with employees they are comfortable with. They choose characteristics in these employees that they want to enhance in themselves. These employees' unique characteristics make the workplace stronger.

■ FIGURE 3-2

Workplace diversity brings a variety of viewpoints to the problem-solving process.

© 2014 Cengage Learning

Several challenges to workplace diversity exist in almost any work environment. Working through these issues can help strengthen an employer, but ignoring them can ruin one.

Some challenges to workplace diversity are presented here:

- **Communication** is essential to a diverse workplace because each person and cultural group communicates differently. The wrong message can be sent through body language, e-mails, a failure to interact properly, or by misinterpreting tone of voice.
- **Change** is easier for some and more difficult for others who have different experiences. Change can be more effective when brought about slowly rather than by trying to adapt all at once.
- **Management** by supportive managers and supervisors is essential in supporting a diversity plan to get maximum results.

Workplace diversity challenges weigh considerably on an employer's success. These challenges serve to improve employee relationships and to promote diversity when handled correctly. These challenges can help their employees to decrease workplace tension, foster a positive place to work and grow as an employee, improve productivity, and improve their relationships with management and fellow employees.

Management through strategic planning can create a diverse and productive workforce. Managers must couple ongoing diversity training with continuous learning opportunities for all employees. Management must work toward creating a culture of diversity that follows the vision set forth by the employer.

3-4   QUESTIONS

1. What is diversity?
2. How did diversity in the workplace come about?
3. What are five things that workplace diversity is associated with?
4. What issues are associated with workplace diversity?
5. How can diversity be promoted in the workplace?

3-4   WORKPLACE DISCRIMINATION

Discrimination means unequal treatment through prejudice by an employer. Title VII of The 1964 Civil Rights Act *(Public Law 88-352)* states that *no person employed or seeking employment by a business with more than 15 employees may be discriminated against due to his or her color, national origin, race, religion, or sex.* It guarantees equal opportunity for an individual in employment.

Employment discrimination occurs when an employer treats an applicant or worker unfairly with one or more of the protected characteristics of Title VII of the 1964 Civil Rights Act. Anyone who experiences discrimination at work or is retaliated against for exercising his or her employment rights can seek legal protection.

Federal and state laws prohibit discrimination in the workplace. State laws may provide different solutions than federal laws. They may, in certain circumstances, be favored over federal laws. The types of employment discrimination are listed here:

- **Age:** The Age Discrimination in Employment Act of 1967 (ADEA) provides that no person shall be discriminated against because of his or her age. If an employer terminates an individual at a certain age and the termination appears to be for no other reason than the employee's age, the former employee may have a valid complaint against the employer for age discrimination under this law.
- **Color:** It is unlawful to discriminate against any employee or applicant for employment based upon skin color with regard to compensation, hiring, job training, promotion, termination, or any other term, condition, or privilege of employment.
- **Disability:** Title I of the Americans Disabilities Act of 1990 (ADA) removes the barriers preventing qualified individuals with disabilities from being employed, and provides transportation, communication, and cultural events that are available to persons without disabilities. It is unlawful for an employer with 15 or more employees to discriminate when recruiting, hiring, promoting, training, laying-off, paying, firing, assigning jobs, giving leave, and/or providing other benefits because of an employee's disability.
- **Gender:** The Equal Pay Act of 1963 prohibits the paying of uneven wages for men and women if they hold the same position in the workplace. Paying a salary to men and women of the same qualifications, responsibility, skill and position, employers are forbidden to discriminate on the basis of gender. Pregnancy-based discrimination is illegal. Employers are required to handle pregnancy

as a temporary condition that requires special consideration.

- **National Origin:** Treating someone less favorably because they or their ancestors are from a certain place or belongs to a particular national origin group is prohibited under Title VII of the Civil Rights Act. It also prohibits discrimination against a person associated with an individual of a particular national origin.
- **Race:** Under Title VII of the Civil Rights Act, it is unlawful to discriminate against any employee or applicant because of his or her race; the act also prohibits employment decisions based on stereotypes and assumptions about abilities, traits, or the performance of individuals of certain racial groups. Title VII is also violated when minority employees are physically isolated from other employees or from customer contacts.
- **Religion:** It is illegal for employers to discriminate based on an individual's religious beliefs. Employers are required to reasonably accommodate an employee's religious beliefs, as long as doing so does not have excessive negative consequences on the workplace.

Under federal laws, employers are prohibited from treating employees unfairly or discriminating against employees based on the characteristics that are legally protected. It is also illegal for an employer to retaliate against an employee who has filed a discrimination complaint or who has participated in an investigation regarding a discrimination complaint.

Remember that not all unfavorable treatment is discrimination. If an employee believes that he or she has experienced workplace discrimination, the employee can file a complaint with the Equal Employment Opportunity Commission (EEOC).

3-4 QUESTIONS

1. What is workplace discrimination?
2. What does Title VII of the Civil Rights Act protect?
3. What types of discrimination does Title VII prohibit in the workplace?
4. From what types of things does the federal law protect an individual who files a discrimination complaint?
5. With whom does an individual file a discrimination complaint?

3-5 WORKPLACE HARASSMENT

Harassment is continually annoying an individual by creating an unpleasant or hostile environment through verbal or physical conduct. Any behavior that is unwelcomed or discriminatory is considered harassment and can have a negative impact on performance and productivity in the workplace. Harassment can also affect an employee's advancement opportunities or pay status.

> **NOTE** The two parts of any type of harassment are unwanted and continued behavior. If no one says to stop the behavior or comments, or if the behavior does not continue, no harassment has taken place. Single instances or comments toward an individual or group may be in bad taste or show poor judgment by an individual but do not constitute harassment in the workplace.

Harassment in the workplace is illegal under the Age Discrimination in Employment Act (ADEA), Americans with Disabilities Act (ADA) of 1990, Civil Rights Act of 1991, Federal Equal Employment Opportunity laws (EEO), and Title VII of the Civil Rights Act of 1964 (Title VII).

Various types of harassment that could occur in a place of work relate to the following issues:

- **Age:** May occur in the form of jokes regarding age and other remarks made by managers or coworkers or in the form of discriminatory practices that are intended to force older workers out of the workplace. Age discrimination is covered by the Age Discrimination in Employment Act (ADEA).
- **Bullying:** Bullying occurs in the workplace when the employer or employees abuse any employee, knowing that the individual will be afraid to stand up to them, and refuse to take responsibility that they are the source of the problem. This type of behavior is considered harassment and is not tolerated.
- **Disability:** Negative comments or behavior toward an individual or group that is differently abled are not allowed. Comments can be regarding special procedures or allowances or the right to work. The Americans with Disabilities Act (ADA)

prohibits this type of harassment by employers and coworkers.

- **Ethnic or National Origin:** Similar to racial harassment, this detrimental behavior targets ethnic groups. Employers or coworkers may not treat any employee differently because of ethnic heritage.
- **Gender:** Employers and coworkers may not belittle or make suggestive comments or jokes about gender or about the ability of members of a gender to perform well at work. This is prohibited in state or local statutes or codes.
- **Gender Identity:** Detrimental comments or actions toward an individual's gender identification, including cross-dressing and transsexualism, is considered harassment.
- **Politics:** Employers and coworkers may not treat any employee differently because of political beliefs and affiliation, including making insulting comments or suggestive remarks regarding political affiliation or associations.
- **Language:** An employee's ability or inability to speak English is a protected category. The language spoken is not a requirement for the right to work and is covered in the legal authority through either citizenship or legal alien status with a work visa.
- **Marital Status:** Any discrimination toward marital status—single, divorced, widowed, or domestic partnership—is considered harassment if it is protected by legal statute
- **Race:** Detrimental comments or actions regarding race toward an individual or a group is considered harassment. Employers or coworkers may not treat any employee differently because of race and must refrain from making either negative or positive comments, using stereotyped language, or making jokes about race and ethnicity, as they may be viewed as offensive and therefore harassing. Other examples or prohibited comments and actions include those relating to appearance, education or intelligence, inferiority of ability, or race comparisons.
- **Religion:** Employers and coworkers may not treat any employee differently because of religious, moral, or ethical beliefs or affiliations with religious groups. They must make reasonable accommodations for observance of religious days and religious dress codes. Some religions require members to adhere to certain modes of dress, such as head coverings or footwear, and to certain styles of facial hair. Employers must allow these unless they materially interfere with safety and productivity of others in the workplace. Negative comments or behavior regarding a particular religious faith, credo, or practice is harassment.

- **Retaliation:** When an employer or supervisor demotes, terminates, verbally harasses, or singles out an employee because this person would not submit to sexual advances or because he or she filed a discrimination complaint, the supervisor is retaliating against the person. Retaliation can also occur as a result of an employee acting as witness or participating in a discrimination complaint. Laws prohibit employers from creating a hostile work environment through retaliation.
- **Sexual Issues:** These can take many forms such as coworkers making jokes or remarks of a suggestive or sexual nature, photos or computer screensavers of a suggestive or sexual nature, requests for sexual favors, or unwanted sexual advances. A person may find these behaviors offensive and may feel uncomfortable in the presence of the person who makes the comments.
- **Skin Color:** This is similar to racial harassment, with the detrimental behavior targeting an individual's skin color. Employers or coworkers may not treat any employee differently because of the color of his or her skin.
- **Verbal:** All harassment can be verbal; however, this category specifically pertains to verbal harassment not included in the categories covered. Verbal abuse involves constant detrimental comments meant to degrade an individual, often a coworker, with the intention of demeaning the person or lowering his or her self-esteem.
- **Whistle-Blower:** This is a new area that protects individual from any action intended to cause harm to a person who reports a business or individual within a business environment of wrongful acts. Any retaliatory action or behavior constitutes harassment and will not be tolerated.

Harassment can be about any personal characteristic and between any two people. It can occur between employees or even employees and non-employees. The victim of harassment is anyone affected by an offensive behavior; this isn't necessarily the person targeted by the harassment.

Any harassment is a form of discrimination and violates Title VII of the Civil Rights Act of 1964 and other federal authority. The federal laws do not prohibit simple teasing, offhand comments, or isolated incidents that are not serious. To be classified as harassment, the conduct must be so offensive that it creates a hostile work environment for an individual.

An individual should report any workplace harassment immediately to a supervisor. If the individual's supervisor is the harasser, the complaint should be taken to a higher level.

An employee should be ensured that any acts of retaliation will not be tolerated. If an employee did not report harassment for fear of retaliation, which in itself is a form of harassment, there would be delays in an investigation. Intimidation could also occur because the harassment was late in being reported. An employer can be held liable for any harassment in the workplace. Managers and supervisors should take immediate action to eliminate any workplace harassment as soon as it is reported and maintain a zero-tolerance harassment policy.

It is important for all employees to know what constitutes harassment. Hence, it is necessary to instruct all employees and supervisors on what behaviors are considered harassment. Most employers have an established procedure for filing any case involving harassment in the workplace, and it is important to make these procedures known to everyone.

3-5   QUESTIONS

1. What is harassment?
2. How is an employee protected from harassment?
3. List five types of harassment in the workplace.
4. When is teasing considered harassment?
5. Who is liable for harassment in the workplace?

3-6   WORKPLACE ETHICS

A **work ethic** is a set of established standards based on core values such as honesty, respect, and trust. Having a good work ethic can help an employee be selected for a promotion, with more responsibility and higher pay (Figure 3-3). An employee who does not exhibit a good work ethic is regarded as failing to provide good value for the pay, and this will not lead to

■ FIGURE 3-3

A good work ethic can be rewarded with more responsibility and higher pay.

© 2014 Cengage Learning

promotions. Employees who demonstrate a poor work ethic are often the first to be let go.

Workplace ethics serve as moral guidelines that are specific to the work environment. Knowing proper workplace ethics and following them in the workplace contributes to an individual's value as an employee, which leads to the longevity of a business.

Employee workplace ethics cover a variety of personal values, all of which an employer should be aware of. Some of the ethical issues that employers and employees alike should consider include these:

- **Behavior:** Do not engage in abusive behavior such as bullying other employees, displaying pornography in their work space or on computer screens, stealing from other employees or the employer, or telling inappropriate or offensive jokes, which is strictly forbidden in the workplace.
- **Fairness:** Refrain from cheating by sneaking around and hiding the real reasons for coming to work late or being absent.
- **Confidentiality:** Refrain from gossip about fellow employees' private issues.
- **Honesty:** Know the difference between what is right and wrong and how to deal with situations that provide opportunities to be dishonest.
- **Initiative:** Accomplish tasks without being told, finding out what is needed and continuing tasks when the going gets tough.

- **Integrity:** Have the courage to do what is right despite popular opinion.
- **Reliability:** Know how to accomplish an assigned task correctly and in a timely manner, without constant supervision.
- **Self-control:** Prevent a difficult situation from growing into a conflict that disrupts workflow.
- **Self-starting:** Take action on tasks without waiting to be told what to do.
- **Credit:** Avoid taking undue credit for work that another employee actually performed.
- **Working with others:** Treat all clients and customers fairly and honestly in all situations.

Ethics programs are offered by employers and are intended to affect how people think about and address ethical issues that arise on the job. An employer's established ethical guidelines benefit the company by steering employees away from any ethical risk-taking situations to a more appropriate and productive type of action. An ethics program allows employees to be sensitive to ethical issues on the job.

Here are some potential benefits from an effective ethics program:

- Aligns the workforce with the employer's mission and vision
- Provides guidance and resources for employees making difficult decisions
- Creates a productive working environment
- Helps in hiring and keeping highly skilled employees

Employees who work for an employer with an ethics program and see their supervisors modeling ethical behavior, and who see honesty, respect, and trust applied in the workplace, generally report the following:

- Greater overall satisfaction with their employer
- The feeling of being valued by their employer
- Feeling less pressured to compromise ethic standards
- Belief that there is less misconduct in the workplace
- A willingness to report any misconduct
- Satisfaction with responses to misconduct being reported

Most supervisors agree that high ethical standards are important in the workplace. A good work ethic provides a natural desire to get the work done.

3-6   QUESTIONS

1. What is a work ethic?
2. What are work ethics based on?
3. List the steps for building key habits into good work ethics.
4. How does a work ethic program transfer to employees?
5. What does good work ethics promote?

3-7   TEAMWORK

A **team** has more than one person, with each team member having different responsibilities. The team working toward a common goal is referred to as **teamwork**.

NOTE   Teamwork—The process of a group of people working together to achieve a common goal.

A challenge of teamwork is to get the team members to become comfortable enough with each other to work together. It requires team members working with each other to learn their jobs (Figure 3-4).

■ FIGURE 3-4

Team-building exercises can help members learn about each others' roles and responsibilities.

© 2014 Cengage Learning

An electronics project requires each team member to be assigned a specific task that he or she is responsible for completing. With multiple tasks, each team member might have more than one team assignment. Task assignments and roles for the fabrication of a printed circuit board (PCB) could be set up like this:

- **Team Leader:** In many ways, the team leader is the anchor for the team. The position brings a high level of responsibility. The team leader is the best person to help the team fulfill its objectives. A strong leadership leads to better teamwork. The team leader's duties include the following:
  - Acts as the liaison between the instructor and the team
  - Is responsible for making sure resources are available
  - Helps the team plan daily what is to be accomplished
  - Motivates the team with clear communications
  - Is responsible for keeping the team on task and for everything related to and done by the team
  - Serves the group, but does not control or dominate it
- **Computer Simulator:** This person verifies that the circuit design is valid by using an electronic simulation program and confirms the circuit does what it is designed to do. The computer simulator's responsibilities include these:
  - Inputs the circuit into an electronic simulation program
  - Confirms that the circuit does what it is designed to do
  - Verifies the components' values
- **PCB Designer:** Using the schematic diagram, the PCB designer develops the PCB layout once the computer simulator verifies the circuit. The PCB designer's duties are listed here:
  - Is responsible for obtaining all the components for the PCB
  - Uses the components to make paper dolls
  - Uses the paper dolls to lay out the top of the PCB
  - Routes the traces on the copper side of the PCB
- **PCB Fabricator:** Using the layout designed by the PCB designer, the PCB fabricator fabricates the PCB and is responsible for the following tasks:
  - Transfers the layout design to a copper-clad board

- Etches the copper-clad board
- Cleans the PCB and drills all holes
- Verifies that the PCB is complete
- **Assembler:** Using the etched PCB and the components used by the PCB designer, the assembler completes the PCB. The assembler's tasks include these:
  - Installs components into the PCB in a logical pattern
  - Solders and trims all component leads to the PCB
  - Inspects the completed board
  - Tests the completed board
  - Troubleshoots the PCB if necessary
- **Recorder:** This individual provides the documentation necessary at all steps of the PCB fabrication. He or she is responsible for providing any information needed, as outlined here:
  - Works with each team member to have accurate daily logs
  - Maintains a folder to provide feedback to the team
  - Provides information during daily planning
- **Media Presentation:** This person documents daily activities using a camera, video camera, and daily notes in order to prepare a media presentation such as PowerPoint, video, or audio. The media presenter's duties follow:
  - Is responsible for capturing all the moments of the team
  - Develops a meaningful presentation for the class
  - Actively involves all team members in the presentation
  - Provides a chronological order to the presentation
  - Is able to explain how the project works

Working on teams can be difficult and frustrating. Poor **communications** by team members can lead to confusion or misunderstanding. Both the team members and team leader need to effectively communicate to create a successful team.

Techniques to become a better team member:

- **Accept responsibility:** People in a group tend to lose respect for an individual who is constantly blaming others for not meeting deadlines. Team members know who is not pulling his or her weight on the team. Encourage nonsupporting team members to get their tasks completed.

- **Be a team member:** Don't act like a know-it-all and burden the team by fishing for compliments every time a task is completed. Doing this creates tension within the group. The team will recognize when good work is being done, and they'll let it be known.

- **Communicate:** If a problem exists with someone in the group, talk to that person about it. Do not let bad feelings grow, as they will make team members uncomfortable and make them want to isolate themselves from the group.

- **Get involved:** Share ideas, suggestions, solutions, and proposals with team members.

- **Listen actively:** Respect the person speaking, ask probing questions, and acknowledge what's said by repeating points that have been made. If something is unclear, ask for more information to clarify any confusion before moving on. Listening is a vital activity of any team and is the other half of communication.

- **Support the team:** Keep an open mind when a team member suggests an idea; respond positively, and consider the idea. Considering the team member's ideas shows an interest in all ideas.

3-7 QUESTIONS

1. What is the definition of teamwork?
2. What is a challenge of teamwork?
3. What are the task assignments for building an electronics project?
4. What is one of the major problems with teamwork?
5. What are the techniques to become a better team member?

SUMMARY

- Good work habits are important for improved productivity and quality.
- The leading cause of poor productivity is procrastination.
- Reasons for procrastination include that the task is unpleasant, overwhelming, or may lead to a negative consequence.
- Developing good work habits is a matter of developing proper attitudes and values.
- Everyone is responsible for health and safety at the workplace.
- Young workers are injured at the workplace because they might not be aware of their rights to be properly trained, supervised, cared for, and provided a safe and healthy environment.
- Hazards identification is a key step in preventing injury at the workplace.
- Discrimination is the unequal treatment of an employee through prejudices by an employer.
- Discrimination complaints can be filed with the Employment Opportunity Commission.
- Harassment is continuing to annoy an individual when asked to stop.
- Harassment is a form of discrimination and is in violation of Title VII of the Civil Rights Act of 1964.
- A work ethic is a set of established standards that is based on core values of honesty, respect, and trust.
- An ethics training program benefits the employer by helping to provide a higher level of productivity.
- Teamwork is team members working toward a common goal.
- Techniques to become a better team member include working toward good communication, not blaming, supporting members' ideas, listening actively, and getting involved.

CHAPTER 3 SELF-TEST

1. Identify how good work habits make successful people.
2. What good work habit steps are needed to be successful?
3. What work habits help produce results for a job?
4. What is the worker responsible for in the workplace?
5. What is involved in working safely at the workplace?
6. What causes electrical accidents?
7. How did diversity in the workplace originate?
8. What acts prevent discrimination in the workplace?
9. What identifies a behavior as harassment?
10. How can work ethics enhance advancement in the workplace?
11. What is the challenge of putting a team together to work on a project?
12. What are some task assignments for a team working on a project?
13. Generate a simplified list for becoming a better team member.

CHAPTER 4

Calculators for Electricity and Electronics

OBJECTIVES

After completing this chapter, the student will be able to:

- Identify why graphing and scientific calculators are popular choices when selecting a calculator.
- Discuss the advantages of a graphing calculator.
- Describe the difference between calculators that have reverse Polish notation and those that do not.
- Describe what the requirements are for a calculator in electronics.
- Discuss basic data-processing requirements for a calculator.
- Describe what must be understood when using a calculator.
- Identify what mathematical skills are still necessary when using a calculator.
- Describe the rounding process.

KEY TERMS

4-1	function
4-1	graphing calculator
4-1	reverse Polish notation (RPN)
4-1	scientific calculator
4-1	stack
4-1	stack based
4-1	trigonometry
4-2	exponents
4-2	memory function
4-2	power
4-3	mathematical skills
4-3	reciprocals
4-3	rounding

A calculator is great time-saving device—but only if the operator is skilled at using it. Therefore, it is important when selecting a calculator to gain the mathematical skills to operate it to its fullest potential.

Many students have rejoiced that all of their mathematical work can be mastered with a calculator. With just a few keystrokes, a calculator gives the correct answer.

However, some students fail to realize that a calculator is just a tool to perform calculations very quickly, but with no guarantees for a correct answer unless these conditions are met:

- The correct numbers are entered.
- The numbers are entered in the correct order.
- The correct function keys are used at the appropriate time.

4-1   TYPES OF CALCULATORS

Graphing calculators and **scientific calculators** are popular choices when selecting a calculator. Although they are not needed for this textbook, they typically include formulas and functions used in **trigonometry**, which is the study of triangles and the relationships of the sides and the angles between the sides, and statistics. With these calculators, it is important to study the manual carefully to be able to use the calculator to its fullest potential.

Graphing calculators are the most powerful and versatile of the calculators. They often are equipped with special features for solving engineering problems. Graphing calculators not only display graphs but are also programmable.

Figure 4-1 shows examples from three of the most popular graphing calculator manufacturers: Casio, Hewlett Packard, and Texas Instruments. Check out more on calculators by visiting the manufacturers' websites.

Casio has been a successful competitor in many consumer electronics field. Casio offers an array of graphing calculators, such as the example shown in Figure 4-2.

Hewlett-Packard, another popular calculator manufacturer, offers some of its calculators with **reverse Polish notation (RPN)** (Figure 4-3).

Texas Instruments offers a variety of models; an example is shown in Figure 4-4. Some of the calculators interface with a computer for downloading and uploading data and come with a calculator-based laboratory.

■ FIGURE 4-1

Examples of graphing calculators from the more popular calculator manufacturers.

■ FIGURE 4-2

An example of Casio's graphing calculator.

■ FIGURE 4-3

An example of Hewlett Packard's graphing calculator, which can use reverse Polish notation (RPN).

■ FIGURE 4-4

Texas Instrument's example of a graphing calculator, which can interface with the computer.

© 2014 Cengage Learning

and therefore RPN, have the advantage of being easy to implement and are very fast.

Scientific calculators are used to solve advanced problems in physics, mathematics, and engineering. They have provisions for handling exponential, trigonometric, and sometimes other special **functions** in addition to basic arithmetic operations.

Figure 4-5 shows some examples of some of the more popular scientific calculators manufactured. Check out more of these calculators by visiting the manufacturers' websites.

Whichever calculator you choose, read the owner's manual thoroughly. Practice with a variety of different problems with known outcomes. Working with a calculator is a great way to review and learn the mathematics of electronics.

The criteria for selecting a calculator should include cost, documentation, durability, ease of use, features, and support.

An alternate system commonly used by Hewlett-Packard and some other manufacturers is reverse Polish notation (RPN). RPN is a convenient method to enter formulas as algebraic equations, using the equal key (=) to submit an expression and get an answer.

If you wanted to find out the value of 53 cubed using a standard algebraic notation calculator, it would be entered as follows:

- 53
- y^x
- 3
- ENTER

To find the value of 53 cubed using a RPN calculator, it would be entered like this:

- 53
- ENTER
- 3
- y^x

The decoders of RPN are often **stack based**; that is, the operands are pushed onto a **stack**. Think of the stack as similar to putting plates into one of those spring-loaded, plate-stacking trolleys seen in cafeterias. The plates are removed, used, cleaned, and put back on the trolley. When an operation is performed, the operands are taken from a stack, the mathematical function is performed, and the result is pushed back on the stack, ready for the next operation. The value of the RPN expression is on the top of the stack. Stacks,

4–1   QUESTIONS

1. What are the two choices of calculators that could be used for electronics?
2. What are the advantages of a graphing calculator?
3. How do the Hewlett Packard graphing calculators differ from other graphing calculators.
4. Describe how to enter a number and operation in an RPN calculator.
5. How does a scientific calculator differ from a basic calculator?

■ FIGURE 4-5

Examples of calculators from three popular calculator manufacturers.

© 2014 Cengage Learning

4-2 CALCULATOR REQUIREMENTS FOR ELECTRICITY AND ELECTRONICS

Selecting a calculator appropriate for electronics is an important decision. The marketplace is flooded with many makes and models. Which is the right one? Which functions will prove to be the most useful? For this textbook, choose a calculator that has the following functions: $+$, $-$, $\times$, $\div$, $\frac{1}{x}$ or X^{-1}, x^2, and $\sqrt{\ }$. A **memory function** is optional but can decrease data entry time.

It is important to know the sequence for entering data into a scientific calculator. These are some basic data entry requirements that should be practiced:

- Use the proper keystroke sequence to apply a formula.
- Store and recall values in memory.
- Calculate using exponents, reciprocals, powers, and roots. A base number and an exponent represent a **power**. The base number is the number to be multiplied. The **exponent**, a small number written above and to the right of the base number, tells how many times the base number is being multiplied. Exponents are shorthand for repeated multiplication of a number by itself.
- Determine the trigonometric functions of a given phase relationship.

All calculators generally come with a manual, which should be kept handy.

4-2 QUESTIONS

1. What calculator functions are required with the textbook?
2. What purpose does a memory function serve with a calculator?
3. When using a scientific calculator, what is important?
4. What basic data entry requirement should be practiced?
5. What do all calculators come with that should be kept handy?

4-3 USING THE CALCULATOR FOR ELECTRICITY AND ELECTRONICS

If operators do not understand principles of the mathematical process, they will not be able to properly enter data into a calculator, nor will they be able to correctly interpret the results. **Mathematical skills** still count. Even when all data is entered correctly, the answer may be incorrect due to battery failure or other conditions.

The following examples show how a calculator is used to solve various types of problems in electronics. Turn on your calculator. Examine the keyboard. Let's do some calculating.

Addition

EXAMPLE 1: Add 39,857 + 19,733.

Solution:

Enter	Display
39857	39857
+	39857
19733	19733
=	59590

Subtraction

EXAMPLE 2: Subtract 30,102 − 15,249.

Solution:

Enter	Display
30102	30102
−	30102
15249	15249
=	14853

Multiplication

EXAMPLE 3: Multiply 33,545 × 981.

Solution:

Enter	Display
33545	33545
×	33545
981	981
=	32907645

Division

EXAMPLE 4: Divide 36,980 ÷ 43, or 36,980/43.

Solution:

Enter	Display
36980	36980
÷	36980
43	43
=	860

Square Root

EXAMPLE 5: Find the square root of 35,721.

Solution:

Enter	Display
35721	35721
√	189

Total Resistance (Parallel Circuit)

The total resistance of a parallel circuit is calculated by first computing the reciprocal of each branch and then taking the reciprocal of the branch total.

Parallel circuits are made up of resistors that are sold in resistance values of ohms. Calculating parallel circuit total resistance involves the use of **reciprocals** (1/R), as shown in the parallel circuit formula:

$$\frac{1}{R_T} = \frac{1}{R_1} + \frac{1}{R_2} + \frac{1}{R_3} \cdots + \frac{1}{R_n}$$

A calculator gives the reciprocal of a number by simply pressing the 1/X or X^{-1} key. If the calculator does not have a 1/X or X^{-1} key, then each reciprocal value must be found separately by dividing 1 by the resistance value.

EXAMPLE 6: Calculate the total equivalent resistance of the parallel circuit shown.

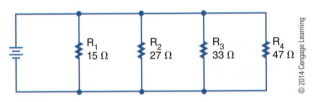

© 2014 Cengage Learning

Solution:

	Enter	Display
Reciprocal of R_1	15	15
	1/X	0.0666667
Reciprocal of R_2	27	27
	1/X	0.037037
Reciprocal of R_3	33	33
	1/X	0.030303
Reciprocal of R_4	47	47
	1/X	0.0212766

	Enter	Display
Totals of reciprocals	0.0666667	0.0666667
	+	
	0.037037	0.1037037
	+	
	0.030303	0.1340067
	+	
	0.0212766	0.1552833

	Enter	Display
Reciprocal of totals	0.1552833	0.1552833
	1/X	6.4398425
Round answer to		6.44 Ω

EXAMPLE 7: Use a calculator with memory function.

Solution:

	Enter	Display
Reciprocal of R_1	15	15
	1/X	0.0666667
	M+	0.0666667 M
	C	0
Reciprocal of R_2	27	27
	1/X	0.037037
	M+	0.037037 M
	C	0
Reciprocal of R_3	33	33
	1/X	0.030303
	M+	0.030303 M
	C	0
Reciprocal of R_4	47	47
	1/X	0.0212766
	M+	0.0212766 M
	C	0
Totals of reciprocals	RM	0.0.155283329
Reciprocal of totals	1/X	6.439841299
Round answer to		6.44 Ω

Rounding

> **NOTE**   **Rounding is not a calculator function and must be done manually.**

Rounding a number can reduce the number of significant digits. This means dropping the least significant digits until the desired number of digits remains. The new least significant digit may be changed using the following rules:

If the highest significant digit dropped is

- less than 5, the new significant digit is not changed.
- greater than 5, the new significant digit is increased by 1.
- 5, the new significant digit is not changed if it is even.
- 5, the new significant digit is increased by 1 if it is odd.

EXAMPLE 8: Round 352.580.

Rounded to the nearest tenth:	352.6
Rounded to the nearest whole number:	352
Rounded to the nearest hundred:	400

These rules result in a rounding off technique that on the average gives the most consistent reliability.

4-3 QUESTIONS

1. Why are mathematical skills necessary when using a calculator?
2. Using a calculator, verify the answers for the seven examples shown in this chapter.
3. What is the function of X^{-1}?
4. How can the X^{-1} be used to multiply the numerator and the denominator?
5. Explain why Examples 6 and 7 have different answers.

SUMMARY

- Graphing and scientific calculators are popular choices when selecting a calculator.
- Graphing calculators are the most powerful and versatile.
- Graphing calculators can display graphs and are programmable.
- RPN calculators allow entering of formulas as algebraic equations.
- RPN places the operands on a stack, which makes them fast.
- Engineering uses scientific calculators.
- The calculator needed for this textbook includes: $\times$, $\div$, $\frac{1}{X}$ or X^{-1}, x^2, and $\sqrt{\ }$, with a memory function optional.
- Using a calculator requires knowing the sequence for entering the data.
- Math skills are necessary to correctly interpret the results of the calculator.
- Calculators do not round the final answer of the calculator.
- Rounding is a way of dropping least significant digits.

CHAPTER 4 SELF-TEST

1. Using the Internet, identify an affordable graphing calculator and a scientific calculator. List the advantages of each.
2. Using conventional notation and RPN, show how to notate 24 squared.
3. What are the basic requirements for entering data into a calculator?
4. Use a calculator to find the answers for the following:
   a. 47,386 + 37,654
   b. 31,256 − 29,345
   c. 29,576 × 241
   d. 24,562/364
   e. ([(24 + 76)/36] × 24) − 46
5. Round 546.51683 to the indicated values:
   a. Hundredths
   b. Tenths
   c. Whole number
   d. Hundreds

Electronic Circuit Diagrams

OBJECTIVES

After completing this chapter, the student will be able to:

- Understand why schematic symbols were developed.
- Explain how schematic symbols are used.
- Identify differences in symbols used in electrical wiring diagrams and electronic schematic diagrams.
- Describe how a reference designator is configured with schematic symbols.
- Discuss the purpose of a schematic diagram.
- Describe the difference between a block diagram and a schematic diagram.
- Explain the difference between an analog schematic diagram and a digital schematic diagram.
- List the techniques for making a schematic diagram.
- Explain the purpose of breadboarding.
- Describe the difference between physical and virtual breadboarding.

KEY TERMS

5-1 electrical symbols

5-1 International Electrotechnical Commission (IEC)

5-1 reference designator

5-1 schematic symbols

5-2 analog schematic diagram

5-2 block diagram

5-2 common signals

5-2 computer aided design (CAD)

5-2 digital schematic diagram

5-2 discrete components

5-2 schematic diagram

5-2 virtual test equipment

5-3 breadboard

5-3 solderless breadboard

5-3 virtual breadboarding

chematic symbols are drawings or pictograms used to construct a schematic diagram. Schematic diagrams are used to show how a circuit is connected. The schematic diagram is different from the actual components layout on a printed circuit board.

Breadboards are used to test a circuit design before building a printed circuit board. Current breadboards do not involve any soldering in the construction of a circuit.

5-1   SCHEMATIC SYMBOLS

Schematic symbols are symbols used to represent electrical and electronic devices in a schematic diagram for an electrical or electronic circuit. The **International Electrotechnical Commission (IEC)** is responsible for keeping the symbols current, removing old symbols from use and adding new symbols as new technological devices emerge. These symbols can vary from country to country but are now, to a large extent, internationally standardized. There are several international standards for schematic symbols in schematic diagrams. Appendix 13—Digital Logic Symbols shows several standards. This textbook uses American (MIL/ANSI) symbols for schematic diagrams.

Different symbols may be used depending on the use of the drawing. For example, **electrical symbols** used in architectural drawings (Figure 5-1) are different from symbols used in schematic diagram electronics (Figure 5-2). In wiring a house, look for symbols prominent in house wiring. Any electrical fixture found in a house has a symbol that coincides to it on an electrical drawing (Figure 5-3). These drawings allow an electrical contractor to easily read the diagram during the construction phase.

Schematic symbols are followed with a **reference designator** that identifies a component in an electrical schematic diagram or on a printed circuit board (PCB). The reference designator usually consists of one or two letters followed by a subscript number, for example, R_2, C_1, S_4, I_{R_1}, and so on. Reference designators followed by a subscript number and letter indicate a component with several sections tied to a common point, for example, S_{1_A}, R_{2_B}, and so forth.

Some of the more common electronic components and reference designators used by industry are shown in Figure 5-4.

■ FIGURE 5-1

Architectural electrical symbols.

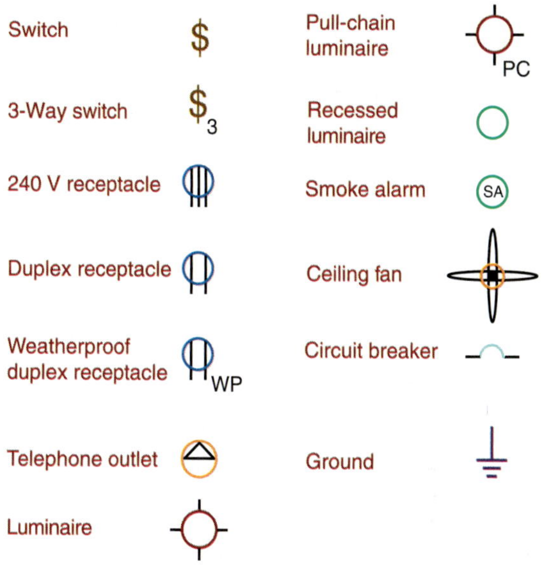

Switch	$	Pull-chain luminaire	
3-Way switch	$_3	Recessed luminaire	
240 V receptacle		Smoke alarm	SA
Duplex receptacle		Ceiling fan	
Weatherproof duplex receptacle	WP	Circuit breaker	
Telephone outlet		Ground	
Luminaire			

© 2014 Cengage Learning

5-1   QUESTIONS

1. Why are schematic symbols used in the electrical and electronic field?
2. Why is there more than one schematic symbol for a device?
3. Is there a difference between schematic symbols used for electrical wiring diagrams and those used for electronic schematic diagram symbols?
4. What is the function of a reference designator?
5. What do letters following a component number in a reference designator signify?

5-2   SCHEMATIC DIAGRAM

The key to an effective PCB design is a **schematic diagram** that becomes the basic reference for a circuit and gives the necessary specifications needed to design a PCB. The diagram should exhibit good signal flow and complete reference data.

The best way to preview a schematic diagram is to lay out or view a **block diagram** of the circuit. The block diagram represents how the component blocks

■ FIGURE 5-2

Symbols used for an electronics schematic diagram.

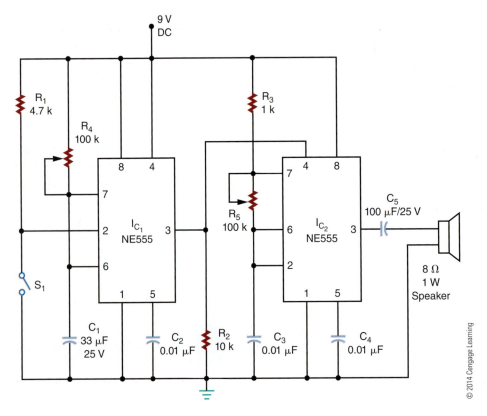

■ FIGURE 5-3

Residential electrical wiring drawing using architectural symbols.

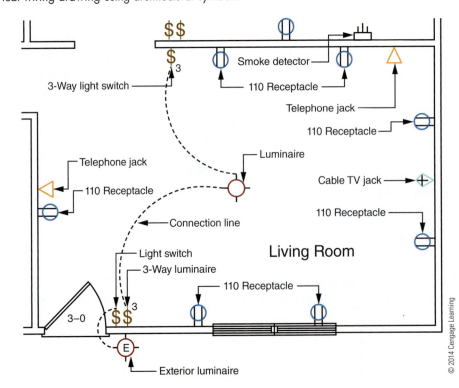

■ FIGURE 5-4

Reference designators for electronic components.

Battery	**B**
Bridge rectifier	**BR**
Capacitor	**C**
Cathode-Ray Tube	**CRT**
Diode	**D**
Fuse	**F**
Integrated Circuit	**IC**
Jacfc Connector	**J**
Inductor	**L**
Liquid Crystal Display	**LCD**
Light-Dependent Resistor	**LDR**
Light-Emitting Diode	**LED**
Speaker	**SPK**
Motor	**M**
Microphone	**MIC**
Neon Lamp	**NE**
Operational Amplifier	**OP**
Plug	**P**
Printed Circuit Boaid	**PCB**
Transistor	**Q**
Resistor	**R**
Relay	**K**
Silicon-Controlled Rectifier	**SCR**
Switch	**S**
Transformer	**T**
Test Point	**TP**
Integrated Circuit	**U**
Vacuum Tube	**V**
Transducer	**X**
Crystal	**Y**

© 2014 Cengage Learning

will be connected. As an example, one block may represent a stage of signal amplification with one signal entering and one signal leaving and proceeding to the next block (Figure 5-5).

■ FIGURE 5-5

Block diagram.

Analog circuits use **discrete components** such as resistors, capacitors, and semiconductor devices. **Analog schematic diagrams** present a different appearance from digital schematics.

Discrete analog components may require close physical proximity to the PCB surface to maintain signal paths and to prevent coupling or delays between stages. It is important to show tight component grouping in the schematic diagram, so the PCB layout can follow it precisely.

Digital schematic diagrams appear quite different from analog schematic diagrams because of differing interconnection patterns. Digital circuits tend to have many **common signals** that occur throughout the schematic. This results in many long signal lines that transverse the length and width of the drawing. To avoid confusion, it is better to label the common signals, as shown in Figure 5-6, rather than draw every connection as a line.

A schematic diagram should show the entire circuit in as few drawings as possible and should use the following techniques:

- Group subcircuit components together in the drawing.
- Signal flow should proceed from left to right, with input on the left and output on the right (Figure 5-7).

■ FIGURE 5-6

Label common signals in digital circuits.

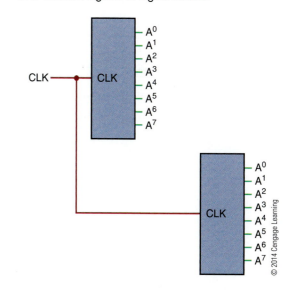

■ FIGURE 5-7

Signal flows from left to right; voltage potential has highest potential at top.

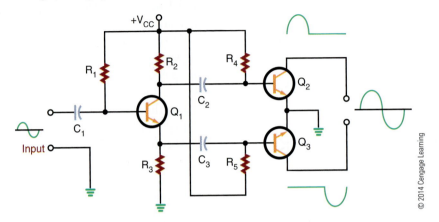

- Voltage potentials are drawn with the highest voltage at the top of the diagram and the lowest at the bottom, as shown in Figure 5-7.
- Signal lines should cross each other as little as possible. On digital schematic diagrams, label the connections with the appropriate signal abbreviation rather than draw a complete maze of lines (Figure 5-8).

- Label components starting on the top left on the input side of the schematic diagram, and move down and then back to the top, repeating this process across the schematic diagram (Figure 5-9).
- Critical leads should be short or isolated from other signals.
- All external connectors and components should be clearly indicated.

■ FIGURE 5-8

Use signal abbreviations rather than drawing a maze of lines.

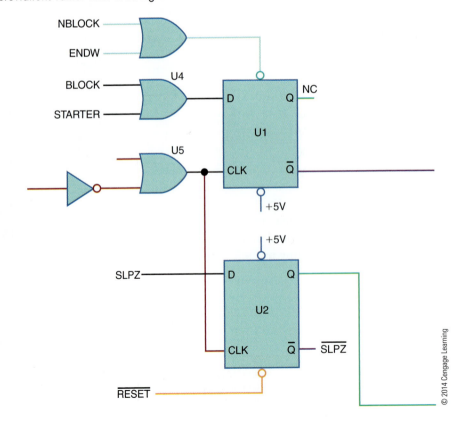

■ FIGURE 5-9

Label components, starting at the left side, and moving top to bottom, repeating across the schematic.

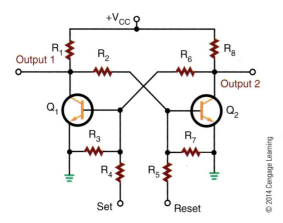

- All integrated circuit pins should be labeled, including the power supply inputs.
- Any unused integrated circuit logic gates or extra subcircuit inputs should be tied to the appropriate power supply level to provide stability in the circuit.
- Extra components, such as bypass capacitors, that are installed during the PCB construction process should be added to the final schematic.

Figure 5-10 shows the schematic symbol for common components used in the electronics field. Keep

in mind the following points while reviewing the symbols:

1. Components that are variable include an arrow as part of their symbol. For example, a potentiometer is a variable resistor.
2. Arrows also indicate the emitting or receiving of various forms of energy. Arrows pointing away from a symbol indicate it is giving off energy, such as a light-emitting diode (LED). If the arrow is pointing toward a symbol, it indicates it is receiving energy, such as a photocell.
3. A component's lead may be identified with letter symbols. For example, SCR leads are identified as K (cathode), A (anode), and G (gate). Transistor leads are identified as E (emitter), B (base), and C (collector).

Figure 5-11 shows two schematic symbols for ground. The most commonly used symbol is shown in Figure 5-11A. It represents chassis or earth ground, may be used multiple times in the same schematic diagram, and represents a common return point for a circuit. The ground symbol shown in Figure 5-11B represents chassis–ground only. If the schematic drawing uses this symbol, it is connected to the ground prong of an AC line cord, and the chassis will be at earth ground potential.

Schematic drawing lines represent wires or component leads. Letting lines cross on a diagram does not mean a connection occurs (Figure 5-12A). If a connection is intended, as shown in Figure 5-12B, a dot

■ FIGURE 5-10

Schematic symbols for common electronic components.

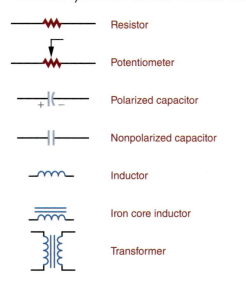
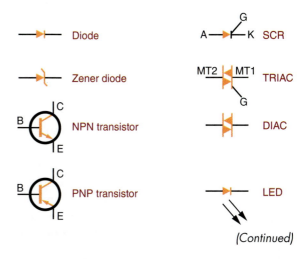

(Continued)

■ FIGURE 5-10 *(Continued)*

Schematic symbols for common electronic components.

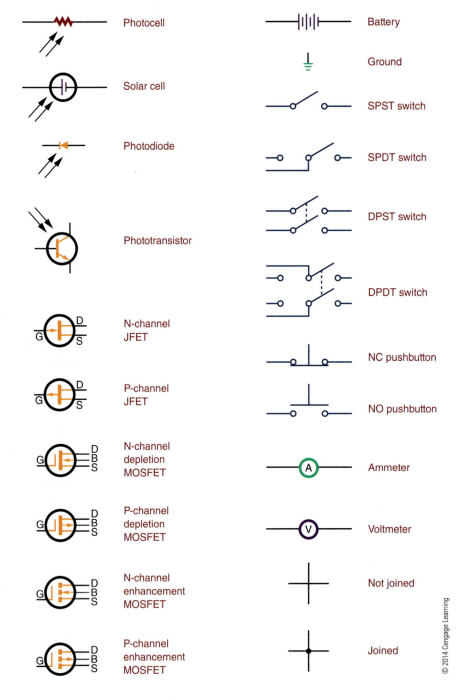

is placed at the junction. A connection dot is also used when three or more lines come together.

Common drafting tools can be used in the generation of schematic diagrams. Using schematic symbol templates speeds up the process. However, computer software programs are now used to speed up the generation of a schematic diagram. In addition to **computer-aided design (CAD)** programs, specialty programs are also available and include electronic circuit simulation programs.

Electronic circuit simulation programs include National Instrument's Multisim and New Wave Concepts' Circuit Wizard. These programs not only enable the user to create a schematic diagram but can simulate current through the circuit and allow the user to troubleshoot on the computer, using **virtual test equipment**. Additionally, the computer program Circuit Wizard shows animated current flow throughout the circuit represented by the schematic diagram.

© 2014 Cengage Learning

■ FIGURE 5-11
Ground symbols.

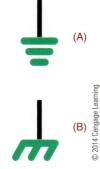

(A)

(B)

© 2014 Cengage Learning

■ FIGURE 5-12
Schematic drawing lines.

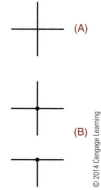

(A)

(B)

© 2014 Cengage Learning

5-2 QUESTIONS

1. What is the purpose of a schematic diagram?
2. What function does a block diagram serve?
3. Why is tight component grouping shown on a schematic diagram?
4. What techniques are used when drawing a schematic diagram?
5. What can be used to speed up the diagram drawing process?

5-3 BREADBOARDING

A **breadboard** is a platform for building a prototype of an electronic circuit. In the early days of radio in the Navy, radio operators would nail copper wires or terminal strips to a wooden board (often the bread

■ FIGURE 5-13
Solderless breadboard with spacing 0.1 (0.25 cm) inches apart in both directions. Metal clips below the perforations make the electrical connection.

© 2014 Cengage Learning

cutting board from the Mess deck) and solder electronic components to them. Sometimes they would glue a paper schematic diagram to act as a guide to locating terminals. Then they would solder components and wires over the symbols on the schematic.

Breadboarding is essential to proving that a circuit design works. *Breadboard* is the term that is now used for all types of devices used for prototyping. In 1971, Ronald J. Portugal of EI Instruments Inc. designed the **solderless breadboard** most commonly used today. It is usually manufactured from white plastic, with perforation and numerous metal clips to plug wires and components in (Figure 5-13). The spacing between the perforations is typically 0.1 inch (0.25 cm), the same as the spacing of the leads on an integrated circuit.

Solderless breadboards were developed to allow circuits to be assembled and altered quickly without soldering. The circuit is constructed using the actual components that would be used in assembling the circuit on a PCB. A PCB is not easy to modify once the circuits are assembled. PCBs only allow minor alterations following final assembly, such as changing resistor or capacitor values or replacing integrated circuits with the same type of devices.

There are many electronic trainers constructed using solderless breadboards. Two such electronic trainers are Digilent's Electronic Explorer (Figure 5-14) and National Instruments' education design and prototyping platform—ELVIS (Figure 5-15). Both of these units include virtual test equipment that displays signals on a computer.

■ **FIGURE 5-14**

An electronic explorer (electronic trainer) with solderless breadboarding and virtual test equipment by Digilent Inc.

Photo Courtesy of Digilent, Inc.

■ **FIGURE 5-15**

An education design and prototyping platform (electronic trainer) with solderless breadboarding and virtual test equipment by National Instruments.

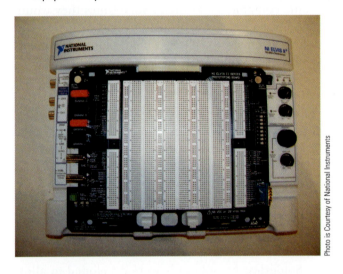

Photo is Courtesy of National Instruments

Computer programs such as National Instruments' Multisim, New Wave Concepts' Circuit Wizard, Cadence's OrCAD PSpice, and Spectrum Software's Micro-Cap 7 allow **virtual breadboarding**, or testing of a circuit and troubleshooting using virtual test equipment. Several of these programs shows the current flow in a circuit to aid in understanding how a circuit performs. These programs are the wave of the future.

5-3   QUESTIONS

1. What function does breadboarding serve?
2. Why were solderless breadboards developed?
3. Why should a circuit be breadboarded?
4. What are some computer programs that allow virtual breadboarding?
5. Why is a program such as Circuit Wizard important to the study of electronics?

SUMMARY

- Schematic symbols are symbols used to represent electrical and electronic schematic diagrams.
- Schematic diagrams are used to show how a circuit is connected.
- Reference designators identify components in a schematic diagram.
- A block diagram represents how the component blocks of a circuit are connected.
- A schematic diagram shows how the components are connected in a circuit.
- Analog schematic diagrams show how the discrete analog components are connected.
- Digital schematic diagrams use integrated circuits and have many common signals throughout the schematic; therefore, they use common signals rather than draw every line.
- A good schematic diagram should show the entire circuit in as few drawings as possible.
- Breadboarding allows circuits to be assembled and altered quickly.
- Breadboarding uses discrete components to build a circuit.
- Virtual breadboarding uses the computer to build and test the circuit without any analog components.

CHAPTER 5 SELF-TEST

1. Draw schematic symbols for the following:
   a. Battery
   b. Resistor
   c. Capacitor
   d. Wires not connecting
   e. Wires connecting
   f. Diode
   g. NPN transistor
   h. LED
   i. Ground
2. In problem 1, identify each symbol, with reference designators where appropriate. Increase the component number for each item that has a reference designator.
3. Where should the battery be placed in an electronic schematic diagram?
4. Redraw the following sketch as a proper schematic diagram.

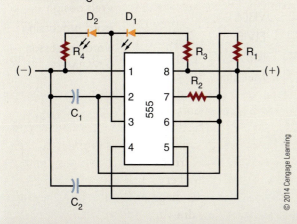

© 2014 Cengage Learning

5. Redraw the following sketch as a block diagram.

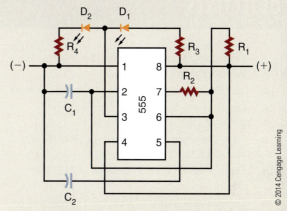

© 2014 Cengage Learning

6. Properly relabel the reference designators in problem 4.
7. What purpose does an electronic circuit simulation program, such as Multisim, serve with reference to schematic diagrams?
8. What purpose does breadboarding serve when constructing an electronic circuit?
9. When would solderless breadboarding be used?
10. What advantages do solderless breadboards serve?

Software for Electronics

OBJECTIVES

After completing this chapter, the student will be able to:

- Identify how electronics circuit simulations work.
- Discuss how electronic circuit simulation enhances the learning process.
- Describe when electronic circuit simulation should be used.
- Explain the difference between analog and digital simulations.
- Demonstrate National Instrument's Multisim, an electronic circuit simulator.
- Explain the function of National Instrument's Ultiboard with National Instrument's Multisim.
- Demonstrate New Wave Concepts' Circuit Wizard as an electronic circuit simulator.
- Explain the differences between the virtual test instruments used with New Wave Concepts' Circuit Wizard.
- Identify four areas of productivity software for electronics.
- Discuss the requirements for word processing.
- Describe the components of a presentation program.
- Explain the difference between a word processor and a page layout program.
- Describe the application of illustration programs and photo editors.
- Identify the purpose of web design.
- Explain the purpose of a spreadsheet program.

KEY TERMS

6-1 autoplace
6-1 autorouting
6-1 electronic circuit simulation
6-1 IBIS (input/output buffer information specification) model
6-1 mixed-mode simulators
6-1 SPICE
6-1 Verilog
6-1 VHDL
6-2 ELVIS
6-2 LabVIEW
6-2 Multisim
6-2 Ultiboard
6-2 virtual instruments
6-3 Circuit Wizard
6-4 desktop publishing (DTP)
6-4 graphical images
6-4 HTML
6-4 presentation programs
6-4 spreadsheet
6-4 Web design
6-4 word processing
6-4 word processor
6-4 WYSIWYG

Electronics is all around us, touching all levels of society, from everyday life to the workplace. With new emerging technologies, new products and services continue to be developed. Within industry, electronics plays an important role for innovation, leading to success in the global marketplace.

The success of industry also depends on having a skilled and motivated workforce. Young people studying electronics are learning skills that will allow them to follow electronics-related careers. This chapter focuses on the use of the computer to use simulation software to design and test electronics circuits in a virtual environment. It also covers several computer productivity programs that are important in the support of electronics.

6-1   SOFTWARE FOR ELECTRICITY AND ELECTRONICS

Electronic circuit simulation software can model electronic circuit performance and is a vital resource for checking circuit proper operation. *Electronic circuit simulation* software uses mathematical algorithms to duplicate the operation of electronic circuits. Due to its highly accurate modeling capability, many educational programs use simulation software for teaching electronics for various programs, including electronics engineering and electronic technology.

Using electronic circuit simulation software allows users to become part of the learning process by actively engaging them in the creation of electronic circuits, studying circuit operation, and evaluating their performance. Being able to simulate an electronic circuit's behavior before constructing it can improve a circuit's design by helping the creator identify poor circuit designs. Knowing a circuit operates poorly provides an opportunity to improve performance through redesigning.

Electronic circuit simulation offers the ability to improve a circuit performance when:

- Breadboarding is not practical.
- It is difficult to look at the internal signals.
- Tooling is expensive for building a circuit.

The most well-known analog simulator is Berkeley **SPICE**. SPICE is industry-standard nomenclature for electronic circuit simulation and is the most advanced in the world. Probably the best-known digital simulators are those based on **Verilog**, a hardware description language and **VHDL**, **V**ery High-Speed Integrated Circuit **H**ardware **D**escription **L**anguage.

Event-driven digital circuit simulations keep track of internal states of the circuit by allowing the study of digital problems. In addition, they can present a complete time diagram of a digital circuit. A logic analyzer can display digital signals, each of which is shown in separate synchronized displays.

There are programs that are analog electronic circuit simulators only. The more popular simulators include both analog and event-driven digital simulation capabilities. These programs, known as **mixed-mode simulators**, can simulate analog, event-driven digital, or a combination of both. An entire mixed signal analysis can be driven from one integrated schematic. All the digital models in mixed-mode simulators provide accurate information.

Printed circuit board (PCB) design process requires specific component models as well as autorouting for the traces. **Autorouting** is a technique for saving time and money when designing a PCB. Historically, autorouting is faster than manual routing, but the perception is that the results are not always satisfactory or do not look the same as manual routing.

Another feature is the ability to **autoplace** the components on a PCB. When autoplace is used to place the components on a board layout, it confirms that the parts will actually fit on the board. To confirm that the component placement can be routed, autoroute the traces. Autorouting can often get the board finished in a fraction of the time needed to manually route it with adequate results.

An **IBIS (input/output buffer information specification) model** is generally used in placing SPICE models for high-speed circuits. IBIS models enable a technique to provide information about a product to potential customers without having to reveal the intellectual property.

6-1   QUESTIONS

1. What is the purpose of electronic circuit simulation?
2. How does electronic circuit simulation engage learning?
3. In what situations would electronic circuit simulation be used?
4. What is the difference between SPICE and Verilog?
5. When would a mixed-mode electronic circuit simulation be used?

6-2 MULTISIM

National Instrument's **Multisim** (Figure 6-1) is an electronic schematic capture and simulation program that is part of a suite of electronic circuit design programs. Multisim uses the original Berkeley SPICE–based software simulation. It includes a microcontroller simulation as well as integrated import and export features to the printed circuit board layout software Ultiboard.

Multisim is widely used in education and industry to provide education on circuit design, electronic schematic design, and SPICE simulation. National Instruments (NI) has maintained this educational component, with a specific version of Multisim that includes features developed for teaching electronics.

Multisim includes a number of measurement instruments that can be used to interactively measure the behavior of circuits. It provides several **virtual instruments** for taking circuit measurements, including these:

- Analog and digital multimeters
- Bode plotter
- Logic analyzer
- Oscilloscope
- Signal generator
- Wattmeter

These instruments behave like their real-world equivalents to set, use, and read the simulated measurements. Using measurement instruments provides an easy way to examine a circuit's behavior through simulated results.

Multisim provides the user with the tools to analyze circuit behavior. The intuitive and easy-to-use software platform combines schematic capture and industry-standard SPICE simulation into a single integrated environment. Multisim abstracts the complexities and difficulties of traditional syntax-based simulation, so users no longer have to be an expert in SPICE to simulate and analyze circuits. Multisim is available in an educational version and a professional version.

Educators use Multisim to guide learning and exploration of circuit behavior by using *what-if* experiments and simulation-driven instruments. Multisim and industry-standard LabVIEW measurement software by National Instrument (NI) were combined with the NI **ELVIS**, a prototyping workstation, to form a complete development platform. **LabVIEW** (Laboratory Virtual Instrumentation Engineering Workbench) is a design and development system using a visual programming language from National Instruments. Multisim and NI Ultiboard combine powerful SPICE simulation and analysis with flexible layout tools to quickly and professionally complete a wide variety of PCB designs.

■ **FIGURE 6-1**

Screenshot of Multisim 11 with virtual oscilloscope present.

Source: National Instruments Multisim™ software © 2014 Cengage Learning

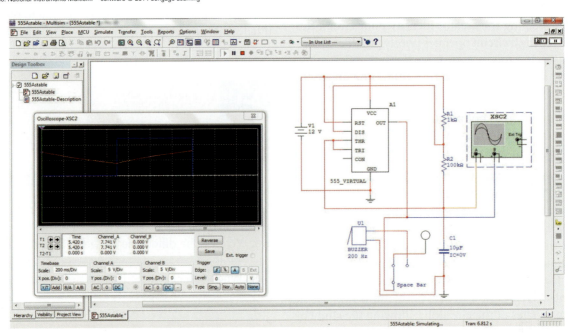

Electronic engineers and researchers use Multisim for schematic capture, SPICE simulation, and circuit design. Multisim is used to identify errors, confirm circuit designs, and build faster prototypes. Schematics are then transferred to NI **Ultiboard** layout to prototype completed PCBs. Ultiboard enables engineers to integrate design and test with a single, unified software platform that includes NI Multisim and NI LabVIEW. Design engineers can now easily compare simulated and real-world measurements to benchmark performance, resulting in fewer design problems.

6-2   QUESTIONS

1. What is Multisim?
2. What is the basis of Multisim?
3. What is the function of the measurement instruments in Multisim?
4. What is the difference between Multisim education version and the Multisim's professional version?
5. What is the function of Ultiboard?

6-3   CIRCUIT WIZARD

Circuit Wizard (Figure 6-2) is a software system that combines schematic circuit design, simulation, PCB design, and CAD/CAM manufacture by New Wave Concepts. By integrating the entire design process, Circuit Wizard provides all the tools necessary to produce electronics projects from start to finish, including on-screen testing of the PCB before assembly.

Designing with Circuit Wizard requires selecting components needed for the circuit from the library and connecting them together using the intelligent wiring tool. Once the circuit is completed, pressing Play starts the simulation.

Circuit Wizard features a simulation engine based on Berkeley SPICE. The component library boasts a library of over 1,500 simulated components covering everything from simple resistors through ICs. For each component there is a choice of either an ideal model for teaching theory or a real-world model (similar to those found in data books) for circuit design.

Circuit Wizard provides several virtual instruments (Figure 6-3) for taking circuit measurements, including the following:

- Analog and digital multimeters
- Logic analyzer

■ **FIGURE 6-2**

Screenshot of Circuit Wizard shown with graph, similar to oscilloscope results.

Source: New Wave Concepts Limited Circuit Wizard™ © 2014 Cengage Learning

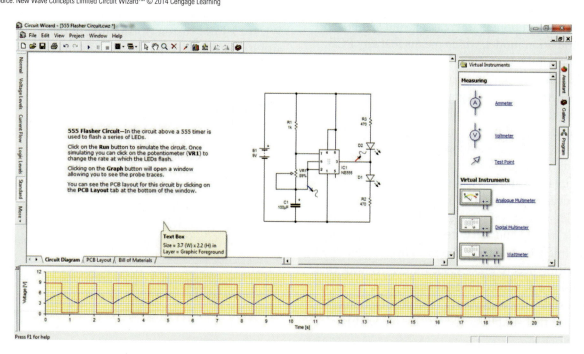

■ FIGURE 6-3

Some of the virtual instruments available in Circuit Wizard.

Source: New Wave Concepts Limited Circuit Wizard™ © 2014 Cengage Learning

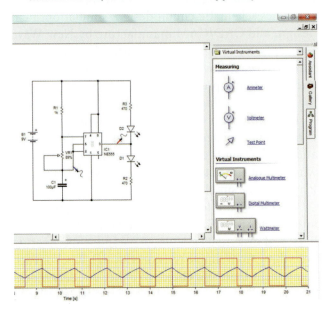

- Oscilloscope
- Signal generator
- Wattmeter

Circuit Wizard allows the operator to change the circuit appearance when it is running a simulation. Features of Circuit Wizard PCB fabrication (Figure 6-4) include these:

- PCB components—Circuit Wizard has over 1500 PCB components covering all the typical devices. For each component, there is a layout symbol and a simulation model, which means that any PCB drawn can be simulated.
- Designing PCBs—Creating a PCB starts with selecting the schematic circuit diagram to be transferred. Clicking Convert to PCB allows Circuit Wizard to complete the layout. Alternatively, the PCBs can be designed using the traditional manual technique using Circuit Wizard's full-featured drafting tools.
- Automatic routing—Circuit Wizard includes an advanced automatic router that can route single-sided and double-sided boards. Using the latest "rip-up and retry" technology, it achieves extremely high completion rates by going back and rerouting nets to make space for connections that could not be routed on a previous pass.
- PCB style views—Styles is a feature of Circuit Wizard that simplifies the process of viewing

PCBs. It combines various display options to show how a particular design would look at different stages in the production process. Several preset styles are shipped with Circuit Wizard, and you can also create your own.

- SPICE PCB simulation—Circuit Wizard can simulate PCBs on a screen before the board is manufactured. To check the PCB: Wire a battery to the board, add any discrete components, and click Play.
- PCB virtual instruments—Circuit Wizard offers several virtual test instruments for analyzing the performance of a PCB: a DC power supply to provide variable power output; a function generator to produce waveforms; a multimeter for voltage, current, and resistance measurement; an oscilloscope to record waveforms; and a probe for simple voltage measurements.
- Linking to other CAD packages—Circuit Wizard offers the ability to export PCB masks to third-party CAD software, so the outlines can be used as a template for correctly aligning screw holes, batteries, switches, and so on.
- CAM support—Circuit Wizard supports CAM outputs and can produce Excellon files for numerically control drill machines. It can generate files for third-party software so PCBs can be machined without the need for chemical etching. It also produces Gerber files for professional board manufacture.

Circuit Wizard can also produce a Bill of Materials spreadsheet (Figure 6-5). The spreadsheet contains the component properties assigned in the design. Circuit Wizard allows custom formatting of the spreadsheet, and it is possible to produce totals in order to determine the cost of producing a PCB.

6-3   QUESTIONS

1. What functions does Circuit Wizard combine into the program?
2. What is the basis of Circuit Wizard?
3. What virtual test instruments are included with Circuit Wizard?
4. How does Circuit Wizard Autorouter route the traces?
5. How can Circuit Wizard help with budgeting the cost of a PCB?

■ FIGURE 6-4

Circuit Wizard PCB fabrication layouts.

Source: New Wave Concepts Limited Circuit Wizard™ © 2014 Cengage Learning

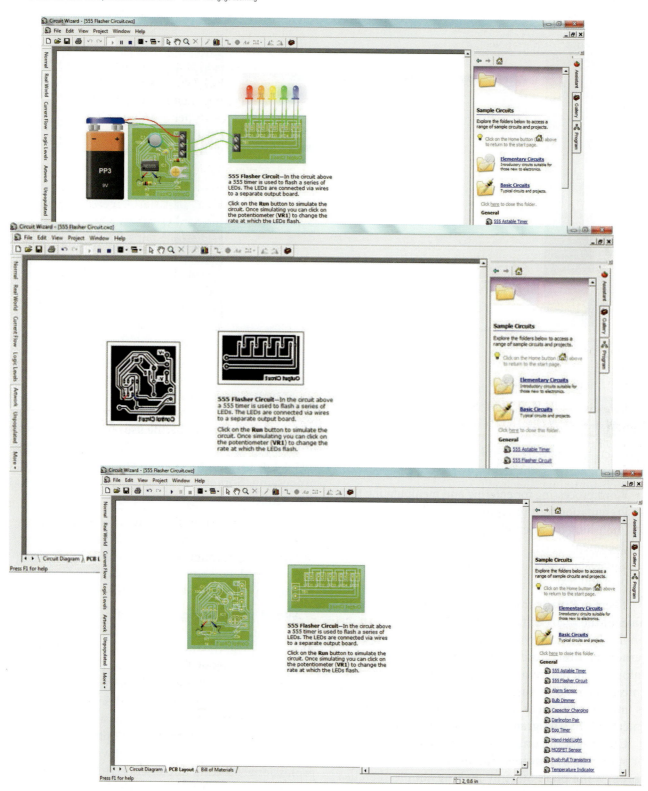

■ **FIGURE 6-5**

Circuit Wizard Bill of Materials based on component properties of the circuit.

Source: New Wave Concepts Limited Circuit Wizard™ © 2014 Cengage Learning

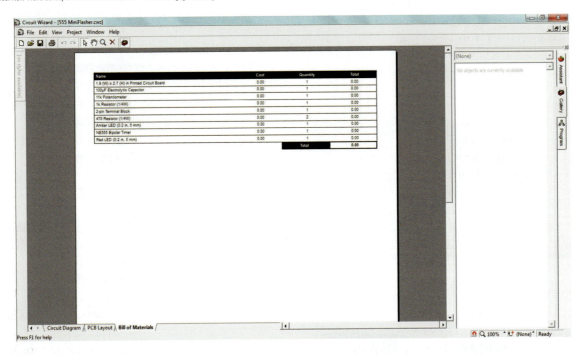

<table>
<tr><th>Name</th><th>Cost</th><th>Quantity</th><th>Total</th></tr>
<tr><td>1.9 (W) x 2.7 (H) in Printed Circuit Board</td><td>0.00</td><td>1</td><td>0.00</td></tr>
<tr><td>100µF Electrolytic Capacitor</td><td>0.00</td><td>1</td><td>0.00</td></tr>
<tr><td>11k Potentiometer</td><td>0.00</td><td>1</td><td>0.00</td></tr>
<tr><td>1k Resistor (1/4W)</td><td>0.00</td><td>1</td><td>0.00</td></tr>
<tr><td>2-pin Terminal Block</td><td>0.00</td><td>1</td><td>0.00</td></tr>
<tr><td>470 Resistor (1/4W)</td><td>0.00</td><td>2</td><td>0.00</td></tr>
<tr><td>Amber LED (0.2 in, 5 mm)</td><td>0.00</td><td>1</td><td>0.00</td></tr>
<tr><td>NE555 Bipolar Timer</td><td>0.00</td><td>1</td><td>0.00</td></tr>
<tr><td>Red LED (0.2 in, 5 mm)</td><td>0.00</td><td>1</td><td>0.00</td></tr>
<tr><td></td><td></td><td>Total</td><td>0.00</td></tr>
</table>

6-4 PRODUCTIVITY SOFTWARE FOR ELECTRONICS

Productivity software useful for electronics includes these programs:

- Word processing for entering and manipulating text
- Presentation for presenting data and images
- Publishing for assembling text and images into a document
- Spreadsheet for entering and manipulating numbers and text

There are many free productivity programs available for the many different types of operating systems on the Internet. Anything downloaded from the Internet requires a word of caution regarding virus. Any information, program, or image downloaded from the Internet should be suspected of containing a virus. It is important to check anything downloaded from the Internet with a program designed to check for viruses.

Creating a document, presentation, publication, or spreadsheet requires planning and then working through the plan. Start by making simple sketches on paper, showing the various pages or elements to be created. Include images and diagrams on how the pages are interconnected or flow.

Word Processing

Word processing involves the use of a word processor program, such as Microsoft's Word® or Apple's Notes®. Just using a word processor to enter data is not word processing. Someone who is a word processor has these skills:

- Identify the components that make up a word processing document.
- Format a document to given specification.
- Style the document to given specifications.
- Explain how he or she created the document.

Word processing can be defined as using software to enter text, establish the formatting, make revisions, and print the final document. Word processing is not limited to text; it also enables the user to insert and edit graphics, clip-art, and photos.

Word processing can accomplish text manipulation functions that extend beyond the basic ability to enter and change text. Using the default options available in most word processing programs allows the creation of professional-looking documents. However, a poorly designed document can result when the user changes the default options and applies special formats, making it less appealing to a reader.

When formatting a document, use basic word processing rules:

- Use the software's default margins to provide the optimum amount of white space, as the white space allows the eyes to rest.
- Use only one space between all sentences. This means no double space with word processing for any reason.
- Use only one font, such as a serif font like Times New Roman, for the text body because it makes the document easier to read.
- For titles or section headings, use a sans serif font such as Helvetica. There is no reason to use more than two fonts in any word processing document.
- Avoid using all the formatting options available in a document; they make the document difficult to read and distract from the content.
- Avoid overusing special formatting such as bold, italic, and underline; save them for emphasis, headings, or citations. Use only one formatting function per application.
- Use the header to put a title and name on each page of the document, if desired, to identify each page in a document.
- Use the footer to number the pages on the bottom right-hand corner.
- Avoid centering titles, page numbers, and so on.

Stick to these basic rules and proofread the document carefully for spelling and grammatical errors that a spell-checker did not catch, to produce a well-formatted, easy-to-read document. Save the document, using a name that reflects its contents.

Presentation

Presentation programs, such as Microsoft's PowerPoint® or Apple's Keynote®, are used to display information, normally in a slide format; the presentation is referred to as a slide show and is often paired with an oral presentation. A presentation program typically includes three major functions:

- An editor to allow text to be inserted and formatted
- A process for inserting and altering the graphic images and photos
- A slide-show arrangement to display the presentation

A slide show can be used to support a speech, help an audience visualize a complicated process or concept, or focus attention on a specific topic. A well-organized slide show allows a presenter to fit text, visual images, and videos to an oral presentation. Keep in mind that a bad presentation can achieve the opposite. Poorly designed slides with too much text or bad graphics can distract and annoy the audience. The slides are only for support, not to replace any talking. The slides act as placeholder, tell a story, describe important data, explain a circumstance, and provide the keywords for the speaker. If the slides are read, the audience will tune out the speaker. To create a presentation for a professional look and concise content use, the following techniques:

- Create a logical flow for the presentation.
- An audience's attention is limited, so avoid details that will not be remembered.
- Use a sans serif font such as Helvetica for titles and a serif font such as Times New Roman for bullets or body text.
- Avoid using title capitalization except for titles, because sentence capitalization is much easier to read.
- Keep the design very basic and simple.
- Use color to highlight the message sparingly.
- Minimize the number of colors on a slide.
- Use colors for design and contrast, to highlight the message.
- Avoid using sentences; instead, use keywords, with only one idea per slide.
- Use good images to help the audience understand the message.
- Use no more than one graphic image or chart per slide
- Do not use images just to "decorate."
- Clearly make the final message stand out.
- Provide a summary of the information presented.
- Spell-check several times, because spelling mistakes attract the wrong kind of attention.
- Keep things simple and only use a basic dissolve to move from one slide to another.

A well-prepared and enthusiastic presentation helps convince an audience of the message and maintains their attention. Some key points that define a good presentation are listed here:

- Know the arrangement and order of the slides.
- Avoid talking too fast.
- Speak aloud, clearly and with confidence
- Move around to help you connect with the audience.
- Maintain eye contact with the audience.
- Do not distribute a handout before you begin the presentation, because the audience will start reading instead of listening. Inform the audience that a handout of the slides will be distributed when the presentation is finished.

Remember, presentation programs are tools designed to enhance a presentation, not to *be* the presentation. The message should be the focus of the presentation, not the slides.

Publishing

There are four types of software used for computer publishing, also referred to as desktop publishing, including these:

- Word processor
- Page layout
- Graphic images
- Web design

These are the main programs for desktop publishing. There are add-ons, extra specialty software, and utility programs that are not discussed here but that can help with desktop publishing. These programs all use similar design elements when designing a document:

- **Alignment:** Each element on the page should have a visual connection to another element to create a clean, refined appearance.
- **Contrast:** This is the most effective way to add visual interest to a page by creating two elements that are really different.
- **Proximity:** This is the grouping of related elements close together to create one visual unity with a clean, fresh appearance.
- **Repetition:** Visual elements should be repeated throughout a page to help develop, organize, and unite appearance through the use of a bold font, design element, format, spatial relationship, and so on.
- **Page Layout:** This is the process of laying out the elements on a page and includes these considerations:
  - Text that can be keyed or placed in the layout with styles applied to text automatically, using style sheets
  - Images that can either be linked and modified as an external source or embedded and modified with the layout software
  - Graphic styles that can be applied to layout elements and include color, filters, transparency, and a parameter that designates the way text flows around the images

A **word processor** is used to capture keystrokes (enter text) and edit text, which also includes spelling and grammar checking. A person using a word processor is word processing and is able to format specific elements into the text, which leads to reducing the formatting time when the text is imported into a page layout program. Word processors provide a wide variety of templates for use in special formatting.

Simple page layouts can be done in a word processor, but word processors are best used for working with text and not for page layout. Additionally, word processing file formats are generally not suitable for printing commercially. A word processor can import and export a variety of formats for compatibility with other programs.

Desktop publishing (DTP) uses page layout software on a personal computer to create a wide range of printed matter. Desktop publishing allows more control over layout and design than word processing does.

When skillfully used, desktop publishing can produce printed literature with an attractive layout and a type quality comparable to traditional typesetting and printing. Desktop publishing allows an individual or small business to publish without the expense of using a commercial printer.

Desktop publishing programs have different strengths (and weaknesses) that should be matched to the publication. Desktop publishing publications fall into three categories:

- **General Publications:** brochures, flyers, newsletters, posters, and small booklets.
- **Multichapter Documents:** academic publications, books, journals, and manuals.
- **Publications with Tables:** data-intensive publications, scientific publications, and statistical and technical publications.

Desktop publishing software allows for combining text and images on a page, manipulation of the page elements, creative layouts, and multipage documents. High-end or professional-level desktop publishing programs include more prepress features, whereas software for home or small business use have more templates and clip art. Prepress refers to the process of preparing the digital files for the printing press. Templates are used in publishing to automatically copy or link elements and graphic design styles to some or all the pages of a multipage document. Linked elements can be modified in all locations or pages without having to change each element on each page in a number of pages that have the same element. Master pages are templates that are used to apply a graphic design style to all the pages in a document and add automatic page numbering if required.

In desktop publishing, **graphical images** are handled by an illustration or drawing program and an image editor to prepare graphic images for inclusion in a document.

- **Illustration programs** and drawing programs use a vector graphics format that allows a more flexible technique for creating artwork that can be resized without losing detail.

- **Image editors** are also referred to as *photo editors* and are used with images such as photos and scanned images. Photo editors are also better for developing images for the web and offer many special photo effects.

Web design contains some overlap with desktop publishing programs and what is known as Hypertext Markup Language, or **HTML**, to create a website. HTML is the main markup language used for displaying web pages on a web browser. To create a website and place it on a web server involves assembling a group of HTML pages, referred to as web pages, with the home page identified as the *index*. A web page is a virtual electronic page that is not limited by paper parameters. Most web pages can be resized either by rescaling the content of the page with new page parameters or by reflowing the content of the page.

The website is designed by using a web authoring program that provides a graphical layout interface similar to that used in desktop publishing or by manually encoding the pages in HTML or by a combination of both. A good solution is to be familiar with both techniques and be able to move back and forth between them.

Many graphical HTML editors such as Microsoft FrontPage® and Adobe Dreamweaver® use a layout engine similar to that in a DTP program and use a **WYSIWYG** (What You See Is What You Get) editor. Serious web designers prefer to write in HTML code for their website because it offers greater control and because graphic editors result in extra codes that are not required to run the website.

Spreadsheet

A computer **spreadsheet** is a software program used for inputting, manipulating, organizing, and storing data, which are placed in cells. Each cell is identified by a grid placement with letters across the top to identify columns and numbers down the left side to identify rows. The cells are identified by column and row designation such as A1, B12, ZZ123, and so on. These cells can contain numbers, formulas, and text. Functions are built-in formulas that are used for common tasks.

Originally, spreadsheets existed only in a paper format; today, they are created and maintained using a computer. Spreadsheets are used in any situation that requires working with numbers and are commonly found in the accounting, analysis, budgeting, and forecasting fields. They are also used for any situation that requires tables and lists to be built and sorted. The advantage of using a computerized spreadsheet is the ability to easily update the data and to increase productivity by the ability to recalculate an entire sheet automatically after a change is made.

A spreadsheet file may consist of multiple worksheets, which are usually referred to as *sheets*. The sheets combine together to form a workbook. A cell from any sheet is capable of referencing cells on other sheets.

A spreadsheet should serve a purpose, and it should be well documented and easy to understand. Here are some basic rules for creating a spreadsheet:

- Create a title for the spreadsheet that is self-explanatory and confirms its purpose, so it is easy for readers to identify what they are looking at.
- Use the footnote section of the spreadsheet to place the filename, date, and creator's name.
- Make the spreadsheet read from left to right and top to bottom. Numerical data is entered on the left because the reader always reads from left to right, with the output data on the right.
- Use formulas rather than inputting numerical values, to assist in following the logic of the spreadsheet. It makes updating information faster.
- Mathematical assumptions should be identified in separate cells, not buried in the formulas, to assist in reading.
- Document any and all assumptions to the right side of a cell, to save time at a later date.
- Present a format that can be easily followed from the process to the results.
- Add notes, procedures, steps, or any other explanations to help the reader.
- Use the intermediate-level functions in the spreadsheet program (e.g., sum, sorting, auto formatting, formula auditing, etc.)

By following these few simple rules, you will create a spreadsheet that is easier to work with, more professional in appearance, and easier for others to read and evaluate. Some guiding principles of spreadsheet design are listed here:

- The spreadsheet's value lies in the information presented, not extra design elements; therefore, do not use logos, graphics, and so forth.
- Do not use any extra formatting or color. Color should only be used to distinguish and highlight data.
- Everything in the spreadsheet should be obvious and clearly presented, so avoid hidden cells, formulas, and sheets.

- The data are the most critical part of any spreadsheet, so make sure they are correctly entered.
- Use a spell-checker to check all text every time it is entered or changed.
- Format cells with some of the formatting options available in most spreadsheets:
  - Add bold or italics, borders, or formulas.
  - Align or wrap text within a cell.
  - Change the color of the text or background.
  - Format cell contents with the appropriate symbols for currency, percentages, commas for thousands, and so on.
  - Merge cells horizontally in selected rows.
- Use the text wrap capability on column headings so that the length of the data entry determines the width of the column.
- Use Print Preview to see how the printed document will look. If any rows or columns are outside the printed area, consider either reorganizing the data to fit or moving the page break to a more logical place. Do not shrink the page to fit everything on a single page.
- A benefit to entering data in a spreadsheet is the ability to sort the information either alphabetically or numerically, in ascending or descending order.
- Charts are a way to analyze spreadsheet data visually and are updated whenever the spreadsheet is modified. Remember, quality graphics give the viewer the greatest impression in the shortest time and with the least amount of space.

Never assume a spreadsheet is correct when entering data. Data must be added carefully and need to be checked and rechecked several times for accuracy.

Do not attempt to simplify a spreadsheet by using lots of formatting, bold lines, and lots of color. Rather, use the fewest number of cells necessary to produce the result that flows from top to bottom and left to right, and position related cells close together on one sheet. Above all, keep the spreadsheet simple.

6-4 QUESTIONS

1. What are the different areas of productivity software?
2. What is a word processor used for?
3. When would a presentation program be used?
4. What are the elements of a page layout program?
5. What function does a spreadsheet serve?

SUMMARY

- Electronic circuit simulation uses mathematical algorithms to duplicate real electronics circuits.
- Electronic circuit simulation allows the user to become part of the learning experience by engaging him or her in these tasks:
  - Creating electronic circuits
  - Studying the operation of the electronics circuit
  - Evaluating the performance of the electronics circuit
- Electronic circuit simulation helps to improve circuit performance under these conditions:
  - When breadboarding is not possible
  - When it is difficult to look at the internal signals
  - When tooling is expensive for building a circuit
- An analog simulation uses the Berkeley SPICE modeling simulation, and digital uses Verilog or VHDL simulations.
- Multisim is an electronics schematic capture and simulation program by National Instrument.
- National Instrument's Ultiboard takes the schematic diagram and makes a printed circuit board layout from it.
- Electronic engineers use Multisim for schematic capture, SPICE simulation, and to design electronic circuits.
- Circuit Wizard combines schematic circuit design, circuit simulation, PCB design, and CAD/CAM fabrication of PCB.
- Circuit Wizard includes the following virtual instruments:
  - Analog and digital multimeter
  - Logic analyzer
  - Oscilloscope
  - Signal generator
  - Wattmeter
- Productivity software includes the areas of word processing, presentation, desktop publishing, and spreadsheets.
- Word processing uses a word processor program for entering text, formatting, making revisions, and printing the final document.
- Presentation programs include a text editor, a process for entering and editing graphics and photos, and an arrangement for a slide show.
- A page layout program is used for desktop publishing and allows the combining of text and images on a page, manipulating the page elements

for creative layouts, and producing multipage documents easily.

- An illustration program uses vector graphics for resizing artwork.
- A photo editor works with photos and scanned images.

- Web design is used to create web pages to be applied to a website on a server, with the home page identified as the index.
- A computer spreadsheet is used for inputting, manipulating, organizing, and storing numbers and other data.

CHAPTER 6 SELF-TEST

1. What does the success of electronic circuit simulation rely on?

2. How does electronic circuit simulation involve the user?

3. What is SPICE?

4. What are Verilog and VDHL?

5. How does Multisim enhance the educational experience?

6. What is National Instrument's ELVIS?

7. What is the National Instrument's LabVIEW?

8. How does Circuit Wizard enhance the learning experience?

9. What is unique about Circuit Wizard's PCB design process?

10. Use an electronic circuit simulation program to determine the voltage across resistor R_5 for the schematic diagram in the figure shown.

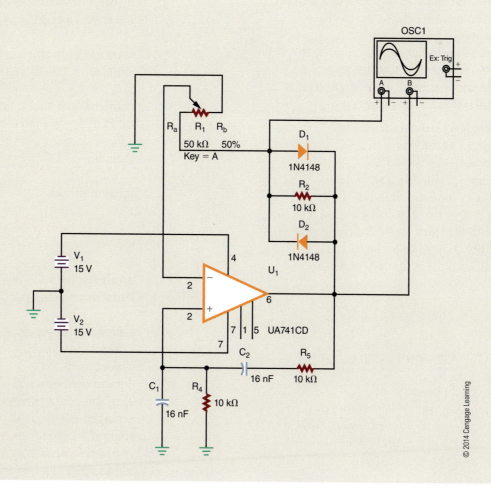

11. Using a word processor, develop an article on the advantages of using a SPICE program.

12. Develop a presentation on using an electronics simulation program.

13. Use a page layout program to develop an instruction manual with text and images for an electronic simulation program.

14. Use an illustration or drawing program to create a schematic of an electronics circuit of choice.

15. Use a photo editor to develop photos or scanned images for a slide show on a physical breadboard of a circuit of choice.

16. Create a personal website on the benefits of using an electronics simulation program.

17. Develop a spreadsheet on the parts list for an electronics circuit, with prices for each part and a total cost.

CHAPTER 7

Safety

OBJECTIVES

After completing this chapter, the student will be able to:

- Identify the dangers of working with electricity.
- Understand how to handle a situation when someone is getting shocked.
- Work confidently on electrical circuits and equipment, following guidelines.
- Identify two techniques to prevent electrical shock.
- Understand the purpose of a ground fault circuit interrupter (GFCI).
- Know what labels to look for when using electrical tools and equipment.
- Identify the dangers of static electricity with semiconductor devices.
- List precautions required to minimize electrostatic discharge (ESD).
- Know how to work safely with electricity.
- Understand the purpose of a material safety data sheet (MSDS).

KEY TERMS

7-1 body resistance

7-1 cardiopulmonary resuscitation (CPR)

7-2 Canadian Standards Association (CSA)

7-2 ground fault circuit interrupter (GFCI)

7-2 grounding

7-2 insulating

7-2 *National Electrical Code (NEC)*

7-2 *NEC* guidebook

7-2 Underwriters Laboratories (UL)

7-3 antistatic workstation

7-3 electrostatic discharge (ESD)

7-3 static electricity

When working with electricity, it is the technician's responsibility to prevent electrical shock and on-the-job injuries and know how to identify the dangers of working with electricity.

Static electricity is an electrical charge at rest on a surface. The static charge becomes larger through the action of contact and separation or by motion. The electrostatic discharge takes place when the charged body comes near or touches a neutral surface.

7–1   DANGERS OF ELECTRICITY

Nothing you learn in this book is more important than learning to work safely with electricity. Work smart and safe when dealing with electricity. You can help protect yourself and everyone around you from painful or even lethal shock. You can also help prevent fires.

During the winter months, the skin dries out, offering some insulation. When the skin becomes wet through water or sweat, the **body resistance** drops. The low resistance allows more current to pass through the body, resulting in impairment.

Chapter 11 discusses the flow of current that is measured in amperes. It is the flow of current through the body that causes pain and even death from electrical shock. Voltage is the pressure behind electrical current, but the current flow is more important than voltage when discussing electrical shock. Figure 7-1 shows that the body resistance is the inverse of current through the body. As body resistance decreases, the current through the body increases. When the body resistance increases, the current flow decreases through the body.

Electrical shock occurs when an electric current flows through the body when a complete circuit exists.

■ FIGURE 7-1

As the body's resistance decreases, the current through the body increases. As the body's resistance increases, the current through the body decreases.

© 2014 Cengage Learning

Different levels of current produce different results, as shown in Figure 7-2.

One technique to reduce current flow is to increase body resistance. Body resistance is high when the skin moisture content is low, and the skin has no cuts or abrasions at the point of electrical contact. In these situations, very little current flows, with a mild shock resulting.

If the situation were reversed, with high skin moisture content lowering the body resistance, a large current would flow. If the current flows through the chest region, the heart could go into ventricular fibrillation, resulting in rapid and irregular muscle contractions and leading to cardiac arrest and respiratory failure.

Factors that influence the effects of electrical shock are listed here:

- Intensity of the current
- Frequency of the current
- Current path through the body
- Length of time current passes through the body

Remember, it is the amount of current flow through the body, not the amount of voltage contacted, which

■ FIGURE 7-2

This table lists the results that occur as different levels of current flow through the body.

Current	Sensation Felt
0.001 Ampere (1 mA)	A mild tingling sensation that can be felt.
0.010 Ampere (10 mA)	Start to lose muscular control.
0.030 Ampere (30 mA)	Breathing becomes upset and labored. Muscular paralysis.
0.100 Ampere (100 mA)	Death if the current lasts for more than a second.
0.200 Ampere (200 mA)	Severe burns, breathing stops. Death.

© 2014 Cengage Learning

determines the severity of a shock. The larger the current flow through the body, the greater the effect of the shock.

When dealing with someone who has experienced severe electrical shock, do not become part of the problem. First, send for help; then remove the source of power. Do not attempt to touch or pull the victim away without removing the power source, or you will also get yourself shocked.

If the power source cannot be secured, use a nonconducting material to remove the victim from the circuit. Once the victim is free, check for signs of breathing and pulse. If trained, begin **cardiopulmonary resuscitation (CPR)** if necessary. CPR is an emergency procedure performed by trained personnel to manually pump blood to the brain until further action can be taken by a qualified medically trained team to restore blood circulation and breathing for a person in cardiac arrest and respiratory failure.

2. Do not wear loose or flapping clothing. Not only might it get caught, but it might also serve as a path for the conduction of electricity.
3. Wear only nonconductive shoes. This reduces the chance of electrical shock.
4. Remove all rings, wristwatches, bracelets, ID chains and tags, and similar metal items. Avoid clothing that contains exposed metal zippers, buttons, or other types of metal fasteners. The metal can act as a conductor, heat up, and cause a bad burn.
5. Do not use bare hands to remove hot parts.
6. Use a shorting stick to remove high-voltage charges on capacitors. Capacitors can hold a charge for long periods of time and are frequently overlooked.
7. Make certain that the equipment being used is properly grounded with polarized plugs. Ground all test equipment to the circuit and/or equipment under test (Figure 7-3).

7-1 QUESTIONS

1. What causes an electrical shock: voltage or current?
2. What does voltage result in when dealing with electrical shock?
3. How much current can result in death from an electrical shock?
4. What technique can reduce current through the body?
5. What factors influence the effects of an electrical shock?

7-2 PREVENTIVE MEASURES

Take time to be safe when working on electrical and electronic circuits. Do not work on any circuits or equipment unless the power is secured.

1. Work only in clean, dry areas. Avoid working in damp or wet locations because the resistance of the skin will be lower, increasing the chance of electrical shock.

■ **FIGURE 7-3**
All equipment should use polarized plugs.

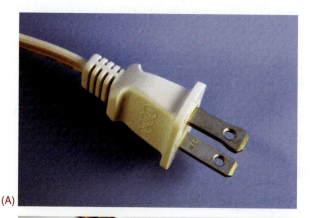

(A)

(B)

8. Remove power to a circuit before connecting alligator clips. Handling uninsulated alligator clips could cause potential shock hazards.

9. When measuring voltages over 300 volts, do not hold the test prods. This eliminates the possibility of shock from leakage on the probes.

> **NOTE** Safety is everyone's responsibility. It is the job of everybody in and out of class to exercise proper precautions to ensure that no one will be injured and no equipment will be damaged.

There are two techniques to prevent electrical shock when using electrical tools and equipment. They are **insulating** and grounding. Insulating refers to using a material that does not support current flow. Copper wire conductors are normally insulated. Hand and power tools and appliances contain layers of insulating material between the interior parts and the exterior housing.

Grounding helps to provide a path to carry the current to ground. All appliances are grounded. Three-prong plugs are an efficient way to ground tools or appliances, as shown in Figure 7-4.

A **ground fault circuit interrupter (GFCI)** (Figure 7-5) is a fast-acting circuit breaker that is sensitive to very low levels of current leakage to ground. GFCIs are designed to limit electrical shock to prevent serious injury. The disadvantage with GFCIs is that they operate only on line-to-ground fault currents.

■ FIGURE 7-4

The ground prong on a grounded plug is an efficient way to ground tools.

© 2014 Cengage Learning

■ FIGURE 7-5

A ground fault circuit interrupter (GFCI) is a fast-acting circuit breaker.

© 2014 Cengage Learning

■ FIGURE 7-6

A portable GFCI can be used on a job site to provide protection to the workers.

© 2014 Cengage Learning

Typically, these are currents leaking through insulation or current that flows during an accidental contact with a hot wire. A portable GFCI circuit is shown in Figure 7-6.

An **Underwriters Laboratories (UL)** label (Figure 7-7) on a device implies that the product bearing the label is safe for use as intended. Tests completed by Underwriters Laboratories determine whether a product meets the minimum safety standards. When purchasing a product, check to determine whether it has the UL label on it. The UL label has nothing to do with the quality of a product, only its safety.

A CSA certification mark indicates that a product has been tested and meets the requirements of

■ **FIGURE 7-7**

The Underwriter Laboratories label implies a product is safe for its intended use.

UL (Underwriters Laboratories), working for a safer world since 1894

recognized standards used as a basis for certification. Only when a product has been certified to an applicable standard does it bear the appropriate CSA certification mark. Certification marks appear on a product's packaging as well as on, or adhered to, the product itself. For consumers, the mark provides increased assurance of safety.

A CSA Group Mark (Figure 7-8) with the indicators "C" and "US" or "NRTL/C" means that the product is certified for both the U.S. and Canadian markets, to the applicable U.S. and Canadian standards. A **Canadian Standards Association (CSA)** mark with the indicator "US" or "NRTL" means that the product is certified for the U.S. market only to the applicable U.S. standards.

A number of insurance companies have formed a group known as the National Fire Protection Association. Every few years, this group publishes a summary of electrical-wiring codes under the general heading

■ **FIGURE 7-8**

The Canadian Standards Association safety test is similiar to the UL safety test, with very strict safety codes.

CSA Group

of the **National Electrical Code (NEC)**. The purpose of this code is to provide guidelines for safe wiring practices in residential and commercial buildings. State and local municipalities may require even more stringent codes than the *NEC* that must be followed. In many states, all wiring must be done or approved by a master electrician. These codes are published for both your own and your neighbor's protection. Electrical fires can and do happen, and they can spread to adjacent homes or apartments. The rules in the **NEC guidebook** help to minimize electrical fires and to provide safety in electrical wiring.

7-2   QUESTIONS

1. List five ways to prevent electrical shock.
2. Who is responsible for safety in the shop?
3. What is the purpose of insulation?
4. What is meant by grounding?
5. What organization checks an appliance or power tool to ensure it is safe to operate?

7-3   ELECTROSTATIC DISCHARGE

Static electricity is created when two substances are rubbed together or separated. The substances can be solid or fluid. The rubbing or separating causes the transfer of electrons from one substance to the other. This results in one substance being positively charged and the other substance being negatively charged. When either substance comes in contact with a conductor, an electrical current flows until it is at the same electrical potential as ground. This is referred to as **electrostatic discharge (ESD)**.

Static electricity is commonly experienced during the winter months when the environment is dry. Synthetics, especially plastic, are excellent generators of static electricity. When a person walks across a vinyl or carpeted floor and touches a metal doorknob or other conductor, an electrical arc to ground may result and a slight shock is felt. For a person to feel a shock, the electrostatic potential must be 3500 to 4000 volts. It typically takes 5000 volts to jump one-quarter of an inch. Lesser voltages are not apparent to a person's nervous system even though they are present.

Inside an integrated circuit, a static discharge can destroy oxide layers or junctions. These static discharges can cause an ESD latent defect. The device may appear to operate properly, but it is damaged. It has been weakened, and failure will result at a later time. There is no method for testing for ESD latent defect.

Figure 7-9 displays potential sources for generating an electrostatic discharge. Figure 7-10 shows

FIGURE 7-9

Potential electrostatic sources.

Object/Process	Material/Activity
Clothes	• Synthetic garments • Nonconductive shoes • Virgin cotton at low humidity levels
Chairs	• Finished wood • Vinyl • Fiberglass
Floors	• Sealed concrete • Waxed finished wood • Vinyl tile or sheeting
Work Surface	• Waxed, painted, or varnished surfaces • Vinyl or plastic
Work Area	• Spray cleaners • Plastic solder suckers • Soldering iron with ungrounded tip • Solvent brushes with synthetic bristles • Heat guns
Personal Items	• Styrofoam or plastic drinking cups • Plastic/rubber hair combs or brushes • Cellophane or plastic candy/gum wrappers • Vinyl purses
Packing and Handling	• Regular plastic bags, wraps, envelopes • Bubble wrap or foam packing material • Plastic trays, tote boxes, parts bin

© 2014 Cengage Learning

FIGURE 7-10

Electrostatic charges in the work environment.

Process	Relative Humidity	
	Low (10–20%)	High (65–90%)
Walking on carpet	35,000 V	1,500 V
Walking on vinyl floor	12,000 V	250 V
Working at workbench	6,000 V	100 V
Work stool padded with urethane foam	18,000 V	1,500 V
Plastic bag picked up from bench	20,000 V	1,200 V
Opening vinyl envelope used for instructions	7,000 V	600 V

© 2014 Cengage Learning

FIGURE 7-11

Semiconductor devices that can be damaged by electrostatic discharge.

Device Type	Sensitivity (Volts)
Bipolar transistors	380–7000
CMOS	250–2000
ECL	500
JFET	140–10,000
MOSFET	100–200
Schottky diodes, TTL	300–2500
SCR	680–1000

© 2014 Cengage Learning

typical measured electrostatic charges in a work environment. Figure 7-11 presents a list of electronic parts that may be damaged by sensitivity to electrostatic discharge.

Antistatic workstations (Figure 7-12) are designed to provide a ground path for static charges that could damage a component. They have a conductive or antistatic work surface that is connected to both a ground and the worker's skin through a wrist strap. The wrist strap has a minimum of 500 kΩ resistances to prevent shock in case of contact with a live circuit. When direct grounding is impractical, an ionized air blower is required.

The following list identifies precautions required to minimize electrostatic discharge:

1. Before starting work on sensitive electronic equipment or circuits, the electronics technician should be grounded using a wrist strap to discharge any static electric charge built up on the body.
2. Always check manuals and package materials for ESD warnings and instructions.
3. Always discharge the package of an ESD-sensitive device before removing it. Keep the package grounded until the device is placed in the circuit.
4. Minimize the handling of ESD devices. Handle an ESD device only when ready to place it in the circuit.
5. When handling an ESD device, minimize physical movement such as scuffing feet.
6. When removing and replacing an ESD device, avoid touching the component leads.
7. Do not permit an ESD device to come in contact with clothing or other ungrounded materials that could have an electrostatic discharge.
8. Before touching an ESD device, always touch the surface on which it rests for a minimum of 1 second to provide a discharge path.
9. When working on a circuit containing an ESD device, do not touch any material that will create a static charge.
10. Use a soldering iron with the tip grounded. Do not use plastic solder suckers with ESD devices.
11. Ground the leads of test equipment momentarily before energizing the test equipment and before probing an ESD device.

FIGURE 7-12

Antistatic workstation.

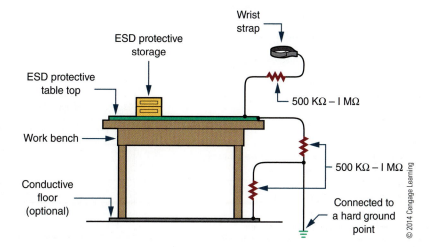

© 2014 Cengage Learning

7-3 QUESTIONS

1. How is an electrostatic charge generated?
2. What voltage discharge potential can be felt by humans?
3. What is meant by ESD latent defect?
4. An electronics technician who is working on ESD-sensitive devices would prepare by doing what first?
5. Is it safe to remove an ESD device before starting to work on a circuit?

7-4 SAFETY PRACTICES

Effective safety habits are necessary to succeed in the electronics field. Practicing good safety habits promotes a safe work environment. Everyone needs to be aware of and support a safe work environment.

A primary concern in the field of electronics is electrical shock, which is the passage of current through the body. Current can flow through the body just as it does through an electrical circuit. How much current flows depends on these factors:

- Body resistance is high if the skin is dry and there are no cuts or abrasions at the point of electrical contact. This results in little current flow with only a mild shock.
- Body resistance is low if the skin is moist. This results in a high current flow, and if the current flow is through the chest region, the heart can receive a lethal dose of current. The heart may go into fibrillation and breathing may be stopped (Figure 7-13).

Small amounts of current through the body can be dangerous to one's health. Large amounts of current through the body can be fatal. A shock of 1 to 20 mA can cause a painful sensation. At a shock of approximately 20 to 30 mA, breathing stops. A shock above 100 mA results in electrocution.

The work environment should be reviewed to identify problem areas. Maintaining a safe work environment is easy when you observe the following guidelines:

1. Keep the work area neat and orderly.
2. Be alert and attentive at all times.

■ **FIGURE 7-13**

Hazards of current flow through the human body.

Sensation Felt	Current in Amperes at 60 Hertz	
	Women	Men
Cannot feel anything	0.0003	0.0004
Tingling—mild sensation	0.0007	0.0011
Shock, not painful, can let go	0.0012	0.0018
Shock, painful, can let go	0.006	0.009
Shock, painful, can barely let go (threshold), may be thrown clear	0.0105	0.016
Shock, painful, severe, cannot let go, muscles immobilized, breathing stops	0.015	0.023
Ventricular fibrillation, fatal		
Length of time—0.03 seconds	1.0	1.0
Length of time—3.0 seconds	0.1	0.1
Heart stops for the length of time current flows; heart may start again if current flows a short time	4	4
Burning of skin, usually not fatal unless heart or other vital organs are burned	5 or more	5 or more
Cardiac arrest, severe burns, and probable death	10,000	10,000

3. Know the correct operation and safety procedures of test equipment, tools, and machinery.
4. If an accident occurs, know what action to take.

> **NOTE** Basic first-aid and cardiopulmonary resuscitation (CPR) training is essential for working in the field of electronics.

Personal safety when working with electricity requires the following:

- Work with one hand in your back pocket or behind your back when working on live circuits. This technique avoids a complete path for current flow that passes through the heart region.
- Use an isolation transformer when working on AC-powered equipment. The transformer isolates the powered equipment from the power source.
- Make sure all capacitors are discharged before beginning troubleshooting. Use a capacitor discharge tool to discharge the capacitor.
- Use grounded line cords and polarized plugs with AC-operated equipment and circuits. They help to reduce the danger from hot chassis.
- Keep hands off live circuits. Test all circuits with a voltmeter or test lamp before working on them.

Before handling toxic and hazardous chemicals, read the material safety data sheet (MSDS) (Chapter 9) to know the best protection and first-aid procedures in case of emergency. When around these types of chemicals, be concerned about inhaling vapors, swallowing liquids, acid burns on the skin, contact with the eyes, and danger of fire and explosions. Follow these guidelines to minimize risk:

- Read the labels of all chemicals being used, paying special attention to the printed warnings.
- Work in well-ventilated areas, especially when working with paint and chemical sprays.
- Wear safety glasses when working with hazardous chemicals.
- Wear rubber gloves when working with acids and acid solutions.
- Use tongs or rubber gloves when placing or removing printed circuit boards from the acid solution.
- Clean all tools that contact hazardous chemicals in case anyone accidentally touches them.

Safety is the number-one priority when soldering. Always follow these safety guidelines:

1. Always wear safety glasses when soldering or cutting leads. When clipping leads on the printed circuit boards, flying leads can be an eye hazard. When clipping leads, aim down and away from the face and anyone close.
2. Never touch the soldering iron tip or heating element. Only pick up the iron by the handle. Never change the tip when the iron is hot.
3. Beware of fingers slipping over the iron handle and contacting the heating element.
4. Beware of pressing the soldering iron too hard against the leads when heating. The soldering iron may slip.
5. Beware of molten solder dripping from the connection. Cotton and wool clothing are the safest to wear when soldering. Many synthetics melt on the skin if molten solder gets splashed on them, causing a greater burn.
6. Use care when handling printed circuit boards. Leads can cut or puncture the skin. Always have cuts or punctures cleaned and treated immediately.
7. Avoid breathing in flux fumes when soldering.
8. Cleaning solvents dry out the skin. Avoid breathing the vapors.
9. Wash hands before eating or drinking. The lead from the solder remaining on the hands can be ingested with food.

Certain components used on printed circuit boards can be damaged by static electricity referred to as electrostatic discharge (ESD). These static charges are created when nonconductive materials, such as plastic from bags or synthetic materials from clothing, are separated. These charges can be induced on a conductor, such as skin, and are delivered in the form of a spark that passes to another conductor. By touching a printed circuit board or semiconductor device, a static discharge can be delivered through it that may damage a sensitive component or device.

Most static charges that can damage a printed circuit board component or semiconductor device are too weak to feel—less than 3000 volts. Most components of printed circuit board assemblies have warnings, such as the labels shown in Figure 7-14, printed on them, noting that they are sensitive to damage by

■ FIGURE 7-14

Warning labels for components sensitive to static electricity.

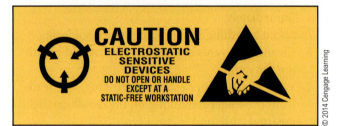

© 2014 Cengage Learning

7-4 QUESTIONS

1. Why are safe habits essential when working in the electronics field?
2. What is the primary concern in the field of electronics?
3. How many milliamperes of current through the body will stop a person's breathing?
4. List procedures that should be observed when working with live electricity.
5. What should be the concern when working around chemicals?
6. List safety guidelines to follow when soldering.
7. List protective measures to prevent ESD damage to components in the work environment.

static electricity. Keep in mind that just because a component does not have a symbol or label does not mean it is not ESD sensitive.

The following protective guidelines help to prevent damage to components:

- Keep sensitive components enclosed in antistatic bags, boxes, or wraps when not working on them.
- Remove sensitive components from their protective wrapping only when in the antistatic work area.
- Keep antistatic workstations free of static-generating materials such as Styrofoam, plastic desoldering tools, plastic notebook folders, combs, and so on.

SUMMARY

- Voltage is the pressure behind electrical current, which can result in electrical shock.
- The higher the body resistance, the less current will flow through it.
- When dealing with someone who is experiencing severe electrical shock, send for help; locate and remove the source of power.
- Do not work on any circuit or equipment unless the power is secured.
- Insulating and grounding are two techniques for preventing electrical shock when working with electrical tools and equipment.
- A GFCI is a fast acting circuit breaker that is sensitive to very low levels of current leakage to ground.
- An Underwriters Laboratories label and a Canadian Standards Association label on a product indicate that it is safe to use as intended without any response to its quality.
- When a charged substance discharges through a conductor, an electric current flows to ground potential and is referred to as electrostatic discharge (ESD).
- A person can feel a shock from an electrostatic discharge of 3500 to 4000 volts.

- An ESD latent discharge results in a device working properly but failing later.
- An electronics technician should be grounded to discharge static electricity buildup on the body before working on a sensitive electronic circuit.

- Components on printed circuit boards can be damaged by static electricity, referred to as electrostatic discharge (ESD).
- Antistatic workstations are designed to provide a ground path for static charges that could damage a component.

CHAPTER 7 SELF-TEST

1. What are the dangers of electrical shock?
2. What has the electrical industry developed that results in preventing electrical shock?
3. Draw a diagram to demonstrate how a grounding circuit works.
4. How does a GFCI protect an individual from electrical shock?
5. Draw a picture and label all components of an ESD workstation.
6. Develop a set of personal guidelines for handling ESD-sensitive materials.
7. What is CPR and where does it fit in electronics?
8. Make a list on how to prevent electrical shock.
9. What steps must be taken when soldering?
10. How are components protected from electrostatic discharge?

Tools and Equipment

After completing this chapter, the student will be able to:

- Understand the precautions for using hand and power tools.
- Describe how to store hand and power tools.
- Demonstrate how to properly use hand and power tools.
- Discuss using test equipment when assembling an electronic circuit.
- Explain who is responsible for safety in the workshop.
- Discuss safety practices when using hand tools.
- Demonstrate safe practices when using power tools.
- Generate a list of safety practices for hand tools, power tools, and working in the workshop on electronic circuits and equipment.

8-1	hand tools	8-3	portable
8-1	power tools	8-3	safety switch
8-2	toolbox	8-3	stationary
8-3	commercial	8-3	torque
8-3	consumer	8-5	safety
8-3	guard	8-5	shorting stick
8-3	insulated tools		

Most accidents that happen in the shop could be avoided by practicing safe work habits. Safe work habits include learning to follow instructions and never taking chances. To stay safe as an electrician, you must learn how to handle and work with tools, machines, and materials correctly. Always read and follow the directions of the tool or equipment manual.

Learn the hazards of the shop and what to avoid. When using hand and power tools, always dress appropriately and keep the workspace clear.

8-1 HANDLING HAND AND POWER TOOLS

Hand tools are tools that are powered manually. Their greatest hazards result from misuse and improper maintenance. When handling hand tools, always observe the following precautions:

1. Always use the proper tool for the job. Use the right type and size tool for each application.
2. When carrying tools, always keep the cutting edge down.
3. Keep hands clean when using tools. Avoid grease, dirt, or oil on hands when using any tool.
4. Clamp small pieces when using a hacksaw, screwdriver, or soldering iron.
5. Avoid using chisels and punches with mushroomed heads. Pieces of metal can break off, causing injuries.
6. Never use a file without a handle.
7. Never use plastic-handled tools near an open flame.
8. Keep metal tools clear of electrical circuits.
9. When cutting wire, always cut one wire at a time to avoid damaging the cutting tool.

Power tools used in electronics are normally powered with electricity. When handling the power tools, always observe the following precautions:

1. Ensure all power tools have grounded three-prong plugs or an insulated housing approved by UL (Underwriters Laboratories).
2. Make all adjustments to the machine before turning it on.
3. Double-check any special setups before applying power.
4. Keep all safety guards in proper position at all times.
5. Clean the work area before using a power tool and keep it clean.
6. Keep power tools clean, and put away them when they are not in use.

7. Avoid treating power tools roughly, dropping them, or hitting them against things.
8. Keep all power tools in good condition with regular maintenance.
9. When finished using a power tool in the workshop, find a safe place to set it down when working.

8-1 QUESTIONS

1. What precautions are directed at the tool user?
2. Why would it be dangerous to use tools with mushroom heads?
3. How should electric power cords be unplugged?
4. When should safety guards be adjusted?
5. What is the function of the three-prong plug used on power tool?

8-2 STORING HAND AND POWER TOOLS

Most workshops have hand tools. Hand tools are classified as any tools that don't require batteries or need to be plugged in to operate. They are typically the easiest tools to misplace and lose; therefore, it is important to keep them organized. How to organize them depends on several factors: the available storage space, the number of tools, and how often they are used.

One simple solution to organizing tools is to use a **toolbox**. A toolbox is a box or container for tool storage. Toolboxes come in a variety of shapes and sizes, so choose one that's proportional in size to the number of tools in the workshop. The best toolboxes will have several compartments to help keep everything organized (Figure 8-1). Put the tools you use most often near the top compartments and those you use less frequently down below.

If the electronics shop does not have a toolbox at each workspace, then the hand tools will be organized in a separate storage cabinet on a tool panel. Many times the tools will be outlined on the tool panel where they are to be stored (Figure 8-2). Keep in mind that the hand tools used on a regularly basis should be easily accessible.

Power tools are generally larger than hand tools. As with your hand tools, the organization of your power tools will depend on how many tools you have, how much storage space is available, and how often you use the tools. If you have a large number of power tools,

■ **FIGURE 8-1**

A good toolbox will have several compartments to help keep things organized.

iStockphoto/hoyaboy

■ **FIGURE 8-2**

Shop tools are often organized on a pegboard using outlines to mark their places.

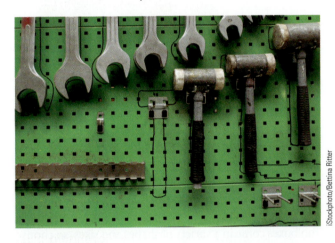

iStockphoto/Bettina Ritter

■ **FIGURE 8-3**

Protect valuable power tools by organizing and storing them carefully.

iStockphoto/Chandlerphoto

■ **FIGURE 8-4**

The proper way to store an electric cord.

© 2014 Cengage Learning

storage cabinets with drawers and shelves are the best solution (Figure 8-3).

One of the most common ways to organize power tools is in bins or containers in the storage cabinet. Buying containers that can seal might be a good idea, because they'll protect your power tools from dampness or any other hazard that could potentially damage them. Avoid storing the power tools too low to prevent back strain.

When storing power tools, do not wrap the power cord around the power tool. Wrapping can damage the power cord where it enters the power tool. Extension cords should never be wound around an arm to store them as it provides an opportunity for the individual wires within the power cord to twist. A better solution is shown in Figure 8-4. This method can prevent strain on the wires in the power cord.

8-2 QUESTIONS

1. What factors affect the storage of hand tools?
2. What is the simplest way to store hand tools?
3. Which tools are stored at the least accessible location?
4. What determines the storage of power tools?
5. What is the best method for storing power tools?

■ FIGURE 8-5A

The official international 1000-volt rating symbol.

1000 V

© 2014 Cengage Learning

■ FIGURE 8-5B

Electricians should have a set of insulated tools.

iStockphoto/sinankocaslan

8-3 USING HAND AND POWER TOOLS

Hand tools are necessary for building electronics circuits. In the electronics workshop, hand tools are of great importance in what they are, what they will be used for, and how to use them. Most of the tools in the electronics workshop can be divided into groups by their use. Each tool has a special purpose and it is important for the electronics technician to know its purpose. Every prospective electronics technician should also learn how to care for the tools they use so each tool can function at its full potential.

Hand tools should always be kept clean, in good working condition, and free of rust. Using or storing the tools in a damp environment creates rust. The formation of rust can be prevented or kept to a minimum by occasionally spraying tools with an oil product or wiping tools with an oily rag. When rust does form, it can be removed easily with fine steel wool or a fine abrasive.

Technicians should also lubricate the moving parts of various tools to keep them working freely. Before using any tool, check to be sure it is in good condition. Never use a tool that is not in good condition.

Using the correct tool for the job is the first step in safe hand tool use. Tools are designed for specific needs. That's why screwdrivers are made with various lengths and tip styles and pliers with different head shapes. The longer the screwdriver is, the more torque it can provide. To avoid personal injury and tool damage, select the proper tool to do the job well and safely.

Many jobs require the use of **insulated tools** to protect the user in the event a circuit is live. While you should always try to work on circuits that are not live,

there are times when a circuit cannot be shut down. All professionals need to have some insulated tools in their collection. Insulated hand tools must be clearly marked with the official, international 1000-volt rating symbol (Figure 8-5). These tools are designed to reduce the chance of injury if the tool should make contact with an energized source.

> **NOTE** Tools with plastic-dipped or slip-on plastic handles are not insulated and are for comfort only. Wrapping a tool with electric tape does not provide adequate insulation.

Inspect insulated tools frequently, watching for any wear or cracking of the insulation. Keep tools clean, dry, and free of surface contaminants, which can compromise their insulating properties. If cutting, wear, or a burn has breached the dielectric insulation, the tool should not be used to ensure safety.

Power tools are classified as **commercial** or **consumer**, with commercial power tools used in business and industry, and consumer power tools used in a household. Power tools are useful in the following tasks: drilling, cutting, fastening, grinding, heating, painting, polishing, routing, sanding, shaping, and more. Power tools are also classified as either **portable** or **stationary**. Portable power tools are handheld and offer mobility. Stationary power tools cannot be moved but offer accuracy and speed. Stationary power tools are available for both metalworking and woodworking. Drill presses and band saws are used for both woodworking and metalworking.

Always protect your eyes, lungs and extremities when using power tools. Wear the protective gear shown in Figure 8-6. The Occupational Safety and Health Administration, OSHA, requires that any workers exposed to hazards of falling, flying, splashing objects or anything that can be inhaled, must use personal protective gear when in the work space. It doesn't take much for an object in the eye to become a major irritation. It is better to cut a glove than skin. When working around fumes, wear the appropriate mask. Never wear long hair, loose clothing, ties or jewelry around power tools as they can get caught in the actions of the power tools.

Any power tool, such as a drill, saw, or impact wrench, generates **torque** when it is turned on. Torque is the tendency of a power tool to rotate around its axis. Before using a power tool, determine how the torque influences the device as it is moved around, before applying it to the task it was designed to do. Make sure to apply a good grip on the tool and balance yourself before using it as intended. When you are finished with the tool, disconnect it and store it properly.

If you are using any power tools that could throw off sparks, make sure that the workshop is vented for flammable gases. Keep the work area clean and dry to prevent accidental slipping with the tool.

Electric power tools present a hazard in that electricity always wants to make a path to ground. Power tools should always have a grounded three-wire cord. Never plug a three-prong cord into a two-prong receptacle, which removes the ground protection. All cords and extensions should be in original condition, with no cracks or frayed ends where they enter the plug or tool. If there is any sign of wear on the cord, discard it or repair the tool or extension cord.

Regulations require that all power tools be fitted with proper **guards** and **safety switches**. Guards protect the operator of power tools from exposed moving parts. Safety switches are used to prevent a tool from accidentally turning on and may require a minimum of two actions to turn a power tool on. Before using any power tool, make sure the guards are in place and the switches are working properly.

Power tools are extremely dangerous when used improperly. Never use a power tool for anything other than its intended use.

■ **FIGURE 8-6**

When using power tool, always wear protective gear. Leather gloves can be substituted for rubber or vinyl gloves.

© 2014 Cengage Learning

8–3   QUESTIONS

1. Why must hand tools be kept in good working condition?
2. What groups can the hand tools in Figure 8-7 be divided into?
3. Why are screwdrivers made in different lengths?
4. Who specifies that technicians exposed to hazards in the workshop must wear protective gear?
5. How are power tools classified?

■ FIGURE 8-7

A. Spudgers for lifting wires and aligning leads.

B. Surgical scissors.

C. Tweezers.

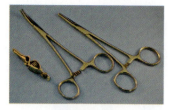

D. Heat sinks.

E. Wire strippers.

F. Needlenose pliers.

G. Screwdrivers.

H. Nut drivers.

I. Soldering iron.

J. Manual desoldering tool.

K. Low-wattage pencil soldering iron.

L. Desoldering bulb and soldering iron with hollow tip.

M. Component lead former tool.

N. Analog voltmeter.

O. Digital voltmeter.

P. Temperature-controlled soldering station.

8–4   USING TEST EQUIPMENT

Electronic test equipment, shown in Figure 8-8, is used to generate signals and capture responses from electronic circuits. Using the test equipment can prove that an electronic circuit is operating properly or show that there are faults in the circuit that can be traced and repaired. Use of electronic test equipment is essential to any serious work on electronics circuits, devices, or systems. It is important to understand how each piece of test equipment operates.

When an electronics engineer designs and assembles an electronic circuit, many different types of electronic test equipment are used to test the circuit for proper operation. Some types of testing equipment can be simulated in a virtual environment. A virtual environment is a computer-simulated environment that is displayed on a computer monitor. The main advantage in using simulation tools for circuit design

■ **FIGURE 8-8**

A. Analog multimeter.

B. Digital multimeter.

C. Analog oscilloscope.

D. Power supply.

E. Frequency counter.

F. Signal generator.

G. Function generator.

H. Logic probe.

I. Logic pulser.

is that the virtual environment lets the user test and evaluate the circuit without damaging test equipment or the circuit itself.

Some of the test equipment we discuss in this textbook are listed here: frequency counter, function generator, logic probes, multimeter, oscilloscope, power supply, signal generator, and virtual test equipment.

Usually, more complex test equipment is necessary when developing electronics circuits and systems than is needed during the production phase and testing or when troubleshooting existing electronics circuits and systems in the field.

8-4   QUESTIONS

1. What is electronics test equipment used for?
2. Where would electronics test equipment be used?
3. When would an electronics engineer use electronics test equipment?
4. What is a virtual environment?
5. What are some of the test equipment required for this textbook?

8-5 SAFETY ISSUES

Each user is responsible for the safe condition of tools and test equipment before usage. Learn how to use hand and power tools properly.

Hand tools used with electronics circuits involve a variety of nonpowered tools such as diagonal cutters, hammers, reamers, screwdrivers, wrenches, and several types of pliers (Figure 8-7). Hand tools may seem harmless, but they are the cause of many injuries.

The following is a list of safe practices related to hand tools:

- Most common hazards associated with the use of hand tools are misuse and improper maintenance.
- Misuse occurs when a hand tool is used for something other than its intended purpose.
- Improper maintenance allows hand tools to deteriorate into an unsafe condition.
- Specially designed tools may be needed in hazardous environments.
- The type of personal protective equipment needed when using a hand tool depends on the nature of the task. At all times, eye protection must be worn.
- The use of hand protection may also be appropriate to provide protection against cuts, abrasion, and repeated impact. Hand protection should include vinyl gloves when acids and other solvents are used.

When working with hand and power tools, always wear proper clothing. Do not wear loose clothing or dangling objects, and tie back long hair when operating power tools. Check the appropriate tool manuals for proper operation.

The following is a list of safe practices related to power tools.

- Always wear eye protection and hearing protection when working with power tools.
- Never carry portable tools by the cord, use the power cord to raise or lower a tool, or yank the power cord to disconnect it.
- Disconnect tools when changing accessories, before servicing and cleaning, and when not in use.
- Never hold the switch button engaged when carrying a tool that is plugged in.

- Keep everyone not involved with the work away from the workspace.
- Fasten work with clamps or place it in a vise, to free both hands to operate tools.
- Always keep all tools sharp and clean.
- To protect against electrical shock, all tools must have a three-wire cord plugged into a grounded receptacle and must be double insulated.
- Never remove a guard from a tool or use the tool without a guard.
- Never adjust a tool when it is running.
- Machine guards protect the operator and others from the point of operation, rotating parts, and flying chips and sparks.

The following is a list of safe practices when working on electrical and electronic circuits.

1. Do not work on any circuits or equipment unless the power is secured.
2. Work only in clean, dry areas. The resistance of the skin is lower in a damp or wet environment, increasing the chance of electrical shock.
3. Do not wear loose or flapping clothing. Not only may it get caught, but it might also serve as a path for the conduction of electricity.
4. Wear only nonconductive shoes. This reduces the chance of electrical shock.
5. Remove all rings, wristwatches, bracelets, ID chains and tags, and similar metal items. Metal can act as a conductor, heat up, and cause a bad burn or short out a circuit.
6. Do not use bare hands to remove hot parts.
7. Use a **shorting stick** to remove high-voltage charges on capacitors. The shorting stick can be a commercial unit or a screwdriver with a clip attached to it. Capacitors and CRTs (cathode-ray tubes) can hold a charge for long periods of time.
8. Make certain that the equipment being used is properly grounded with three-prong or polarized plugs. Ground all test equipment and all equipment and electronic circuits being tested and repaired.
9. Remove power to a circuit before connecting alligator clips. Handling uninsulated alligator clips could cause potential shock hazards.
10. When measuring voltages over 300 volts, do not hold the test prods. This eliminates the possibility of shock from leakage on the probes.

Safety is everyone's responsibility. It is the job of everybody in and out of class to exercise proper precautions to ensure that no one is injured and no equipment is damaged.

8-5 QUESTIONS

1. Who is responsible for safety in the workspace?
2. When do accidents occur with hand tools?
3. What must be worn at all times when using hand and power tools?
4. How should power tools never be carried?
5. Why should nonconductive shoes be worn when using power tools?

SUMMARY

- Precautions should always be observed when using hand and power tools.
- Hand and power tools should always be stored when not in use.
- Always use hand and power tools properly.
- Test equipment is used when assembling a circuit to ensure it is working properly.
- When working in the workshop, everyone is responsible for safety.
- Safe practices should be observed when using hand and power tools.
- Safe practices should be observed when working on any electronic circuit or equipment.

CHAPTER 8 SELF-TEST

1. Generate a list of precautions that should be observed when handling hand tools.
2. Describe how to handle power tools in the workshop.
3. Make a list for handling hand and power tools in the workshop.
4. Referring to Figure 8-7, identify a storage method for each of the hand tools identified.
5. Make a safety poster for using hand and/or power tools.
6. List personal protection devices identified by OSHA when using power tools.
7. Make a drawing identifying how torque impacts the operation of a power tool.
8. Where can information be found on how a piece of test equipment operates?
9. Make a chart regarding hand tools and safe practices.
10. Make a list of don'ts when using power tools.
11. When working on electrical and electronic circuits, what is the first thing to do?

CHAPTER 9

Hazardous Materials

A material safety data sheets (MSDS) is required to be maintained and available on request by persons using a chemical or hazardous material.

The purpose of this chapter is to provide the reader with information on regulations for the safe and proper use, storage, and handling of hazardous materials.

Although this chapter does not cover all the requirements concerning the handling and storage of hazardous materials, it is a good starting point.

9-1   MATERIAL SAFETY DATA SHEET (MSDS)

On December 29, 1970, President Richard M. Nixon signed the Occupational Safety and Health Act, which established the **Occupational Safety and Health Administration**, known as **OSHA**.

OSHA's mission is twofold: one is to prevent work-related illnesses, injuries, and death by issuing and enforcing workplace safety and health standards. The second is to create a better work environment for the workforce ensuring the safety of everyone. OSHA accomplished this mission by making and enforcing certain standards to protect workers.

OSHA requires that a **material safety data sheet (MSDS)** be supplied when a chemical or hazardous substance is purchased. The law requires that a copy of the MSDS be maintained and available on request by the person(s) using this material.

When unpacking a shipment of chemicals, look for the MSDS in the box. File the MSDS in a folder in the location where the chemical is being used.

MSDSs should be kept on file for two reasons:

1. To help protect personnel from injury and exposure hazards.
2. To abide by the law.

Read each MSDS before handling or using hazardous material. Everyone who uses the chemical product must understand its dangers and the precautions to be taken while using it.

Figure 9-1 provides a brief description of the sections of an MSDS. Figure 9-2 shows an MSDS for ferric chloride.

Recipients of MSDSs should consult the OSHA Safety and Health Standards for guidance on control of potential occupational health and safety hazards.

9-1   QUESTIONS

1. What is the purpose of an MSDS?
2. What is the first thing a person should do when unpacking a shipment of chemicals?
3. What happens when a school or employer fails to provide students or employees the MSDS to be read when they are working with chemicals?
4. What are the sections of an MSDS?
5. Who should be contacted regarding general guidance of potential health and safety hazards?

9-2   CLASSIFICATION OF HAZARDOUS MATERIALS

A **hazardous material** is any material that is capable of harming physically or causing an illness to an individual.

On October 15, 1966, the **Department of Transportation (DOT)** was established by an act of Congress. April 1, 1967, was the DOT's first official day of operation.

The DOT was given broad authority to regulate the transportation of hazardous materials. This authority includes the discretion to determine what materials are classified as hazardous.

Hazardous materials are broken into nine categories based on their chemical and physical properties. Figure 9-3 is a chart for classifying hazardous materials by the DOT.

Based on the classification of the hazardous material, the DOT is responsible for determining the appropriate packaging materials for shipping or transport. **Packing groups** are used to indicate the degree of risk a hazardous material may pose when in transport. Packing groups are always represented by roman numerals that indicate the type of packaging required for the hazardous materials as well as quantity limits allowed. The groups are shown below:

- Packing Group I—High danger
- Packing Group II—Moderate danger
- Packing Group III—Low danger

Also, based on the material classification, there are strict guidelines for proper labeling or marking of packages with hazardous materials to be transported and for signs on the transport vehicles (Figure 9-4).

■ FIGURE 9-1A

Sections of a material safety data sheet.

Top Portion

- Name of Manufacturer
- Manufacturer's Address
- Emergency Phone Number
- Trade name of Product(s) with stock number(s)

Section 1 Hazardous Materials

- Exposure recommendations for individual components.
  - *Components are listed if present at or above 1% in the mixture and present a physical or health hazard.*
  - *Components identified as carcinogens by NTP, IARC, and OSHA are listed if present at or above 0.1% in the mixture.*
- All exposure limits are those that OSHA and ACGIH (American Conference of Governmental Industrial Hygienists) concur on unless specifically listed as an OSHA Permissible Exposure Limit (PEL) and/or an ACGIH Threshold Limit Value (TLV).

Section 2 Physical and Chemical Properties

- **Boiling Point**—Of product, if known. If unknown, the lowest value of the component is listed for mixtures.
- **Vapor Pressure**—Of product, if known. If unknown, the lowest value of the component is listed for mixtures.
- **Vapor Density**—Compared to Air = 1. If specific vapor density of product is not known, the value is expressed as lighter or heavier than air.
- **Evaporation Rate**—Indicated as faster or slower than ethyl ether or butyl acetate.
- **Specific Gravity**—Compared to water = 1. If specific gravity of product is not known, the value is expressed as less than or greater than water.
- **Appearance and Odor**—Describes the physical appearance, color, and smell of the product.

Section 3 Fire and Explosion Hazard Data

- **Flash Point**—Method identified.
- **Flammability Limits**—For product, if known. The lowest and highest values of the components are listed for mixtures.
- **Extinguishing Media**—Following National Fire Protection Association criteria.
- **Special Fire Fighting**—Minimum equipment to protect firefighters from toxic products of vaporization, combustion, or decomposition in fire situations.
- **Unusual Fire Hazard**—Known or expected hazardous products resulting from heating, burning, or other reactions.
- **DOT Category**—Describes the classification for shipping by road.

■ **FIGURE 9-1B** *(Continued)*
Sections of a material safety data sheet.

Section 4 Reactivity Hazard Data

- **Stability**—If known, conditions to avoid preventing hazardous or violent decomposition.
- **Conditions to Avoid**—If known, conditions to avoid preventing hazardous reactions.
- **Hazardous Polymerization**—If known, conditions to avoid preventing hazardous polymerization resulting in a large release of energy.

Section 5 Health Hazard Data

- **Primary Route(s) of Entry**—Based on properties and expected use.
- **Effects of Overexposure (Acute)**—Potential local and systemic effects due to single or short-term overexposure to the eyes and skin or through inhalation or ingestion.
- **Signs and Symptoms of Exposure**—If known, warning signs which may indicate exposure to the skin or eyes, or through inhalation or ingestion.
- **Medical Conditions Generally Aggravated by Overexposure**—If known, preexisting conditions which may contribute to the effects of overexposure to the eyes and skin or through inhalation or ingestion.
- **Emergency and First-Aid Procedures**—Procedures to be followed when dealing with accidental overexposures.

Section 6 Control and Protective Measures

- **Protective Equipment**—Identifies equipment that may be needed when handling the product.
- **Ventilation**—Describes the need for forced mechanical ventilation if required.

Section 7 Spill or Leak Procedures

- Identifies precautions to be taken and methods of containment, cleanup, and disposal.
- RCRA hazardous wastes and CERCLA hazardous substances are listed in this section.

Section 8 Hazardous Material Identification

- Provides hazardous materials identification system rating based on ratings of individual components.
- DOT emergency response codes are provided in this section.

Section 9 Special Precautions and Comments

- Covers any relevant points not previously mentioned.
- Reinforces pertinent points previously covered.

FIGURE 9-2A

Material safety data sheet for ferric chloride.

Company Logo

Material Safety Data Sheet

Revision Date	**Prepared by**	**Technical Information**
Date Sheet Prepared	Who prepared MSDS	Toll-Free Phone Number or email address
Home Office		**Emergency**
Company Address		Toll-Free Phone Number

Section 1: Product Identification

Name: Ferric Chloride
MSDS Code: XXX - Liquid;   **Related Part Numbers:** XXX-500ML; XXX-1L; XXX-4L; XXX-20L
Use: For etching printed circuit boards.

Section 2: Hazardous Ingredients

CAS#	Chemical Name	Percentage by Weight	ACGIH TWA	OSHA Pel	OSHA Stel
7705-08-0	Ferric Chloride	40	$1mg/m^3$	N/E	N/E
7647-01-0	Hydrochloric Acid	1	5ppm	N/E	N/E

Section 3: Hazards Identification

Eyes:	Corrosive. Contact of liquid will cause severe eye burns and corneal damage.
Skin:	Corrosive. May cause severe skin irritation with possible burns.
Inhalation:	May cause irritation of mucous membranes and respiratory tract.
Ingestion:	May be corrosive to the gastrointestinal tract.
	May cause chemical burns in the mouth, throat, esophagus, and stomach.
Chronic:	Repeated exposure may cause an increased body load of iron, with possible chronic systemic effects.

Section 4: First-Aid Measures

Eyes:	Remove contact lenses. Flush with water or saline for 15 minutes. Get medical aid.
Skin:	Wash skin with soap and water for 15 minutes. Get medical aid if symptoms persist.
Inhalation:	Immediately remove from exposure to fresh air. If not breathing, give artificial respiration. If breathing is difficult, give oxygen. Get medical aid.
Ingestion:	Do not induce vomiting. If conscious, give 1–2 glasses of water. Get medical aid.

Section 5: Firefighting Measures

Auto Ignition Temperature: N/A   **Flash Point:** N/A   **LEL/UEL:** N/A
Extinguishing Media: N/A
General Information: Will not burn.

NFPA Ratings:	Health 2	Flammability 0	Reactivity 2
HMIS Ratings:	Health 2	Flammability 0	Reactivity 2	WHMIS Codes E

■ **FIGURE 9-2B** *(Continued)*
Material safety data sheet for ferric chloride.

Section 6: Accidental Release Measures

Spill Procedure:	Provide adequate ventilation. Wear appropriate personal protection. Sprinkle absorbent compound onto spill, then sweep into a plastic container. Wipe up further residue with paper towel and place into container. Wash spill area with soap and water.

Section 7: Handling and Storage

Handling:	Wash thoroughly after handling. Avoid contact with eyes, skin, and clothing. Do not ingest or inhale. Do not expose container to heat or flame.
Storage:	Keep from freezing. Store in a cool, dry, well ventilated area, away from incompatible substances.

Section 8: Exposure Controls

Routes of entry:	Eyes, ingestion, inhalation, and skin.
Ventilation:	Use adequate general or local exhaust ventilation to keep airborne concentrations below exposure limits.
Personal Protection:	Wear appropriate protective eyeglasses or chemical safety goggles. Wear appropriate protective clothing to prevent skin contact. Use a NIOSH-approved respirator when necessary.

Section 9: Physical and Chemical Properties

Physical State: Liquid	**Odor:** Odorless	**Solubility:** Miscible
Boiling Point: 110°C	**Vapor Pressure:** 1PSI@20°C	**Vapor Density:** N/E
Specific Gravity: 1.43	**Evaporation Rate:** Slow	**pH:** 1

Section 10: Stability and Reactivity

Stability:	Stable at normal temperatures and pressures.
Conditions to Avoid:	Incompatible materials.
Incompatibilities:	Alkalis, oxidizers, corrosive to metals.
Polymerization:	Will not occur.
Decomposition:	Prolonged contact with metals may produce flammable hydrogen gas.

Section 11: Toxicological Information

Sensitization (effects of repeated exposure):	No		
Carcinogenicity (risk of cancer):	No		
Teratogenicity (risk of malformation in an unborn fetus):	No		
Reproductive Toxicity (risk of sterility):	TDLo: 29mg/kg (rat)		
Mutangenicity (risk of heritable genetic effects):	No		
LethalExposure (ingestion):	**Inhalation**	**Skin**	
Concentrations (LD50): 4500 mg/kg (rat)	**(LC50):** N/E	**(LD50):** N/E	

■ **FIGURE 9-2C** *(Continued)*
Material safety data sheet for ferric chloride.

Section 12: Ecological Information

General Information: Avoid runoff into storms and sewers that lead into waterways. Water runoff can cause environmental damage.

Environmental Impact Data: (percentage by weight)

CFC: 0   **HFC:** 0   **Cl.Solv.:** 0   **VOC:** 0   **HCFC:** 0   **ODP:** 0

Section 13: Disposal Information

General Information: Dispose of in accordance with all local, provincial, state, and federal regulations. Water runoff can cause environmental damage.

Section 14: Transportation Information

Ground: (All sizes 4L or smaller)
Consumer commodity

Ground: (All sizes larger than 4L)
Shipping name ferric chloride solution, packing group 3, class 8. Shipper must be trained and certified.

Air: Do not ship by air.

Sea: Shipper must be trained and certified.

Section 15: Regulatory Information

SARA (Superfund Amendments and Reauthorization Act of 1986, USA, 40 CFR 372.4)
None of the chemicals in this product have a reportable quantity.

EPCRA (Emergency Planning and Right to Know Act, USA, 40 CFR 372.45)
This product does not contain any chemicals subject to the reporting requirements of section 313 Title III of the SARA of 1986 and 40 CFR part 372.

TSCA (Toxic Substances Control Act of 1976, USA)
All substances are TSCA listed.

CAA (Clean Air Act, USA)
This product does not contain any class 1 ozone depletors.
This product does not contain any class 2 ozone depletors.
This product does not contain any chemicals listed as hazardous air pollutants.

California Proposition 65 (chemicals known to cause cancer or reproductive toxicity, May 1, 1997 revision, USA)
This product does not contain any chemicals listed.

Section 16: Other Information

Definitions: N/A = not applicable, N/E = not established

Disclaimer: This material safety data sheet is provided as an information resource only. The information contained herein is accurate and compiled from reliable sources. It is the responsibility of the user to verify its validity. The buyer assumes all responsibility of using and handling the product in accordance with federal, state, and local regulations.

■ **FIGURE 9-3**
The DOT provides charts that classify hazardous materials.

Class 1: Explosives
Division 1.1 Explosives with a mass explosion hazard
Division 1.2 Explosives with a projection hazard
Division 1.3 Explosives with predominantly a fire hazard
Division 1.4 Explosives with no significant blast hazard
Division 1.5 Very insensitive explosives
Division 1.6 Extremely insensitive explosive articles
Class 2: Gases
Division 2.1 Flammable gases
Division 2.2 Nonflammable gases
Division 2.3 Poison gas
Division 2.4 Corrosive gases
Class 3: Flammable liquids.
Division 3.1 Flashpoint below −18°C (0°F)
Division 3.2 Flashpoint −18°C and above, but less than 23°C (73°F)
Division 3.3 Flashpoint 23°C and up to 61°C (141°F)
Class 4: Flammable solids, spontaneously combustible materials, and materials that are dangerous when wet.
Division 4.1 Flammable solids
Division 4.2 Spontaneously combustible materials
Division 4.3 Materials that are dangerous when wet
Class 5: Oxidizers and organic peroxides
Division 5.1 Oxidizers
Division 5.2 Organic peroxides
Class 6: Poisons and etiologic materials
Division 6.1 Poisonous materials
Division 6.2 Etiologic (infectious) materials
Class 7: Radioactive materials
Any material or combination of materials that spontaneously gives off ionizing radiation.
Class 8: Corrosives
A material, liquid or solid, that causes visible destruction or irreversible alteration to human skin or a liquid that has a severe corrosion rate on steel or aluminum.
Class 9: Miscellaneous
A material that presents a hazard during transport, but which is not included in any other hazard class.
ORM-D: Other regulated material
A material that, although otherwise subjected to regulations, presents a limited hazard during transportation due to its form, quantity, and packaging.

■ FIGURE 9-4

Labeling for hazardous materials for transporting, as identified by DOT.

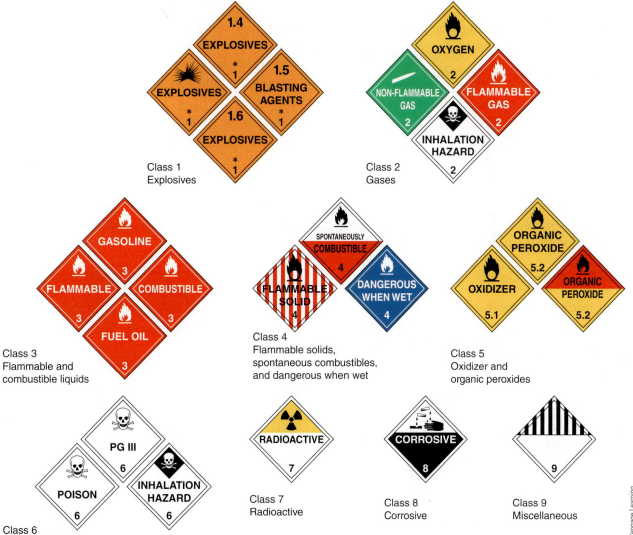

Class 1
Explosives

Class 2
Gases

Class 3
Flammable and
combustible liquids

Class 4
Flammable solids,
spontaneous combustibles,
and dangerous when wet

Class 5
Oxidizer and
organic peroxides

Class 6
Poison (toxic) and
poison inhalation hazard

Class 7
Radioactive

Class 8
Corrosive

Class 9
Miscellaneous

9–2   QUESTIONS

1. What is a hazardous material?
2. Who regulates hazardous materials that are transported?
3. List the nine categories of hazardous materials.
4. Who is responsible for determining the appropriate packaging materials for shipping or transport of hazardous materials?
5. What are packing groups used for?

9–3   HANDLING HAZARDOUS MATERIALS

Material safety data sheets should be retained and consulted often when using a hazardous material. All personnel should know where they are filed.

Chemicals that are generally considered to be nonthreatening could be hazardous under specific circumstances. **Reactions** between two chemicals could prove fatal. Chemicals and other hazardous materials should be stored properly to prevent spills. Those responsible should ensure guidelines for an organization are followed regarding the storage of hazardous materials.

Several chemicals are used in electronics manufacturing and repair. Many of these chemicals are used in PCB fabrication and soldering. Components that explode can emit harmful vapors.

Chemical Storage in the Workspace

- Maintain chemicals per manufacturer requirements on MSDS.
- Ensure containers are labeled in accordance with OSHA.
- Ensure chemical containers are closed when not in use.
- Use secondary containment such as acid carriers when transporting hazardous liquids, for example, ferric chloride, for more than a very short distance.

Flammable and Combustible Liquids

- **Flammable** and **combustible** liquids should be stored in a metal container designed for them.
- All flammable and combustible liquids should be stored in a flammable storage cabinet.
- Flammable storage cabinets should be properly vented to the outside of the building.

9-3   QUESTIONS

1. Why should the MSDS be retained in the workspace?
2. What would happen if two chemicals combine in a spill?
3. How are chemicals typically used in electronics?
4. How should chemicals be stored in the workspace?
5. How should flammable and combustible liquids be stored in the workspace?

9-4   DISPOSING OF HAZARDOUS MATERIALS

Federal law strictly regulates **disposal** of most hazardous materials. Do not dispose of any hazardous material in the sewer, on the ground, or in the trash.

Identify who is responsible for collecting **hazardous waste** and treating it or preparing it for shipment to an approved hazardous waste disposal facility.

Waste products must be clearly labeled with the complete names of the contents, and they must be stored in nonleaking, safe containers.

By law, any facility is required to reduce the amount of hazardous waste generated; therefore, follow these measures:

- Buy only the amounts of hazardous materials that can be used before the expiration date of the material.
- Use the hazardous material only for the purpose for which it is intended.
- Determine whether someone else has a legitimate use for the product and transfer any extra material to this person.

9-4   QUESTIONS

1. Who regulates the disposal of hazardous materials?
2. Is it approved to dump chemical liquids in the sewer?
3. Identify who is responsible in the workspace for disposing of hazardous materials.
4. When disposing of hazardous materials, what must be done?
5. How can the amount of hazardous waste be reduced?

SUMMARY

- The Occupational Safety and Health Administration (OSHA) mission is to prevent work-related illnesses, injuries, and death by issuing and enforcing workplace safety and health standards.
- OSHA accomplishes its mission by making and enforcing certain standards needed to protect workers.
- OSHA requires a material safety data sheet (MSDS) be supplied when a chemical or hazardous substance is purchased.
- MSDS provides information on a hazardous material product identifying its dangers and the precautions to be taken while using it.

- The Department of Transportation (DOT) was given broad authority to regulate hazardous materials if transported by Congress in 1966.
- DOT broke hazardous materials into nine categories based on chemical and physical properties.
- DOT identified appropriate packaging materials for shipping or transport of hazardous materials.
- Packing groups are used to indicate the degree of risk a hazardous material may pose when in transport.
- In every workspace with hazardous materials, someone is responsible for the hazards of each chemical present.
- Chemicals that are generally considered to be non-threatening could be hazardous if two chemicals react during a spill.
- There are proper methods for storage of chemicals in the workspace.
- All flammable and combustible liquids should be stored in metal containers in a flammable storage cabinet.
- Federal law strictly regulates the disposal of most hazardous materials.
- Federal law also requires any facility to reduce the amount of hazardous waste generated.

CHAPTER 9 SELF-TEST

1. What is OSHA and when was it formed?
2. What is OSHA's mission?
3. Identify the parts of a material safety data sheet (MSDS).
4. What are the storage requirements for ferric chloride?
5. List the nine categories of hazardous materials.
6. How should ferric chloride be transported?
7. State how more than 4 liters of ferric chloride should be packed.
8. Who is physically responsible for the hazardous material in the electronic workspace?
9. How should combustible liquids be stored?
10. Explain how to properly dispose of ferric chloride.

PRACTICAL APPLICATIONS

Practical application activities are provided at the end of each section in this text book to support the material covered in the section. Always observe all safety precautions provided by your instructor and included with components. When using power and hand tools, always follow all safety rules for each piece of equipment. Never use a tool for a purpose other than its intended purpose. Always wear safety glasses in the lab area.

Before starting a project, it is important to ascertain that the following guidelines have been completed:

- Obtain all parts.
- Develop a timeline for how long it will take to finish the project. (Be realistic.)
- Review all safety precautions.
- Breadboard the circuit before making the PCB. Breadboarding involves assembling the circuit using a specially designed board.
- Lay out a PCB using the guidelines laid out in Chapter 48.

Section 1 Practical Applications

1. Identify a possible career choice and research it using DOT, the Occupational Outlook handbook, and the Internet. Prepare a PowerPoint presentation on the career to present to an audience.

2. Download and take a Computerized Associate Practice Exam from the Internet.

3. Form a group, determine assignments, and identify a task to complete with a timeline of two weeks.

4. Use a calculator to solve the following problems:

    a. $R_T = 470 + 1000 + 560$

    b. $I_T = 24 / 467$

    c. $P_{R2} = (0.0046)(23)$

    d. $1/R_A = 1/3300 + 1/5600$

    e. $X_C = 1/ (2)(3.13)(400)(0.000010)$

    f. $X_L = (2)(3.14)(400)(0.015)$

    g. $Z = \sqrt{180^2 + 560^2}$

5. Redraw correctly the following schematic diagram.

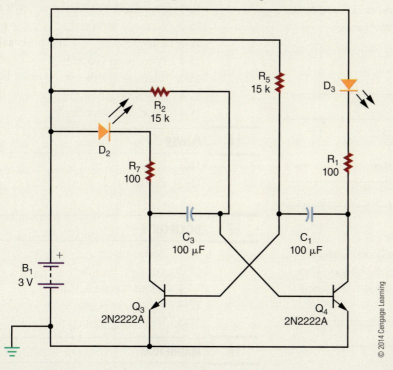

© 2014 Cengage Learning

6. Enter the redrawn circuit from problem 5 into a circuit simulation program.

7. Make a poster identifying the dangers of electrical shock.

8. Review how hand tools are stored in the workshop, and design an alternate storage technique.

9. Review all power tools, both portable and stationary, for any electrical shock hazards or safety issues.

10. Obtain the MSDSs for all hazardous materials in the workshop and review them to find any discrepancies between the MSDS and workshop practice.

DC CIRCUITS

PEOPLE IN ELECTRONICS

André Marie Ampère (1775–1836)
In 1820, Ampère demonstrated the principles of electromagnetism.

Samuel Hunter Christie (1764–1865)
In 1833, Christie invented a measuring technique for comparing wire resistance that later became known as the Wheatstone bridge.

Charles Augustin de Coulomb (1736–1806)
De Coulomb worked with electricity and magnetism and is known best for the formation of Coulomb's law.

Michael Faraday (1791–1867)
Faraday invented and developed electromagnetic rotary devices, which formed the foundation of electric motors.

Joseph Henry (1797–1878)
Henry discovered the electromagnetic phenomenon of self-inductance and mutual inductance.

Gustav Robert Kirchhoff (1824–1887)
In 1847, Kirchhoff extended Ohm's law with Kirchhoff's current law and Kirchhoff's voltage law.

Heinrich Friedrich Emil Lenz (1804–1865)
In 1833, Lenz formulated Lenz's law and also discovered Joule's law.

George Simon Ohm (1787–1854)
In 1827, Ohm was the first to observe the relationship between current, voltage and resistance.

Robert Jemison Van de Graaf (1901–1967)
In 1929, de Graaf invented the Van de Graaf generator that produces high voltage.

Count Alessandro Volta (1745–1827)
In 1800, Volta developed the first electric cell.

James Watt (1736–1819)
Watt improved the Newcomen steam engine that was fundamental to the industrial revolution and developed the concept of horsepower.

Sir Charles Wheatstone (1802–1875)
In 1843, Wheatstone improved on the design of the original Wheatstone bridge.

Fundamentals of Electricity

OBJECTIVES

After completing this chapter, the student will be able to:

- Define atom, matter, element, and molecule.
- List the parts of an atom.
- Define the valence shell of an atom.
- Identify the unit for measuring current.
- Draw the symbol used to represent current flow in a circuit.
- Describe the differences among conductors, insulators, and semiconductors.
- Explain the differences among potential, electromotive force, and voltage.
- Draw the symbol used to represent voltage.
- Identify the unit used to measure voltage.
- Define resistance, and identify the unit for measuring resistance.
- Identify characteristics of resistance in a circuit.
- Draw the symbol used to represent resistance in a circuit.

KEY TERMS

10-1 atom
10-1 compound
10-1 element
10-1 matter
10-1 mixture
10-1 molecule
10-2 atomic number
10-2 atomic weight
10-2 conductor
10-2 electron
10-2 insulator
10-2 ionization
10-2 negative ion
10-2 neutron
10-2 nucleus
10-2 positive ion

10-2 proton
10-2 semiconductor
10-2 shell
10-2 valence
10-2 valence shell
10-3 ampere (A)
10-3 coulomb (C)
10-3 current (I)
10-4 difference of potential
10-4 electromotive force (emf)
10-4 potential
10-4 volt (V)
10-4 voltage (E)
10-5 ohm (Ω)
10-5 resistance (R)

Everything, whether natural or artificial, can be broken down into either an element or a compound. However, the smallest part of each of these is the atom.

The atom is made up of protons, neutrons, and electrons. The protons and neutrons group together to form the center of the atom, called the nucleus. The electrons orbit the nucleus in shells located at various distances from the nucleus.

When appropriate external force is applied to electrons in the outermost shell, they are knocked loose and become free electrons. The movement of free electrons is called current. The external force needed to create this current is called voltage. As it travels along its path, the current encounters some opposition, called resistance.

This chapter looks at how current, voltage, and resistance collectively form the fundamentals of electricity.

10-1 MATTER, ELEMENTS, AND COMPOUNDS

Matter is anything that occupies space and has weight. It may be found in any one of three states: solid, liquid, or gas. Examples of matter include the air we breathe, the water we drink, the clothing we wear, and ourselves. Matter may be either an element or a compound.

An **element** is the basic building block of nature. It is a substance that cannot be reduced to a simpler substance by chemical means. There are now over 100 known elements (Appendix 1). Examples of elements are gold, silver, copper, and oxygen.

The chemical combination of two or more elements is called a **compound** (Figure 10-1). A compound can be separated by chemical, but not by physical means. Examples of compounds are water, which consists of hydrogen and oxygen, and salt, which consists of sodium and chlorine. The smallest part of the compound that still retains the properties of the compound is called a **molecule**. A molecule is the chemical combination of two or more atoms. An **atom** is the smallest particle of an element that retains the characteristic of the element. The physical combination of elements and compounds is called a **mixture**. Examples of mixtures include air, which is made up of oxygen, nitrogen, carbon dioxide, and other gases, and salt water, which consists of salt and water.

© 2014 Cengage Learning

■ FIGURE 10-1

The chemical combination of two or more elements is called a compound. A molecule is the chemical combination of two or more atoms. The example here is water (H_2O) and salt (NaCl).

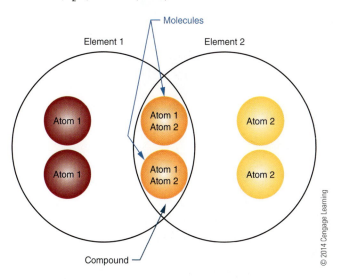

10-1 QUESTIONS

1. In what forms can matter be found?
2. What is a substance called that cannot be reduced to a simpler substance by chemical means?
3. What is the smallest possible particle that retains the characteristic of a compound?
4. What is the smallest possible particle that retains the characteristic of an element?
5. What is the combination of elements and compounds called?

10-2 A CLOSER LOOK AT ATOMS

As previously stated, an atom is the smallest particle of an element. Atoms of different elements differ from each other. If there are over 100 known elements, then there are over 100 known atoms.

Every atom has a **nucleus**. The nucleus is located at the center of the atom. It contains positively charged particles called **protons** and uncharged particles called **neutrons**. Negatively charged particles called **electrons** orbit around the nucleus (Figure 10-2).

■ **FIGURE 10-2**

Parts of an atom.

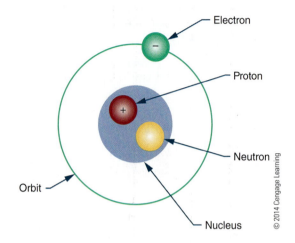

© 2014 Cengage Learning

■ **FIGURE 10-3**

The electrons are held in shells around the nucleus.

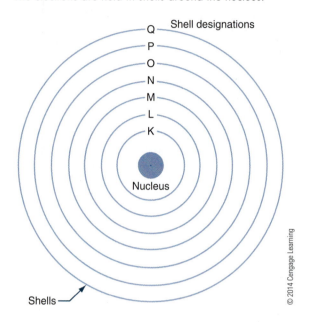

© 2014 Cengage Learning

The number of protons in the nucleus of the atom is called the element's **atomic number**. Atomic numbers distinguish one element from another.

Each element also has an **atomic weight**. The atomic weight is the mass of the atom and is determined by the total number of protons and neutrons in the nucleus. Electrons do not contribute to the total mass of the atom; an electron's mass is only 1/1845 that of a proton and is not significant enough to consider.

The electrons orbit in concentric circles about the nucleus. Each orbit is called a **shell**. These shells are filled in sequence; K is filled first, then L, M, N, and so on (Figure 10-3). The maximum number of electrons that each shell can accommodate is shown in Figure 10-4.

The outer shell is called the **valence shell**, and the number of electrons it contains is the **valence**. The farther the valence shell is from the nucleus, the less attraction the nucleus has on each valence electron. Thus the potential for the atom to gain or lose electrons increases if the valence shell is not full and is located far enough away from the nucleus. Conductivity of an atom depends on its valence band. The greater the number of electrons in the valence shell, the less it conducts. For example, an atom having seven electrons in the valence shell is less conductive than an atom having three electrons in the valence shell.

Electrons in the valence shell can gain energy. If these electrons gain enough energy from an external force, they can leave the atom and become free electrons, moving randomly from atom to atom. Materials that contain a large number of free electrons are called **conductors**. Figure 10-5 compares the conductivity of various metals used as conductors.

■ **FIGURE 10-4**

The number of electrons each shell can accommodate.

Shell Designation	Total Number of Electrons
K	2
L	8
M	18
N	32
O	18
P	12
Q	2

© 2014 Cengage Learning

On the chart, silver, copper, and gold have a valence of 1 (Figure 10-6). However, silver is the best conductor because its free electron is more loosely bonded.

Insulators, the opposite of conductors, prevent the flow of electricity. Insulators are stabilized by absorbing valence electrons from other atoms to fill their valence shells, thus eliminating free electrons. Materials classified as insulators are compared in Figure 10-7. Mica is the best insulator because it has the fewest free electrons in its valence shell. A perfect insulator has atoms with full valence shells. This means it cannot gain electrons.

■ **FIGURE 10-5**

Conductivity of various metals used as conductors.

Material	Conductance
Silver	High
Copper	
Gold	
Aluminum	
Tungsten	
Iron	
Nichrome	Low

© 2014 Cengage Learning

■ **FIGURE 10-6**

Copper has a valence of 1.

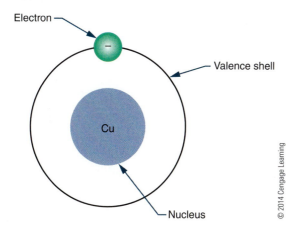

© 2014 Cengage Learning

■ **FIGURE 10-7**

Insulation properties of various materials used as insulators.

Material	Insulation Properties
Mica	High
Glass	
Teflon	
Paper (paraffin)	
Rubber	
Bakelite	
Oils	
Porcelain	
Air	Low

© 2014 Cengage Learning

Halfway between conductors and insulators are **semiconductors**. Semiconductors are neither good conductors nor good insulators but are important because they can be altered to function as conductors or insulators. Silicon and germanium are two semiconductor materials.

An atom that has the same number of electrons and protons is identified as an electrically balanced atom. A balanced atom that receives one or more electrons is no longer balanced. It is said to be negatively charged and is called a **negative ion**. A balanced atom that loses one or more electrons is said to be positively charged and is called a **positive ion**. The process of gaining or losing electrons is called **ionization**. Ionization is significant in current flow.

10–2 **QUESTIONS**

1. What atomic particle has a positive charge and a large mass?
2. What atomic particle has no charge at all?
3. What atomic particle has a negative charge and a small mass?
4. What does the number of electrons in the outermost shell determine?
5. What is the term for describing the gaining or losing of electrons?

10–3 **CURRENT**

Given an appropriate external force, the movement of electrons is from negatively charged atoms to positively charged atoms. This flow of electrons is called **current (I)**. The symbol "I" is used to represent current. The amount of current is the sum of the charges of the moving electrons past a given point.

An electron has a very small charge, so the charge of 6.24×10^{18} electrons is added together and called a **coulomb (C)**. When 1 coulomb of charge moves past a single point in 1 second, it is called an **ampere (A)**. The ampere is named for the French physicist André Marie Ampère (1775–1836), one of the early founders of the science of electromagnetism, which he called electrodynamics. Current is measured in amperes.

1. What action causes current in an electric circuit?
2. What action results in an ampere of current?
3. What symbol is used to represent current?
4. What symbol is used to represent the unit ampere?
5. What unit is used to measure current?

10-4 VOLTAGE

When there is an excess of electrons (negative charge) at one end of a conductor and a deficiency of electrons (positive charge) at the opposite end, a current flows between the two ends as long as this condition persists. The source that creates this excess of electrons at one end and the deficiency at the other end represents the **potential**, or the ability of the source to perform electrical work.

The actual work accomplished in a circuit is a result of the **difference of potential** available at the two ends of a conductor. This difference of potential causes electrons to move or flow in a circuit (Figure 10-8) and is referred to as **electromotive force (emf)** or **voltage**. Voltage is the force that moves the electrons in the circuit. Think of voltage as the pressure or pump that moves the electrons.

The symbol **E** is used in electronics to represent voltage. The unit for measuring voltage is the **volt (V)**, named for Count Alessandro Volta (1745–1827), inventor of the first cell to produce electricity.

1. What force moves electrons in a circuit?
2. What is the term that represents the potential between the two ends of a conductor?
3. What symbol is used to represent voltage?
4. What symbol is used to represent the unit volt?
5. Who invented the first cell to produce electricity?

10-5 RESISTANCE

As the free electrons move through the circuit, they encounter atoms that do not readily give up electrons. This opposition to the flow of electrons (the current) is called **resistance (R)**.

Every material offers some resistance or opposition to current flow. The degree of resistance of a material depends on its size, shape, and temperature.

Materials with a low resistance are called conductors. Conductors have many free electrons and offer little resistance to current flow. As previously mentioned, silver, copper, gold, and aluminum are examples of good conductors.

Materials with a high resistance are called *insulators*. Insulators have few free electrons and offer a high resistance to current flow. As previously mentioned, glass, rubber, and plastic are examples of good insulators.

Resistance is measured in **ohms**, a unit named for the German physicist and mathematician George Simon Ohm (1787–1854). In 1827, Ohm published a paper acknowledging the mathematical relationship among current, voltage, and resistance known as Ohm's law. The symbol for the ohm is the Greek letter omega (Ω).

■ **FIGURE 10-8**

Electrons flow in a circuit because of the difference of potential.

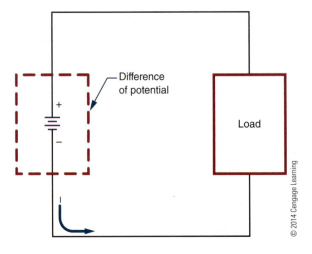

Difference of potential

+

−

Load

I

© 2014 Cengage Learning

10–5 QUESTIONS

1. What is the term used to describe opposition to current flow?
2. What is the main difference between conductors and insulators?
3. What letter is used to represent resistance?
4. What is the symbol used to represent the unit of resistance?
5. What is the unit of resistance?

SUMMARY

- Matter is anything that occupies space.
- Matter can be an element or compound.
- An element is the basic building block of nature.
- A compound is a chemical combination of two or more elements.
- A molecule is the smallest unit of a compound that retains the properties of the compound.
- An atom is the smallest unit of matter that retains the structure of the element.
- An atom consists of a nucleus, which contains protons and neutrons. It also has one or more electrons that orbit around the nucleus.
- Protons have a positive charge, electrons have a negative charge, and neutrons have no charge.
- The atomic number of an element is the number of protons in the nucleus.
- The atomic weight of an atom is the sum of protons and neutrons.
- The orbits of the electrons are called shells.
- The outer shell of an atom is called the valence shell.
- The number of electrons in the valence shell is called the valence.
- An atom that has the same number of protons as electrons is electrically balanced.
- The process by which atoms gain or lose electrons is called ionization.
- The flow of electrons is called current.
- Current is represented by the symbol I.
- The charge of 6,240,000,000,000,000,000 (or 6.24×10^{18}) electrons is called a coulomb.
- An ampere of current is measured when 1 coulomb of charge moves past a given point in 1 second.
- Ampere is represented by the symbol A.
- Current is measured in amperes.
- An electric current flows through a conductor when there is an excess of electrons at one end and a deficiency at the other end.
- A source that supplies excess electrons represents a potential or electromotive force.
- The potential or electromotive force is referred to as voltage.
- Voltage is the force that moves electrons in a circuit.
- The symbol E is used to represent voltage.
- A volt (V) is the unit for measuring voltage.
- Resistance is the opposition to current flow.
- Resistance is represented by the symbol R.
- All materials offer some resistance to current flow.
- The resistance of a material is dependent on the material's size, shape, and temperature.
- Conductors are materials with low resistance.
- Insulators are materials with high resistance.
- Resistance is measured in ohms.
- The Greek letter omega (Ω) is used to represent ohms.

CHAPTER 10 SELF-TEST

1. What are the differences among an element, an atom, a molecule, and a compound?
2. What criteria determine whether an atom is a good conductor?
3. What determines whether a material is a conductor, a semiconductor, or an insulator?
4. Why is it essential to understand the relationship among conductors, semiconductors, and insulators?
5. How many electrons need to move past a point to be defined as an ampere?

6. In a circuit, what is doing the actual work?
7. What is resistance in a circuit really doing?
8. Make a chart comparing current, voltage, and resistance. Include the symbol and unit for each.
9. Explain the differences among current, voltage and resistance.
10. Describe how the resistance of a material is determined.

CHAPTER 11

Current

OBJECTIVES

After completing this chapter, the student will be able to:

- State the two laws of electrostatic charges.
- Define coulomb.
- Identify the unit used to measure current flow.
- Define the relationship of amperes, coulombs, and time through a formula.
- Describe how current flows in a circuit.
- Describe how electrons travel in a conductor.
- Define and use scientific notation.
- Identify commonly used prefixes for powers of 10.

KEY TERMS

11-1 repelling

11-2 current (I)

11-2 hole

11-2 voltage source

11-3 microampere (μA)

11-3 milliampere (mA)

11-3 scientific notation

The atom has been defined as the smallest particle of an element. It is composed of electrons, protons, and neutrons.

Electrons breaking away from atoms and flowing through a conductor produce an electric current.

This chapter examines how electrons break free from atoms to produce a current flow and how to use scientific notation. Scientific notation expresses very large and small numbers in a form of mathematical shorthand.

11-1 ELECTRICAL CHARGE

Two electrons together or two protons together represent "like" charges. Like charges resist being brought together and instead move away from each other. This movement is called **repelling**. This is the first law of electrostatic charges: Like charges repel each other (Figure 11-1). According to the second law of electrostatic charges: Unlike charges attract each other.

The negative electrons are drawn toward the positive protons in the nucleus of an atom. This attractive force is balanced by the centrifugal force caused by the electron's rotation about the nucleus. As a result, the electrons remain in orbit and are not drawn into the nucleus.

The amount of attracting or repelling force that acts between two electrically charged bodies depend on two factors: their charge and the distance between them.

Single electrons have a charge too small for practical use. The unit adopted for measuring charges is the coulomb (C), named for Charles Coulomb. The electrical

charge (Q) carried by 6,240,000,000,000,000,000 electrons (six quintillion two hundred forty quadrillion, or 6.24×10^{18}) represents one coulomb.

$$1 \text{ C} = 6.24 \times 10^{18} \text{ electrons}$$

Electrical charges are created by the displacement of electrons. When there is an excess of electrons at one point and a deficiency of electrons at another point, a difference of potential exists between the two points. When a difference of potential exists between two charged bodies connected by a conductor, electrons will flow along the conductor. This flow of electrons is called current.

11-2 CURRENT FLOW

An electric **current (I)** consists of the drift of electrons from an area of negative charge to an area of positive charge. The unit of measurement for current flow is the ampere (A). An ampere represents the amount of current in a conductor when 1 coulomb of charge moves past a point in 1 second. The relationship between amperes and coulombs per second can be expressed as follows:

$$I = \frac{Q}{t}$$

Where: I = current measured in amperes
 Q = quantity of electrical charge in coulombs
 t = time in seconds

EXAMPLE: What is the current in amperes if 9 coulombs of charge flow past a point in an electric circuit in 3 seconds?

Given	Solution
I = ?	$I = \dfrac{Q}{t}$
Q = 9 coulombs	$I = \dfrac{9}{3}$
t = 3 seconds	I = 3 amperes

■ FIGURE 11-1

Basic laws of electrostatic charges.

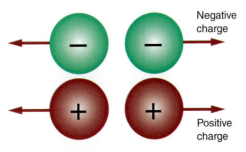

Negative charge

Positive charge

Like charges repel each other

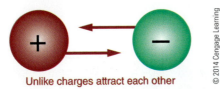

Unlike charges attract each other

© 2014 Cengage Learning

EXAMPLE: A circuit has a current of 5 amperes. How long does it take for 1 coulomb to pass a given point in the circuit?

Given

$I = 5$ amperes

$Q = 1$ coulomb

$t = ?$

Solution

$$I = \frac{Q}{t}$$

$$5 = \frac{1}{t}$$

$$\frac{5}{1} \underset{\diagdown}{\diagup} \frac{1}{t} \quad \text{(cross-multiply)}$$

$$(1)(1) = (5)(t)$$

$$1 = 5t$$

$$\frac{1}{5} = \frac{5t}{5} \quad \text{(divide both sides by 5)}$$

$$\frac{1}{5} = t$$

$$0.2 \text{ second} = t$$

Electrons, with their negative charge, represent the charge carrier in an electric circuit. Therefore, electric current is the flow of negative charges. Scientists and engineers once thought that current flowed in a direction opposite to electron flow. Later work revealed that the movement of an electron from one atom to the next created the appearance of a positive charge, called a **hole**, moving in the opposite direction (Figure 11-2 and Figure 11-3). Electron movement and current were found to be the same.

If electrons are added to one end of a conductor and provision is made to take electrons from the other end, an electric current flows through the conductor. As free electrons move slowly through the conductor, they collide with atoms, knocking other electrons free. These new free electrons travel toward the positive end of the conductor and collide with other atoms. The

■ **FIGURE 11-2**

As electrons move from one atom to another, they create the appearance of a positive charge, called a hole.

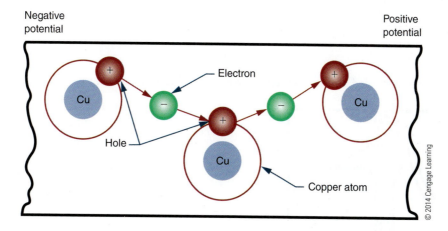

■ **FIGURE 11-3**

Electron movement occurs in the opposite direction to hole movement.

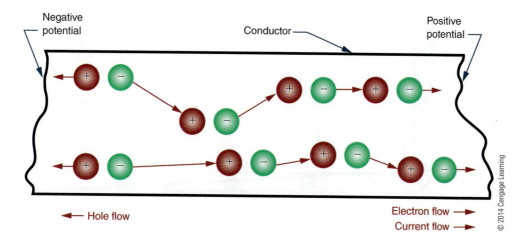

■ FIGURE 11-4

Electrons in a conductor react like Ping-Pong balls in a hollow tube.

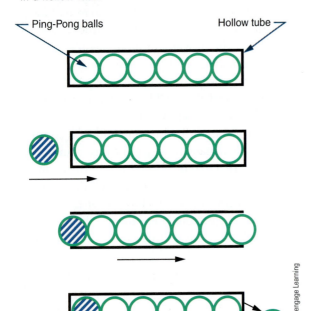

electrons drift from the negative to the positive end of the conductor because like charges repel. In addition, the positive end of the conductor, which represents a deficiency in electrons, attracts the free electrons because unlike charges attract.

The drift of electrons is slow (approximately an eighth of an inch per second), but individual electrons ricochet off atoms, knocking other electrons loose, at the speed of light (186,000 miles per second). For example, visualize a long, hollow tube filled with Ping-Pong balls (Figure 11-4). As a ball is added to one end of the tube, a ball is forced out the other end of the tube. Although an individual ball takes time to travel down the tube, the speed of its impact can be far greater.

The device that supplies electrons from one end of a conductor (the negative terminal) and removes them from the other end of the conductor (the positive terminal) is called the **voltage source**. It can be thought of as a kind of pump (Figure 11-5).

■ FIGURE 11-5

A voltage source can be considered a pump that supplies electrons to the load and recycles the excess electrons.

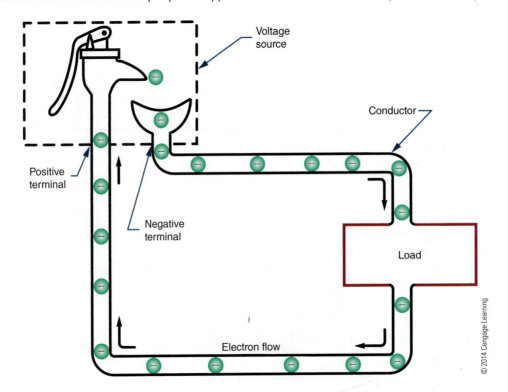

11-2 QUESTIONS

1. Define electric current.
2. What is the unit for measuring flow?
3. What is the relationship among current, coulombs, and time?
4. What is the current if 15 coulombs of charge flow past a point in a circuit in 5 seconds?
5. How long does it take for 3 coulombs to move past a point in a circuit if the circuit has 3 amperes of current flow?
6. What makes electrons move through a conductor in only one direction?

■ **FIGURE 11-6**

Prefixes commonly used in electronics.

Prefix	Symbol	Value	Decimal Value
Tetra-	T	10^{12}	1,000,000,000,000
Giga-	G	10^{9}	1,000,000,000
Mega-	M	10^{6}	1,000,000
Kilo-	k	10^{3}	1,000
Milli-	m	10^{-3}	0.001
Micro-	μ	10^{-6}	0.000001
Nano-	n	10^{-9}	0.000000001
Pico-	p	10^{-12}	0.000000000001

11-3 SCIENTIFIC NOTATION

In electronics, it is common to encounter very small and very large numbers. **Scientific notation** is a means of using single-digit numbers plus powers of 10 to express large and small numbers. For example, 300 in scientific notation is 3×10^{2}.

The exponent indicates the number of decimal places to the right or left of the decimal point in the number. If the power is positive, the decimal point is moved to the right. For example:

$$3 \times 10^{3} = 3.0 \times 10^{3} = 3.000 = 3000$$

3 places

If the power is negative, the decimal point is moved to the left. For example:

$$3 \times 10^{-6} = 3.0 \times 10^{-6} = 0\ 000003. = 0.000003$$

6 places

Figure 11-6 lists some commonly used powers of 10, both positive and negative, and the prefixes and symbols associated with them. For example, an ampere (A) is a large unit of current that is not often found in low-power electronic circuits. More frequently used units are the **milliampere (mA)** and the **microampere (μA)**. A milliampere is equal to one-thousandth (1/1000) of an ampere, or 0.001 A. In other words, it takes 1000 milliamperes to equal one ampere. A microampere is equal to one-millionth

(1/1,000,000) of an ampere, or 0.000001A; it takes 1,000,000 microamperes to equal 1 ampere.

EXAMPLE: How many milliamperes are there in 2 amperes?

Solution:

$$\frac{1000 \text{ mA}}{1 \text{ A}} = \frac{X \text{ mA}}{2 \text{ A}} \quad (1000 \text{ mA} = 1 \text{ A})$$

$$\frac{1000}{1} = \frac{X}{2}$$

$$(1)(X) = (1000)(2)$$

$$X = 2000 \text{ mA}$$

EXAMPLE: How many amperes are there in 50 microamperes?

Solution:

$$\frac{1,000,000 \text{ μA}}{1 \text{ A}} = \frac{50 \text{ μA}}{X \text{ A}}$$

$$\frac{1,000,000}{1} = \frac{50}{X}$$

$$(1)(50) = (1,000,000)(X)$$

$$\frac{50}{1,000,000} = X$$

$$0.00005 = X$$

$$0.00005 \text{ A} = X$$

11-3 QUESTIONS

1. Define scientific notation.
2. In scientific notation:
   a. What does a positive exponent mean?
   b. What does a negative exponent mean?
3. Convert the following numbers to scientific notation:
   a. 500
   b. 3768
   c. 0.0056
   d. 0.105
   e. 356.78
4. Define the following prefixes:
   a. Milli-
   b. Micro-
5. Perform the following conversions:
   a. 1.5 A = _____ mA
   b. 1.5 A = _____ μA
   c. 150 mA = _____ A
   d. 750 μA = _____ A

SUMMARY

- Laws of electrostatic charges: Like charges repel; unlike charges attract.
- Electrical charge (Q) is measured in coulombs (C).
- One coulomb is equal to 6.24×10^{18} electrons.
- An electric current is the slow drift of electrons from an area of negative charge to an area of positive charge.
- Current flow is measured in amperes.
- One ampere (A) is the amount of current that flows in a conductor when 1 coulomb of charge moves past a point in 1 second.
- The relationship among current, electrical charge, and time is represented by the formula:

$$I = \frac{Q}{t}$$

- Electrons (negative charge) represent the charge carrier in an electrical circuit.
- Hole movement (positive charge) occurs in the opposite direction to electron movement.
- Current flow in a circuit is from negative to positive.
- Electrons travel very slowly through a conductor, but individual electrons move at the speed of light.
- Scientific notation expresses a very large or small number as a numeral from 1 to 9 to a power of 10.
- If the power-of-10 exponent is positive, the decimal point is moved to the right.
- If the power-of-10 exponent is negative, the decimal point is moved to the left.
- The prefix *milli-* means one-thousandth.
- The prefix *micro-* means one-millionth.

CHAPTER 11 SELF-TEST

1. How much current is in a circuit if it takes 5 seconds for 7 coulombs to flow past a given point?
2. Describe how electrons flow in a circuit with reference to the potential in the circuit.
3. Convert the following numbers to scientific notation:
   a. 235
   b. 0.002376
   c. 56323.786
4. What do the following prefixes represent?
   a. Milli-
   b. Micro-
5. Make a table and convert the following:
   a. 305 mA to _____ A
   b. 6 μA to _____ mA
   c. 17 volts to _____ mV
   d. 0.023 mV to _____ μV
   e. 0.013 kΩ to _____ Ω
   f. 170 MΩ to _____ Ω

Voltage

OBJECTIVES

After completing this chapter, the student will be able to:

- Identify the six most common voltage sources.
- Describe six different methods of producing electricity.
- Define a cell and a battery.
- Describe the difference between primary and secondary cells.
- Describe how cells and batteries are rated.
- Identify ways to connect cells or batteries to increase current or voltage output or both.
- Define voltage rise and voltage drop.
- Identify the two types of grounds associated with electrical circuits.

KEY TERMS

12-1 alkaline

12-1 alternating current (AC)

12-1 cell

12-1 direct current (DC)

12-1 generator

12-1 microphone

12-1 photovoltaic cell (solar cell)

12-1 piezoelectric effect

12-1 thermocouple

12-1 Van de Graaf generator

12-2 alkaline cell

12-2 battery

12-2 dry cell

12-2 Leclanche cell

12-2 lithium cell

12-2 nickel cadmium (Ni-Cad) cell

12-2 primary cell

12-2 secondary cell

12-3 parallel

12-3 series

12-3 series aiding

12-3 series opposing

12-3 series-parallel

12-4 voltage drop

12-4 voltage rise

12-5 ground

In a piece of copper wire, the electrons are in random motion with no direction. To produce a current flow, the electrons must all move in the same direction. To produce motion in a given direction, energy must be imparted to the electrons in the copper wire. This energy comes from a source connected to the wire.

The force that causes the electrons to move in a common direction is referred to as *difference of potential*, or *voltage*. This chapter examines how voltage is produced.

12-1   VOLTAGE SOURCES

A current is produced when an electron is forced from its orbit around an atom. Any form of energy that dislodges electrons from atoms can be used to produce current. It is important to note that energy is not created; rather, there is simply a transfer of energy from one form to another. The source supplying the voltage is not simply a source of electrical energy. Instead, it is the means of converting some other form of energy into electrical energy. The six most common voltage sources are friction, magnetism, chemicals, light, heat, and pressure.

Friction is the oldest known method of producing electricity. A glass rod can become charged when rubbed with a piece of fur or silk. This is similar to the charge you can generate by scuffing your feet across a carpet in a dry room. A **Van de Graaf generator** is a device that operates using the same principles as the glass rod and is capable of producing millions of volts (Figure 12-1).

Magnetism is the most common method of producing electrical energy today. If a wire is passed through a magnetic field, voltage is produced as long as there is motion between the magnetic field and the conductor. A device based on this principle is called a **generator** (Figure 12-2). A generator can produce either direct current or alternating current, depending on how it is wired. When electrons flow in only one direction, the current is called **direct current (DC)**. When electrons flow in one direction and then in the opposite direction, the current is called **alternating current (AC)**. A generator may be powered by steam from nuclear power or coal, water, wind, or gasoline or diesel engines. The schematic symbol for a DC generator and an AC generator is shown in Figure 12-3. The second most common method of producing electrical energy today is by the use of a chemical **cell**. The cell

FIGURE 12-1

A Van de Graaf generator is capable of producing millions of volts.

Hollow metal ball

Support tower (non conductive; houses latex belt)

Base to house drive motor

© 2014 Cengage Learning

consists of two dissimilar metals, copper and zinc, immersed in a salt, acid, or **alkaline** solution. The metals, copper and zinc, are the electrodes. The electrodes

FIGURE 12-2

A generator uses magnetism to produce electricity.

Field   Battery

© 2014 Cengage Learning

■ **FIGURE 12-3**

Schematic symbol for A.) direct current generator and B.) alternating current generator.

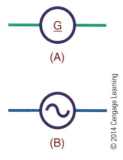

(A)

(B)

© 2014 Cengage Learning

establish contact with the electrolyte (the salt, acid, or alkaline solution) and the circuit. The electrolyte pulls the free electrons from the copper electrode, leaving it with a positive charge. The zinc electrode attracts free electrons in the electrolyte and thus acquires a negative charge. Several of these cells can be connected together to form a battery. Figure 12-4 shows the schematic symbol for a cell and battery. Many types of cells and batteries are in use today (Figure 12-5).

■ **FIGURE 12-4**

Schematic symbol for a cell and a battery. The combination of two or more cells forms a battery.

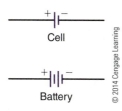

Cell

Battery

© 2014 Cengage Learning

■ **FIGURE 12-5**

Some of the more common chemical batteries and cells in use today.

© 2014 Cengage Learning

Light energy can be converted directly to electrical energy by light striking a photosensitive (light-sensitive) substance in a **photovoltaic cell (solar cell)** (Figure 12-6). A solar cell consists of photosensitive materials mounted between metal contacts. When the surface of the photosensitive material is exposed to light, it dislodges electrons from their orbits around the surface atoms of the material. This occurs because light has energy. A single cell can produce a small voltage. Figure 12-7 shows the schematic symbol for a solar cell. Many cells must be linked together to produce a usable voltage and current. Solar cells are used primarily in satellites and cameras. The high cost of construction has limited their general application. However, the price of solar cells is coming down.

Heat can be converted directly to electricity with a device called a **thermocouple** (Figure 12-8). The schematic symbol for a thermocouple is shown in Figure 12-9. A thermocouple consists of two dissimilar metal wires twisted together. One wire is copper and the other wire is zinc or iron. When heat is applied to the twisted connection, the copper wire readily gives up free electrons, which are transferred to the other wire. Thus the copper wire develops a positive charge and the other wire develops a negative charge, and a

■ **FIGURE 12-6**

A photovoltaic cell can convert sunlight directly into electricity.

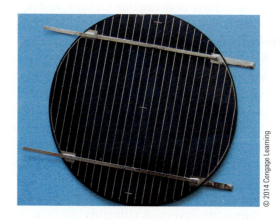

© 2014 Cengage Learning

■ **FIGURE 12-7**

Schematic symbol of a photovoltaic cell (solar cell).

© 2014 Cengage Learning

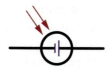

■ **FIGURE 12-8**

Thermocouples convert heat energy directly into electrical energy.

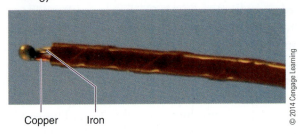

Copper   Iron

© 2014 Cengage Learning

■ **FIGURE 12-9**

Schematic symbol for a thermocouple.

© 2014 Cengage Learning

small voltage occurs. The voltage is directly proportional to the amount of heat applied. One application of the thermocouple is as a thermometer. Also called a pyrometer, these devices are often used in high-temperature kilns and foundries.

When pressure is applied to certain crystalline materials such as quartz, tourmaline, Rochelle salts, or barium titanate, a small voltage is produced. This is referred to as the **piezoelectric effect**. Initially, negative and positive charges are distributed randomly throughout a piece of crystalline material, and no overall charge can be measured. However, when pressure is applied, electrons leave one side of the material and accumulate on the other side. A charge is produced as long as the pressure remains. When the pressure is removed, the charge is again distributed, so no overall charge exists. The voltage produced is small and must be amplified to be useful. Uses of the piezoelectric effect include crystal **microphones**, phonograph pickups (crystal cartridges), and precision oscillators (Figure 12-10 and Figure 12-11).

Note that although a voltage can be produced by these means, the reverse is also true; that is, a voltage can be used to produce magnetism, chemicals, light, heat, and pressure. Magnetism is evident in motors, speakers, solenoids, and relays. Chemical activities can be produced through electrolysis and electroplating. Light is produced with lightbulbs and other optoelectric devices. Heat is produced with heating elements in stoves, irons, and soldering irons. And voltage can be applied to bend or twist a crystal.

■ **FIGURE 12-10**

Crystal Microphone and how it is constructed.

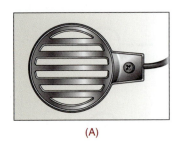

(A)

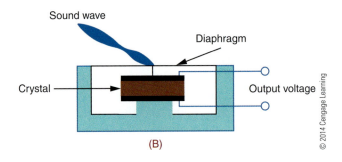

Sound wave

Diaphragm

Crystal

Output voltage

(B)

© 2014 Cengage Learning

■ **FIGURE 12-11**

Schematic diagram of a piezoelectric crystal.

© 2014 Cengage Learning

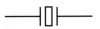

12–1 QUESTIONS

1. What are the six most common voltage sources?
2. What is the most common method for producing a voltage?
3. What is the second most common method for producing a voltage?
4. Why are solar cells not used more for producing a voltage?
5. What determines current flow with the piezoelectric effect?
6. What are the results when voltage is applied to produce magnetism, chemical, light, heat, and pressure?

12–2 CELLS AND BATTERIES

As mentioned in the previous unit, a cell contains a positive and a negative electrode separated by an electrolytic solution. A **battery** is a combination of two or more cells. There are two basic types of cells. Cells that

■ FIGURE 12-12

Diagram of a dry cell (shown cut away).

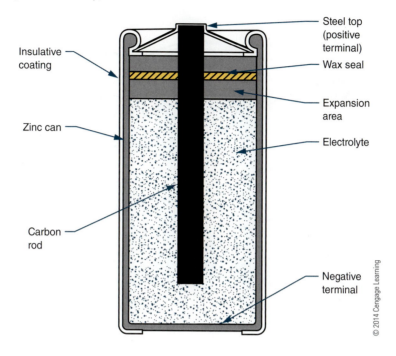

Insulative coating

Zinc can

Carbon rod

Steel top (positive terminal)

Wax seal

Expansion area

Electrolyte

Negative terminal

© 2014 Cengage Learning

cannot be recharged are called **primary cells**. Cells that can be recharged are called **secondary cells**.

An example of a primary cell is a **Leclanche cell**, also called a **dry cell** (Figure 12-12). This type of cell is not actually dry. It contains a moist paste as the electrolyte. A seal prevents the paste from leaking out when the cell is turned sideways or upside down. The electrolyte in a dry cell is a solution of ammonium chloride and manganese dioxide. The electrolyte dissolves the zinc electrode (the case of the cell), leaving an excess of electrons with the zinc. As the current is removed from the cell, the zinc, ammonium chloride, and manganese dioxide produce manganese dioxide, water, ammonia, and zinc chloride. The carbon rod (center electrode) gives up the extra electrons that accumulate on the zinc electrode. This type of cell produces as much as 1.75 to 1.8 volts when new. A typical Leclanche cell has an energy density of approximately 120 watt-hours per pound. As the cell is used, the chemical action decreases, and eventually the resulting current ceases. If the cell is not used, the electrolytic paste eventually dries out. The cell has a shelf life of about two years. The output voltage of this type of cell is determined entirely by the materials used for the electrolyte and the electrodes. The AAA cell, AA cell, C cell, and D cell (Figure 12-13) are all constructed of the same materials and therefore produce the same

voltage. It should be noted that although the Leclanche cell is frequently referred to as a carbon-zinc (or zinc carbon) cell, the carbon does not take any part in the chemical reaction that produces electricity.

The **alkaline cell** is named because of the highly caustic base, potassium hydroxide (KOH), used as the electrolyte. The design of an alkaline cell on the outside is very similar to that of a carbon-zinc cell. However, the inside of the alkaline cell is significantly different (Figure 12-14). Alkaline cells have an

■ FIGURE 12-13

Some common examples of dry cells.

© 2014 Cengage Learning

■ **FIGURE 12-14**

Alkaline cells are constructed inside out. The cathode surrounds the anode.

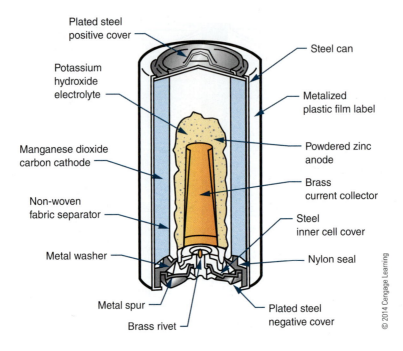

open-circuit rating of approximately 1.52 volts, and an energy density of about 45 watt-hours per pound. The alkaline cell performs much better over temperature extremes than carbon-zinc cells. Alkaline cells perform best where moderate to high currents are drawn over extended periods of time.

Lithium cells (Figure 12-15) have overcome the inherent properties associated with lithium. Lithium is extremely reactive with water. Lithium cell formation uses lithium, manganese dioxide (MnO_2), and a lithium perchlorate ($LiClO_4$) in an organic solvent (water cannot be used). The output of a lithium cell is approximately 3 volts. Lithium cells are very efficient, with energy densities of about 90 watt-hours per pound. The greatest benefit of lithium cells is their extremely long shelf life of 5 to 10 years.

A secondary cell is a cell that can be recharged by applying a reverse voltage. An example is the lead-acid battery used in automobiles (Figure 12-16). It is made of six 2-volt secondary cells connected in series. Each cell has a positive electrode of lead peroxide (PbO_2) and a negative electrode of spongy lead (Pb). The electrodes are separated by plastic or rubber and immersed in an electrolytic solution of sulfuric acid (H_2SO_4) and distilled water (H_2O). As the cell is discharged, the sulfuric acid combines with the lead sulfate, and the electrolyte

converts to water. Recharging the cell involves applying a source of DC voltage greater than that produced by the cell. As the current flows through the cell, it changes the electrode back to lead peroxide and spongy lead and converts the electrolyte back to sulfuric acid and water. This type of cell is also referred to as a wet cell.

Another type of secondary cell is the **nickel cadmium (Ni-Cad) cell** (Figure 12-17). This is a dry cell that can be recharged many times and can hold its charge for long periods of time. It consists of a positive and a negative electrode, a separator, an electrolyte, and a package. The electrodes are comprised of a deposit of powdered nickel on a nickel wire screen, which is coated with a nickel salt solution for the positive electrode and a cadmium salt solution for a negative electrode. The separator is made of an absorbent insulating material. The electrolyte is potassium hydroxide. A steel can forms the package and is sealed tightly. A typical voltage from this type of cell is 1.2 volts.

The ability of a battery to deliver power continuously is expressed in ampere-hours. A battery rated at 100 ampere-hours can continuously supply any of the following: 100 amperes for 1 hour ($100 \times 1 = 100$ ampere-hours), 10 amperes for 10 hours ($10 \times 10 = 100$ ampere-hours), or 1 ampere for 100 hours ($1 \times 100 = 100$ ampere-hours).

■ FIGURE 12-15

Lithium cells have extremely high energy density.

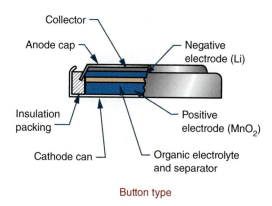

Collector
Anode cap
Negative electrode (Li)
Insulation packing
Positive electrode (MnO_2)
Cathode can
Organic electrolyte and separator

Button type

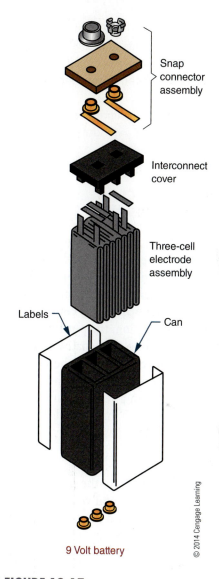

Snap connector assembly

Interconnect cover

Three-cell electrode assembly

Labels

Can

9 Volt battery

© 2014 Cengage Learning

■ FIGURE 12-16

An example of a secondary cell (shown cut away).

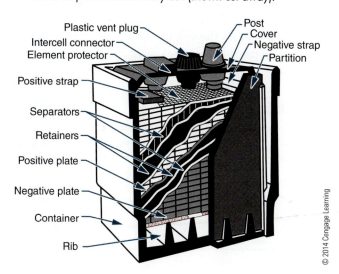

Plastic vent plug
Intercell connector
Element protector
Positive strap
Separators
Retainers
Positive plate
Negative plate
Container
Rib
Post
Cover
Negative strap
Partition

© 2014 Cengage Learning

■ FIGURE 12-17

A nickel-cadmium (Ni-Cad) battery is another example of a secondary cell.

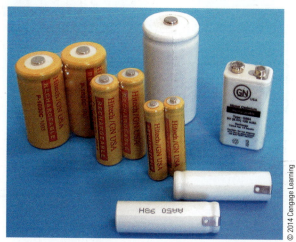

© 2014 Cengage Learning

12–2   QUESTIONS

1. What are the components of a cell?
2. What are the two basic types of cells?
3. What is the major difference between the two types of cells?
4. List some examples of a primary cell.
5. List some examples of a secondary cell.

12–3   CONNECTING CELLS AND BATTERIES

Cells and batteries can be connected together to increase voltage and/or current. They can be connected in **series**, in **parallel**, or in **series-parallel**. In series, cells or batteries can be connected in either **series-aiding** or **series-opposing** configurations. In a series-aiding configuration, the positive terminal of the first cell is connected to the negative terminal of the second cell; the positive terminal of the second cell is connected to the negative terminal of the third cell; and so on (Figure 12-18). In a series-aiding configuration, the same current flows through all the cells or batteries. This can be expressed as

$$I_T = I_1 = I_2 = I_3$$

The subscript numbers refer to the number of each individual cell or battery. The total voltage is the sum of the individual cell voltages and can be expressed as

$$E_T = E_1 + E_2 + E_3$$

■ FIGURE 12-18

Cells or batteries can be connected in series to increase voltage.

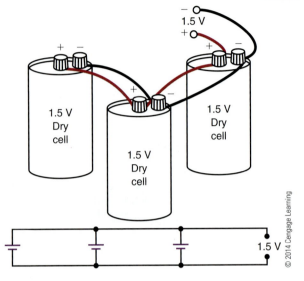

■ FIGURE 12-19

Cells or batteries can be connected in parallel to increase current flow.

In a series-opposing configuration, the cells or batteries are connected with like terminals together, negative to negative or positive to positive. However, this configuration has little practical application.

In a parallel configuration, all the positive terminals are connected together and all the negative terminals are connected together (Figure 12-19). The total current available is the sum of the individual currents of each cell or battery. This can be expressed as

$$I_T = I_1 + I_2 + I_3$$

The total voltage is the same as the voltage of each individual cell or battery. This can be expressed as

$$E_T = E_1 = E_2 = E_3$$

If both a higher voltage and a higher current are desired, the cells or batteries can be connected in a series-parallel configuration. Remember, connecting cells or batteries in series increases the voltage, and connecting cells or batteries in parallel increases the current. Figure 12-20 shows four 3-volt batteries connected in a series-parallel

■ FIGURE 12-20

Cells and batteries can be connected in series-parallel to increase current and voltage output.

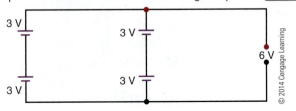

■ **FIGURE 12-21**

The voltage increases when cells are connected in series.

■ **FIGURE 12-22**

Connecting the series-connected cells in parallel increases the output current. The net result is a series-parallel configuration.

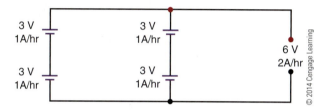

configuration. This configuration produces a total voltage of 6 volts with a current twice that of an individual battery. It is necessary to connect the two 3-volt batteries in series to get the 6 volts (Figure 12-21). To increase the current, a second pair of 3-volt batteries is connected in series, and the resulting series-connected batteries are connected in parallel (Figure 12-22). The overall result is a series-parallel configuration.

| 12–3 | **QUESTIONS** |

1. Draw three cells connected in a series-aiding configuration.
2. What effect does a series-aiding configuration have on current and voltage?
3. Draw three cells connected in parallel.
4. What effect does connecting cells in parallel have on current and voltage?
5. How can cells or batteries be connected to increase both current and voltage?

| 12–4 | **VOLTAGE RISES AND VOLTAGE DROPS** |

In electric and electronic circuits, there are two types of voltage: **voltage rise** and **voltage drop**. Potential energy, or voltage, introduced into a circuit is

■ **FIGURE 12-23**

A potential applied to a circuit is a called a voltage rise.

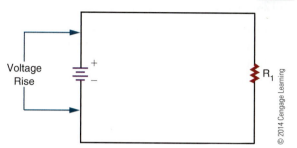

called a voltage rise (Figure 12-23). The voltage is connected to the circuit so that the current flows from the negative terminal of the voltage source and returns to the positive terminal of the voltage source. A 12-volt battery connected to a circuit gives a voltage rise of 12 volts to the circuit.

As the electrons flow through the circuit, they encounter a load, some resistance to the flow of electrons. As electrons flow through the load, they give up their energy. The energy given up is called a voltage drop (Figure 12-24). The energy is given up in most cases as heat. The energy that the electrons give up to a circuit is the energy given to them by the source.

To repeat, energy introduced into a circuit is called a voltage rise. Energy used up in a circuit by the load is called a voltage drop. A voltage drop occurs when there is a current flow in the circuit. Current moves through a circuit from the negative polarity to the positive polarity. The voltage source establishes the voltage rise from the negative terminal to the positive terminal.

The voltage drop in a circuit equals the voltage rise of the circuit, because energy cannot be created or destroyed, only changed to another form. If a 12-volt

■ **FIGURE 12-24**

The energy used by the circuit in a passing current through the load (resistance) is called a voltage drop. A voltage drop occurs when current flows in the circuit.

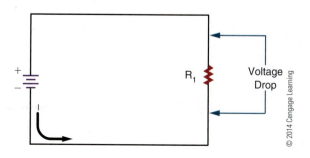

■ **FIGURE 12-25**

Two identical 6-volt lamps each produce a 6-volt drop when connected in series to a 12-volt source.

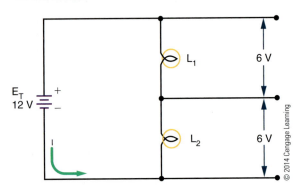

■ **FIGURE 12-26**

When two lamps of different voltage are connected in series to a 12-volt source, the voltage drop across each lamp differs, based on the voltage requirement of the lamp.

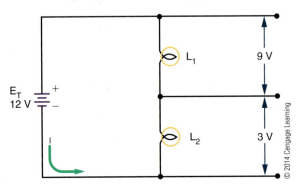

source is connected to a 12-volt lamp, the source supplies a 12-volt voltage rise, and the lamp produces a 12-volt voltage drop. All the energy is consumed in the circuit. If two identical 6-volt lamps are connected in series to the same 12-volt source (Figure 12-25), each lamp produces a 6-volt drop, for a total of 12 volts. If two different lamps are connected in series, such as a 9-volt lamp and a 12-volt lamp (Figure 12-26), the 9-volt lamp produces a voltage drop of 9 volts, and the 3-volt lamp produces a voltage drop of 3 volts. The sum of the voltage drops equals the voltage rise of 12 volts.

12-4 QUESTIONS

1. What is a voltage rise?
2. What is a voltage drop?
3. Where did the energy that is given up by the electrons in a circuit come from?
4. If two identical resistors are connected in series to a voltage source, what is the voltage drop across each of the resistors?
5. If a 6-volt lamp and a 3-volt lamp were connected in series with a 9-volt battery, which lamp would be brighter?

12-5 GROUND AS A VOLTAGE REFERENCE LEVEL

Ground is a term used to identify zero potential. All other potentials are either positive or negative with respect to ground. There are two types of grounds: earth and electrical.

In the home, all electrical circuits and appliances are earth grounded. Consequently, no difference of potential exists between any two appliances or circuits. All circuits are tied to a common point in the circuit panel box (circuit breaker or fuse box) (Figure 12-27). This common point (the neutral bus) is then connected by a heavy copper wire to a copper rod driven

■ **FIGURE 12-27**

In a residential circuit panel box, all circuits are tied to a common point (the neutral bus).

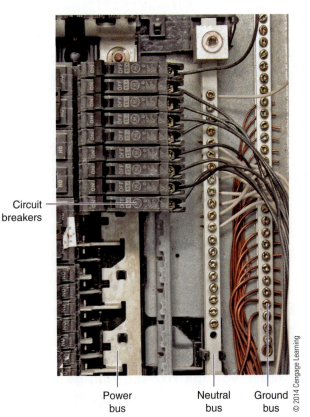

Circuit breakers

Power bus   Neutral bus   Ground bus

into the earth or fastened to the metal pipe that supplies water to the home. This protects the user from electrical shock in case of a faulty connection.

Electrical grounding is used in automobiles. Here the chassis of the automobile is used as a ground. This can be verified by seeing where the battery cables are attached to the automobile. Generally, the negative terminal is bolted directly to the frame of the automobile. This point or any other point on the frame of the automobile is considered to be ground. Ground serves as part of the complete circuit.

In electronics, electrical ground serves a different purpose: Ground is defined as the zero reference point against which all voltages are measured. Therefore, the voltage at any point in a circuit may be measured with reference to ground. The voltage measured may be positive or negative with respect to ground.

In larger pieces of electronic equipment, the chassis or metal frame serves as the ground point (reference point), as in the automobile. In smaller electronic devices, plastic is used for the chassis, and all components are connected to a printed circuit board. In this case, ground is a copper pad on the circuit board that acts as a common point to the circuit.

12-5 QUESTIONS

1. What are the two types of grounds?
2. What is the purpose of earth grounding?
3. How is an electrical ground used in an automobile?
4. How is an electrical ground used in a piece of electronic equipment?
5. In electronics, what function does ground serve when taking voltage measurements?

SUMMARY

- Current is produced when an electron is forced from its orbit.
- Voltage provides the energy to dislodge electrons from their orbit.
- A voltage source provides a means of converting some other form of energy into electrical energy.
- Six common voltage sources are friction, magnetism, chemicals, light, heat, and pressure.
- Voltage can be used to produce magnetism, chemicals, light, heat, and pressure.
- Magnetism is the most common method used to produce a voltage.
- Chemical cells are the second most common means of producing a voltage.
- A cell contains positive and negative electrodes separated by an electrolytic solution.
- A battery is a combination of two or more cells.
- Cells that cannot be recharged are called primary cells.
- Cells that can be recharged are called secondary cells.
- Dry cells are primary cells.
- Lead-acid batteries and nickel-cadmium (Ni-Cad) cells are examples of secondary cells.
- Cells and batteries can be connected in series, in parallel, or in series-parallel to increase voltage, current, or both.
- When cells or batteries are connected in a series-aiding configuration, the output current remains the same, but the output voltage increases.

$$I_T = I_1 = I_2 = I_3 \qquad E_T = E_1 + E_2 + E_3$$

- When cells or batteries are connected in parallel, the voltage output remains the same but the output current available increases.

$$I_T + I_1 + I_2 + I_3 \qquad E_T = E_1 = E_2 = E_3$$

- A series-parallel combination increases both the output voltage and the output current.
- Voltage applied to a circuit is referred to as a voltage rise.
- The energy used by a circuit is referred to as a voltage drop.
- The voltage drop in a circuit equals the voltage rise.
- Two types of ground are earth and electrical.
- Earth grounding is used to prevent electric shock by keeping all appliances and equipment at the same potential.
- Electrical grounding provides a common reference point.

CHAPTER 12 SELF-TEST

1. Does current or voltage perform the work in a circuit?
2. List six forms of energy that can be used to produce electricity.
3. How are secondary cells rated?
4. Draw a series-parallel combination that will supply 9 volts at 1 ampere. Use 1½-volt cells rated at 250 milliamperes.
5. What is the voltage drop across three lamps of 3 volts, 3 volts, and 6 volts, with 9 volts applied?

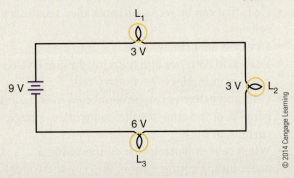

6. What is the voltage and current of two 12-volt, 600-ampere batteries connected in series, and then connected in parallel with two more 12-volt, 600-ampere batteries connected in series?
7. What is one way that a circuit uses the excess energy flowing in a circuit?
8. How does the current flow in a circuit from a battery?
9. What is the difference between the use of ground in home, automobiles, and electronics?
10. Make a chart showing the formulas used for series and parallel for voltage and current in a circuit.

© 2014 Cengage Learning

Resistance

OBJECTIVES

After completing this chapter, the student will be able to:

■ Define resistance and explain its affect in a circuit.

■ Determine the tolerance range of a resistor.

■ Identify carbon composition, wirewound, and film resistors.

■ Identify potentiometers and rheostats.

■ Describe how a variable resistor operates.

■ Decode a resistor's value using the color code or alphanumeric code.

■ Identify the three types of resistor circuits.

■ Calculate total resistance in series, parallel, and series-parallel circuits.

KEY TERMS

13-2 conductance (G)

13-2 mho (℧)

13-2 Siemens (S)

13-3 carbon composition resistor

13-3 carbon film resistor

13-3 metal film resistor

13-3 potentiometer, or pot

13-3 resistor

13-3 rheostat

13-3 surface-mount resistor

13-3 tolerance

13-3 variable resistor

13-3 wirewound resistor

13-4 resistor color code

13-4 parallel circuit

13-4 series circuit

13-4 series-parallel circuit

13-7 branch

Resistance is opposition to the flow of current. Some materials such as glass and rubber offer great opposition to current flow. Other materials such as silver and copper offer little opposition to current flow. This chapter examines the characteristics of resistance, types of resistance, and the effects of connecting resistors together by a conductor to form a circuit.

13-1 RESISTANCE

As previously mentioned, every material offers some resistance or opposition to the flow of current. Some conductors such as silver, copper, and aluminum offer very little resistance to current flow. Insulators such as glass, wood, and paper offer high resistance to current flow.

The size and type of wires in an electric circuit are chosen to keep the electrical resistance as low as possible. This allows the current to flow easily through the conductor. In an electric circuit, the larger the diameter of the wire, the lower the electrical resistance to current flow.

Temperature also affects the resistance of an electrical conductor. In most conductors (copper, aluminum, and so on), resistance increases with temperature. Carbon is an exception because the resistance decreases as temperature increases. Certain alloys of metals (Manganin and Constantan) have resistance that does not change with temperature.

The relative resistance of several conductors of the same length and cross section is shown in Figure 13-1. Silver is used as a standard of 1, and the remaining metals are arranged in order of ascending resistance.

■ FIGURE 13-1

Resistance of several conductors of the same length and cross-section area.

Conductor Material	Resistivity
Silver	1.000
Copper	1.0625
Lead	1.3750
Gold	1.5000
Aluminum	1.6875
Iron	6.2500
Platinum	6.2500

© 2014 Cengage Learning

The resistance of an electric circuit is expressed by the symbol "R." Manufactured circuit parts containing definite amounts of resistance are called resistors. Resistance (R) is measured in ohms (Ω). One ohm is the resistance of a circuit or circuit element that permits a steady current flow of 1 ampere (1 coulomb per second) when 1 volt is applied to the circuit.

13-1 QUESTIONS

1. What is the difference between conductors and insulators?
2. How does the diameter of a piece of wire affect its resistance?
3. What factors affect the resistance of a conductor?
4. What material makes the best conductor?
5. Why is copper, rather than silver, used for wire?

13-2 CONDUCTANCE

The term in electricity that is the opposite of resistance is **conductance (G)**. Conductance is the ability of a material to pass electrons. The unit of conductance is a **mho**, ohm spelled backwards. The symbol used to represent conductance is the inverted Greek letter omega ($\mho$). Conductance is the reciprocal of resistance and is measured in **Siemens (S)**. A reciprocal is obtained by dividing the number into 1.

$$R = 1/G$$
$$G = 1/R$$

If the resistance of a material is known, dividing its value into 1 gives its conductance. Similarly, if the conductance is known, dividing its value into 1 gives its resistance.

13-2 QUESTIONS

1. Define the term *conductance*.
2. What is the significance of conductance in a circuit?
3. What symbol is used to represent conductance?
4. What is the unit of conductance?
5. What is the conductance of a 100-ohm resistor?

13-3 RESISTORS

Resistance is a property of all electrical components. Sometimes the effect of resistance is undesirable; other times it is constructive. **Resistors** are components manufactured to possess a specific value of resistance to the flow of current. A resistor is the most commonly used component in an electronic circuit. Resistors are available with fixed or variable resistance values. They are available in a variety of shapes and sizes to meet specific circuit, space, and operating requirements (Figure 13-2 and Figure 13-3). Resistors are drawn schematically as a series of jagged lines, as shown in Figure 13-4. A resistor's **tolerance** is the amount that the resistor may vary and still be acceptable. It is expensive for a manufacturer to hold a resistor to a certain value when an exact value is not needed. Therefore, the larger the tolerance, the cheaper it is to manufacture. Resistors are available with tolerances of ±20%, ±10%, ±5%, ±2%, and ±1%. Precision resistors are available with even smaller tolerances. In most electronic circuits, resistors of 10% tolerance are satisfactory.

EXAMPLE: How much can a 1000-ohm resistor with a 20% tolerance vary and still be acceptable?

Solution:

$$1000 \times 0.20 = \pm 200 \text{ ohms}$$

The tolerance is ±200 ohms. Therefore, the 1000-ohm resistor may vary from 800 to 1200 ohms and still be satisfactory. For the sake of uniformity, electronic manufacturers offer a number of standard

■ **FIGURE 13-2**

Fixed resistors come in various sizes and shapes.

© 2014 Cengage Learning

■ **FIGURE 13-3**

Variable resistors come in many styles to meet the needs of manufacturers of electronic equipment.

Potentiometers

© 2014 Cengage Learning

Trimming potentiometers

© 2014 Cengage Learning

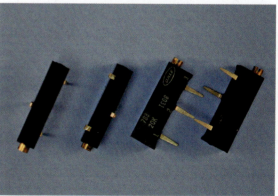

Trim potentiometers

© 2014 Cengage Learning

■ **FIGURE 13-4**

Schematic diagram of a fixed resistor.

© 2014 Cengage Learning

■ FIGURE 13-5

EIA standard resistor values (not including multiplier band).

±2% and ±5% Tolerance	±10% Tolerance	±20% Tolerance
1.0	1.0	1.0
1.1		
1.2	1.2	
1.3		
1.5	1.5	1.5
1.6		
1.8	1.8	
2.0		
2.2	2.2	2.2
2.4		
2.7	2.7	
3.0		
3.3	3.3	3.3
3.6		
3.9	3.9	
4.3		
4.7	4.7	4.7
5.1		
5.6	5.6	
6.2		
6.8	6.8	6.8
7.5		
8.2	8.2	
9.1		

© 2014 Cengage Learning

■ FIGURE 13-6

Carbon composition resistors were the most widely used resistors in electronic circuits.

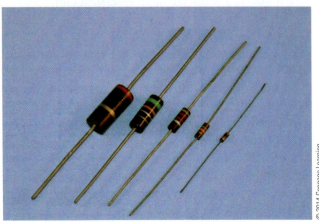

© 2014 Cengage Learning

The wirewound resistor is constructed of a nickel-chromium alloy (nichrome) wire wound on a ceramic form (Figure 13-7). Leads are attached and the entire resistor is sealed with a coating. Wirewound resistors are used in high-current circuits where precision is necessary. The resistance range varies from a fraction of an ohm to several thousand ohms.

Film resistors have become popular (Figure 13-8) because they offer the small size of the composition resistor with the accuracy of the wirewound resistor. A thin film of carbon is deposited on a cylindrical ceramic core and sealed in an epoxy or glass coating to form a carbon film resistor. Cutting a spiral groove

■ FIGURE 13-7

Wirewound resistors are available in many different styles.

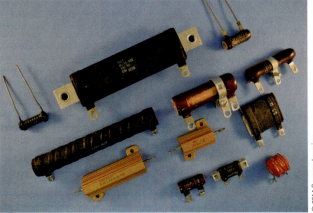

© 2014 Cengage Learning

resistor values. Figure 13-5 is a list of standard values for resistors with ±5 , ±10%, and ±20% tolerance. After the value on the chart is obtained, it is multiplied by the value associated with the color of the multiplier band.

Resistors fall into three major categories, named for the material they are made of: molded **carbon composition resistors, wirewound resistors**, and **carbon film resistors**.

Until recently, the molded carbon composition resistor was the most commonly used resistor in electronic circuits (Figure 13-6). It is manufactured in standard resistor values.

■ **FIGURE 13-8**

The film resistor offers the size of the carbon resistor with the accuracy of the wirewound resistor.

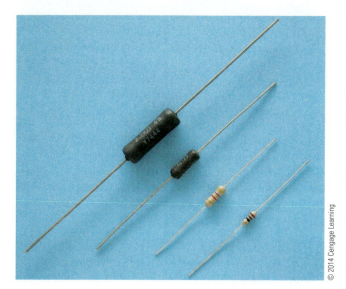

© 2014 Cengage Learning

■ **FIGURE 13-9**

Tin oxide resistors.

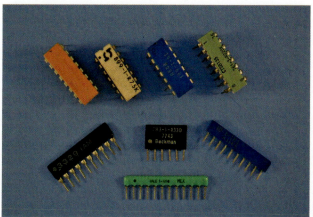

© 2014 Cengage Learning

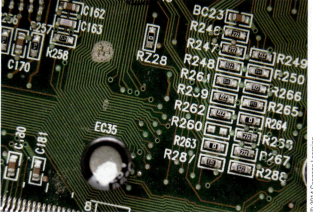

© 2014 Cengage Learning

through the film, the length of the resistor, sets the value of the resistor. The closer the pitch of the spiral, the higher the resistance. Carbon film resistors are available from 10 ohms to 2.5 megohms at a ±1% tolerance. **Metal film resistors** are physically similar to carbon film resistors but use a metal alloy and are more expensive. They are available from 10 ohms to 1.5 megohms at a ±1% tolerance, although tolerances down to ±0.1% are available. Another type of film resistor is the tin oxide resistor (Figure 13-9). It consists of a tin oxide film on a ceramic substrate and is also available in a single inline format or dual-inline pin format.

Surface-mount resistors are ideal for small circuit applications. A thin film of carbon or metal alloy is deposited on a ceramic base or substrate. Contact from the resistive element to the printed circuit board is via metal end caps, or terminals, resulting in zero lead length. In application, these end caps are soldered directly to the circuit board conductive trace using an automated soldering process. The lack of long leads to solder into the printed circuit board yields several advantages. Among them are lightweight, smaller printed circuit board sizes and the use of automated assembly processes. Surface-mount resistors are available in both thick and thin films. They are available from 0 ohms to 10 megohms at tolerance of ±5%

down to ±0.1%, with power handling capabilities of 1/16 watt to 1 watt.

Variable resistors allow the resistance to vary. They have a resistive element of either carbon composition or wire that is connected to two terminals. A third terminal is attached to a movable wiper, which is connected to a shaft. The wiper slides along the resistive element when the shaft is rotated. As the shaft is rotated, the resistance between the center terminal and one outer terminal increases, whereas the resistance between the center terminal and the other outer terminal decreases (Figure 13-10). Variable resistors are available with resistance that varies linearly (a linear taper) or logarithmically (an audio taper).

A variable resistor used to control voltage is called a **potentiometer**, or **pot**. A variable resistor used to control current is called a **rheostat** (Figure 13-11).

■ **FIGURE 13-10**

Variable resistors allow the resistance value to increase or decrease at random.

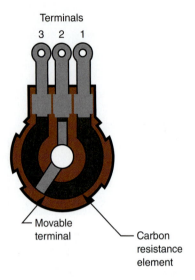

Terminals

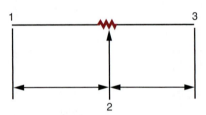

■ **FIGURE 13-11**

A rheostat is a variable resistor used to control current.

© 2014 Cengage Learning

13-3   QUESTIONS

1. What is the purpose of specifying the tolerance of a resistor?
2. What are the three major types of fixed resistors?
3. What is the advantage of film resistors over carbon composition resistors?
4. Explain how a variable resistor works.
5. What is the difference between a potentiometer and a rheostat?

13-4   RESISTOR IDENTIFICATION

The small size of the resistor prevents printing the resistance value and tolerance on its case. Therefore, a color-coded strip system is used to display the resistor value. The strips can be seen and read in any position that the resistor is placed. The Electronic Industries Association (EIA) color code is shown in Figure 13-12.

The meaning of the colored bands on a resistor is as follows. The first band, closest to the end of the resistor, represents the first digit of the resistor value. The second band represents the second digit of the resistor value. The third band represents the number of zeros to be added to the first two digits. The fourth band represents the tolerance of the resistor (Figure 13-13).

For example, the resistor shown in Figure 13-14 has a resistance value of 1500 ohms. The brown band (first

■ FIGURE 13-12

The Electronic Industries Association (EIA) two-significant digit color code.

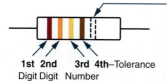

Note: A fifth band may be present, which represents reliability

1st 2nd 3rd 4th—Tolerance
Digit Digit Number
of zeros

Two-significant-figure color code

	1st Band 1st Digit	2nd Band 2nd Digit	3rd Band Number of Zeros	4th Band Tolerance
Black	0	0	—	—
Brown	1	1	0	1%
Red	2	2	00	2%
Orange	3	3	000	—
Yellow	4	4	0,000	—
Green	5	5	00,000	0.5%
Blue	6	6	000,000	0.25%
Violet	7	7		0.10%
Gray	8	8		0.05%
White	9	9		—
Gold			×0.1	5%
Silver			×0.01	10%
No color				20%

© 2014 Cengage Learning

■ FIGURE 13-13

Meaning of the colored bands on a carbon composition resistor.

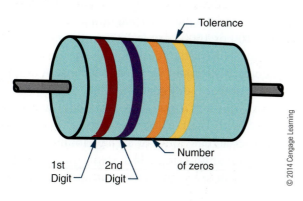

Tolerance

1st Digit 2nd Digit Number of zeros

© 2014 Cengage Learning

■ FIGURE 13-14

This resistor has a resistance value of 1500 ohms.

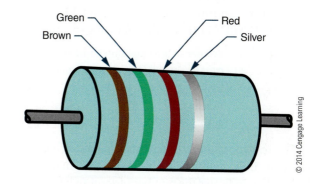

Green Red
Brown Silver

© 2014 Cengage Learning

■ **FIGURE 13-15**

The fifth band on a resistor indicates the resistor's reliability.

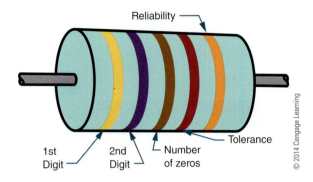

Reliability

1st Digit · 2nd Digit · Number of zeros · Tolerance

© 2014 Cengage Learning

■ **FIGURE 13-16**

Resistors may also be identified by a letter-and-number system.

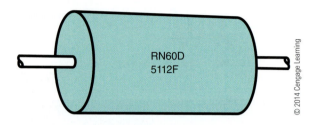

RN60D 5112F

© 2014 Cengage Learning

band) represents the first digit (1). The green band (second band) represents the second digit (5). The red band (third band) represents the number of zeros (2 zeros—00) to be added to the first two digits. The silver band (fourth band) indicates a resistance tolerance of 10%. Therefore, this is a 1500-ohm resistor with a ±10% tolerance.

A resistor may have a fifth band (Figure 13-15). This band indicates the reliability of the resistor. It tells how many of the resistors (per thousand) will fail after 100 hours of operation. Generally, when there are five bands on a resistor, the same amount of body color shows at each end. In this case, look for the tolerance band, position it on the right, and read the resistor as described previously.

There are two instances where the third band does not mean the number of zeros. For resistor values of less than 10 ohms, the third band is gold. This means that the first two digits should be multiplied by 0.1. For resistor values of less than 1 ohm, the third band is silver. This means the first two digits are multiplied by 0.01.

A resistor may also be identified by a letter-and-number (alphanumeric) system (Figure 13-16). For example, RN60D5112F has the following meaning:

RN60	Resistor style (composition, wirewound, film)
D	Characteristic (effects of temperature)
5112	Resistance value (2 represents the number of zeros)
F	Tolerance

The resistor value is the primary concern. Three to five digits indicate the value of the resistor. In all cases, the last digit indicates the number of zeros to be added to the preceding digits. In the example given, the last digit

(2) indicates the number of zeros to be added to the first three digits (511). So 5112 translates to 51,100 ohms.

In some cases an R may be inserted into the number. The R represents a decimal point and is used when the resistor value is less than 10 ohms. For example, 4R7 represents 4.7 ohms.

The five-digit numbering system is similar to the three- and four-digit systems. The first four digits represent significant digits, and the last digit indicates the number of zeros to be added. For values of less than 1000 ohms, the R is used to designate a decimal point.

Surface-mount resistors are identified similar to the letter-and-number system. The part number of the resistor is interpreted as follows; the number varies between manufacturers. For example, RC0402J103T has the following meaning:

RC	**Chip resistor**
0402	Size (0. 04" × 0.02")
J	Tolerance (J = ±5%, F = ±1%, D = ±0.5%, B = ±0.1%)
103	Resistance (three- or four-digit code available)
T	Packaging method

The resistance value is indicated by three or four digits. In either case, the last digit indicates the number of zeros to be added to the preceding digits. In the example given, the first two digits are 1 and 0, followed by 3 zeroes, for a value of 10,000 ohms. For values of less than 100 or 1000 (depending on the number of digits used), D is used to designate the decimal point. For example, 3D9 would represent 3.9 ohms. A 0-ohm resistor or jumper is designated as 000.

Potentiometers (variable resistors) are also imprinted with their values (Figure 13-17). These may be their actual values or an alphanumeric code. With the alphanumeric code system, the resistance value is determined from the last part of the code. For example, in MTC253L4, the number 253 means 25

■ FIGURE 13-17

Potentiometers (variable resistors) are also labeled with their values.

© 2014 Cengage Learning

■ FIGURE 13-18

Three types of resistive circuits: (A) series circuit, (B) parallel circuit, (C) series-parallel circuit.

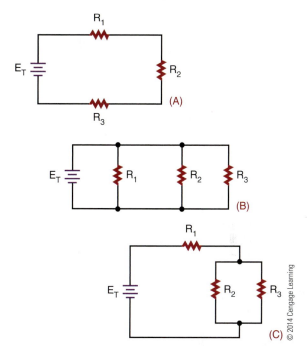

© 2014 Cengage Learning

followed by three 0s, or 25,000 ohms. The L4 indicates the resistor construction and body type.

1. Write the color code from memory.
2. What do the four bands on a carbon composition resistor represent?
3. Decode the following resistors:

	1st Band	2nd Band	3rd Band	4th Band
a.	Brown	Black	Red	Silver
b.	Blue	Green	Orange	Gold
c.	Orange	White	Yellow	(None)
d.	Red	Red	Red	Silver
e.	Yellow	Violet	Brown	Gold

4. What does a fifth band on a resistor indicate?
5. What does a gold or silver third band represent?

13-5	**CONNECTING RESISTORS**

There are three important types of resistive circuits: the **series circuit**, the **parallel circuit**, and the **series-parallel circuit** (Figure 13-18). A *series circuit* provides a single path for current flow. A *parallel* circuit provides two or more paths for current flow. A *series-parallel* circuit is a combination of a series circuit and a parallel circuit.

13-5	**QUESTIONS**

1. What are the three basic types of circuit configurations?
2. What is the difference between the three circuit types?

13-6	**CONNECTING RESISTORS IN SERIES**

A series circuit contains two or more resistors and provides one path for current to flow. The current flows from the negative side of the voltage source through each resistor to the positive side of the voltage source. If there is only one path for current to flow between two points in a circuit, the circuit is a series circuit.

The more resistors connected in series, the more opposition there is to current flow. The more opposition there is to current flow, the higher the resistance in the circuit. In other words, when a resistor is added in series to a circuit, the total resistance in the circuit increases. The total resistance in a series circuit is the

■ **FIGURE 13-19**

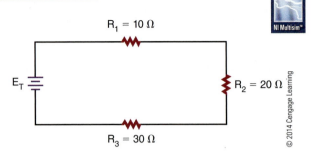

$R_1 = 10\ \Omega$

E_T

$R_2 = 20\ \Omega$

$R_3 = 30\ \Omega$

© 2014 Cengage Learning

sum of the individual resistances in the circuit. This can be expressed as:

$$R_T = R_1 + R_2 + R_3 \ldots + R_n$$

The numerical subscripts refer to the individual resistors in the circuit. R_n is the last resistor in the circuit. The symbol R_T represents the total resistance in the circuit.

EXAMPLE: What is the total resistance of the circuit shown in Figure 13-19?

Given	Solution
$R_T = ?$	$R_T = R_1 + R_2 + R_3$
$R_1 = 10\ \Omega$	$R_T = 10 + 20 + 30$
$R_2 = 20\ \Omega$	$R_T = 60\ \Omega$
$R_3 = 30\ \Omega$	

EXAMPLE: Calculate the total resistance for the circuit shown in Figure 13-20.

Given	Solution
$R_T = ?$	$R_T = R_1 + R_2 + R_3 + R_4 + R_5$
$R_1 = 1\ k\Omega$	$R_T = 1\,k + 4.7\,k + 3.9\,k + 0.82\,k$
$R_2 = 4.7\ k\Omega$	$+ 10\,k$
$R_3 = 3.9\ k\Omega$	$R_T = 1000 + 4700 + 3900 + 820$
$R_4 = 820\ \Omega$	$+ 10,000$
$R_5 = 10\ k\Omega$	$R_T = 20,420\ \Omega$

■ **FIGURE 13-20**

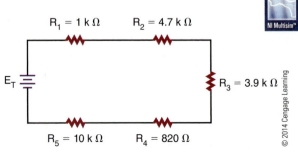

$R_1 = 1\ k\Omega$ $R_2 = 4.7\ k\Omega$

E_T

$R_3 = 3.9\ k\Omega$

$R_5 = 10\ k\Omega$ $R_4 = 820\ \Omega$

© 2014 Cengage Learning

13-6 **QUESTIONS**

1. Write the formula for determining total resistance in a series circuit.
2. What is the total resistance of a series circuit with the following resistors? *(Make a drawing of each series circuit.)*
   a. $R_T = ?,\quad R_1 = 1500\ \Omega,\quad R_2 = 3300\ \Omega,$
      $R_3 = 4700\ \Omega$
   b. $R_T = ?,\quad R_1 = 100\ \Omega,\quad R_2 = 10\ k\Omega,$
      $R_3 = 5.6\ M\Omega$
   c. $R_T = ?,\quad R_1 = 4.7\ k\Omega,\quad R_2 = 8.2\ k\Omega,$
      $R_3 = 330\ \Omega$
   d. $R_T = ?,\quad R_1 = 5.6\ M\Omega,\quad R_2 = 1.8\ M\Omega,$
      $R_3 = 8.2\ M\Omega$

13-7 **CONNECTING RESISTORS IN PARALLEL**

A parallel circuit contains two or more resistors and provides two or more paths for current to flow. Each current path in a parallel circuit is called a **branch**. The current flows from the negative side of the voltage source, through each branch of the parallel circuit, to the positive side of the voltage source. If there is more than one path for current to flow between two points in a circuit with two or more resistors, the circuit is a parallel circuit.

The more resistors are connected in parallel, the less opposition there is to current flow. The less opposition there is to current flow, the lower the resistance in the circuit. In other words, when a resistor is added in parallel to a circuit, the total resistance in the circuit decreases because additional paths for current flow are provided. In a parallel circuit, the total resistance is always less than the resistance of any branch.

The total resistance in a parallel circuit is given by the formula:

$$\frac{1}{R_T} = \frac{1}{R_1} + \frac{1}{R_2} + \frac{1}{R_3} \ldots + \frac{1}{R_n}$$

Again, R_T is the total resistance, R_1, R_2, and R_3 are the individual (branch) resistors, and R_n is the number of the last resistor in the circuit.

EXAMPLE: What is the total resistance of the circuit shown in Figure 13-21?

■ FIGURE 13-21

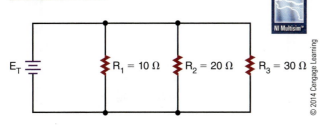

Given

$$R_T = ?$$
$$R_1 = 10 \ \Omega$$
$$R_2 = 20 \ \Omega$$
$$R_3 = 30 \ \Omega$$

Solution

$$\frac{1}{R_T} = \frac{1}{R_1} + \frac{1}{R_2} + \frac{1}{R_3}$$

$$\frac{1}{R_T} = \frac{1}{10} + \frac{1}{20} + \frac{1}{30} \quad \text{(common denominator is 60)}$$

$$\frac{1}{R_T} = \frac{6}{60} + \frac{3}{60} + \frac{2}{60}$$

$$\frac{1}{R_T} = \frac{11}{60}$$

$$\frac{1}{R_T} \diagdown \frac{11}{60}$$

$$(11)(R_T) = (1)(60)$$

$$11R_T = 60$$

$$\frac{11R_T}{11} = \frac{60}{11} \quad \text{(divide both sides by 11)}$$

$$1R_T = \frac{60}{11}$$

$$R_T = 5.45 \ \Omega$$

The circuit shown in Figure 13-21 could be replaced with one 5.45 Ω resistor.

NOTE **The total resistance in a parallel circuit is always less than the smallest resistor.**

EXAMPLE: Calculate the total resistance for the circuit shown in Figure 13-22.

Given

$$R_T = ?$$
$$R_1 = 1 \ \text{k}\Omega \ (1000 \ \text{ohms})$$
$$R_2 = 4.7 \ \text{k}\Omega \ (4700 \ \text{ohms})$$
$$R_3 = 3.9 \ \text{k}\Omega \ (3900 \ \text{ohms})$$
$$R_4 = 820 \ \Omega$$
$$R_5 = 10 \ \text{k}\Omega \ (10{,}000 \ \text{ohms})$$

Solution

$$\frac{1}{R_T} = \frac{1}{R_1} + \frac{1}{R_2} + \frac{1}{R_3} + \frac{1}{R_4} + \frac{1}{R_5}$$

$$\frac{1}{R_T} = \frac{1}{1000} + \frac{1}{4700} + \frac{1}{3900} + \frac{1}{820} + \frac{1}{10{,}000}$$

It is too complicated to find a common denominator, so work with decimals.

$$\frac{1}{R_T} = 0.001 + 0.000213 + 0.000256 + 0.00122 + 0.0001$$

$$\frac{1}{R_T} = 0.002789$$

$$\frac{1}{R_T} \diagdown \frac{0.002789}{1} \quad \text{(Cross-multiply)}$$

$$(0.002789)(R_T) = (1)(1)$$

$$0.002789R_T = 1$$

$$\frac{0.002789R_T}{0.002789} = \frac{1}{0.002789} \quad \text{(Divide both sides by 0.00289.)}$$

$$1R_T = \frac{1}{0.002789}$$

$$R_T = 358.55 \ \Omega$$

■ FIGURE 13-22

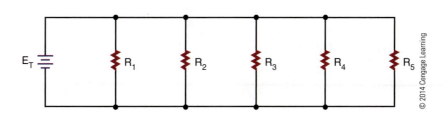

How many places each number is rounded off significantly affects the accuracy of the final answer.

EXAMPLE: What resistor value must be connected in parallel with a 47-ohm resistor to provide a total resistance of 27 ohms? See Figure 13-23.

Given

$$R_T = 27\ \Omega$$
$$R_1 = 47\ \Omega$$
$$R_2 = ?$$

Solution

$$\frac{1}{R_T} = \frac{1}{R_1} + \frac{1}{R_2}$$

$$\frac{1}{27} = \frac{1}{47} + \frac{1}{R_2}$$

$$\frac{1}{27} - \frac{1}{47} = \frac{1}{47} - \frac{1}{47} + \frac{1}{R_2} \quad \textit{Subtract } \frac{1}{47} \textit{ from both sides.}$$

$$\frac{1}{27} - \frac{1}{47} = \frac{1}{R_2} \quad \text{(easier to work with decimals)}$$

$$0.00370 - 0.0213 = \frac{1}{R_2}$$

$$0.0157 = \frac{1}{R_2}$$

$$63.69\ \Omega = R_2$$

Note that 63.69 ohms is not a standard resistor value. Use the closest standard resistor value, which is 62 ohms.

13-7   QUESTIONS

1. Write the formula for determining total resistance in a parallel circuit.
2. What is the total resistance of a parallel circuit with the following resistors? *(Make a drawing of each parallel circuit.)*
   a. $R_T = ?$, $R_1 = 1500\ \Omega$, $R_2 = 3300\ \Omega$, $R_3 = 4700\Omega$
   b. $R_T = ?$, $R_1 = 100\ \Omega$, $R_2 = 10\ k\Omega$, $R_3 = 5.6\ M\Omega$
   c. $R_T = ?$, $R_1 = 4.7\ k\Omega$, $R_2 = 8.2\ k\Omega$, $R_3 = 330\Omega$
   d. $R_T = ?$, $R_1 = 5.6\ M\Omega$, $R_2 = 1.8\ M\Omega$, $R_3 = 8.2\ M\Omega$

■ FIGURE 13-23

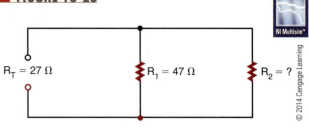

$R_T = 27\ \Omega$ $R_1 = 47\ \Omega$ $R_2 = ?$

© 2014 Cengage Learning

13-8   CONNECTING RESISTORS IN SERIES AND PARALLEL

A series-parallel circuit is a combination of a series and a parallel circuit. Figure 13-24 shows a simple series-parallel circuit with resistors. Notice that R_2 and R_3 are in parallel and that this parallel combination is in series with R_1 and R_4. The current flows from the negative side of the voltage source through resistor R_4 and divides at point A to flow through the two branches, R_2 and R_3. At point B, the current recombines and flows through R_1.

The total resistance for a series-parallel circuit or compound circuit is computed using the series formula:

$$R_T = R_1 + R_2 + R_3 \ldots + R_n$$

and the parallel formula:

$$\frac{1}{R_T} = \frac{1}{R_1} + \frac{1}{R_2} + \frac{1}{R_3} \ldots + \frac{1}{R_n}$$

Most circuits can be broken down to a simple parallel or series circuit. The procedure is as follows:

1. Calculate the parallel portion of the circuit first to determine the equivalent resistance.
2. If there are series components within the parallel portion of the circuit, determine the equivalent resistance for the series components first.
3. After the equivalent resistance is determined, redraw the circuit, substituting the equivalent resistance for the parallel portion of the circuit.
4. Do final calculations.

EXAMPLE: What is the total resistance for the circuit shown in Figure 13-24?

The first step is to determine the equivalent resistance (R_A) for R_2 and R_3.

■ **FIGURE 13-24**

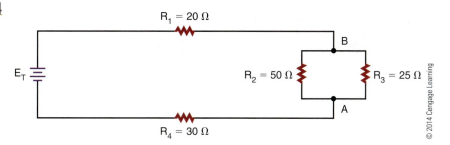

■ **FIGURE 13-25**

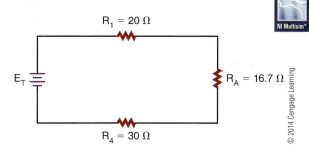

Given

$$R_A = ?$$
$$R_2 = 50 \ \Omega$$
$$R_3 = 25 \ \Omega$$

Solution

$$\frac{1}{R_A} = \frac{1}{R_2} + \frac{1}{R_3}$$

$$\frac{1}{R_A} = \frac{1}{50} + \frac{1}{25}$$

$$\frac{1}{R_A} = \frac{1}{R_1} + \frac{1}{R_2}$$

$$R_A = 16.7 \ \Omega$$

Redraw the circuit, substituting the equivalent resistance for the parallel portion of the circuit. See Figure 13-25.

Now determine the total series resistance for the redrawn circuit.

Given

$$R_1 = 20 \ \Omega$$
$$R_A = 16.7 \ \Omega$$
$$R_4 = 30 \ \Omega$$

Solution

$$R_T = R_1 + R_A + R_4$$
$$R_T = 20 + 16.7 + 30$$
$$R_T = 66.7 \ \Omega$$

EXAMPLE: Calculate the total resistance for the circuit shown in Figure 13-26.

First find the equivalent resistance (R_A) for parallel resistors R_2 and R_3. Then find the equivalent resistance (R_B) for resistors R_5, R_6, and R_7.

Given

$$R_A = ?$$
$$R_2 = 47 \ \Omega$$
$$R_3 = 62 \ \Omega$$

Solution

$$\frac{1}{R_A} = \frac{1}{R_2} + \frac{1}{R_3}$$

$$\frac{1}{R_A} = \frac{1}{47} + \frac{1}{62}$$

$$R_A = 26.7 \ \Omega$$

■ **FIGURE 13-26**

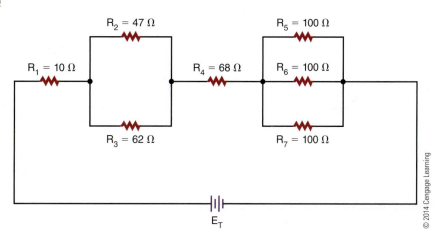

■ **FIGURE 13-27**

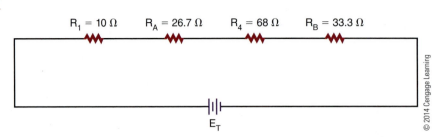

Given

$R_B = ?$
$R_5 = 100\ \Omega$
$R_6 = 100\ \Omega$
$R_7 = 100\ \Omega$

Solution

$$\frac{1}{R_B} = \frac{1}{R_5} + \frac{1}{R_6} + \frac{1}{R_7}$$

$$\frac{1}{R_B} = \frac{1}{100} + \frac{1}{100} + \frac{1}{100}$$

$$R_B = 33.3\ \Omega$$

Now redraw the circuit using equivalent resistance R_A and R_B, and determine the total series resistance for the redrawn circuit. See Figure 13-27.

Given

$R_T = ?$
$R_1 = 10\ \Omega$
$R_A = 26.7\ \Omega$
$R_4 = 68\ \Omega$
$R_B = 33.3\ \Omega$

Solution

$R_T = R_1 + R_A + R_4 + R_B$
$R_T = 10 + 26.7 + 68 + 33.3$
$R_T = 138\ \Omega$

The circuit shown in Figure 13-26 could be replaced with a single resistor of 138 ohms. (Figure 13-28).

■ **FIGURE 13-28**

EXAMPLE: Find the total resistance for the circuit shown in Figure 13-29.

The equivalent resistance of the series in the parallel portion of the circuit must be determined first. This is labeled R_S.

Given

$R_S = ?$
$R_2 = 180\ \Omega$
$R_3 = 200\ \Omega$
$R_4 = 620\ \Omega$

Solution

$R_S = R_2 + R_3 + R_4$
$R_S = 180 + 200 + 620$
$R_S = 1000\ \Omega$

Redraw the circuit, substituting equivalent resistance R_S for the series resistors R_2, R_3, and R_4. See Figure 13-30.

Determine the equivalent parallel resistance R_A for R_S and R_5.

Given

$R_A = ?$
$R_S = 1000\ \Omega$
$R_5 = 1000\ \Omega$

Solution

$$\frac{1}{R_A} = \frac{1}{R_S} + \frac{1}{R_5}$$

$$\frac{1}{R_A} = \frac{1}{1000} + \frac{1}{1000}$$

$$R_A = 500\ \Omega$$

■ **FIGURE 13-29**

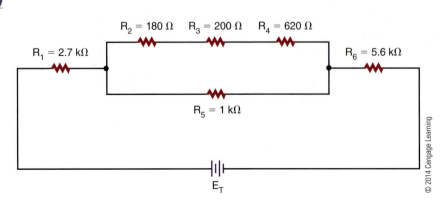

FIGURE 13-30

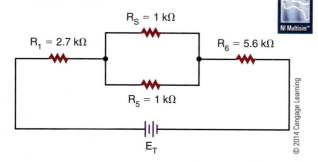

FIGURE 13-31

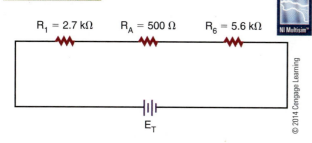

Redraw the circuit again, substituting equivalent resistance R_A for parallel resistors R_S and R_5, and determine the total series resistance for the redrawn circuit. See Figure 13-31.

Given	Solution
$R_T = ?$	$R_T = R_1 + R_A + R_6$
$R_1 = 2700 \ \Omega$	$R_T = 2700 + 500 + 5600$
$R_A = 500 \ \Omega$	$R_T = 8800 \ \Omega$
$R_6 = 5600 \ \Omega$	

The circuit shown in Figure 13-29 can be replaced with a single resistor of 8800 ohms (Figure 13-32).

13–8 QUESTIONS

1. Describe the process for finding the total resistance in a series-parallel circuit.
2. Which is solved first: the parallel resistance or the series resistance within the parallel resistance?
3. What is the total resistance of series-parallel circuits with the following resistors? *(Draw a diagram of each series circuit.)*

|——————PARALLEL——————| |—SERIES—|

a. $R_T = ?$, $R_1 = 1500 \ \Omega$, $R_2 = 3300 \ \Omega$, $R_3 = 4700 \ \Omega$
b. $R_T = ?$, $R_1 = 100 \ \Omega$, $R_2 = 10 \ k\Omega$, $R_3 = 5.6 \ M\Omega$
c. $R_T = ?$, $R_1 = 4.7 \ k\Omega$, $R_2 = 8.2 \ k\Omega$, $R_3 = 330 \ \Omega$
d. $R_T = ?$, $R_1 = 5.6 \ M\Omega$, $R_2 = 1.8 \ M\Omega$, $R_3 = 8.2 \ M\Omega$

FIGURE 13-32

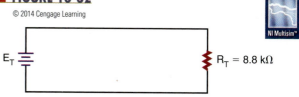
© 2014 Cengage Learning

SUMMARY

- Resistors are either fixed or variable.
- The tolerance of a resistor is the amount that its resistance can vary and still be acceptable.
- Resistors are either carbon composition, wirewound, or film.
- Carbon composition resistors were the most commonly used resistors.
- Wirewound resistors are used in high-current circuits that must dissipate large amounts of heat.
- Film resistors offer small size with high accuracy.
- Variable resistors used to control voltage are called potentiometers.
- Variable resistors used to control current are called rheostats.
- Resistor values may be identified by colored bands:
  - The first band represents the first digit.
  - The second band represents the second digit.
  - The third band represents the number of zeros to be added to the first two digits.
  - The fourth band represents the tolerance.
  - A fifth band may be added to represent reliability.
- Resistor values of less than 100 ohms are shown with a black third band.
- Resistors may be placed in three configurations— series, parallel, and compound.
- Resistor values of less than 10 ohms are shown with a gold third band.
- Resistor values of less than 1 ohm are shown with a silver third band.
- Resistor values for 1% tolerance resistors are shown with the fourth band as the multiplier.
- Resistor values may also be identified by an alphanumeric system.
- The total resistance in a series circuit can be found by the formula:

$$R_T = R_1 + R_2 + R_3 \ldots + R_n$$

- The total resistance in a parallel circuit can be found by the formula:

$$\frac{1}{R_T} = \frac{1}{R_1} + \frac{1}{R_2} + \frac{1}{R_3} \cdots + \frac{1}{R_n}$$

- The total resistance in a series-parallel circuit is determined by both series and parallel.

CHAPTER 13 SELF-TEST

1. Describe how the resistance of a material is determined.
2. What is the tolerance range of a 2200-ohm resistor with a 10% tolerance?
3. Write the color codes for the following resistors:
   a. 5600 ohms ±5%
   b. 1.5 megohms ±10%
   c. 2.7 ohms ±5%
   d. 100 ohms ±20%
   e. 470 kilohms ±10%
4. Decode the following code for a chip resistor: RC0402D104T.
5. Describe how potentiometers are labeled.
6. Make a chart for calculating total resistance for resistors in series, parallel, and series-parallel circuits.
7. Determine the total resistance for four 8-ohm resistors in parallel.

8. Describe the steps for solving the total resistance in question 7.
9. Determine the total resistance for the circuit shown.

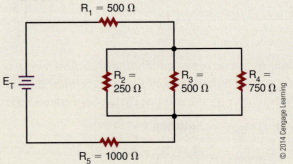

$R_1 = 500\ \Omega$
$R_2 = 250\ \Omega$
$R_3 = 500\ \Omega$
$R_4 = 750\ \Omega$
$R_5 = 1000\ \Omega$
E_T

© 2014 Cengage Learning

10. Describe how current flows through a series-parallel circuit.

Ohm's Law

OBJECTIVES

After completing this chapter, the student will be able to:

- Identify the three basic parts of a circuit.
- Identify three types of circuit configurations.
- Describe how current flow can be varied in a circuit.
- State Ohm's law with reference to current, voltage, and resistance.
- Solve problems using Ohm's law for current, resistance, or voltage in series, parallel, and series-parallel circuits.
- Describe how the total current flow differs between series and parallel circuits.
- Describe how the total voltage drop differs between series and parallel circuits.
- Describe how the total resistance differs between series and parallel circuits.
- State and apply Kirchhoff's current and voltage laws.
- Verify answers using Ohm's law with Kirchhoff's law.

KEY TERMS

14-1 electric circuit

14-1 load

14-2 Ohm's law

14-4 Kirchhoff's current law

14-5 Kirchhoff's voltage law

Ohm's law defines the relationship among three fundamental quantities: current, voltage, and resistance. It states that current is directly proportional to voltage and inversely proportional to resistance.

This chapter examines Ohm's law and how it is applied to a circuit. Some of the concepts were introduced in previous chapters.

14–1   ELECTRIC CIRCUITS

As stated earlier, current flows from a point with an excess of electrons to a point with a deficiency of electrons. The path that the current follows is called an **electric circuit**. All electric circuits consist of a voltage source, a load, and a conductor. The voltage source establishes a difference of potential that forces the current to flow. The source can be a battery, a generator, or another of the devices described in Chapter 12: Voltage. The **load** consists of some type of resistance to current flow. The resistance may be high or low, depending on the purpose of the circuit. The current in the circuit flows through a conductor from the source to the load. The conductor must give up electrons easily. Copper is used for most conductors.

The path the electric current takes to the load may be through any of three types of circuits: a series circuit, a parallel circuit, or a series-parallel circuit. A series circuit (Figure 14-1) offers a single continuous path for the current flow, going from the source to the load. A parallel circuit (Figure 14-2) offers more than one path for current flow. It allows the source to apply voltage to more than one load. It also allows several sources to be connected to a single load. A series-parallel circuit (Figure 14-3) is a combination of the series and parallel circuits.

Current in an electric circuit flows from the negative side of the voltage source through the load to the positive side of the voltage source (Figure 14-4). As long as the path is not broken, it is a closed circuit and current flows (Figure 14-5). However, if the path is broken, it is an open circuit and no current can flow (Figure 14-6).

■ FIGURE 14-1

A series circuit offers a single path for current flow.

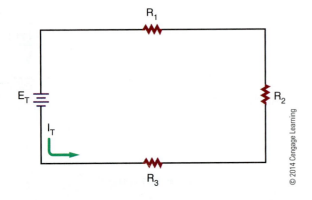

■ FIGURE 14-2

A parallel circuit offers more than one path for current flow.

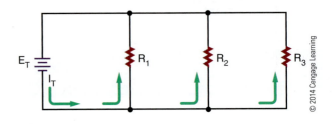

■ FIGURE 14-3

A series-parallel circuit is a combination of a series circuit and a parallel circuit.

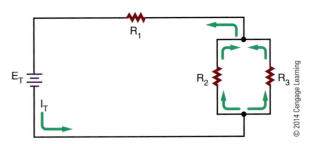

■ FIGURE 14-4

Current flow in an electric circuit flows from the negative side of the voltage source, through the load, and returns to the voltage source through the positive terminal.

■ FIGURE 14-5

A closed circuit supports current flow.

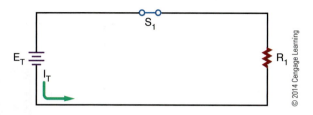

■ FIGURE 14-6

An open circuit does not support current flow.

Changing either the voltage applied to the circuit or the resistance in the circuit can vary the current flow in an electric circuit. The current changes in exact proportion to the change in the voltage or resistance. If the voltage is increased, the current also increases. If the voltage is decreased, the current also decreases (Figure 14-7). On the other hand, if the resistance is increased, the current decreases (Figure 14-8).

■ FIGURE 14-7

Current flow in an electric circuit can be changed by varying the voltage.

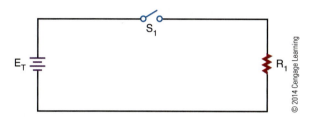

■ FIGURE 14-8

Current flow in an electric circuit can also be changed by varying the resistance in the circuit.

14–1 QUESTIONS

1. What are the three basic parts of an electric circuit?
2. Define the following:
   a. Series circuit
   b. Parallel circuit
   c. Series-parallel circuit
3. Draw a diagram of a circuit showing how current would flow through the circuit. *(Use arrows to indicate current flow.)*
4. What is the difference between an open circuit and a closed circuit?
5. What happens to the current in an electric circuit when the voltage is increased? When it is decreased? When the resistance is increased? When it is decreased?

14–2 OHM'S LAW

In 1827, George Ohm first observed **Ohm's law**, or the relationship among current, voltage, and resistance. Ohm's law states that the current in an electric circuit is directly proportional to the voltage and inversely proportional to the resistance in a circuit. This may be expressed as follows:

$$\text{Current} = \frac{\text{Voltage}}{\text{Resistance}}$$

or

$$I = \frac{E}{R}$$

Where: I = current in amperes
E = voltage in volts
R = resistance in ohms

Whenever two of the three quantities are known, the third quantity can always be determined.

EXAMPLE: How much current flows in the circuit shown in Figure 14-9?

Given

$$I_T = ?$$
$$E_T = 12 \text{ volts}$$
$$R_T = 1000 \text{ ohms}$$

■ FIGURE 14-9
© 2014 Cengage Learning

■ FIGURE 14-10
© 2014 Cengage Learning

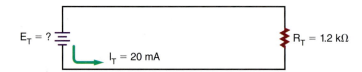

■ FIGURE 14-11
© 2014 Cengage Learning

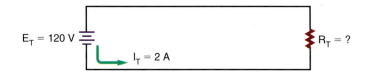

Solution

$$I_T = \frac{E_T}{R_T}$$

$$I_T = \frac{12}{1000}$$

$$I_T = 0.012 \, \text{amp, or 12 milliamps}$$

EXAMPLE: In the circuit shown in Figure 14-10, how much voltage is required to produce 20 milliamps of current flow?

Given

$$I_T = 20 \, \text{mA} = 0.02 \, \text{amp}$$
$$E_T = \, ?$$
$$R_T = 1.2 \, \text{k}\Omega = 1200 \, \text{ohms}$$

Solution

$$I_T = \frac{E_T}{R_T}$$

$$0.02 = \frac{E_T}{1200}$$

$$E_T = 24 \, \text{volts}$$

EXAMPLE: What resistance value is needed for the circuit shown in Figure 14-11 to draw 2 amperes of current?

Given

$$I_T = 2 \, \text{amps}$$
$$E_T = 120 \, \text{volts}$$
$$R_T = \, ?$$

Solution

$$I_T = \frac{E_T}{R_T}$$

$$2 = \frac{120}{R_T}$$

$$60 \, \text{ohms} = R_T$$

14-2 QUESTIONS

1. State Ohm's law as a formula.
2. How much current flows in a circuit with 12 volts applied and a resistance of 2400 ohms?
3. How much resistance is required to limit current flow to 20 milliamperes with 24 volts applied?
4. How much voltage is needed to produce 3 amperes of current flow through a resistance of 100 ohms?
5. Complete the following table:

	E	I	R
a.	9 V		180 Ω
b.	12 V	20 mA	
c.	120 V	10 A	
d.	120 V		10 MΩ
e.	1.5 V	20 μA	
f.	20 V		560 Ω

© 2014 Cengage Learning

14-3 APPLICATION OF OHM'S LAW

In a series circuit (Figure 14-12), the same current flows throughout the circuit.

$$I_T = I_{R_1} = I_{R_2} = I_{R_3} \ldots = I_{R_n}$$

FIGURE 14-12

In a series circuit, the current flow is the same throughout the circuit.

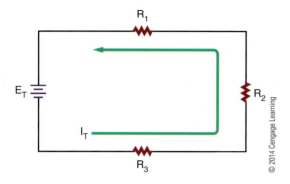

The total voltage in a series circuit is equal to the voltage drop across the individual loads (resistance) in the circuit.

$$E_T = E_{R_1} + E_{R_2} + E_{R_3} \ldots + E_{R_n}$$

The total resistance in a series circuit is equal to the sum of the individual resistances in the circuit.

$$R_T = R_1 + R_2 + R_3 \ldots + R_n$$

In a parallel circuit (Figure 14-13), the same voltage is applied to each branch in the circuit.

$$E_T = E_{R_1} = E_{R_2} = E_{R_3} \ldots = E_{R_n}$$

FIGURE 14-13

In a parallel circuit, the current divides among the branches of the circuit and recombines on returning to the voltage source.

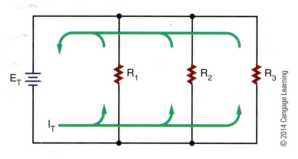

The total current in a parallel circuit is equal to the sum of the individual branch currents in the circuit.

$$I_T = I_{R_1} + I_{R_2} + I_{R_3} \ldots + I_{R_n}$$

The reciprocal of the total resistance is equal to the sum of the reciprocals of the individual branch resistances.

$$\frac{1}{R_T} = \frac{1}{R_1} + \frac{1}{R_2} + \frac{1}{R_3} \ldots + \frac{1}{R_n}$$

The total resistance in a parallel circuit will always be smaller than the smallest branch resistance.

Ohm's law states that the current in a circuit (series, parallel, or series-parallel) is directly proportional to the voltage and inversely proportional to the resistance.

$$I = \frac{E}{R}$$

In determining unknown quantities in a circuit, follow these steps:

1. Draw a schematic of the circuit and label all known quantities.
2. Solve for equivalent circuits and redraw the circuit.
3. Solve for the unknown quantities.

NOTE Ohm's Law is true for any point in a circuit and can be applied at any time. The same current flows throughout a series circuit, and the same voltage is present at any branch of a parallel circuit.

EXAMPLE: What is the total current flow in the circuit shown in Figure 14-14?

FIGURE 14-14

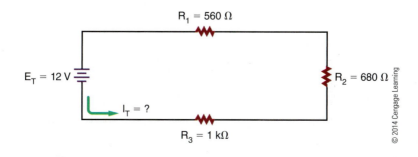

■ FIGURE 14-15

© 2014 Cengage Learning

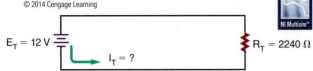

$E_T = 12\,V$ $R_T = 2240\,\Omega$ $I_T = ?$

Given

$I_T = ?$

$E_T = 12\,\text{volts}$

$R_T = ?$

$R_1 = 560\,\text{ohms}$

$R_2 = 680\,\text{ohms}$

$R_3 = 1\,k\Omega = 1000\,\text{ohms}$

Solution

First solve for the total resistance of the circuit:

$R_T = R_1 + R_2 + R_3$

$R_T = 560 + 680 + 1000$

$R_T = 2240\,\text{ohms}$

Draw an equivalent circuit. See Figure 14-15. Now solve for the total current flow:

$$I_T = \frac{E_T}{R_T}$$

$$I_T = \frac{12}{2240}$$

$$I_T = 0.0054\,\text{amp, or} \quad 5.4\,\text{milliamp}$$

EXAMPLE: How much voltage is dropped across resistor R_2 in the circuit shown in Figure 14-16?

Given

$I_T = ?$

$E_T = 48\,\text{volts}$

$R_T = ?$

$R_1 = 1.2\,k\Omega = 1200\,\text{ohms}$

$R_2 = 3.9\,k\Omega = 3900\,\text{ohms}$

$R_3 = 5.6\,k\Omega = 5600\,\text{ohms}$

Solution

First solve for the total circuit resistance:

$R_T = R_1 + R_2 + R_3$

$R_T = 1200 + 3900 + 5600$

$R_T = 10,700\,\text{ohms}$

■ FIGURE 14-16

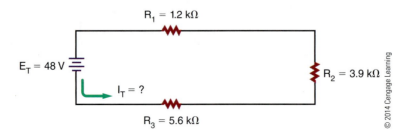

$R_1 = 1.2\,k\Omega$

$E_T = 48\,V$ $I_T = ?$ $R_2 = 3.9\,k\Omega$

$R_3 = 5.6\,k\Omega$

© 2014 Cengage Learning

■ FIGURE 14-17

© 2014 Cengage Learning

$E_T = 48\,V$ $R_T = 10,700\,\Omega$ $I_T = ?$

Draw the equivalent circuit. See Figure 14-17. Solve for the total current in the circuit:

$$I_T = \frac{E_T}{R_T}$$

$$I_T = \frac{48}{10,700}$$

$$I_T = 0.0045\,\text{amp, or 4.5 milliamps}$$

Remember, in a series circuit, the same current flows throughout the circuit. Therefore, $I_{R_2} = I_T$.

$$I_{R_2} = \frac{E_{R_2}}{R_2}.$$

$$0.0045 = \frac{E_{R_2}}{3900}$$

$$E_{R_2} = 17.55\,\text{volts}$$

EXAMPLE: What is the value of R_2 in the circuit shown in Figure 14-18?

First solve for the current that flows through R_1 and R_3. Because the voltage is the same in each branch of a parallel circuit, each branch voltage is equal to the source voltage of 120 volts.

Given

$I_{R_1} = ?$

$E_{R_1} = 120\,\text{volts}$

$R_1 = 1000\,\text{ohms}$

Solution

$$I_{R_1} = \frac{E_{R_1}}{R_1}$$

$$I_{R_1} = \frac{120}{1000}$$

$$I_{R_1} = 0.12\,\text{amp}$$

■ FIGURE 14-18

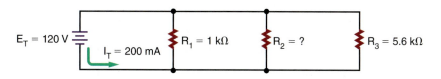

$E_T = 120\,V$ $I_T = 200\,mA$ $R_1 = 1\,k\Omega$ $R_2 = ?$ $R_3 = 5.6\,k\Omega$

Given

$I_{R_3} = ?$

$E_{R_3} = 120\,volts$

$R_3 = 1000\,ohms$

Solution

$$I_{R_3} = \frac{E_{R_3}}{R_3}$$

$$I_{R_3} = \frac{120}{5600}$$

$$I_{R_3} = 0.021\,amp$$

In a parallel circuit, the total current is equal to the sum of the currents in the branch currents.

Given

$$I_T = 0.200\,amp$$
$$I_{R_1} = 0120\,amp$$
$$I_{R_2} = ?$$
$$I_{R_3} = 0.021\,amp$$

Solution

$$I_T = I_{R_1} + I_{R_2} + I_{R_3}$$

$$0.200 = 0.120 + I_{R_2} + 0.021$$
$$0.200 = 0.141 + I_{R_2}$$
$$0.200 - 0.141 = I_{R_2}$$
$$0.059\,amp = I_{R_2}$$

Resistor R_2 can now be determined using Ohm's law.

Given

$I_{R_2} = 0.059\,amp$

$E_{R_2} = 120\,volts$

$R_2 = ?$

Solution

$$I_{R_2} = \frac{E_{R_2}}{R_2}$$

$$0.059 = \frac{120}{R_2}$$

$$R_2 = 2033.9\,ohms$$

EXAMPLE: What is the current through R_3 in the circuit shown in Figure 14-19?

First determine the equivalent resistance (R_A) for resistors R_1 and R_2.

Given

$R_A = ?$

$R_1 = 1000\,ohms$

$R_2 = 2000\,ohms$

Solution

$$\frac{1}{R_A} = \frac{1}{R_1} + \frac{1}{R_2}$$

$$\frac{1}{R_A} = \frac{1}{1000} + \frac{1}{2000}$$

$$R_A = 666.67\,ohms$$

Then determine the equivalent resistance (R_B) for resistors R_4, R_5 and R_6. First, find the total series resistance (R_S) for resistors R_5 and R_6.

Given

$R_S = ?$

$R_5 = 1500\,ohms$

$R_6 = 3300\,ohms$

Solution

$$R_S = R_5 + R_6$$
$$R_S = 1500 + 3300$$
$$R_S = 4800\,ohms$$

Given

$R_B = ?$

$R_4 = 4700\,ohms$

$R_S = 4800\,ohms$

■ FIGURE 14-19

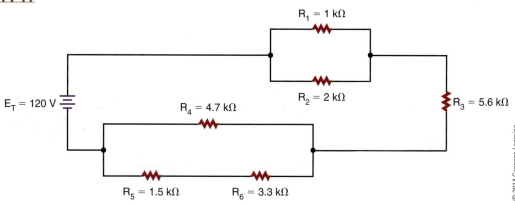

$R_1 = 1\,k\Omega$

$E_T = 120\,V$ $R_4 = 4.7\,k\Omega$ $R_2 = 2\,k\Omega$ $R_3 = 5.6\,k\Omega$

$R_5 = 1.5\,k\Omega$ $R_6 = 3.3\,k\Omega$

■ **FIGURE 14-20**

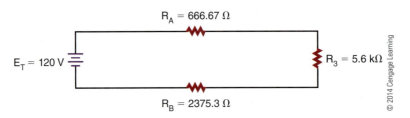

$R_A = 666.67\ \Omega$

$E_T = 120\ V$

$R_3 = 5.6\ k\Omega$

$R_B = 2375.3\ \Omega$

© 2014 Cengage Learning

Solution

$$\frac{1}{R_B} = \frac{1}{R_4} + \frac{1}{R_S}$$

$$\frac{1}{R_B} = \frac{1}{4700} + \frac{1}{4800}$$

$$R_B = 2374.74\ \text{ohms}$$

Redraw the equivalent circuit substituting R_A and R_B, and find the total series resistance of the equivalent circuit. See Figure 14-20.

Given

$$R_T = ?$$

$$R_A = 666.67\ \text{ohms}$$

$$R_3 = 5600\ \text{ohms}$$

$$R_B = 2374.74\ \text{ohms}$$

Solution

$$R_T = R_A + R_3 + R_B$$

$$R_T = 666.67 + 5600 + 2374.74$$

$$R_T = 8641.41\ \text{ohms}$$

Now solve for the total current through the equivalent circuit, using Ohm's law.

Given

$$I_T = ?$$

$$E_T = 120\ \text{volts}$$

$$R_T = 8641.41\ \text{ohms}$$

Solution

$$I_T = \frac{E_T}{R_T}$$

$$I_T = \frac{120}{8641.41}$$

$$I_T = 0.0139\ \text{amp, or } 13.9\ \text{milliamps}$$

In a series circuit, the same current flows throughout the circuit. Therefore, the current flowing through R_3 is equal to the total current in the circuit.

$$I_{R_3} = I_T$$

$$I_{R_3} = 13.9\ \text{milliamps}$$

14-3 QUESTIONS

1. State the formulas necessary for determining total current in series and parallel circuits when the current flow through the individual components is known.
2. State the formulas necessary for determining total voltage in series and parallel circuits when the individual voltage drops are known.
3. State the formulas for determining total resistance in series and parallel circuits when the individual resistances are known.
4. State the formula to solve for total current, voltage, or resistance in a series or parallel circuit when at least two of the three values (current, voltage, and resistance) are known.
5. What is the total circuit current in Figure 14-21?
   $I_T = ?$
   $E_T = 12\ \text{volts}$
   $R_1 = 500\ \text{ohms}$
   $R_2 = 1200\ \text{ohms}$
   $R_3 = 2200\ \text{ohms}$

■ **FIGURE 14-21**

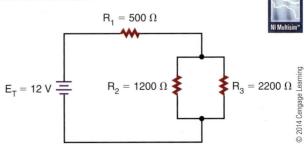

$R_1 = 500\ \Omega$

$E_T = 12\ V$ $R_2 = 1200\ \Omega$ $R_3 = 2200\ \Omega$

© 2014 Cengage Learning

14-4 KIRCHHOFF'S CURRENT LAW

In 1847, G. R. Kirchhoff extended Ohm's law with two important statements that are referred to as Kirchhoff's laws. The first law, known as **Kirchhoff's current law**, states the following:

- The algebraic sum of all the currents entering and leaving a junction is equal to 0.

Here is another way of stating Kirchhoff's current law:

- The total current flowing into a junction is equal to the sum of the current flowing out of that junction.

A junction is defined as any point of a circuit at which two or more current paths meet. In a parallel circuit, the junction is where the parallel branches of the circuit connect.

In Figure 14-22, point A is one junction and point B is the second junction. Following the current in the circuit, I_T flows from the voltage source into the junction at point A. There the current splits among the three branches as shown. Each of the three branch currents (I_1, I_2, and I_3) flows out of junction A. According to Kirchhoff's current law, which states that the total current into a junction is equal to the total current out of the junction, the current can be stated as

$$I_T = I_1 + I_2 + I_3$$

Following the current through each of the three branches finds them coming back together at point B. Currents I_1, I_2, and I_3 flows out into junction B, and I_T flows out. Kirchhoff's current law formula at this junction is the same as at junction A:

$$I_1 + I_2 + I_3 = I_T$$

14-4 QUESTIONS

1. State Kirchhoff's current law.
2. A total of 3 A flows into a junction of three parallel branches. What is the sum of the three branch currents?
3. If 1 mA and 5 mA of current flow into a junction, what is the amount of current flowing out of the junction?
4. In a parallel circuit with two branches, one branch has 2 mA of current flowing through it. The total current is 5 mA. What is the current through the other branch?
5. Refer to Figure 14-23. What are the values of I_2 and I_3?

14-5 KIRCHHOFF'S VOLTAGE LAW

Kirchhoff's second law is referred to as **Kirchhoff's voltage law**, and it states the following:

- The algebraic sum of all the voltages around a closed circuit equals 0.

Here is another way of stating Kirchhoff's voltage law:

- The sum of all the voltage drops in a closed circuit equals the voltage source.

■ **FIGURE 14-22**

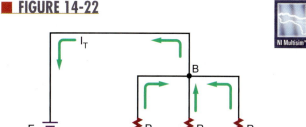

© 2014 Cengage Learning

■ **FIGURE 14-23**

© 2014 Cengage Learning

■ **FIGURE 14-24**

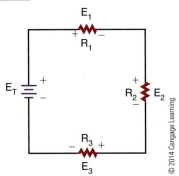

■ **FIGURE 14-25**

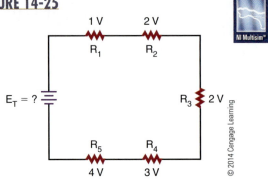

In Figure 14-24 there are three voltage drops and one voltage source (voltage rise) in the circuit. If the voltages are summed around the circuit as shown, they equal 0.

$$E_T - E_1 - E_2 - E_3 = 0$$

Notice that the voltage source (E_T) has a sign opposite that of the voltage drops. Therefore the algebraic sum equals 0.

Looking at this another way, the sum of all the voltage drops equals the voltage source.

$$E_T = E_1 + E_2 + E_3$$

Both of the formulas shown are stating the same thing and are equivalent ways of expressing Kirchhoff's voltage law.

The key to remember is that the polarity of the voltage source in the circuit is opposite to that of the voltage drops.

14–5 QUESTIONS

1. State Kirchhoff's voltage law in two different ways.
2. A series resistive circuit is connected to a 12-volt voltage source. What is the total voltage drop in the circuit?
3. A series circuit has two identical resistors connected in series with a 9-volt battery. What is the voltage drop across each resistor?
4. A series circuit is connected to a 12-volt voltage source with three resistors. One resistor drops 3 V and another resistor drops 5 V. What is the voltage drop across the third resistor?
5. Refer to Figure 14-25. What is the total voltage applied to the circuit?

SUMMARY

- An electric circuit consists of a voltage source, a load, and a conductor.
- The current path in an electric circuit can be series, parallel, or series-parallel.
- A series circuit offers only one path for current to flow.
- A parallel circuit offers several paths for the flow of current.
- A series-parallel circuit provides a combination of series and parallel paths for the flow of current.
- Current flows from the negative side of the voltage source through the load to the positive side of the voltage source.
- Changing either the voltage or the resistance can vary current flow in an electric circuit.
- Ohm's law gives the relationship of current, voltage, and resistance.
- Ohm's law states that the current in an electric circuit is directly proportional to the voltage applied and inversely proportional to the resistance in the circuit.

$$I = \frac{E}{R}$$

- Ohm's law applies to all series, parallel, and series-parallel circuits.
- To determine unknown quantities in a circuit:
  - Draw a schematic of the circuit and label all quantities.
  - Solve for equivalent circuits and redraw the circuit.
  - Solve for all unknown quantities.

- Kirchhoff's current law: The algebraic sum of all the currents entering and leaving a junction is equal to 0; it may be restated as the total current flowing into a junction is equal to the sum of the current flowing out of that junction.

- Kirchhoff's voltage law: The algebraic sum of all the voltages around a closed circuit equals 0; it may be restated as follows: The sum of all the voltage drops in a closed circuit equals the voltage source.

CHAPTER 14 SELF-TEST

Using Ohm's law, find the unknown values for the following:

1. I = ?   E = 9 V   R = 4500 ohms
2. I = 250 mA   E = ?   R = 470 ohms
3. I = 10 A   E = 240 V   R = ?
4. Find the current and voltage drop through each component for the following circuits:

5. Use Kirchhoff's laws to verify answers for question 4.

(A)

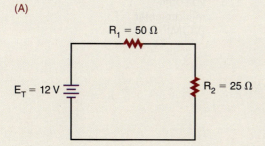

(B)

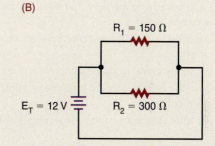

(C)

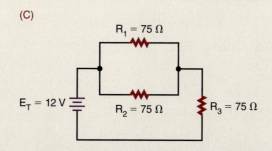

Electrical Measurements — Meters

OBJECTIVES

After completing this chapter, the student will be able to:

- Identify the two types of meter movements available.
- Describe how a voltmeter is used in a circuit.
- Describe how an ammeter is used in a circuit.
- Identify the functions of a multimeter.
- Identify the advantages and disadvantages of DMMs and VOMs.
- Describe how to use a multimeter to measure voltage, current, and resistance.
- Describe how to measure current using an ammeter.
- Describe how to connect an ammeter into a circuit.
- List safety precautions for using an ammeter.
- Describe how to connect a voltmeter to an electrical circuit.
- List safety precautions for connecting a voltmeter to a circuit.
- Describe how resistance values are measured using an ohmmeter.
- Define continuity check.
- Describe how an ohmmeter is used to check open, short, or closed circuits.

KEY TERMS

15-1	analog meter
15-1	digital meter
15-1	digital multimeter (DMM)
15-2	ammeter
15-2	ohmmeter
15-2	voltmeter
15-3	multimeter
15-3	volt-ohm-milliammeter (VOM)
15-4	ammeter shunt
15-5	polarity
15-5	short circuit
15-6	closed circuit
15-6	continuity test
15-6	open circuit
15-7	full-scale value

n the field of electricity, accurate quantitative measurements are essential. A technician commonly works with current, voltage, and resistance. Ammeters, voltmeters, and ohmmeters are used to provide the essential measurements. A good understanding of the design and operation of electrical measuring meters is important.

This chapter describes the more commonly used analog meters, including the multimeter, or multifunction meter.

15-1 INTRODUCTION TO METERS

Meters are the means by which the invisible action of electrons can be detected and measured. Meters are indispensable in examining the operation of a circuit. Two types of meters are available. One type is the **analog meter**, which uses a graduated scale with a pointer (Figure 15-1). The other type is the **digital meter**, which provides a reading in numbers (Figure 15-2). Digital meters are easier to read and provide a more accurate reading than analog meters. However, analog meters provide a better graphic display of rapid changes in current or voltage.

Most meters are housed in a protective case. Terminals are provided for connecting the meter to the circuit. The polarity of the terminals must be observed

■ FIGURE 15-2

A digital meter.

© 2014 Cengage Learning

for proper connection. A red terminal is positive and a black terminal is negative.

Before using an analog meter, the pointer should be adjusted to 0. A small screw is located on the front of the meter to permit this adjustment (Figure 15-3). To zero the meter, place the meter in the position where it is to be used. If the needle does not point to 0, use a screwdriver to turn the screw until it does. The meter should not be connected to a circuit while this adjustment is being made.

■ FIGURE 15-1

Analog meters.

© 2014 Cengage Learning

■ FIGURE 15-3

Location of the zero adjustment screw on an analog meter.

Zero adjusting screw

© 2014 Cengage Learning

FIGURE 15-6

■ FIGURE 15-6

(A) Schematic symbol for a voltmeter. (B) A voltmeter is connected in parallel in a circuit.

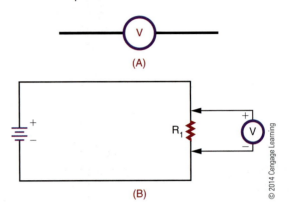

(A)

(B)

© 2014 Cengage Learning

15-1   QUESTIONS

1. What is the purpose of a meter?
2. What are the two types of meters available?
3. Which type of meter, analog or digital, would be easier to read?
4. What colors identify the positive and negative terminals of a meter?
5. What adjustments should be made before using an analog meter?

15-2   TYPES OF METERS

An **ammeter** is used to measure current in a circuit. An ammeter (schematic symbol shown in Figure 15-4) can be considered a flow meter. It measures the number of electrons flowing past a given point in a circuit. The electrons must flow through the ammeter to obtain a reading as shown in Figure 15-5, and this is accomplished by opening the circuit and inserting the ammeter.

A **voltmeter** is used to measure the voltage (difference of potential) between two points in a circuit. A voltmeter can be considered a pressure gauge, used to measure the electrical pressure in a circuit (Figure 15-6).

Resistance is measured with an **ohmmeter**. To measure resistance, a voltage is placed across the device to be measured, inducing a current flow through the device (Figure 15-7). When there is little resistance, a large current flows and the ohmmeter registers a low resistance. When there is great resistance, a small current flows and the ohmmeter registers a high resistance.

15-2   QUESTIONS

1. What is used to measure current?
2. What is used to measure voltage?
3. What meter is used to measure resistance?
4. Describe how to measure current with an ammeter.
5. Describe how to measure voltage with a voltmeter.
6. Describe how to measure resistance with an ohmmeter.

■ FIGURE 15-4

Schematic symbol for an ammeter.

© 2014 Cengage Learning

■ FIGURE 15-5

The placement of an ammeter in a circuit.

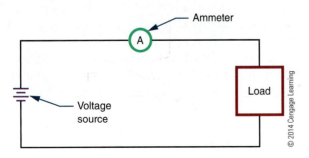

© 2014 Cengage Learning

■ FIGURE 15-7

(A) Schematic symbol for an ohmmeter. (B) An ohmmeter applies a voltage across the component being measured and monitors the current flowing through it.

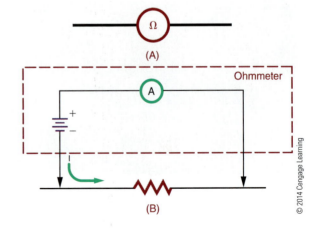

(A)

(B)

© 2014 Cengage Learning

15-3 MULTIMETERS

When working on a piece of equipment, many different measurements must be taken. To eliminate the need for several meters, the voltmeter, ammeter, and ohmmeter can be combined into a single instrument called a **multimeter**. An analog multimeter is referred to as a **volt-ohm-milliammeter (VOM)**, Figure 15-8. A **digital multimeter** is also referred to as a **DMM**.

The meter in Figure 15-8 has five voltage positions, four current positions, and three resistance positions. There are five scales on the meter to accommodate the various ranges and functions. The technician selects the switch on the multimeter for the desired voltage, current, or resistance range. The zero-ohm control adjusts the ohmmeter circuit to compensate for variations in the voltage of the internal batteries. The function switch has three positions: −DC, +DC, and AC. To measure current, DC voltage, and resistance, the function switch is placed at −DC or +DC, according to the polarity of the applied current or voltage. The function switch permits reversal of the test lead connections without removing the leads from the circuit being tested.

■ FIGURE 15-8

Analog multimeter.

To measure DC voltage, set the function switch to +DC. With the function switch set at +DC, the common jack is negative and the plus jack is positive. The voltmeter is connected in parallel with the circuit. When measuring an unknown voltage, always set the meter to the highest range (500 volts). If the measured voltage is lower, a lower position can be selected. This procedure protects the meter from damage. Read the voltage on the scale marked DC. For the 2.5-volt range, use the 0–250 scale and divide by 100. For the 10-volt, 50-volt, and 250-volt ranges, use the scales directly. For the 500-volt range, use the 0–50 scale and multiply by 10.

To measure current, the selector switch is set for the desired current position and the meter is connected in series with the circuit. The DC scale on the meter is used. For the 1 mA range, use the 0–10 scale and divide by 10. For the 10 mA range, use the 0–10 scale directly. For the 100 mA range, use the 0–10 scale and multiply by 10. For the 500 mA range, use the 0–50 scale and multiply by 10.

To measure resistance, set the selector switch to the desired resistance range. Short the test leads together. Rotate the zero-ohm control until the pointer indicates zero ohms. Separate the test leads and connect them across the component being measured. For measuring resistance between 0 and 200 ohms, use the R × 1 range. To measure resistance between 200 and 20,000 ohms, use the R × 100 range. For resistance above 20,000 ohms, use the R × 10,000. When measuring resistance in digital and semiconductor circuits, do not use the R × 1 scale. This scale is energized by a 9 V battery and could damage the circuitry. Observe the reading on the ohm scale at the top of the meter scales. Note that the ohm scale reads from right to left. To determine the actual resistance value, multiply the reading by the factor at the switch position. The letter K equals 1000.

To use the other voltage and current jacks located on the multimeter, refer to the operator's manual.

With a DMM, the meter movement is replaced by a digital readout, which is similar to the readout used in an electronic calculator. The difference is that the display is typically four digits.

The controls of a DMM are similar to those found on an analog meter. The exception is with a digital meter that has an autoranging feature. It does not need to identify the range of the signal being tested.

The readout of the DMM automatically places the decimal point with the appropriate unit prefix. DC measurements are displayed with a + or − sign. This avoids the necessity of reversing the test leads.

The advantages of DMMs over VOMs include these:

- Easier readability, resulting in a higher accuracy with smaller voltage, and results that are repeatable with different readers
- Autoranging capabilities
- Autozeroing for resistance reading
- Autolock of a displayed value

The disadvantages of a DMM include these:

- Requirement for batteries or other source of electricity
- Voltage limitation that if exceeded could seriously damage the meter
- Inability to measure an instantaneous change in a signal that is faster than the sampling time

15-3 QUESTIONS

1. What is a multimeter?
2. What is an analog multimeter called?
3. What is a digital multimeter called?
4. Explain how to use the ohmmeter portion of a VOM.
5. What are the advantages of the DMM over the VOM?

15-4 MEASURING CURRENT

To use an ammeter to measure current, the circuit must be opened and the meter inserted into the circuit in series (Figure 15-9).

When placing the ammeter in the circuit, polarity must be observed. The two terminals on an ammeter are marked: red for positive and black for negative (or common) (Figure 15-10).

■ FIGURE 15-9

An ammeter is connected in series in a circuit.

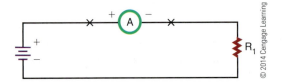

© 2014 Cengage Learning

■ FIGURE 15-10

An ammeter is one part of this VOM. A black negative lead plugs into the common or negative jack. A red positive lead plugs into the jack with the plus symbol.

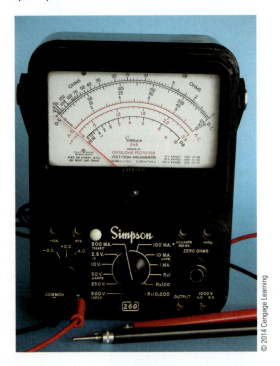

© 2014 Cengage Learning

CAUTION!

Always turn off the power before you connect an ammeter to a circuit.

The negative terminal must be connected to the more negative point in the circuit, and the positive terminal to the more positive point in the circuit (Figure 15-11). When the ammeter is connected, the needle (pointer) of the meter moves from left to right. If the needle moves in the opposite direction, reverse the leads.

Another way to measure current that does not require opening the circuit and placing an ammeter in series is to use a resistor of a known value in the circuit. The resistor is referred to as an **ammeter shunt** and may be installed by the manufacturer or inserted for temporary use. The resistor may be left in for future testing of the circuit and must be of a small enough value to not interfere with the normal operation of the circuit.

The resistor's purpose is to provide a small output voltage for measuring. The best meter to use is a DMM because it is able to accurately measure small voltages.

■ FIGURE 15-11

Connect the positive terminal of the ammeter to the more positive point in the circuit. Connect the negative terminal to the more negative point in the circuit.

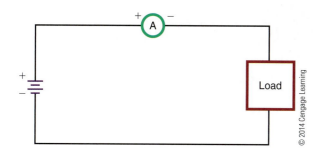

© 2014 Cengage Learning

Ohm's law (I = E/R) is used to calculate the amount of current measured by the shunt. Dividing the voltage read by the resistance value yields the current flowing in the circuit.

Figure 15-12 shows a 0.1 Ω resistor used as a shunt. If the DMM reads 330 mV, then the circuit has 3.3 A flowing through it.

$$I = \frac{E}{R}$$

$$I = \frac{0.330}{0.1}$$

$$I = 3.3 \text{ A}$$

The advantage of using this technique is that the circuit does not have to be opened and an ammeter inserted. Also note that voltage—rather than current—was read.

CAUTION!

An analog ammeter must never be connected in parallel with any of the circuit components. If connected in parallel, the fuse in the ammeter will blow and may seriously damage the meter or the circuit. Also, never connect an ammeter directly to a voltage source.

Before turning on power to the circuit after the ammeter is installed, set the meter to its highest scale (its highest ammeter range before applying power). After the power is applied, the ammeter can be set to the appropriate scale. This prevents the needle of the meter from being driven into its stop.

■ FIGURE 15-12

Determined current through a load.

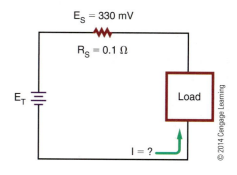

© 2014 Cengage Learning

The internal resistance of the ammeter adds to the circuit and increases the total resistance of the circuit. Therefore, the measured circuit current can be slightly lower than the actual circuit current. However, because the resistance of an ammeter is usually minute compared to the circuit resistance, the error is ignored.

A clip-on ammeter requires no connection to the circuit being measured and uses the electromagnetic field created by the current flow to measure the amount of current in the circuit.

15–4 QUESTIONS

1. How is an ammeter connected to a circuit?
2. What is the first step before connecting an ammeter in a circuit?
3. What should be done if the ammeter reads in the reverse direction?
4. What scale should the ammeter be placed on before applying power?
5. Why should an ammeter never be connected in parallel in a circuit?

15–5 MEASURING VOLTAGE

Voltage exists between two points; it does not flow through a circuit as current does. Therefore, a voltmeter, used to measure voltage, is connected in parallel with the circuit.

■ FIGURE 15-13

When connecting a voltmeter into a circuit, be sure to observe polarity.

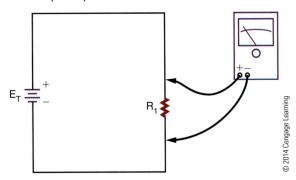

© 2014 Cengage Learning

CAUTION!

If an analog voltmeter is connected in series with a circuit, a large current could flow through the meter and damage it.

Polarity is important with analog meters. The negative terminal of the voltmeter must be connected to the more negative point in the circuit, and the positive terminal to the more positive point in the circuit (Figure 15-13). If the connections are reversed, the pointer deflects toward the left side of the meter, registering no measurement. If this occurs, reverse the meter leads.

A good practice is to remove power from the circuit, connect the voltmeter, and then reapply power. Initially, set the voltmeter for its highest scale. After the voltage is applied to the circuit, set the meter down to the proper scale.

The voltmeter's internal resistance is connected in parallel with the component being measured. The total resistance of resistors in parallel is always less than that of the smallest resistor. As a result, the voltage read by the voltmeter is smaller than the actual voltage across the component. In most cases, the internal resistance of a voltmeter is high, so the error is small and can be ignored. However, if voltage is being measured in a high-resistance circuit, meter resistance may have a noticeable effect. Some voltmeters are designed with extra-high internal resistance for such purposes.

Currently, the DMM is the best all-around meter available. The voltage ranges have very little loading effect on the circuits being measured. The meter's internal resistance is typically up to 10 MΩ on its voltage

range, compared to a quality analog meter with an internal resistance of 20,000 Ω /V, which would yield 20,000 Ω on its 1 V range.

15–5 QUESTIONS

1. How is a voltmeter connected to a circuit?
2. What is the recommended practice for connecting a voltmeter to a circuit?
3. What should be done if the voltmeter reads in the reverse direction?
4. What caution must be taken when measuring a high-resistance circuit?
5. Why should a voltmeter never be connected in series in a circuit?

15–6 MEASURING RESISTANCE

An ohmmeter measures the resistance of a circuit or component by applying a known voltage. Batteries supply the voltage. When a constant voltage is applied to the meter circuit through the component under test, the pointer is deflected based on the current flow. The meter deflection varies with the resistance being measured. To measure the resistance of a circuit or component, the ohmmeter is connected in parallel with the circuit or component.

CAUTION!

Before connecting an ohmmeter to a component in a circuit, make sure the power is turned off.

When measuring a component in a circuit, disconnect one end of the component from the circuit. This eliminates parallel paths, which result in an incorrect resistance reading. The device must be removed from the circuit to obtain an accurate reading. Then, the ohmmeter leads are connected across the device (Figure 15-14).

Because the primary purpose of an ohmmeter is to measure resistance, it can be used to determine whether a circuit is open, shorted, or closed. An **open circuit** has infinite resistance because no current flows

■ FIGURE 15-14

When you are using an ohmmeter to measure resistance, the component being measured must be removed from the circuit.

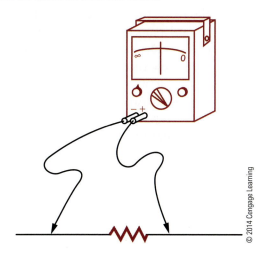

© 2014 Cengage Learning

through it (Figure 15-15). A **short circuit** has 0 ohms of resistance because current flows through it without developing a voltage drop. A **closed circuit** is a complete path for current flow. Its resistance varies depending on the components in the circuit (Figure 15-16).

The testing for an open, short, or closed circuit is called a **continuity test**, or check. It is a check to determine whether a current path is continuous. To determine whether a circuit is open or closed, the lowest scale on the ohmmeter should be used. First,

■ FIGURE 15-15

An ohmmeter can be used to determine whether a circuit is open. An open circuit indicates a high resistance.

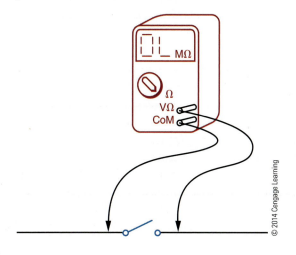

© 2014 Cengage Learning

■ FIGURE 15-16

An ohmmeter can be also be used to determine whether a circuit offers a complete path for current flow. A closed circuit indicates a low resistance.

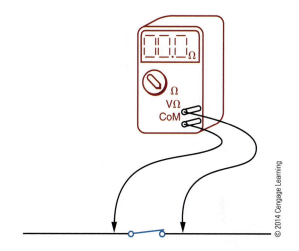

© 2014 Cengage Learning

ensure that there are no components in the circuit that may be damaged by the current flow from the ohmmeter. Then place the leads of the ohmmeter across the points in the circuit to be measured. If a reading occurs, the path is closed or shorted. If no reading occurs, the path is open. This test is useful to determine why a circuit does not work.

The DMM has several advantages over the analog multimeter, including higher accuracy, digital readout, and repeatability of the reading. Unlike the VOM, the DMM uses very little current to test resistances. This allows the testing of semiconductor junctions. In fact, many DMMs have a special range for testing semiconductor junctions. These meters use approximately 1 mA of current through the junction under test.

Another feature of a DMM is an audio signal, or "beeper," for continuity testing. This allows the operator to focus on the circuit rather than on the meter for continuity testing. The use of the beeper function is not to check for resistance, but rather for continuity.

NOTE **When using any meter, it is important to keep fingers from touching the probes to avoid putting body resistance in parallel with the component under test, thereby introducing errors.**

15–6 QUESTIONS

1. How does an ohmmeter work?
2. What caution must be observed before connecting an ohmmeter to a circuit?
3. What is the primary purpose of an ohmmeter?
4. For what other purpose may an ohmmeter be used?
5. Why is a DMM better for testing semiconductors than a VOM?
6. Why is a DMM better for continuity testing than a VOM?

15–7 READING METER SCALES

Voltmeter and ammeter scales are read in the same manner. However, voltmeters measure volts, and ammeters measure amperes. The maximum value indicated by a meter is called the **full-scale value**. In other words, the maximum voltage or current that a meter can read is its full-scale value.

The measured value of voltage or current is read on the scale under the pointer. For example, the pointer in Figure 15-17 is shown deflected one major division, indicating a voltage of 1 volt or a current of 1 ampere.

In Figure 15-18, the meter is shown deflected seven major divisions, indicating a current of 7 amperes or a voltage of 7 volts.

If the pointer of the meter rests between the major divisions of the scale, the smaller divisions are read. Figure 15-19 shows four small lines between each major division of the scale, creating five equally spaced intervals. Each of these small intervals represents one-fifth of the major interval, or 0.2 units.

If the pointer falls between the small lines on a meter scale, the value must be estimated. In Figure 15-20, the pointer falls between the ⅖ (0.4) and ⅗ (0.6) marks. This indicates a value of approximately 2.5 volts or amperes. In Figure 15-21, the pointer is one-fourth

■ FIGURE 15-18
The reading indicates 7 volts or amperes.

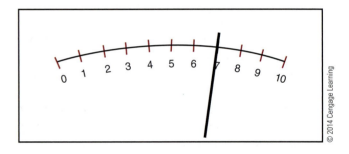

■ FIGURE 15-19
Each small division represents 0.2 volt or ampere.

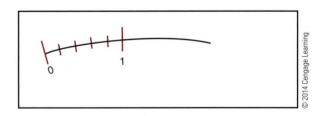

■ FIGURE 15-17
The reading indicates 1 volt or ampere.

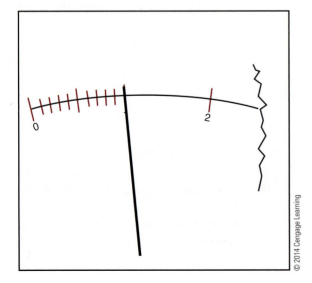

■ FIGURE 15-20
The reading indicates 2.5 volts or amperes.

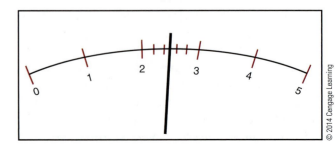

© 2014 Cengage Learning

FIGURE 15-21

The reading indicates 4.65 volts or amperes.

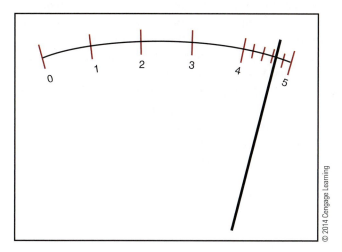

FIGURE 15-22

The ohmmeter scale is read from right to left.

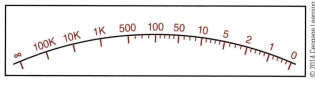

FIGURE 15-23

The reading indicates 1.5 ohms.

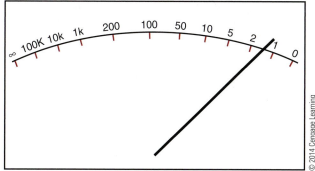

FIGURE 15-24

The reading indicates 200 ohms.

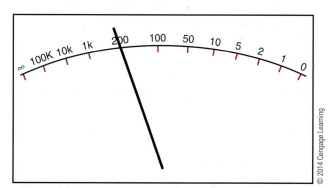

of the distance between the ⅗ (0.6) and ⅘ (0.8) marks. Each small interval represents 0.2. One-fourth of 0.2 is 0.05. Therefore, the pointer indicates a value of about 4.65 volts or amperes.

The number of major and minor divisions on a meter scale depends on the range of the voltage or current that the meter is designed to measure. In all cases, the value of the small intervals can be found by dividing the value of the major interval by the number of spaces it contains.

The ohm scale on a meter is different from most voltage and current scales (Figure 15-22). It is read from right to left instead of from left to right. Also, it is a nonlinear scale, so the number of small spaces between the major intervals is not the same throughout the scale. Between 0 and 1 there are five small spaces, which equal 0.2 unit each. There are four intervals between 5 and 10, representing 1 unit each, and between each of these there is a minor division that represents 0.5 unit. Between the 50 and 100 marks are five small intervals, which each represent 10 units. Between 100 and 500, there are four small intervals, each representing 100 units. The last mark on the left is labeled infinity (∞). If the pointer deflects to this mark, the resistance is beyond the range of the meter. The pointer normally rests on the infinity mark when no resistance is being measured. Figure 15-23 shows the pointer deflected to 1.5 ohms. Figure 15-24 shows the pointer indicating 200 ohms.

Before an ohmmeter is used, the test leads are shorted together and the zero control is adjusted so the pointer rests on the zero mark. This calibrates the meter and compensates for battery deterioration.

15-7 QUESTIONS

1. What determines the maximum value an analog meter can measure?
2. What are the differences between an ohmmeter scale and a voltmeter or ammeter scale?
3. What is really being measured with an ammeter?
4. Estimate the reading of the voltmeter scales in Figure 15-25.
5. When an ohmmeter reads infinity (∞), what is being measured?

■ FIGURE 15-25

Analog multimeter.

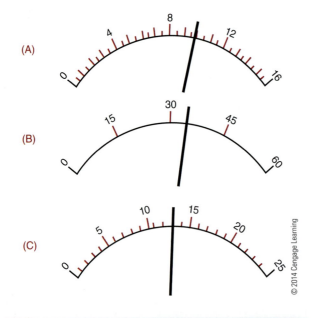

(A)

(B)

(C)

© 2014 Cengage Learning

SUMMARY

- Analog meters use a graduated scale with a pointer.
- Digital meters provide a direct readout.
- On both analog and digital meters, the red terminal is positive and the black terminal is negative.
- Before using an analog meter, check the mechanical zero adjustment.
- A multimeter combines a voltmeter, ammeter, and ohmmeter into one package.
- A VOM is an analog multimeter that measures volts, ohms, and milliamperes.
- A DMM is a digital multimeter.
- On a multimeter, the range selector switch selects the function to be used.
- An ammeter must be connected in series with a circuit.
- A voltmeter is connected in parallel with a circuit.
- An ohmmeter measures resistance by the amount of current flowing through the resistor being measured.
- The maximum value of a meter scale is called the full-scale value.
- The number of divisions on the meter scale depends on the range the meter is designed to measure.
- Ammeters and voltmeters are read from left to right and have a linear scale.
- Ohmmeters are read from right to left and have nonlinear scales.
- An analog ohmmeter must be calibrated before use to compensate for battery deterioration.

CHAPTER 15 SELF-TEST

1. Which type of meter, analog or digital, would you use for an accurate reading?
2. Which type of meter, analog or digital, would you use to gauge rapid changes in a source?
3. Draw a meter scale for each of the following readings, and show where the needle would point.
   a. 23 V
   b. 220 mA
   c. 2700 ohms
4. What are the advantages of using a multimeter?
5. Draw diagrams showing a multimeter connected in a circuit to measure voltage, current, and resistance.
6. What is the difference between a VOM and a DMM?
7. When using a VOM, to what range should the selector switch be set?
8. When using a VOM, if the pointer is driven hard into the pin, what should be done immediately? What action should follow?
9. Why are analog ammeters and voltmeters read from left to right, whereas analog ohmmeters are read from right to left?
10. Does the DMM ohmmeter function have to be calibrated before use?

Power

OBJECTIVES

After completing this chapter, the student will be able to:

- Define power as it relates to electric circuits.
- State the relationship of current and voltage.
- Solve for power consumption in an electrical circuit.
- Determine the total power consumption in a series, parallel, or series-parallel circuit.

KEY TERMS

16-1 electric power rate

16-1 instantaneous rate

16-1 power (P)

16-1 watt

16-1 work

16-2 power dissipated

n addition to current, voltage, and resistance, a fourth quantity is important in circuit analysis. This quantity is called power. Power is the rate at which work is done. Every time a circuit is energized, power is expended. Power is directly proportional to both current and voltage. This chapter looks at circuit applications involving power.

16-1 POWER

Electrical or mechanical power relates to the rate at which work is being done. **Work** is done whenever a force causes motion. If a mechanical force is used to lift or move a weight, work is being done. However, force exerted without causing motion, such as a force of a compressed spring between two fixed objects, does not constitute work.

Voltage is an electrical force that creates current to flow into a closed circuit. When voltage exists between two points and current cannot flow, no work is done. This is similar to a spring under tension that produces no motion. When voltage causes electrons to move in a circuit, work is being done. The **instantaneous rate** at which work is done is called the **electric power rate** and is measured in **watts**. **Power** can be defined at the rate which energy is dissipated in a circuit. The symbol for power is **P**.

The total amount of work done may be accomplished in different lengths of time. For example, a given quantity of electrons may be moved from one location to another in one second, one minute, or one hour, depending on the rate in which they were moved. In all cases, the total amount of work done is the same. However, when the work is done in a short period of time, the instantaneous power rate (wattage) is greater than when the same amount of work is done over a longer period of time.

As mentioned, the basic unit of power is the watt. A watt is equal to the voltage across a circuit multiplied by the current through the circuit. It represents the rate at any given instant in which work is being done, moving electrons through the circuit. The relationship of power, current, and voltage may be expressed as follows:

$$P = I \times E, \text{ or } P = IE$$

I represents the current through the circuit and E represents the voltage applied to the circuit being measured. The amount of power will differ with any change in the voltage or current in the circuit.

■ FIGURE 16-1

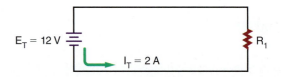

EXAMPLE: Calculate the power consumed in the circuit shown in Figure 16-1.

Given	Solution
P = ?	P = IE
I = 2 A	P = (2)(12)
E = 12 V	P = 24 W

EXAMPLE: What voltage is required to deliver 2 amperes of current at 200 watts?

Given	Solution
P = 200 W	P = IE
I = 2 A	200 = 2(E)
E = ?	100 V = E

EXAMPLE: How much current does a 100-watt, 120-volt light bulb use?

Given	Solution
P = 100 W	P = IE
I = ?	100 = (I)(120)
E = 120 V	0.83 A = I

16-1 QUESTIONS

1. Define power as it relates to electricity.
2. What unit is used to measure power?
3. What represents the rate of work being done at any given instant?
4. When the same work is being done at different lengths of time, is the total amount of work done the same?
5. Calculate the missing value:
   a. P = ?, E = 12 V, I = 1 A
   b. P = 1000 W, E = ?, I = 10 A
   c. P = 150 W, E = 120 V, I = ?

16–2 POWER APPLICATION (CIRCUIT ANALYSIS)

Resistive components in a circuit consume power. To determine the **power dissipated** by a component, multiply the voltage drop across the component by the current flowing through the component.

$$P = IE$$

The total power dissipated in a series or parallel circuit is equal to the sum of the power dissipated by the individual components. This can be expressed as:

$$P_T = P_{R_1} + P_{R_2} + P_{R_3} \ldots + P_{R_n}$$

The power dissipated in a circuit is often less than 1 watt. To facilitate the use of these smaller numbers, the milliwatt (mW) and the microwatt (μW) are used.

$$1000 \text{ milliwatts} = 1 \text{ watt}$$

$$1 \text{ milliwatt} = \frac{1}{1000} \text{ watt}$$

$$1{,}000{,}000 \text{ microwatts} = 1 \text{ watt}$$

$$1 \text{ microwatt} = \frac{1}{1{,}000{,}000} \text{ watt}$$

EXAMPLE: How much power is consumed in the circuit shown in Figure 16-2?

■ FIGURE 16-2

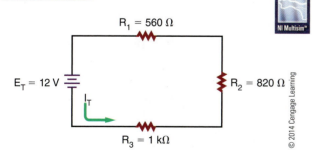

$R_1 = 560\ \Omega$
$E_T = 12\ V$
I_T
$R_2 = 820\ \Omega$
$R_3 = 1\ k\Omega$

© 2014 Cengage Learning

First determine the total resistance for the circuit.

Given	Solution
$R_T = ?$	$R_T = R_1 + R_2 + R_3$
$R_1 = 560\ \Omega$	$R_T = 560 + 820 + 1000$
$R_2 = 820\ \Omega$	$R_T = 2380\ \Omega$
$R_3 = 1000\ \Omega$	

Now determine the total current flowing in the circuit, using Ohm's law.

Given	Solution
$I_T = ?$	$I_T = \dfrac{E_T}{R_T}$
$E_T = 12\ V$	
$R_T = 2380\ \Omega$	$I_T = \dfrac{12}{2380}$
	$I_T = 0.005\ A$

The total power consumption can now be determined using the power formula.

Given	Solution
$P_T = ?$	$P_T = I_T E_T$
$I_T = 0.005\ A$	$P_T = (0.005)(12)$
$E_T = 12\ V$	$P_T = 0.06\ W,$ or $60\ mW$

EXAMPLE: What is the value of resistor R_2 in the circuit shown in Figure 16-3?

First determine the voltage drop across resistor R_1.

Given	Solution
$P_{R_1} = 0.018\ W$	$P_{R_1} = I_{R_1} E_{R_1}$
$I_{R_1} = 0.0015\ A$	$0.018 = (0.0015)(E_{R_1})$
$E_{R_1} = ?$	$E_{R_1} = 12\ V$

Now the current through resistor R_2 can be determined. Remember, in a parallel circuit, the voltage is the same in all branches and $E_T = E_{R_1} = E_{R_2} = E_{R_3}$.

Given	Solution
$P_{R_2} = 0.026\ W$	$P_{R_2} = I_{R_2} E_{R_2}$
$I_{R_2} = ?$	$0.026 = (I_{R_2})(12)$
$E_{R_2} = 12\ V$	$I_{R_2} = 0.00217\ A$

■ FIGURE 16-3

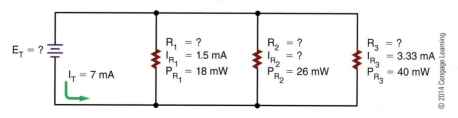

$E_T = ?$
$I_T = 7\ mA$

$R_1 = ?$
$I_{R_1} = 1.5\ mA$
$P_{R_1} = 18\ mW$

$R_2 = ?$
$I_{R_2} = ?$
$P_{R_2} = 26\ mW$

$R_3 = ?$
$I_{R_3} = 3.33\ mA$
$P_{R_3} = 40\ mW$

© 2014 Cengage Learning

The resistance value of R_2 can now be determined using Ohm's law.

Given	Solution
$I_{R_2} = 0.00217\,A$	$I_{R_2} = \dfrac{E_{R_2}}{R_2}$
$E_{R_2} = 12\,V$	
$R_2 = ?$	$0.00217 = \dfrac{12}{R_2}$
	$R_2 = 5530\,\Omega$

EXAMPLE: If a current of 0.05 ampere flows through a 22-ohm resistor, how much power does the resistor dissipate?

The voltage drop across the resistor must first be determined using Ohm's law.

Given	Solution
$I_R = 0.05\,A$	$I_R = \dfrac{E_R}{R}$
$E_R = ?$	
$R = 22\,\Omega$	$0.05 = \dfrac{E_R}{22}$
	$E_R = 1.1\,V$

The power consumed by the resistor can now be determined using the power formula.

Given	Solution
$P_R = ?$	$P_R = I_R E_R$
$E_R = 1.1\,V$	$P_R = (0.05)(1.1)$
$I_R = 0.05\,A$	$P_R = 0.055\,W,$ or $55\,mW$

16-2 QUESTIONS

1. What is the formula for power when both the current and voltage are known?
2. What is the formula for determining the total power in a series circuit? A parallel circuit?
3. Convert the following:
   a. $100\,mW = $ _____ W
   b. $10\,W = $ _____ mW
   c. $10\,\mu W = $ _____ W
   d. $1000\,\mu W = $ _____ mW
   e. $0.025\,W = $ _____ mW
4. What is the power consumed by each resistor in the circuit shown in Figure 16.4?
5. What is the power consumed by each resistor in the circuit shown in Figure 16.5?
6. What is the power consumed by each resistor in the circuit shown in Figure 16.6?

■ **FIGURE 16-4**

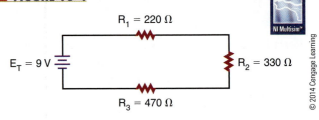

■ **FIGURE 16-5**

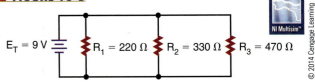

■ **FIGURE 16-6**

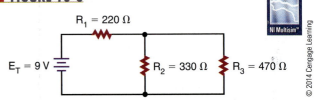

SUMMARY

- Power is the rate at which energy is delivered to a circuit.
- Power is also the rate at which energy (heat) is dissipated by the resistance in a circuit.
- Power is measured in watts.
- Power is the product of current and voltage:

$$P = IE$$

- The total power dissipated in a series or parallel circuit is equal to the sum of the power dissipated by the individual components.

$$P_T = P_1 + P_2 + P_3 \ldots + P_n$$

CHAPTER 16 SELF-TEST

Using Watt's law, find the missing value for the following:

1. $P = ?$ $E = 30\,V$ $I = 40\,mA$
2. $P = 1\,W$ $E = ?$ $I = 10\,mA$
3. $P = 12.3\,W$ $E = 30\,V$ $I = ?$

4. What is the total power consumption for the circuits used in question 5?
5. What is the individual power consumption of each resistor in the following circuits?

(A)

$E_T = 120\,V$ $R_1 = 5.6\,k\Omega$ $R_2 = 5.6\,k\Omega$

(B)

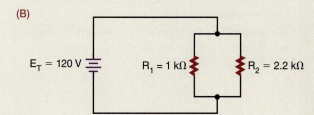

$E_T = 120\,V$ $R_1 = 1\,k\Omega$ $R_2 = 2.2\,k\Omega$

(C)

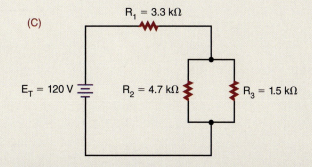

$R_1 = 3.3\,k\Omega$

$E_T = 120\,V$ $R_2 = 4.7\,k\Omega$ $R_3 = 1.5\,k\Omega$

DC Circuits

After completing this chapter, the student will be able to:

- Solve for all unknown values (current, voltage, resistance, and power) in a series, parallel, or series-parallel circuit.
- Understand the importance of voltage dividers.
- Design and solve for all unknown values in a voltage divider circuit.

17	voltage divider	17-5	strain gauge
17-4	inversely proportional	17-5	variometer
17-4	scaling	17-5	Wheatstone bridge
17-5	explosimeter		

In the study of electronics, certain circuits appear again and again. The most commonly used circuits are the series circuit, the parallel circuit, and the series-parallel circuit.

This chapter applies information from the last few chapters to the solving of all unknowns in these three basic types of circuits. **Voltage dividers** can make available several voltages from a single voltage source. Voltage dividers are essentially series circuits with parallel loads. This chapter helps the student understand how significant voltage dividers are as an application of series circuits and how to design one for a specific application.

17-1 SERIES CIRCUITS

A series circuit (Figure 17-1) provides only one path for current flow. The factors governing the operation of a series circuit are as follows:

1. The same current flows through each component in a series circuit.

$$I_T = I_{R_1} = I_{R_2} = I_{R_3} \ldots = I_{R_n}$$

2. The total resistance in a series circuit is equal to the sum of the individual resistances.

$$R_T = R_1 + R_2 + R_3 \ldots + R_n$$

3. The total voltage across a series circuit is equal to the sum of the individual voltage drops.

$$E_T = E_{R_1} + E_{R_2} + E_{R_3} \ldots + E_{R_n}$$

4. The voltage drop across a resistor in a series circuit is proportional to the size of the resistor.

$$(I = E/R)$$

5. The total power dissipated in a series circuit is equal to the sum of the individual power dissipations.

$$P_T = P_{R_1} + P_{R_2} + P_{R_3} \ldots + P_{R_n}$$

EXAMPLE: Three resistors, 47 ohms, 100 ohms, and 150 ohms, are connected in series with a battery rated at 12 volts. Solve for all values in the circuit.

The first step is to draw a schematic of the circuit and list all known variables. See Figure 17-2.

	Given	
$I_T = ?$	$R_1 = 47\,\Omega$	$E_{R1} = ?$
$E_T = 12\,V$	$R_2 = 100\,\Omega$	$E_{R_2} = ?$
$R_T = ?$	$R_3 = 150\,\Omega$	$E_{R_3} = ?$
$P_T = ?$		$P_{R_1} = ?$
		$P_{R_2} = ?$
		$P_{R3} = ?$

In solving for all values in a circuit, the total resistance must be found first. Then the total circuit current can be determined.

Once the current is known, the voltage drops and power dissipation can be determined.

Given	Solution
$R_T = ?$	$R_T = R_1 + R_2 + R_3$
$R_1 = 47\,\Omega$	$R_T = 47 + 100 + 150$
$R_2 = 100\,\Omega$	$R_T = 297\,\Omega$
$R_3 = 150\,\Omega$	

Using Ohm's law, the current is as shown here:

Given	Solution
$I_T = ?$	$I_T = \dfrac{E_T}{R_T}$
$E_T = 12\,V$	
$R_T = 297\,\Omega$	$I_T = \dfrac{12}{297}$
	$I_T = 0.040\,\Omega$

■ **FIGURE 17-1**

Series circuit.

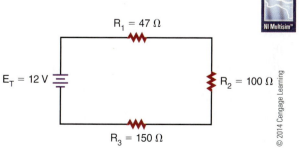

■ **FIGURE 17-2**

Because $I_T = I_{R_1} = I_{R_2} = I_{R_3} \ldots = I_{R_n}$, the voltage drop (E_{R_1}) across resistor R_1 is as follows:

Given	Solution
$I_{R_1} = 0.040$ A	$I_{R_1} = \dfrac{E_{R_1}}{R_1}$
$E_{R_1} = ?$	
$R_1 = 47\ \Omega$	$0.040 = \dfrac{E_{R_1}}{47}$
	$E_{R_1} = 1.88$ V

The voltage drop (E_{R_2}) across resistor R_2 is as shown here:

Given	Solution
$I_{R_2} = 0.040$ A	$I_{R_2} = \dfrac{E_{R_2}}{R_2}$
$E_{R_2} = ?$	
$R_2 = 100\ \Omega$	$0.040 = \dfrac{E_{R_2}}{100}$
	$E_{R_2} = 4$ V

The voltage drop (E_{R_3}) across resistor R_2 is as follows:

Given	Solution
$I_{R_3} = 0.040$ A	$I_{R_3} = \dfrac{E_{R_3}}{R_3}$
$E_{R_3} = ?$	
$R_3 = 150\ \Omega$	$0.040 = \dfrac{E_{R_3}}{150}$
	$E_{R_3} = 6$ V

Verify that the sum of the individual voltages is equal to the total voltage.

Given	Solution
$E_T = 12$ V	$E_T = E_{R_1} + E_{R_2} + E_{R_3}$
$E_{R_1} = 1.88$ V	$E_T = 1.88 + 4 + 6$
$E_{R_2} = 4$ V	$E_T = 11.88$ V
$E_{R_3} = 6$ V	

There is a difference between the calculated and the total given voltage due to the rounding of the total current to three decimal places.

The power dissipated across resistor R_1:

Given	Solution
$P_{R_1} = ?$	$P_{R_1} = I_{R_1} E_{R_1}$
$I_{R_1} = 0.040$ A	$P_{R_1} = (0.040)(1.88)$
$E_{R_1} = 1.88$ V	$P_{R_1} = 0.075$ W

The power dissipated across resistor R_2:

Given	Solution
$P_{R_2} = ?$	$P_{R_2} = I_{R_2} E_{R_2}$
$I_{R_2} = 0.040$ A	$P_{R_2} = (0.040)(4)$
$E_{R_2} = 4$ V	$P_{R_2} = 0.16$ W

The power dissipated across resistor R_3:

Given	Solution
$P_{R_3} = ?$	$P_{R_3} = I_{R_3} E_{R_3}$
$I_{R_3} = 0.040$ A	$P_{R_3} = (0.040)(6)$
$E_{R_3} = 6$ V	$P_{R_3} = 0.24$ W

The total power dissipated:

Given	Solution
$P_T = ?$	$P_T = P_{R_1} + P_{R_2} + P_{R_3}$
$P_{R_1} = 0.075$ W	$P_T = 0.075 + 0.16 + 0.24$
$P_{R_2} = 0.16$ W	$P_T = 0.475$ W, or 475 mW
$P_{R_3} = 0.24$ W	

17–1   QUESTIONS

1. What is the formula for total current in a series circuit?
2. What is the formula for total voltage in a series circuit?
3. What is the formula for total resistance in a series circuit?
4. What is the formula for Ohm's law?
5. Four resistors, 270 ohms, 560 ohms, 1200 ohms, and 1500 ohms, are connected in series with a battery rated at 28 volts. Solve for all values of the circuit.

17–2   PARALLEL CIRCUITS

A parallel circuit (Figure 17-3) is defined as a circuit having more than one current path. The factors governing the operations of a parallel circuit are listed here:

1. The same voltage exists across each branch of the parallel circuit and is equal to that of the voltage source.

$$E_T = E_{R_1} = E_{R_2} = E_{R_3} \ldots = E_{R_n}$$

2. The current through each branch of a parallel circuit is inversely proportional to the amount of resistance of the branch ($I = E/R$).

■ FIGURE 17-3

Parallel circuit.

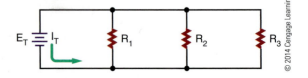

3. The total current in a parallel circuit is the sum of the individual branch currents.

$$I_T = I_{R_1} + I_{R_2} + I_{R_3} \ldots + I_{R_n}$$

4. The reciprocal of the total resistance in a parallel circuit is equal to the sum of the reciprocals of the individual resistances.

$$\frac{1}{R_T} = \frac{1}{R_1} + \frac{1}{R_2} + \frac{1}{R_3} \ldots + \frac{1}{R_n}$$

5. The total power consumed in a parallel circuit is equal to the sum of the power consumed by the individual resistors.

$$P_T = P_{R_1} + P_{R_2} + P_{R_3} \ldots + P_{R_n}$$

EXAMPLE: Three resistors, 100 ohms, 220 ohms, and 470 ohms, are connected in parallel with a battery rated at 48 volts. Solve for all values of the circuit.

First, draw a schematic of the circuit and list all known variables (Figure 17-4).

Given		
$I_T = ?$	$R_1 = 100\ \Omega$	$I_{R_1} = ?$
$E_T = 48\ V$	$R_2 = 220\ \Omega$	$I_{R_2} = ?$
$R_T = ?$	$R_3 = 470\ \Omega$	$I_{R_3} = ?$
$P_T = ?$		$P_{R_1} = ?$
		$P_{R_2} = ?$
		$P_{R_3} = ?$

In solving the circuit for all values, the total resistance must be found first. Then the individual currents flowing through each branch of the circuit can be found.

Once the current is known, the power dissipation across each resistor can be determined.

Given

$R_T = ?$

$R_1 = 100\ \Omega$

$R_2 = 220\ \Omega$

$R_3 = 470\ \Omega$

Solution

$$\frac{1}{R_T} = \frac{1}{R_1} + \frac{1}{R_2} + \frac{1}{R_3}$$

$$\frac{1}{R_T} = \frac{1}{100} + \frac{1}{220} + \frac{1}{470}$$

$$R_T = 59.98\ \Omega$$

The current (I_{R_1}) through resistor R_1:

Given

$I_{R_1} = ?$

$E_{R_1} = 48\ V?$

$R_1 = 100\ \Omega$

Solution

$$I_{R_1} = \frac{E_{R_1}}{R_1}$$

$$I_{R_1} = \frac{48}{100}$$

$$I_{R_1} = 0.48\ A$$

The current (I_{R_2}) through resistor R_2:

Given

$I_{R_2} = ?$

$E_{R_2} = 48\ V$

$R_2 = 220\ \Omega$

Solution

$$I_{R_2} = \frac{E_{R_2}}{R_2}$$

$$I_{R_2} = \frac{48}{220}$$

$$I_{R_2} = 0.218\ A$$

The current (I_{R_3}) through resistor R_3:

Given

$I_{R_3} = ?$

$E_{R_3} = 48\ V$

$R_3 = 470\ \Omega$

Solution

$$I_{R_3} = \frac{E_{R_3}}{R_3}$$

$$I_{R_3} = \frac{48}{470}$$

$$I_{R_3} = 0.102\ A$$

The total current:

Given

$I_T = ?$

$I_{R_1} = 0.48\ A$

$I_{R_2} = 0.218\ A$

$I_{R_3} = 0.102\ A$

Solution

$$I_T = I_{R_1} + I_{R_2} + I_{R_3}$$

$$I_T = 0.48 + 0.218 + 0.102$$

$$I_T = 0.800\ A$$

The total current can also be found using Ohm's law:

Given

$I_T = ?$

$E_T = 48\ V$

$R_T = 59.98\ \Omega$

Solution

$$I_T = \frac{E_T}{R_T}$$

$$I_T = \frac{48}{59.98}$$

$$I_T = 0.800\ A$$

■ **FIGURE 17-4**

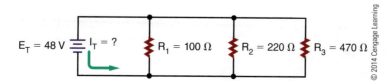

Again, a difference occurs because of rounding.

The power dissipated across resistor R_1:

Given	Solution
$P_{R_1} = ?$	$P_{R_1} = I_{R_1}E_{R_1}$
$I_{R_1} = 0.48\,A$	$P_{R_1} = (0.48)(48)$
$E_{R_1} = 48\,V$	$P_{R_1} = 23.04\,W$

The power dissipated across resistor R_2:

Given	Solution
$P_{R_2} = ?$	$P_{R_2} = I_{R_2}E_{R_2}$
$I_{R_2} = 0.218\,A$	$P_{R_2} = (0.218)(48)$
$E_{R_2} = 48\,V$	$P_{R_2} = 10.46\,W$

The power dissipated across resistor R_3:

Given	Solution
$P_{R_3} = ?$	$P_{R_3} = I_{R_3}E_{R_3}$
$I_{R_3} = 0.102\,A$	$P_{R_3} = (0.102)(48)$
$E_{R_3} = 48\,V$	$P_{R_3} = 4.90\,W$

The total power dissipated:

Given	Solution
$P_T = ?$	$P_T = P_{R_1} + P_{R_2} + P_{R_3}$
$P_{R_1} = 23.04\,W$	$P_T = 23.04 + 10.46 + 4.90$
$P_{R_2} = 10.46\,W$	$P_T = 38.40\,W$
$P_{R_3} = 4.90\,W$	

The total power can also be determined using Ohm's law:

Given	Solution
$P_T = ?$	$P_T = I_T E_T$
$I_T = 0.80\,A$	$P_T = (0.80)(48)$
$E_T = 48\,V$	$P_T = 38.4\,W$

17-2 QUESTIONS

1. What is the formula for total current in a parallel circuit?
2. What is the formula for total voltage in a parallel circuit?
3. What is the formula for total resistance in a parallel circuit?
4. What is the formula for total power in a parallel circuit?
5. Four resistors, 2200 ohms, 2700 ohms, 3300 ohms, and 5600 ohms, are connected in parallel with a battery rated at 9 volts. Solve for all values of the circuit.

■ **FIGURE 17-5**

Series-parallel circuit.

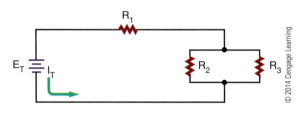

© 2014 Cengage Learning

17-3 SERIES-PARALLEL CIRCUITS

Most circuits consist of both series and parallel circuits. Circuits of this type are referred to as series-parallel circuits (Figure 17-5). The solution of most series-parallel circuits is simply a matter of applying the laws and rules discussed earlier. Series formulas are applied to the series part of the circuit, and parallel formulas are applied to the parallel parts of the circuit.

EXAMPLE: Solve for all unknown quantities in Figure 17-6.

Given			
$R_T = ?$	$I_T = ?$	$E_T = 48\,V$	$P_T = ?$
$R_1 = 820\,\Omega$	$I_{R_1} = ?$	$E_{R_1} = ?$	$P_{R_1} = ?$
$R_2 = 330\,\Omega$	$I_{R_2} = ?$	$E_{R_2} = ?$	$P_{R_2} = ?$
$R_3 = 680\,\Omega$	$I_{R_3} = ?$	$E_{R_3} = ?$	$P_{R_3} = ?$
$R_4 = 470\,\Omega$	$I_{R_4} = ?$	$E_{R_4} = ?$	$P_{R_4} = ?$
$R_5 = 120\,\Omega$	$I_{R_5} = ?$	$E_{R_5} = ?$	$P_{R_5} = ?$
$R_6 = 560\,\Omega$	$I_{R_6} = ?$	$E_{R_6} = ?$	$P_{R_6} = ?$

To solve for total resistance (R_T), first find the equivalent resistance (R_A) for parallel resistors R_2 and R_3. Then solve for the equivalent resistance of resistors R_A and R_4 (identified as R_{S1}) and R_5 and R_6 (identified as R_{S2}). Then the equivalent resistance (R_{BTS}) can be determined for R_{S1} and R_{S2}. Finally, find the total series resistance for R_1 and R_B.

Given	Solution
$R_A = ?$	$\dfrac{1}{R_A} = \dfrac{1}{R_2} + \dfrac{1}{R_3}$
$R_2 = 330\,\Omega$	
$R_3 = 680\,\Omega$	$\dfrac{1}{R_A} = \dfrac{1}{330} + \dfrac{1}{680}$
	$R_A = 222.18\,\Omega$

Redraw the circuit, substituting resistor R_A for resistors R_2 and R_3. See Figure 17-7.

■ FIGURE 17-6

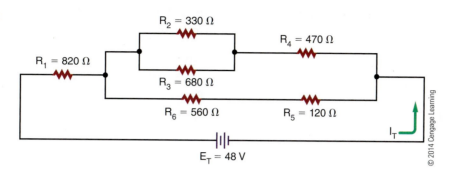

■ FIGURE 17-7

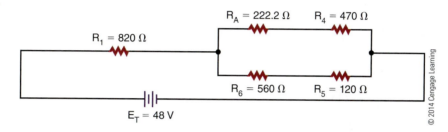

Now determine series resistance R_{S_1} for resistors R_A and R_4.

Given	Solution
$R_{S_1} = ?$	$R_{S_1} = R_A + R_4$
$R_A = 222.18\ \Omega$	$R_{S_1} = 222.18 + 470$
$R_4 = 470\ \Omega$	$R_{S_1} = 692.18\ \Omega$

Determine series resistance R_{S_2} for resistors R_5 and R_6.

Given	Solution
$R_{S_2} = ?$	$R_{S_2} = R_5 + R_6$
$R_5 = 120\ \Omega$	$R_{S_2} = 120 + 560$
$R_6 = 560\ \Omega$	$R_{S_2} = 680\ \Omega$

Redraw the circuit with resistors R_{S_1} and R_{S_2}. See Figure 17-8.

Now determine the parallel resistance (R_B) for resistors R_{S_1} and R_{S_2}.

Given	Solution
$R_B = ?$	$\dfrac{1}{R_B} = \dfrac{1}{R_{S_1}} + \dfrac{1}{R_{S_2}}$
$R_{S_1} = 692.18\ \Omega$	
$R_{S_2} = 680\ \Omega$	$\dfrac{1}{R_B} = \dfrac{1}{692.18} + \dfrac{1}{680}$
	$R_B = 343.02\ \Omega$

Redraw the circuit-using resistor R_B. See Figure 17-9.

Now determine the total resistance in the circuit.

Given	Solution
$R_T = ?$	$R_T = R_1 + R_B$
$R_1 = 820\ \Omega$	$R_T = 820 + 343.02$
$R_B = 343.02\ \Omega$	$R_T = 1163.02\ \Omega$

■ FIGURE 17-8

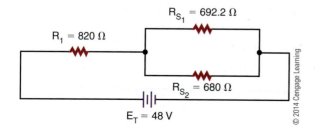

■ FIGURE 17-9

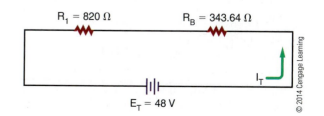

The total current in the circuit can now be determined using Ohm's law.

Given

$I_T = ?$

$E_T = 48\,V$

$R_T = 1163.02\,\Omega$

Solution

$$I_T = \frac{E_T}{R_T}$$

$$I_T = \frac{48}{1163.02}$$

$$I_T = 0.0413\,A \text{ or } 41.3\,mA$$

The voltage drop across resistor R_1 can now be determined.

Given

$I_{R_1} = 0.0413\,A$

$E_{R_1} = ?$

$R_1 = 820\,\Omega$

Solution

$$I_{R_1} = \frac{E_{R_1}}{R_1}$$

$$0.0413 = \frac{E_{R_1}}{820}$$

$$E_{R_1} = 33.87\,V$$

The voltage drop across equivalent resistance R_B:

Given

$I_{R_B} = 0.0413\,A$

$E_{R_B} = ?$

$R_B = 343.02\,\Omega$

Solution

$$I_{R_B} = \frac{E_{R_B}}{R_B}$$

$$0.0413 = \frac{E_{R_B}}{343.02}$$

$$E_{R_B} = 14.167\,V$$

The current through each branch in a parallel circuit has to be calculated separately.

The current through branch resistance R_{S_1}:

Given

$I_{R_{S_1}} = ?$

$E_{R_{S_1}} = 14.166\,V$

$R_{S_1} = 692.18\,\Omega$

Solution

$$I_{R_{S_1}} = \frac{E_{R_{S_1}}}{R_{S_1}}$$

$$I_{R_{S_1}} = \frac{14.167}{692.18}$$

$$I_{R_{S_1}} = 0.0205\,A$$

The current through branch resistance R_{S_1}:

Given

$I_{R_{S_2}} = ?$

$E_{R_{S_2}} = 14.167\,V$

$R_{S_2} = 680\,\Omega$

Solution

$$I_{R_{S_2}} = \frac{E_{R_{S_2}}}{R_{S_2}}$$

$$I_{R_{S_2}} = \frac{14.158}{680}$$

$$I_{R_{S_2}} = 0.0208\,A$$

The voltage drop across resistors R_A and R_4 can now be determined.

Given

$I_{R_A} = 0.0205\,A$

$E_{R_A} = ?$

$R_A = 222.18\,\Omega$

Solution

$$I_{R_A} = \frac{E_{R_A}}{R_A}$$

$$0.0205 = \frac{E_{R_A}}{222.18}$$

$$E_{R_A} = 4.55\,V$$

Given

$I_{R_4} = 0.0205\,A$

$E_{R_4} = ?$

$R_4 = 470\,\Omega$

Solution

$$I_{R_4} = \frac{E_{R_4}}{R_4}$$

$$0.0205 = \frac{E_{R_4}}{470}$$

$$E_{R_4} = 9.64\,V$$

The voltage drop across resistors R_5 and R_6:

Given

$I_{R_5} = 0.0208\,A$

$E_{R_5} = ?$

$R_5 = 120\,\Omega$

Solution

$$I_{R_5} = \frac{E_{R_5}}{R_5}$$

$$0.0208 = \frac{E_{R_5}}{120}$$

$$E_{R_5} = 2.50\,V$$

Given

$I_{R_6} = 0.0208\,A$

$E_{R_6} = ?$

$R_6 = 560\,\Omega$

Solution

$$I_{R_6} = \frac{E_{R_6}}{R_6}$$

$$0.0208 = \frac{E_{R_6}}{560}$$

$$E_{R_6} = 11.65\,V$$

The current through equivalent resistance R_A splits through parallel branches with resistors R_2 and R_3. The current through each resistor has to be calculated separately.

The current through resistor R_2:

Given

$I_{R_2} = ?$

$E_{R_2} = 4.55\,V$

$R_2 = 330\,\Omega$

Solution

$$I_{R_2} = \frac{E_{R_2}}{R_2}$$

$$I_{R_2} = \frac{4.56}{330}$$

$$I_{R_2} = 0.0138\,A$$

Given

$I_{R_3} = ?$

$E_{R_3} = 4.55\,V$

$R_3 = 680\,\Omega$

Solution

$$I_{R_3} = \frac{E_{R_3}}{R_3}$$

$$I_{R_3} = \frac{4.56}{680}$$

$$I_{R_3} = 0.0067\,A$$

Now the power dissipation through each resistor can be determined.

The power consumed by resistor R_1:

Given

$P_{R_1} = ?$

$I_{R_1} = 0.0413\,A$

$E_{R_1} = 33.87\,V$

Solution

$P_{R_1} = I_{R_1}E_{R_1}$

$P_{R_1} = (0.0412)(33.87)$

$P_{R_1} = 1.40\,W$

The power consumed by resistor R_2:

Given

$P_{R_2} = ?$

$I_{R_2} = 0.0138\,A$

$E_{R_2} = 4.55\,V$

Solution

$P_{R_2} = I_{R_2}E_{R_2}$

$P_{R_2} = (0.00138)(4.56)$

$P_{R_2} = 0.063\,W,\ or\ 63\,mW$

The power consumed by resistor R_3:

Given

$P_{R_3} = ?$

$I_{R_3} = 0.0067\,A$

$E_{R_3} = 4.55\,V$

Solution

$P_{R_3} = I_{R_3}E_{R_3}$

$P_{R_3} = (0.0067)(4.55)$

$P_{R_3} = 0.030\,W,\ or\ 30\,mW$

The power consumed by resistor R_4:

Given

$P_{R_4} = ?$

$I_{R_4} = 0.0205\,A$

$E_{R_4} = 9.64\,V$

Solution

$P_{R_4} = I_{R_4}E_{R_4}$

$P_{R_4} = (0.0205)(9.64)$

$P_{R_4} = 0.20\,W,\ or\ 200\,mW$

The power consumed by resistor R_5:

Given

$P_{R_5} = ?$

$I_{R_5} = 0.0208\,A$

$E_{R_5} = 2.50\,V$

Solution

$P_{R_5} = I_{R_5}E_{R_5}$

$P_{R_5} = (0.0208)(2.50)$

$P_{R_5} = 0.052\,W,\ or\ 52\,mW$

The power consumed by resistor R_6:

Given

$P_{R_6} = ?$

$I_{R_6} = 0.0208\,A$

$E_{R_6} = 11.65\,V$

Solution

$P_{R_6} = I_{R_6}E_{R_6}$

$P_{R_6} = (0.0208)(11.65)$

$P_{R_6} = 0.24\,W,\ or\ 240\,mW$

The total power consumption of the circuit:

Given

$P_T = ?$

$P_{R_1} = 1.39\,W$

$P_{R_2} = 0.063\,W$

$P_{R_3} = 0.031\,W$

$P_{R_4} = 0.20\,W$

$P_{R_5} = 0.052\,W$

$P_{R_6} = 0.24\,W$

Solution

$P_T = P_{R_1} + P_{R_2} + P_{R_3} + P_{R_4} + P_{R_5} + P_{R_6}$

$P_T = 1.40 + 0.063 + 0.030 + 0.20 + 0.052 + 0.24$

$P_T = 1.99\,W$

The total power consumption could also be determined by the use of the power formula. Rounding may result in differences in answers.

Given

$P_T = ?$

$I_T = 0.0413\,A$

$E_T = 48\,V$

Solution

$P_T = I_T E_T$

$P_T = (0.0413)(48)$

$P_T = 1.98\,W$

17-3 QUESTION

1. Solve for all unknown values in Figure 17-10.

17-4 VOLTAGE DIVIDERS

One of the more common uses of resistors is voltage division, which is used extensively in electronics. Voltage dividers are often used to set a bias or operating

■ **FIGURE 17-10**

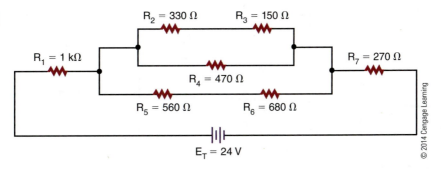

$R_2 = 330\,\Omega$ $R_3 = 150\,\Omega$

$R_1 = 1\,k\Omega$

$R_4 = 470\,\Omega$

$R_7 = 270\,\Omega$

$R_5 = 560\,\Omega$ $R_6 = 680\,\Omega$

$E_T = 24\,V$

point of various active electronic components such as transistors or integrated circuits. A voltage divider is also used to convert a higher voltage to a lower one so that an instrument can read or measure a voltage above its normal range. This often referred to as **scaling**.

In Chapter 13, the concept of connecting resistors in series was presented. Ohm's law (presented in Chapter 14) is one of the most important concepts to understand because power formulas, voltage dividers, current dividers, and so on can all be traced back to and solved by it. Ohm's law states that the current through a circuit is directly proportional to the voltage across the circuit and inversely proportional to the resistance.

$$\text{Current} = \frac{\text{Voltage}}{\text{Resistance}}$$

$$I = \frac{E}{R}$$

That the current is directly proportional to the voltage across the circuit is a very simple yet profound statement. The first part of the statement simply means that as the voltage changes, so does the current, and the change is in the same direction (i.e., if the voltage increases, then the current increases; similarly, if the voltage decreases, then the current decreases). The second part of Ohm's law states that current is inversely proportional to resistance. **Inversely proportional** simply means that when one thing occurs, the opposite happens to the other quantity. In this case, if resistance increases, then current decreases.

Figure 17-11 depicts the simplest of circuits. It has a single voltage source of 10 volts, which is connected to a single 10 kΩ resistor. By Ohm's law, 1 milliampere of current is flowing in this circuit.

$$I = \frac{E}{R}$$

$$I = 10/10000$$

$$I = 0.001 \text{ A or } 1 \text{ mA}$$

In Figure 17-12, the original 10 kΩ resistor in Figure 17-11 is shown divided into two equal resistors, R_1 and R_2. This circuit still has the same total resistance as Figure 17-11 and therefore has the same amount of current flowing through it (i.e., 1 mA). Ohm's law can be used to determine the voltage across each of the resistors.

$$I_{R_1} = \frac{E_{R_1}}{R_1}$$

$$0.001 = \frac{E_{R1}}{5000}$$

$$5 \text{ V} = E_{R_1}$$

Therefore:

$$I_{R_2} = \frac{E_{R_2}}{R_2}$$

$$0.001 = \frac{E_{R2}}{5000}$$

$$5 \text{ V} = E_{R_2}$$

Therefore, the voltage drop across R_1 and R_2 has effectively divided the source voltage of 10 volts in half.

Now, let us take Figure 17-12 another step and divide resistor R_2 into two resistors, R and R_3, which are 4 kΩ and 1 kΩ, as shown in Figure 17-13. As far as the voltage source is concerned, nothing has really changed in the circuit because the total resistance is still 10 kΩ. There is still 1 mA of current flowing in the circuit. The important difference is that now there are multiple voltage drops. The voltage drop at the junction of R_1 and R_2 is still 5 volts.

However, at the junction of R_2 and R_3 to the ground reference point, the voltage drop is 1 volt.

The product of this discussion is depicted in Figure 17-14. It represents a 10-volt voltage source connected to a 10 kΩ potentiometer, or pot. As the wiper moves up and down, the overall value of the pot does not change as in the earlier examples. However, the value of the resistance between the wiper and the top

■ **FIGURE 17-11**

Simple circuit.

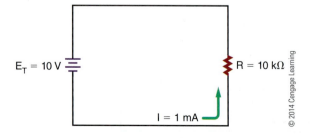

■ **FIGURE 17-12**

Simple voltage divider.

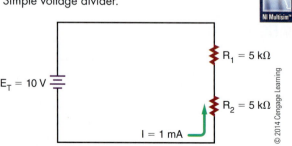

FIGURE 17-13

Voltage divider.

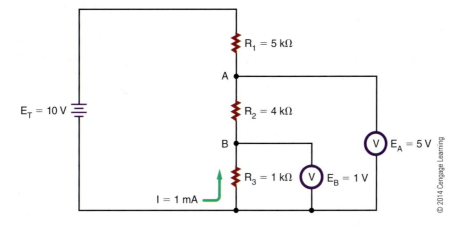

FIGURE 17-14

Voltage divider application.

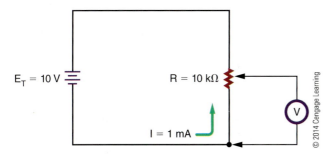

of the pot does, thus effectively creating a two-resistor divider of any value where the resistors can be between zero and the value of the pot. This is a common control used to set voltages, audio levels (volume control), and so on. A word of warning: This voltage divider action

is predictable only when these simple circuits are connected to resistances greater than they are. For example, if the 1-volt drop in Figure 17-13 were supplying a load less than approximately 10 kΩ, then, because it is in parallel with R_3, the load would reduce its value, thus changing the voltage divider action (Figure 17-15).

What may not be obvious is that voltage dividers work on a ratio principle. Look at Figure 17-12. Resistor R_2 is 5 kΩ, which just happens to be half (or 50%) of the total network. It also happens to have half (or 50%) of the total supplied voltage dropped across it. Inspection of Figure 17-13 depicts a similar event occurring. Resistor R3 is 1 kΩ, which just happens to be one-tenth (or 10%) of the total network resistance. In this case, the voltage dropped across it is 10% of the supplied voltage, or 1 volt. The voltage R_2 remains unchanged because the sum of the resistances R_2 and R_3 make up R_2 voltage, and their combined value is still 5 kΩ.

FIGURE 17-15

Voltage divider with load.

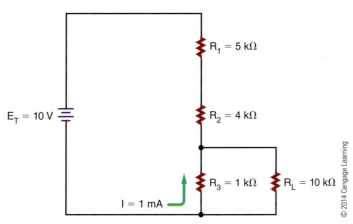

A generalization can be stated for voltage dividers of this nature. The voltage drop is equal to the percentage of the dropping resistor to the sum of the dropping network.

$$E_{Drop} = \frac{E_{source} \times R_{Drop}}{R_{Total}}$$

The next example (Figure 17-16) is a more practical example of a real-world voltage divider. The voltage source is the standard automotive accessory power source of 14.8 volts. What is desired is a voltage-divider network that will divide this 14.8 volt source into a lower-voltage source specified by the manufacturer of a tape or CD player. The first voltage desired is 3 volts, and the load for this voltage will draw 150 mA. The second voltage required is 6 volts, and this voltage's load will draw 400 mA. The final voltage required is 9 volts, and its load draws 600 mA. Reviewing Figure 17-16 indicates that both series and parallel networks are involved. It is necessary to know the load currents for each voltage in advance.

At first glance, the design of this circuit may seem very complex. In reality, it is a simple process. Refer to Figure 17-17. Selecting the amount of current that is desired to flow through resistor R_4 starts the process. In this example, an arbitrary value of 10 mA was chosen.

Next, inspect junction A between resistors R_3 and R_4. At this point, there is current summing or addition. There is 10 mA of current flowing through R_4, and it is joined or summed with 150 mA of current through the 3-volt load, R_{L3}. This means that a total of 160 mA of current is flowing through resistor R_3.

At junction B between resistors R_2 and R_3, there is another summation process. The 400 mA of current flowing through the 6-volt load now joins the 160 mA of current through R_3. The total amount of current flowing through resistor R_2 is this summation, or 560 mA.

Junction C has 560 mA of current entering it through the divider chain, and it is summed with the current from the 9-volt load, which is 600 mA. This summation becomes the total current flowing through the final resistor R_1 to be a value of 1160 mA, or 1.16 A.

The final step in the process is calculating the values of R_1, R_2, R_3, and R_4. These calculations involve nothing more than applying Ohm's law. Again, start at the bottom of the network with resistor R_4. At this point in the circuit, it is determined that there is only the previously selected 10 mA of current flowing. The desired voltage at junction A is 3 volts. Therefore, the following relationships exist:

$$I_{R_4} = \frac{E_{R_4}}{R_4}$$

$$0.01 = \frac{3}{R_4}$$

$$300\,\Omega = R_4$$

The calculation of R_3 is only slightly more difficult. From the previous calculations, the total current flowing through the resistor was determined to be 160 mA. Remember that voltage dividers are ratios, or the relation of one potential point to another. The desired voltage junction B is 6 volts. However, this voltage is not used in the calculation. Junction B is above junction A,

■ **FIGURE 17-16**

Real-world voltage divider.

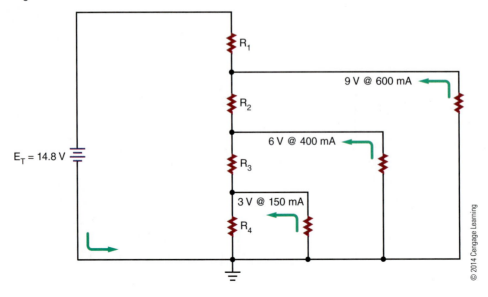

FIGURE 17-17

Analyzing the circuit.

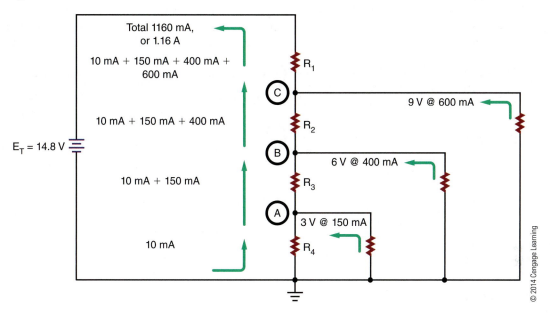

which has been established at 3 volts. Therefore, junction B is 6 V – 3 V, or 3 V greater than junction A. This value is used in the calculation of resistor R_3.

$$I_{R_3} = \frac{E_{R_3}}{R_3}$$

$$0.160 = \frac{3}{R_3}$$

$$18.75\,\Omega = R_3$$

The process is repeated in the calculation of R_2. The current flowing through R_2 is 560 mA. The desired voltage at junction A is 9 volts. Junction A is 9 V to 6 V, or 3 volts more than at junction B.

$$I_{R_2} = \frac{E_{R_2}}{R_2}$$

$$0.560 = \frac{3}{R_2}$$

$$5.36\,\Omega = R_2$$

The final calculation to determine the value of resistor R_1 is somewhat different. In this case, the total current flowing through the circuits is 1.16 A. In this example, the junction A desired voltage is 9 volts. The source voltage is 14.8 volts; therefore, resistor R_1 must drop 14.8 V to 9 V, or drop 5.8 volts.

$$I_{R_1} = \frac{E_{R_1}}{R_1}$$

$$1.16 = \frac{5.8}{R_1}$$

$$5\,\Omega = R_1$$

As in the real world, the resistor values required in this circuit are not exact values but are rounded (Figure 17-18).

NOTE The values of resistors indicated by a computer simulation will not precisely match calculated values. In the real-world the resistors' values are not carried past 1 ohm.

A real-world ammeter has some internal resistance associated with it, and a real-world voltmeter has some current flowing through it. Both of these devices cause the displayed values to be slightly different from those calculated.

Computer simulations use calculated values, and this presents a common problem in electronics. Components are manufactured in standard values, and usually only these values are available when a circuit needs to be constructed.

The only way to obtain a resistor such as resistor R_2 (5.36 Ω) or R_3 (18.75 Ω) is to spend a large amount of money for a custom-made resistor. The student should realize the goal is almost always to design and build practical and economical equipment. The use of precision components often increases costs with little or no benefit.

■ FIGURE 17-18

Determining the resistor values.

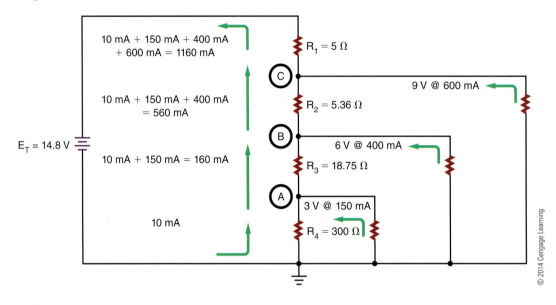

17–4	QUESTIONS

1. What function does a voltage divider perform?
2. What is a common function for voltage dividers?
3. What principle do voltage dividers work on?
4. What generalization can be stated for voltage dividers?
5. Design a voltage divider to use a 14.8 V voltage source to control a 3 V load at 300 mA.

17–5	WHEATSTONE BRIDGE

A **Wheatstone bridge** is a measuring instrument that was invented by Samuel Hunter Christie in 1833. In 1843, a talented and versatile scientist named Sir Charles Wheatstone improved on the original design of the Wheatstone bridge. He did not claim to have invented the circuit named after him, but he was the first to use the circuit effectively for making resistance measurements.

The Wheatstone bridge consists essentially of two voltage dividers (Figure 17-19). It is used to measure an unknown electrical resistance by balancing one leg that includes the unknown component against another leg of known components, with one being variable.

■ FIGURE 17-19

A Wheatstone bridge consists of two voltage dividers.

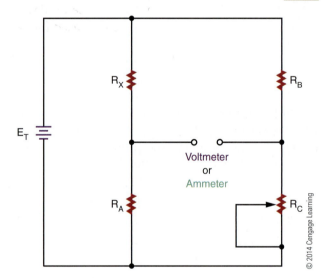

In the circuit, R_X is the unknown resistor value. R_C is a potentiometer and is adjusted until the voltage from the second voltage divider is equal to voltage from the voltage divider containing R_X. When the voltage values are equal, the bridge balanced. The balance point can be detected by connecting either a voltmeter or an ammeter across the output terminals. Both meters will give a zero reading when a balance is achieved.

In a balanced circuit, the ratio R_X/R_A is equal to the ratio R_B/R_C.

$$\frac{R_X}{R_A} = \frac{R_B}{R_C}$$

$\therefore$

$$R_X = \frac{R_A \times R_B}{R_C}$$

In other words, if the values of R_A, R_B, and R_C are known, it is easy to calculate R_X. R_C would have to be measured with a meter to determine its value after adjusting for a zero reading. In an actual Wheatstone bridge test instrument, R_A and R_B are fixed, R_C is adjusted, and R_X can be easily read off a sliding scale connected to R_C. The circuit for a Wheatstone bridge as designed by Wheatstone is shown in Figure 17-20. This is how the circuit is drawn for applications using it.

Variations on the Wheatstone bridge can be used to measure capacitance, inductance, and impedance. Wheatstone bridge circuits are rarely used today to measure resistance values. They are now used for designing sensing circuits such as **strain gauges** for transforming a strain applied to a proportional change of resistance, widely used in industry; **variometers** to detect changes in air pressure for alerting glider pilots to updrafts or thermals so height can be gained allowing for longer flights; and **explosimeters** for sampling the amount of combustible gases in a space.

■ **FIGURE 17-20**

Wheatstone bridge schematic.

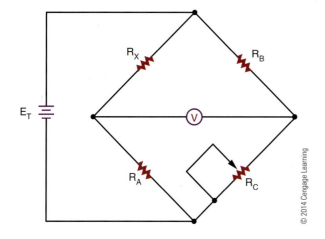

© 2014 Cengage Learning

17-5 QUESTIONS

1. Who invented the Wheatstone bridge?
2. What is the Wheatstone bridge circuit constructed of?
3. When the meter reads 0, what does this mean?
4. What is the value of R_X when R_A and R_B are 10 kΩ each and R_C is 96,432 Ω?
5. What are other quantities that can be measured using the Wheatstone bridge concept?
6. What applications are used today for the Wheatstone bridge?

SUMMARY

- A series circuit provides only one path for current flow.
- Formulas governing the operation of a series circuit include these:

$$I_T = I_{R_1} = I_{R_2} = I_{R_3} \ldots = I_{R_n}$$
$$E_T = E_{R_1} + E_{R_2} + E_{R_3} \ldots + E_{R_n}$$
$$I = E/R$$
$$R_T = R_1 + R_2 + R_3 \ldots + R_n$$
$$P_T = P_{R_1} + P_{R_2} + P_{R_3} \ldots + P_{R_n}$$

- A parallel circuit provides more than one path for current flow.
- Formulas governing the operation of a parallel circuit include the following:

$$I_T = I_{R_1} + I_{R_2} + I_{R_3} \ldots + I_{R_n}$$
$$E_T = E_{R_1} = E_{R_2} = E_{R_3} \ldots = E_{R_n}$$
$$\frac{1}{R_T} = \frac{1}{R_1} + \frac{1}{R_2} + \frac{1}{R_3} \ldots + \frac{1}{R_n}$$
$$I = E/R$$
$$P_T = P_{R_1} + P_{R_2} + P_{R_3} \ldots + P_{R_n}$$

- Series-parallel circuits are solved by using series formulas for the series parts of the circuit and parallel formulas for the parallel parts of the circuit.
- Voltage dividers are used to set the bias or operating point of active electronic components.
- A Wheatstone bridge is used to measure an unknown electrical resistance.

CHAPTER 17 SELF-TEST

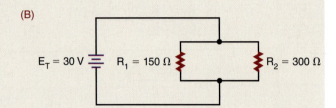

1. Solve for all unknown quantities in the circuits shown.

(A)

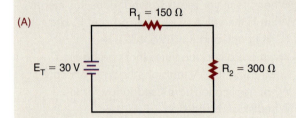

$R_1 = 150 \ \Omega$

$E_T = 30 \ V$

$R_2 = 300 \ \Omega$

(B)

$E_T = 30 \ V$ $R_1 = 150 \ \Omega$ $R_2 = 300 \ \Omega$

(C)

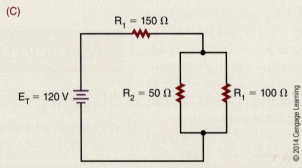

$R_1 = 150 \ \Omega$

$E_T = 120 \ V$ $R_2 = 50 \ \Omega$ $R_1 = 100 \ \Omega$

© 2014 Cengage Learning

2. Design a voltage divider that will provide 1.5 V at 250 mA, 6 V at 750 mA, and 9 V at 500 mA from 12 V.

3. Draw and label all values for the voltage divider in problem 2.

4. Design a voltage divider to produce 9 V at 150 mA from 13.8 volts used in an automobile.

5. Draw and label a Wheatstone bridge and solve for R_X when R_A and R_B are 1 kΩ each and R_C is 7.59 kΩ.

Magnetism

OBJECTIVES

After completing this chapter, the student will be able to:

- Identify three types of magnets.
- Describe the basic shapes of magnets.
- Describe the differences between permanent magnets and temporary magnets.
- Describe how the earth functions as a magnet.
- State the laws of magnetism.
- Explain magnetism based on the theory of atoms and electron spin.
- Explain magnetism based on the domain theory.
- Identify flux lines and their significance.
- Define permeability.
- Describe the magnetic effects of current flowing through a conductor.
- Describe the principle of an electromagnet.
- Describe how to determine the polarity of an electromagnet using the left-hand rule.
- Define magnetic induction.
- Define retentivity and residual magnetism.
- Define a magnetic shield.
- Describe how magnetism is used to generate electricity.
- State the basic law of electromagnetism.
- Describe how the left-hand rule for generators can be used to determine the polarity of induced voltage.
- Describe how AC and DC generators convert mechanical energy into electrical energy.
- Describe how a relay operates as an electromechanical switch.
- Discuss the similarities between a doorbell and a relay.
- Discuss the similarities between a solenoid and a relay.
- Describe how a magnetic phonograph cartridge works.
- Describe how a loudspeaker operates.
- Describe how information can be stored and retrieved using magnetic recording.
- Describe how a DC motor operates.

KEY TERMS

18-1	AC generator
18-1	artificial magnet
18-1	domains
18-1	electromagnet
18-1	ferromagnetic materials
18-1	flux lines
18-1	magnet
18-1	magnetic field
18-1	magnetism
18-1	permanent magnet
18-1	permeability
18-1	temporary magnets
18-2	left-hand rule for coils
18-2	left-hand rule for conductors
18-3	electromagnetic induction
18-3	Faraday's law
18-3	left-hand rule for generators
18-3	magnetic induction
18-3	magnetic shield
18-3	residual magnetism
18-3	retentivity
18-4	alternator
18-4	armature
18-4	commutator
18-4	DC generator
18-4	field
18-4	frequency
18-4	loudspeaker
18-4	relay
18-4	right-hand motor rule
18-4	solenoid

Electricity and magnetism are inseparable. To understand electricity means to understand the relationship that exists between magnetism and electricity.

Electric current always produces some form of magnetism, and magnetism is the most common method for generating electricity. In addition, electricity behaves in specific ways under the influence of magnetism.

This chapter looks at magnetism, electromagnetism, and the relationship between magnetism and electricity.

18-1 MAGNETIC FIELDS

The word **magnet** is derived from magnetite, the name of a mineral found in Magnesia, a part of Asia Minor. This mineral is a natural magnet. Another type of magnet is the **artificial magnet**. This magnet is created by rubbing a piece of soft iron with a piece of magnetite. A third type of magnet is the **electromagnet** created by current flowing through a coil of wire.

Magnets come in various shapes (Figure 18-1). Among the more common shapes are the horseshoe, the bar or rectangle, and the ring.

Magnets that retain their magnetic properties are called **permanent magnets**. Magnets that retain only a small portion of their magnetic properties are called **temporary magnets**.

Magnets are made of metallic or ceramic materials. Alnico (*al*uminum, *ni*ckel, and *co*balt) and Cunife

(*copper* [*Cu*], *ni*ckel, and Iron [*Fe*]) are two metallic alloys used for magnets.

The earth itself is a huge magnet (Figure 18-2). The earth's magnetic north and south poles are situated close to the geographic North and South Poles. If a bar magnet is suspended, the magnet aligns in a north–south direction, with one end pointing toward the North Pole of the earth and the other toward the South Pole. This is the principle behind the compass. It is also the reason the two ends of the magnet are called the north pole and the south pole.

Magnets align in a north–south direction because of laws similar to those of positive and negative charges: Unlike magnetic poles attract each other and like magnetic poles repel each other. The color code for a magnet is red for the north pole and blue for the south pole.

Magnetism, the property of the magnet, can be traced to the atom. As electrons orbit around the nucleus of the atom, they spin on their axis, like the earth as it orbits the sun. This moving electrostatic charge produces a magnetic field. The direction of the magnetic field depends on the electron's direction of spin. Iron, nickel, and cobalt are the only naturally magnetic elements. Each of these materials has two valence

■ FIGURE 18-1

Magnets come in various sizes and shapes.

■ FIGURE 18-2

The earth's magnetic north and south poles are situated close to the geographic North and South Poles.

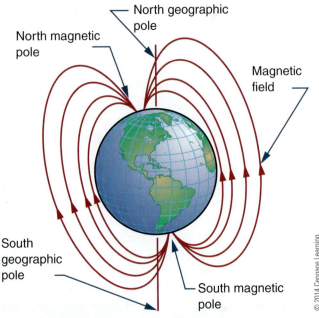

North geographic pole

North magnetic pole

Magnetic field

South geographic pole

South magnetic pole

■ FIGURE 18-3

The domains in unmagnetized material are randomly arranged with no overall magnetic effect.

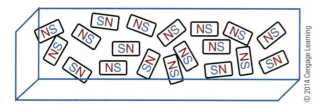

© 2014 Cengage Learning

■ FIGURE 18-4

When material becomes magnetized, all the domains align in a common direction.

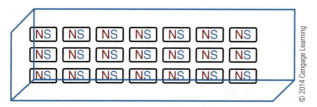

© 2014 Cengage Learning

■ FIGURE 18-5

To prevent loss of magnetism, (A) bar magnets are stacked on top of each other, and (B) keeper bars are placed across horseshoe magnets.

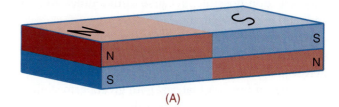

(A)

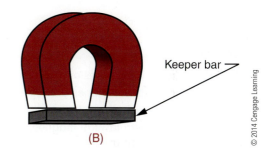

Keeper bar

(B)

© 2014 Cengage Learning

■ FIGURE 18-6

Magnetic lines of flux can be seen in patterns of iron filings.

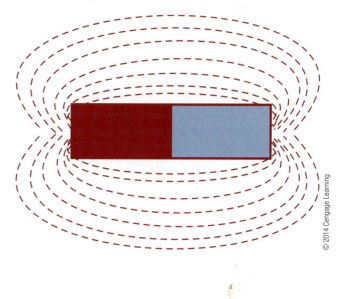

© 2014 Cengage Learning

electrons that spin in the same direction. Electrons in other materials tend to spin in opposite directions and this cancels their magnetic characteristics.

Ferromagnetic materials are materials that respond to magnetic fields. In ferromagnetic materials, the atoms combine into **domains**, or groups of atoms arranged in the form of magnets. In an unmagnetized material, the magnetic domains are randomly arranged, and the net magnetic effect is zero (Figure 18-3). When the material becomes magnetized, the domains align in a common direction and the material becomes a magnet (Figure 18-4). If the magnetized material is divided into smaller pieces, each piece becomes a magnet with its own poles.

Evidence of this "domain theory" is that if a magnet is heated or hit repeatedly with a hammer, it loses its magnetism (the domains are jarred back into a random arrangement). Also, if an artificial magnet is left by itself, it slowly loses its magnetism. To prevent this loss, bar magnets should be stacked on top of each other with opposite poles together; keeper bars should be placed across horseshoe magnets (Figure 18-5). Both methods maintain the magnetic field.

A **magnetic field** consists of invisible lines of force that surround a magnet. These lines of force are called **flux lines**. They can be "seen" by placing a sheet of paper over a magnet and sprinkling iron filings on the paper. When the paper is lightly tapped, the iron filings arrange themselves into a definite pattern that reflects the forces attracting them (Figure 18-6).

Flux lines have several important characteristics: They have polarity from north to south. They always form a complete loop. They do not cross each other because like polarities repel. They tend to form the smallest loop possible, because unlike poles attract and tend to pull together.

The characteristic that determines whether a substance is ferromagnetic is called **permeability**. Permeability is the ability of a material to accept magnetic lines of force. A material with great permeability offers less resistance to flux lines than air.

18–1 QUESTIONS

1. What are the three types of magnets?
2. What are the basic shapes of magnets?
3. How are the ends of a magnet identified?
4. What are the two laws of magnetism?
5. What are flux lines?

18–2 ELECTRICITY AND MAGNETISM

When current flows through a wire, it generates a magnetic field around the wire (Figure 18-7). This can be shown by placing a compass next to a wire that has no current flowing through it. The compass aligns itself with the earth's magnetic field. When current passes through the wire, however, the compass needle is deflected and aligns itself with the magnetic field generated by the current. The north pole of the compass indicates the direction of the flux lines. The direction of the flux lines can also be determined if the direction of the current flow is known, using the **left-hand rule for conductors**. If the wire is grasped with the left hand, with the thumb pointing in the direction of current flow, the fingers point in the direction of the flux lines (Figure 18-8). When the polarity of the voltage source is reversed, the flux lines are also reversed.

If two wires are placed next to each other with current flowing in opposite directions, they create opposing magnetic fields that repel each other (Figure 18-9). If the two wires carry current flowing in the same direction, the fields combine (Figure 18-10).

■ **FIGURE 18-7**

A current flowing through a conductor creates a magnetic field around the conductor.

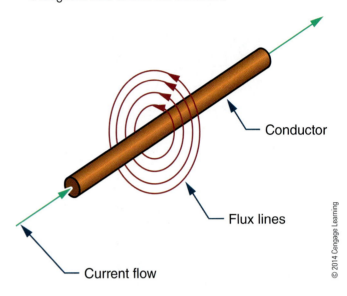

Conductor

Flux lines

Current flow

© 2014 Cengage Learning

A single piece of wire produces a magnetic field with no north or south pole and little strength or practical value. If the wire is twisted into a loop, three things occur: (1) The flux lines are brought together. (2) The flux lines are concentrated at the center of the loop. (3) North and south poles are established. This is the principle of the electromagnet.

An electromagnet is composed of many turns of wire close together. This allows the flux lines to add together when a current flows through the wire. The more turns of wire, the more flux lines are added together. Also, the greater the current, the greater the number of flux lines generated. The strength of the magnetic field, then, is directly proportional to the number of turns in the coil and the amount of current flowing through it.

A third method of increasing the strength of the magnetic field is to insert a ferromagnetic core into the center of the coil. An iron core is typically used because it has a higher permeability (can support more flux lines) than air.

Use the **left-hand rule for coils** to determine the polarity of the electromagnet; grasp the coil with the left hand, with the fingers pointing in the direction of the current flow. The thumb then points in the direction of the north pole (Figure 18-11).

■ FIGURE 18-8

Determining the direction of the flux lines around a conductor when the direction of the current flow is known (left-hand rule for conductors).

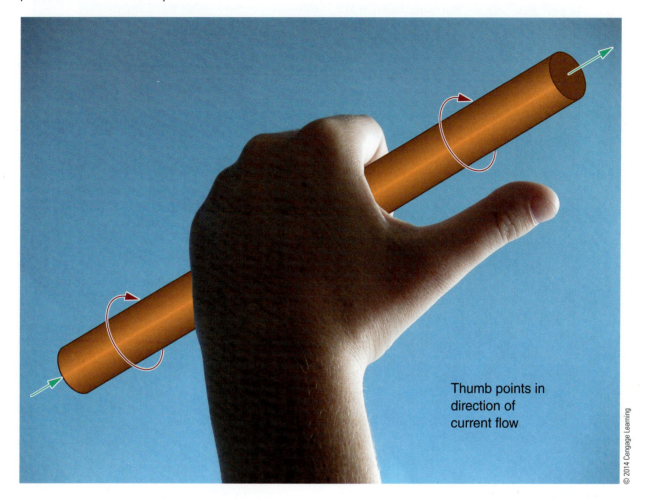

Thumb points in direction of current flow

© 2014 Cengage Learning

■ FIGURE 18-9

When current flows in opposite directions through two conductors placed next to each other, the resulting magnetic fields repel each other.

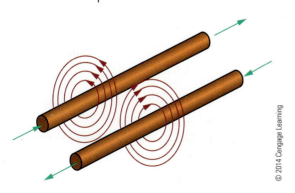

© 2014 Cengage Learning

■ FIGURE 18-10

When current flows in the same direction through two conductors placed next to each other, the magnetic fields combine.

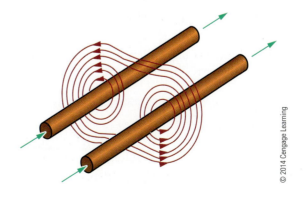

© 2014 Cengage Learning

■ FIGURE 18-11

Determining the polarity of an electromagnet (left-hand rule for coils).

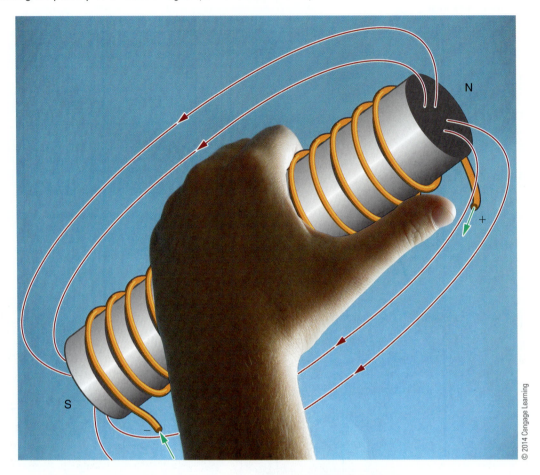

18-2 QUESTIONS

1. How can a magnetic field be shown to exist when a current flows through a wire?
2. How can the direction of flux lines around a wire be determined?
3. What happens when two current-carrying wires are placed next to each other with the current flowing in:
   a. The same direction?
   b. Opposite directions?
4. What are three ways to increase the strength of an electromagnetic field?
5. How can the polarity of an electromagnet be determined?

18-3 MAGNETIC INDUCTION

Magnetic induction is the effect a magnet has on an object without physical contact. For example, a magnet may induce a magnetic field in an iron bar (Figure 18-12). In passing through the iron bar, the magnetic lines of force cause the domains in the iron bar to align in one direction. The iron bar is now a magnet. The domains in the iron bar align themselves with their south pole toward the north pole of the magnet, because opposite poles attract. For the same reason, the iron bar is drawn toward the magnet. The flux lines now leave from the end of the iron bar; the iron bar is an extension of the magnet. This is an effective way to increase the length or change the shape of a magnet without physically altering it.

■ FIGURE 18-12

Placing an iron bar in a magnetic field extends the magnetic field and magnetizes the iron bar.

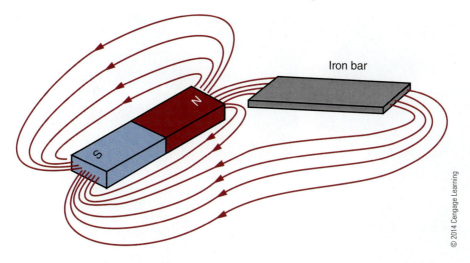

Iron bar

© 2014 Cengage Learning

When the magnet and iron bar are separated, the domains in the iron bar return to their random pattern, although a few domains remain in north–south alignment, giving the iron bar a weak magnetic field. This remaining magnetic field is called **residual magnetism**. The ability of a material to retain its magnetic field after the magnetizing force is removed is called **retentivity**. Soft iron has low retentivity. Alnico, an alloy made of aluminum, nickel, and cobalt, has high retentivity.

Inserting a low reluctance material in front of the magnetic field can bend flux lines. Low-reluctance materials are called **magnetic shields**. An example is the material called Mu-metal. The magnetic shield is placed around the item to be protected. Electronic equipment, especially oscilloscopes, requires shielding from magnetic flux lines.

Electromagnetic induction is the principle behind the generation of electricity: When a conductor passes or is passed by a magnetic field, a current is produced (induced) in the conductor. As the conductor passes through the magnetic field, free electrons are forced to one end of the conductor, leaving a deficiency of electrons at the other end. This results in a difference of potential between the ends of the conductor. This difference of potential exists only when the conductor is passing through a magnetic field. When the conductor is removed from the magnetic field, the free electrons return to their parent atoms.

For electromagnetic induction to occur, either the conductor must move or the magnetic field must move. The voltage produced in the conductor is called *induced voltage*. The strength of the magnetic field, the speed with which the conductor moves through the magnetic field, the angle at which the conductor cuts the magnetic field, and the length of the conductor determine the amount of induced voltage.

The stronger the magnetic field, the greater the induced voltage. The faster the conductor moves through the field, the greater the induced voltage. Moving the conductor, the magnetic field, or both can produce motion between the conductor and the magnetic field. The maximum voltage is induced when the conductor moves at right angles to the field. Angles less than 90° induce less voltage. If a conductor is moved parallel to the flux lines, no voltage is induced. The longer the conductor is moving through the magnetic field, the greater the induced voltage.

Faraday's law, the basic law of electromagnetism, states: The induced voltage in a conductor is directly proportional to the rate at which the conductor cuts the magnetic lines of force.

The polarity of an induced voltage can be determined by the following **left-hand rule for generators**: The thumb, index finger, and middle finger are held at right angles to each other (Figure 18-13). The thumb points in the same direction as the motion of the conductor, the index finger points in the direction of the flux lines (north to south), and the middle finger points toward the negative end of the conductor, the direction of the current flow.

■ **FIGURE 18-13**

The left-hand rule for generators can be used to determine the direction of the induced current flow in a generator.

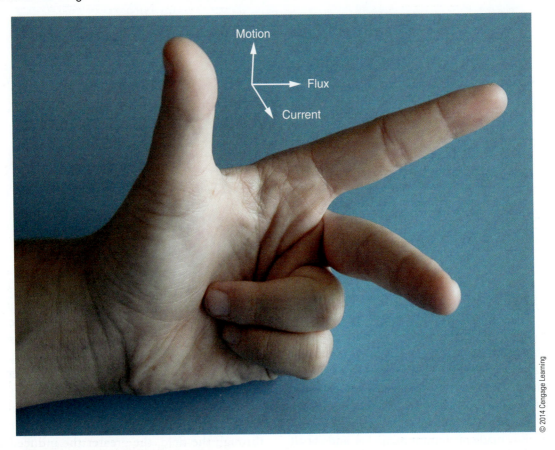

Motion

Flux

Current

© 2014 Cengage Learning

| 18-3 | **QUESTIONS** |

1. How can the length of a magnet be increased without physically altering the magnet?
2. What is residual magnetism?
3. How does a magnetic shield work?
4. How is electromagnetic induction used to generate electricity?
5. What does Faraday's law state?

| 18-4 | **MAGNETIC AND ELECTROMAGNETIC APPLICATIONS** |

An alternating current (AC) generator converts mechanical energy to electrical energy by utilizing the principle of electromagnetic induction. The mechanical energy is needed to produce motion between the magnetic field and the conductor.

Figure 18-14 shows a loop of wire (conductor) being rotated (moved) in a magnetic field. The loop has a light and dark side for ease of explanation. At the point shown in part A, the dark half is parallel to the lines of force, as is the light half. No voltage is induced. As the loop is rotated toward the position shown in part B, the lines of force are cut and a voltage is induced. The induced voltage is greatest at this position, when the motion is at right angles to the magnetic field. As the loop is rotated to position C, fewer lines of force are cut and the induced voltage decreases from the maximum value to zero volts. At this point, the loop has rotated 180°, or half a circle.

The direction of current flow can be determined by applying the left-hand rule for generators. The arrow shows the direction of current flow at position B. When the loop is rotated to position D, the action reverses. As the dark half moves up through the magnetic lines of force and the light half moves down, applying the left-hand rule for generators shows that the induced voltage changes polarities. The voltage reaches a peak

■ **FIGURE 18-14**
Induced voltage in an AC generator.

■ **FIGURE 18-15**
Induced voltage in a DC generator.

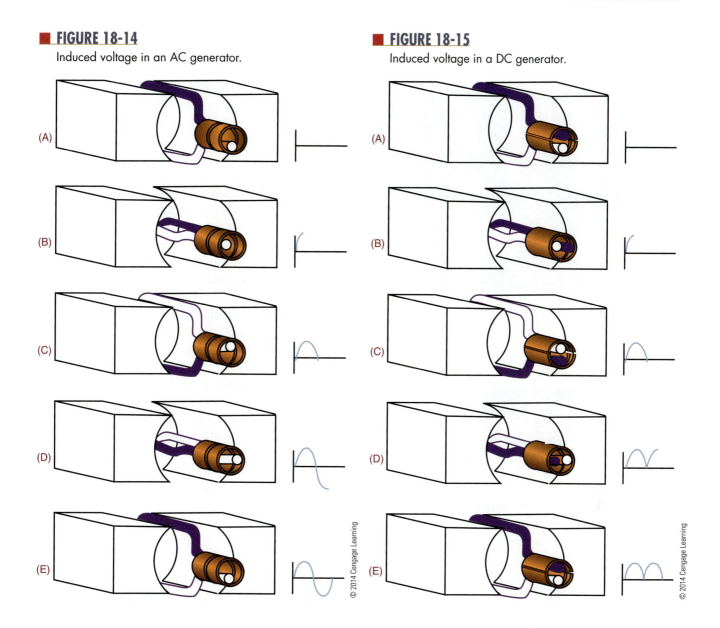

© 2014 Cengage Learning

at position D and decreases until the loop reaches the original position. The induced voltage has completed one cycle of two alternations.

The rotating loop is called an **armature**, and the source of the electromagnetic field is called the **field**. The armature can have any number of loops. The term *armature* refers to the part that rotates in the magnetic field, regardless of whether it consists of one loop or multiple loops. The **frequency** of the alternating current or voltage is the number of complete cycles completed per second. The speed of rotation determines the frequency. An **AC generator** is often called an **alternator** because it produces alternating current.

A DC (direct current) generator also converts mechanical energy into electrical energy. It functions like an AC generator with the exception that it converts the AC voltage to DC voltage. It does this with a device called a **commutator**, as shown in Figure 18-15. The output is taken from the commutator, a split ring. When the loop is rotated from position A to position B, a voltage is induced. The induced voltage is greatest as the motion is at right angles to the magnetic field. As the loop is rotated to position C, the induced voltage decreases from the maximum value to zero. As the loop continues to rotate to position D, a voltage is induced, but the commutator reverses the output polarity so that it remains the same as before.

■ FIGURE 18-16

Examples of various types of relays.

© 2014 Cengage Learning

■ FIGURE 18-17

Examples of types of solenoids.

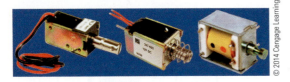

© 2014 Cengage Learning

The loop then returns to the original position, E. The voltage generated from the commutator pulsates, but in one direction only, varying twice during each revolution between zero and maximum.

A **relay** is an electromechanical switch that opens and closes with an electromagnetic coil (Figure 18-16). As a current flows through the coil, it generates a magnetic field that attracts the plunger, pulling it down. As the plunger is pulled down, it closes the switch contacts. When the current through the coil stops, a spring pulls the armature back to its original position and opens the switch.

A relay is used where it is desirable to have one circuit control another circuit. It electrically isolates the two circuits. A small voltage or current can control a large voltage or current. A relay can also be used to control several circuits some distance away.

A doorbell is an application of the relay. The striker to ring the bell is attached to the plunger. As the doorbell is pressed, the relay coil is energized, pulling down the plunger and striking the bell. As the plunger moves down, it opens the circuit, de-energizing the relay. The plunger is pulled back by the spring closing the switch

contacts, energizing the circuit again, and the cycle repeats until the button is released.

A **solenoid** is similar to a relay (Figure 18-17). A coil, when energized, pulls a plunger that does some mechanical work. This is used on some door chimes where the plunger strikes a metal bar. It is also used on automotive starters. The plunger pulls the starter gear in to engage the flywheel to start the engine.

Phonograph pickups use the electromagnetic principle. A magnetic field is produced by a permanent magnet that is attached to a stylus (needle). The permanent magnet is placed inside a small coil. As the stylus is tracked through the groove of a record, it moves up and down and from side to side in response to the audio signal recorded. The movement of the magnet in the coil induces a small voltage that varies at the audio signal response. The induced voltage is then amplified and used to drive a loudspeaker, reproducing the audio signal.

Loudspeakers are used for all types of audio amplification. Most speakers today are constructed of a moving coil around a permanent magnet. The magnet produces a stationary magnetic field. As current is passed through the coil, it produces a magnetic field that varies at the rate of the audio signal. The varying magnetic field of the coil is attracted and repelled by the magnetic field of the permanent magnet. The coil is attached to a cone that moves back and forth in response to the audio signal. The cone moving back and forth moves the air, reproducing the audio sound.

Magnetic recording is used for cassette recorders, video recorders, reel-to-reel recorders, floppy disk drives, and hard disk drives. All these devices use the same electromagnetic principle to store information. A signal is stored on the tape or disk with a record head, to be read back later with a playback head. In some equipment, the record and playback head are combined in one package or they may be the same head. The record and playback head are a coil of wire with a ferromagnetic core. In a tiny gap between the ends of the core is a magnetic field. As the storage medium, a piece of material covered with iron oxide, is pulled across the record head, the magnetic field penetrates the tape, magnetizing it. The information is written on it in a magnetic pattern that corresponds to the original information. To play back or read the information, the medium is moved past the gap in the playback head. The changing magnetic field induces a small voltage into the coil winding. When this voltage is amplified, the original information is reproduced.

The operation of a DC motor (Figure 18-18) depends on the principle that a current-carrying conductor, placed in and at right angles to a magnetic field, tends to move at right angles to the direction of the field. Figure 18-18A shows a magnetic field extending between a north pole and a south pole. Figure 18-18B shows the magnetic field that exists around a current-carrying conductor. The plus sign implies that the current flows inward. The direction of the flux lines can be determined using the left-hand rule. Figure 18-18C shows the conductor placed in the magnetic field. Note that both fields become distorted. Above the wire, the field is weakened, and the conductor tries to move upward. The amount of force exerted upward depends on the strength of the field between the poles and the amount of current flowing through the conductor. If the current through the conductor is reversed (Figure 18-18D), the direction of the magnetic flux around the conductor is reversed. If this occurs, the field below the conductor is weakened and the conductor tends to move downward.

A method of determining the direction of a current-carrying conductor in a magnetic field uses the **right-hand motor rule**: When the thumb, index finger, and middle finger of the right hand are extended at right angles to each other, the middle finger points in the direction of current flow in the conductor; the index finger indicates the magnetic field from the north pole to the south pole; the thumb points in the direction of the motion of the conductor.

The force acting on a current-carrying conductor in a magnetic field depends on the strength of the magnetic field, the length of the conductor, and the amount of current flowing in the conductor. If a loop of wire, free to rotate horizontally, is placed between the two poles of a magnet, the loop spins as the poles repel each other. The current flows in one direction on one side of the loop and in the other direction on the other side of the loop. One side of the loop moves down and the other side of the loop moves upward. The loop rotates in a counterclockwise direction around its axis. A commutator reverses the direction of current flow in the loop every time it reaches the top or zero torque position. This is how a DC motor rotates. The loop or armature rotates in a magnetic field. Permanent magnets

■ **FIGURE 18-18**

Operation of a DC motor.

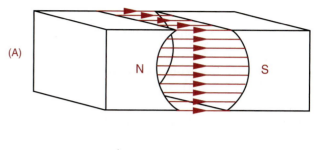
(A)

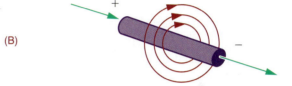
(B)

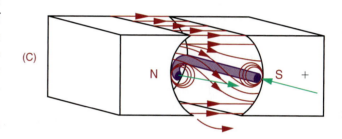
(C)

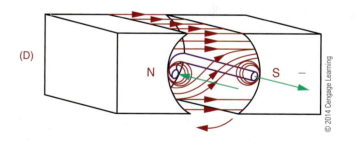
(D)

© 2014 Cengage Learning

or electromagnets may produce this field. The commutator reverses the direction of current through the armature. Note the resemblance between a DC motor and a **DC generator**.

The basic meter movement uses the principle of the DC motor. It consists of a stationary permanent magnet and a moveable coil. When the current flows through the coil, the resulting magnetic field reacts with the field of the permanent magnet and causes the coil to rotate. The greater the current flow through the coil, the stronger the magnetic field produced. The stronger the magnetic field produced, the greater the rotation of the coil. To determine the amount of current flow, a pointer is attached to the rotating coil. As the coil turns, the pointer also turns. The pointer moves across a graduated scale and indicates the amount of current flow. This type of meter movement is used for analog meters such as voltmeters, ammeters, and ohmmeters.

A conductor carrying a current can be deflected (moved) by a magnetic field. It is not the conductor that is deflected, but the electrons traveling in the conductor. The electrons are restricted to the conductor, so the conductor also moves. Electrons can travel through other media. In the case of television picture tubes, electrons travel through a vacuum to strike a phosphor screen where they emit light. The electrons are produced by an electron gun. By varying the electron beam over the surface of the picture screen, a picture can be created. To move the beam back and forth across the screen, two magnetic fields deflect the beam. One magnetic field moves the beam up and down the screen, and the other magnetic field moves the beam from side to side. This method is used in television, radar, oscilloscopes, computer terminals, and other applications where a picture is desired on a screen.

18–4 QUESTIONS

1. What are the differences between an AC and a DC generator?
2. Why are relays important?
3. How does a loudspeaker produce sound?
4. What is the principle behind DC motors and meter movements?
5. How does an electromagnetic field produce an image on a screen?

SUMMARY

- The word *magnet* is derived from the name of magnetite, a mineral that is a natural magnet.
- Rubbing a piece of soft iron with another magnet can create a magnet.
- Current flowing in a coil of wire creates an electromagnet.
- Horseshoe, bar, rectangular, and ring are the most common shapes of magnets.
- Unlike poles attract and like poles repel.
- One theory of magnetism is based on the spin of electrons as they orbit around an atom.
- Another theory of magnetism is based on the alignment of domains.
- Flux lines are invisible lines of force surrounding a magnet.
- Flux lines form the smallest loop possible.
- Permeability is the ability of a material to accept magnetic lines of force.
- A magnetic field surrounds a wire when current flows through it.
- The direction of the flux lines around a wire can be determined by grasping the wire with the left hand, with the thumb pointing in the direction of current flow. The fingers then point in the direction of the flux lines.
- If two current-carrying wires are placed next to each other, with current flowing in the same direction, their magnetic fields combine.
- The strength of an electromagnet is directly proportional to the number of turns in the coil and the amount of current flowing through the coil.
- The polarity of an electromagnet is determined by grasping the coil with the left hand with the fingers in the direction of current flow. The thumb then points toward the north pole.
- Retentivity is the ability of a material to retain a magnetic field.
- Electromagnetic induction occurs when a conductor passes through a magnetic field.
- Faraday's law: Induced voltage is directly proportional to the rate at which the conductor cuts the magnetic lines of force.
- The left-hand rule for generators can be used to determine the direction of induced voltage.
- AC and DC generators convert mechanical energy into electrical energy.
- A relay is an electromechanical switch.
- Electromagnetic principles are applied in the design and manufacture of doorbells, solenoids, phonograph pickups, loudspeakers and magnetic recordings.
- DC motors and meters use the same principles.
- Electron beams can be deflected by an electromagnetic field to produce images on television, radar and oscilloscope screens.

CHAPTER 18 SELF-TEST

1. How can the domain theory of magnetism be verified?

2. What three methods can be used to increase the strength of an electromagnet?

3. Explain the left-hand rule for conductors.

4. Explain how a DC generator operates through one cycle.

5. Show in a drawing how an electromagnet works.

6. Explain how to determine the polarity of an electromagnet.

7. Explain the left-hand rule for generators.

8. Draw and identify the parts of a DC generator.

9. Describe how a DC motor operates.

10. What other devices can use a magnetic field to operate besides DC motors and generators?

Inductance

OBJECTIVES

After completing this chapter, the student will be able to:

- Explain the principles of inductance.
- Identify the basic units of inductance.
- Identify different types of inductors.
- Determine the total inductance in series and parallel circuits.
- Explain L/R time constants and how they relate to inductance.

KEY TERMS

19-1 henry (H)

19-1 inductance (L)

19-1 inductor

19-1 Lenz's law

19-2 air-core inductor

19-2 ferrite-core inductor

19-2 fixed inductor

19-2 laminated iron-core inductor

19-2 shielded inductor

19-2 toroid-core inductor

19-2 variable inductor

19-3 L/R time constant

When a current flows through a conductor, a magnetic field builds up around the conductor. This field contains energy and is the foundation for inductance.

This chapter examines inductance and its application to DC circuits. The effects of inductance on a circuit are discussed in Chapter 25.

19-1 PRINCIPLES OF INDUCTANCE

Inductance is the characteristic of an electrical conductor that opposes a change in current flow. The symbol for inductance is **L**. An **inductor** is a device that stores energy in a magnetic field.

Inductance exhibits the same effect on current in an electric circuit as inertia does on velocity of a mechanical object. It takes more work to start a load moving than it does to keep it moving because the load possesses the property of inertia. Inertia is the characteristic of mass that opposes a change in velocity. Once current is moving through a conductor, inductance helps to keep it moving. The effects of inductance are sometimes desirable and other times undesirable.

As noted in Chapter 18, the basic principle behind inductance states, *when a current flows through a conductor, it generates a magnetic field around the conductor.* As the magnetic flux lines build up, they create an opposition to the flow of current.

When the current changes direction or stops, or the magnetic field changes, an electromotive force (emf) is induced back into the conductor through the collapsing of the magnetic field. The opposition to the changes in current flow is identified as a counter electromotive force (counter emf). **Lenz's law** summarizes this effect—an induced emf in any circuit is always in a direction to oppose the effect that produced it. The amount of counter emf is in proportion to the rate of change: The faster the rate of change, the greater the counter emf.

All conductors have some inductance. The amount of inductance depends on the conductor and the shape of it. Straight wire has small amounts of inductance, whereas coils of wire have much more inductance.

The unit by which inductance is measured is the **henry (H)**, named for Joseph Henry (1797–1878), an American physicist. A henry is the amount of inductance required to induce an emf of 1 volt when the current in a conductor changes at the rate of 1 ampere per second. The henry is a large unit; the millihenry (mH) and microhenry (μH) are more commonly used.

19-2 INDUCTORS

Inductors are devices designed to have a specific inductance. They consist of a conductor coiled around a core and are classified by the type of core material—magnetic or nonmagnetic. Figure 19-1 shows the symbol used for inductors.

Inductors can also be **fixed** or variable. Figure 19-2 shows the symbol for a variable inductor. **Variable inductors** are created with adjustable core material. Figure 19-3 shows several types of inductors used for adjusting the core material. Maximum inductance occurs when the core material is in line with the coil of the wire.

Air-core inductors, or inductors without core material, are used for up to 5 millihenries of inductance. They are wrapped on a ceramic or phenolic core (Figure 19-4).

■ FIGURE 19-1

Schematic symbol for an inductor.
© 2014 Cengage Learning

■ FIGURE 19-2

Schematic symbol for a variable inductor.
© 2014 Cengage Learning

■ FIGURE 19-3

Various types of fixed inductors (A), variable inductors (B).

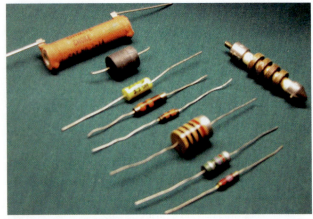

(A)

(B)

© 2014 Cengage Learning

■ FIGURE 19-4

Air-core inductors.

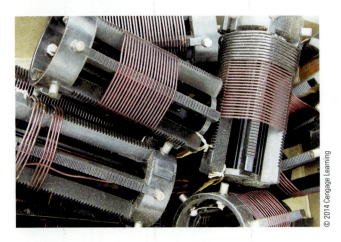

© 2014 Cengage Learning

■ FIGURE 19-5

Schematic symbol for an iron-core inductor.

© 2014 Cengage Learning

■ FIGURE 19-6

Toroid-core inductor.

© 2014 Cengage Learning

Ferrite-core inductors and powdered iron cores inductors are used for up to 200 millihenries. The symbol used for an iron-core inductor is shown in Figure 19-5. An iron-core inductor is also referred to as a choke. Chokes are used in DC and low-frequency AC applications.

Toroid-core inductors are donut-shaped and offer a high inductance for a small size (Figure 19-6). The magnetic field is contained within the core.

Shielded inductors have a shield made of magnetic material to protect them from the influence of other magnetic fields (Figure 19-7).

Laminated iron-core inductors are used for all large inductors (Figure 19-8). These inductors vary from 0.1 to 100 henries, the inductance depending on the amount of current flowing through the inductor. These inductors are sometimes referred to as chokes. They are used in the filtering circuits of power supplies to remove AC components from the DC output. They will be discussed further later on.

Inductors typically have tolerances of ±10%, but tolerances of less than 1% are available. Inductors, like resistors, can be connected in series, parallel, or series-parallel combinations. The total inductance of several inductors connected in series (separated to prevent

FIGURE 19-7

Shielded inductor.

FIGURE 19-8

Laminated iron-core inductor.

their magnetic fields from interacting) is equal to the sum of the individual inductances.

$$L_T = L_1 + L_2 + L_3 \ldots + L_n$$

If two or more inductors are connected in parallel (with no interaction of their magnetic fields), the total inductance is found by using the formula:

$$\frac{1}{L_T} = \frac{1}{L_1} + \frac{1}{L_2} + \frac{1}{L_3} \ldots + \frac{1}{L_n}$$

19-2 QUESTIONS

1. What are inductors?
2. Draw the symbols used to represent fixed and variable inductors.
3. What is another name for a laminated iron-core inductor?
4. What are the formulas for determining total inductance in
   a. Series circuits?
   b. Parallel circuits?
5. What is the total inductance of a circuit with three inductors, 10 H, 3.5 H, and 6 H, connected in parallel?

19-3 L/R TIME CONSTANTS

An **L/R time constant** is the time required for current through a conductor to increase to 63.2% or decrease to 36.8% of the maximum current. An RL circuit is shown in Figure 19-9. L/R is the symbol used

FIGURE 19-9

Circuit used to determine L/R time constant.

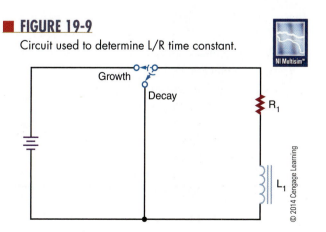

■ **FIGURE 19-10**

Time constants required to build up or collapse the magnetic field in an inductor.

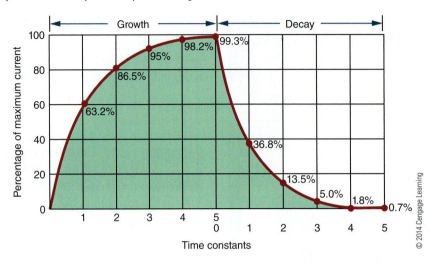

for the time constant of an RL circuit. This can be expressed as follows:

$$t = \frac{L}{R}$$

where: t = time in seconds
        L = inductance in henries
        R = resistance in ohms

Figure 19-10 charts the growth and decay (or increase and decrease) of a magnetic field, in terms of time constants. It takes five time constants to fully transfer all the energy into the magnetic field, or to build up the maximum magnetic field. It takes five full-time constants to completely collapse the magnetic field.

19-3   QUESTIONS

1. What is a time constant for an inductor?
2. How is a time constant determined?
3. How many time constants are required to fully build up a magnetic field for an inductor?
4. How many time constants are required to fully collapse a magnetic field for an inductor?
5. How long does it take to fully build up a magnetic field for a 0.1-henry inductor in series with a 100,000-ohm resistor?

SUMMARY

- Inductance is the ability to store energy in a magnetic field.
- The unit for measuring inductance is the henry (H).
- The letter L represents inductance.
- Inductors are devices designed to have specific inductances.
- The symbol for fixed inductance is

© 2014 Cengage Learning

- The symbol for a variable inductor is

© 2014 Cengage Learning

- Types of inductors include air core, ferrite or powdered iron core, toroid core, shielded, and laminated iron core.
- The total inductance for inductors connected in series is calculated by the formula:

$$L_T = L_1 + L_2 + L_3 \ldots + L_n$$

- The total inductance for inductors connected in parallel is

$$\frac{1}{L_T} = \frac{1}{L_1} + \frac{1}{L_2} + \frac{1}{L_3} \ldots + \frac{1}{L_n}$$

- A time constant is the time required for current to increase to 63.2% or decrease to 36.8% of the maximum current.

• A time constant can be determined by the formula:

$$t = \frac{L}{R}$$

• It takes five time constants to fully build up or collapse the magnetic field of an inductor.

CHAPTER 19 SELF-TEST

1. Explain Lenz's law as it relates to emf.
2. What is the relationship between a conductor and counter emf?
3. Identify types and shapes of inductors.
4. How can the magnetic field be increased for a particular inductance?
5. What is the total inductance for the circuit shown?

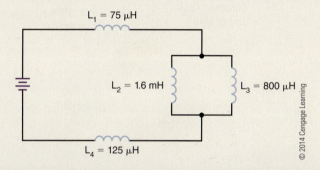

L₁ = 75 µH
L₂ = 1.6 mH
L₃ = 800 µH
L₄ = 125 µH

© 2014 Cengage Learning

6. A 500-mH inductor and a 10-kilohm resistor are connected in series to a 25-volt source. What will be the voltage across the inductor 100 microseconds after energizing the circuit?
7. Make a chart for the growth and decay time constants for the following components:
   a. 1H, 100 Ω
   b. 100 mH, 10 kΩ
   c. 10 mH, 1 kΩ
   d. 10 H, 10 Ω
   e. 1000 mH, 1 kΩ

Capacitance

OBJECTIVES

After completing this chapter, the student will be able to:

- Explain the principles of capacitance.
- Identify the basic units of capacitance.
- Identify different types of capacitors.
- Determine total capacitance in series and parallel circuits.
- Explain RC time constants and how they relate to capacitance.

KEY TERMS

20-1 capacitance (C)
20-1 capacitor
20-1 dielectric
20-1 farad (F)
20-1 microfarad (μF)
20-1 plates
20-1 picofarad (pF)
20-2 ceramic disk capacitor

20-2 dielectric constant
20-2 electrolytic capacitor
20-2 fixed capacitor
20-2 paper capacitor
20-3 RC time constant
20-2 variable capacitor

Capacitance allows for the storage of energy in an electrostatic field. Capacitance is present whenever two conductors are separated by an insulator.

This chapter focuses on capacitance and its application to DC circuits. The effects of capacitance in a circuit are discussed in Chapter 25.

20-1 PRINCIPLES OF CAPACITANCE

Capacitance is the ability of a device to store electrical energy in an electrostatic field. The symbol for capacitance is **C**. A capacitor is a device that possesses a specific amount of capacitance. A **capacitor** is made of two conductors separated by an insulator (Figure 20-1). The conductors are called **plates** and the insulator is called a **dielectric**. Figure 20-2 shows the symbols for capacitors.

When a DC voltage source is connected to a capacitor, a current flows until the capacitor is charged. The capacitor is charged with an excess of electrons on one plate (negative charge) and a deficiency of electrons on the other plate (positive charge). The dielectric prevents electrons from moving between the plates. Once the capacitor is charged, all current stops. The capacitor's voltage is equal to the voltage of the voltage source.

A charged capacitor can be removed from the voltage source and used as an energy source. However, as energy is removed from the capacitor, the voltage diminishes rapidly. In a DC circuit, a capacitor acts as an open circuit after its initial charge. An open circuit is a circuit with an infinite resistance.

CAUTION!

Because a capacitor can be removed from a voltage source and holds the potential of the voltage source indefinitely, treat all capacitors as though they were charged. Never touch both leads of a capacitor with your hand until you have discharged it by shorting both leads together or using a shorting stick. A capacitor in a circuit can hold a potential indefinitely if it does not have a discharge path.

The amount of energy stored in a capacitor is in proportion to the size of the capacitor. The capacitors used in classrooms are usually small and deliver only a small shock if discharged through the body. If a capacitor is large and charged with a high voltage, however, the shock can be fatal. Charged capacitors should be treated like any other voltage source. Use a commercial shorting stick or a screwdriver with an insulated handle and a lead attached to discharge capacitors.

The basic unit of capacitance is the **farad (F)**. A farad is the amount of capacitance that can store 1 coulomb (C) of charge when the capacitor is charged to 1 volt. The farad is too large a unit for ordinary use, so the **microfarad (μF)** and **picofarad (pF)** are used. The letter C stands for capacitance.

$$1 \text{ microfarad} = 0.000,001 \text{ or } \frac{1}{1,000,000} \text{ farad}$$

$$1 \text{ picofarad} = 0.000,000,000,001 \text{ or}$$

$$\frac{1}{1,000,000,000,000} \text{ farad}$$

■ FIGURE 20-1

A capacitor consists of two plates (conductors) separated by a dielectric (insulator or nonconductor).

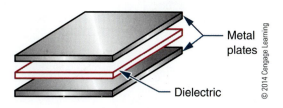

Metal plates

Dielectric

© 2014 Cengage Learning

■ FIGURE 20-2

Schematic symbols for capacitors.

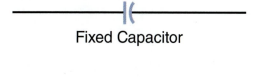

Fixed Capacitor

Variable Capacitor

© 2014 Cengage Learning

20-1 QUESTIONS

1. What is capacitance?
2. Draw the symbol for a fixed and a variable capacitor.
3. What precautions should be observed when handling capacitors?
4. What is the basic unit for measuring capacitance?
5. What units are normally associated with capacitors?

20-2 CAPACITORS

Four factors affect the capacitance of a capacitor:

1. Area of the plate
2. Distance between the plates
3. Type of dielectric material
4. Temperature

A capacitor is either fixed or variable. A **fixed capacitor** has a definite value that cannot be changed. A **variable capacitor** is one whose capacitance can be changed either by varying the space between plates (trimmer capacitor) or by varying the amount of meshing between two sets of plates (tuning capacitor).

Capacitance is directly proportional to the area of the plate. For example, doubling the plate area doubles the capacitance if all other factors remain the same.

Capacitance is inversely proportional to the distance between the plates. In other words, as the plates move apart, the strength of the electric field between the plates decreases.

The ability of a capacitor to store electrical energy depends on the electrostatic field between the plates and the distortion of the orbits of the electrons in the dielectric material. The degree of distortion depends on the nature of dielectric material and is indicated by its dielectric constant. A **dielectric constant** is a measure of the effectiveness of a material as an insulator. The constant compares the material's ability to distort and store energy in an electric field to the ability of air, which has a dielectric constant of 1. Paper has a dielectric constant of 2 to 3; mica has a dielectric constant of 5 to 6; and titanium has a dielectric constant of 90 to 170.

The temperature of a capacitor is the least significant of the four factors. It need not be considered for most general-purpose applications.

Capacitors come in many types and styles to meet the needs of the electronics industry. **Electrolytic capacitors** offer large capacitance for small size and weight (Figure 20-3). Electrolytic capacitors consist of two metal foils separated by fine gauze or other absorbent material that is saturated with a chemical paste called an electrolyte. The electrolyte is a good conductor and serves as part of the negative plate. The dielectric is formed by oxidation of the positive plate. The oxidized layer is thin and a good insulator. An electrolytic capacitor is polarized, having positive and negative leads. Polarity must be observed when connecting an electrolytic capacitor in a circuit.

Paper capacitors and plastic capacitors are constructed by a rolled foil technique (Figure 20-4). A paper dielectric has less resistance than a plastic film

■ FIGURE 20-3

Electrolytic capacitors. (A) Radial capacitor leads (B) Axial capacitor leads.

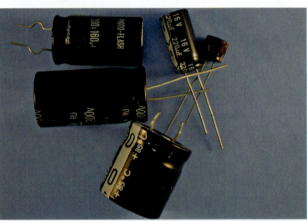

(A)

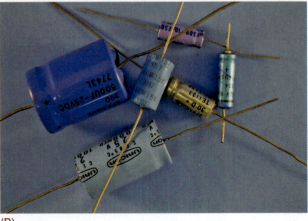

(B)

■ **FIGURE 20-4**

Paper and plastic capacitors.

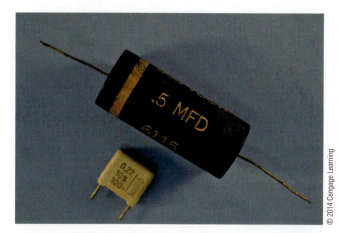

■ **FIGURE 20-5**

Ceramic disk capacitors.

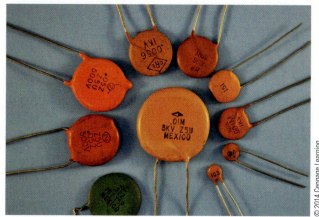

dielectric, and plastic film is now being used more as a result. The plastic film allows a metallized film to be deposited directly on it. This reduces the distance between plates and produces a smaller capacitor.

The **ceramic disk capacitor** is popular because it is inexpensive to produce (Figure 20-5). It is used for capacitors of 0.1 microfarad and smaller. The ceramic material is the dielectric. This is a tough, reliable, general-purpose capacitor.

Variable capacitors also come in all sizes and shapes (Figure 20-6). Types include padders, trimmers, and tuning capacitors. Padder and trimmer capacitors must be adjusted by a technician. The user can adjust tuning capacitors.

Like resistors and inductors, capacitors can be connected in series, parallel, and series-parallel combinations. Placing capacitors in series effectively increases the thickness of the dielectric. This decreases the total capacitance, because capacitance is inversely proportional to the distance between the plates. The total capacitance of capacitors in series is calculated like the total resistance of parallel resistors:

$$\frac{1}{C_T} = \frac{1}{C_1} + \frac{1}{C_2} + \frac{1}{C_3} \ldots + \frac{1}{C_n}$$

When capacitors of different values are connected in series, the smaller capacitors charge up to the highest voltage. Placing capacitors in parallel effectively adds to the plate area. This makes the total capacitance equal to the sum of the individual capacitances:

$$C_T = C_1 + C_2 + C_3 \ldots + C_n$$

Capacitors in parallel all charge to the same voltage.

■ **FIGURE 20-6**

Variable capacitors.

20-2 **QUESTIONS**

1. What four factors affect capacitance?
2. What are the advantages of electrolytic capacitors?
3. What are other names for variable capacitors?
4. What is the formula for total capacitance in series circuits?
5. What is the formula for total capacitance in parallel circuits?

20-3 RC TIME CONSTANTS

An **RC time constant** reflects the relationship among time, resistance, and capacitance. An RC circuit is shown in Figure 20-7. The time it takes a capacitor to charge and discharge is directly proportional to the amount of the resistance and capacitance. The time constant reflects the time required for a capacitor to charge up to 63.2% of the applied voltage or to discharge down to 36.8%. The time constant is expressed as follows:

$$t = RC$$

where: t = time in seconds
         R = resistance in ohms
         C = capacitance in farads

FIGURE 20-7

Circuit used to determine RC time constant.

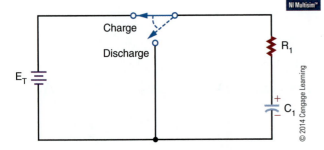

EXAMPLE: What is the time constant of a 1-microfarad capacitor and a 1-megohm resistor?

Given	Solution
C = 1 μF	t = RC
R = 1 MΩ	t = (1,000,000)(0.000001)
t = ?	t = 1 second

The time constant is not the time required to charge or discharge a capacitor fully. Figure 20-8 shows the time constants needed to charge and discharge a capacitor. Note that it takes approximately five time constants to fully charge or discharge a capacitor.

20-3 QUESTIONS

1. What is a time constant for a capacitor?
2. How is the time constant determined for a capacitor?
3. How many time constants are required to fully charge or discharge a capacitor?
4. A 1-microfarad and a 0.1-microfarad capacitor are connected in series. What is the total capacitance for the circuit?
5. A 0.015-microfarad capacitor is charged to 25 volts. What is the voltage 25 milliseconds after placing a 2-megohm resistor across its terminals?

FIGURE 20-8

Chart of time constants required to charge and discharge a capacitor.

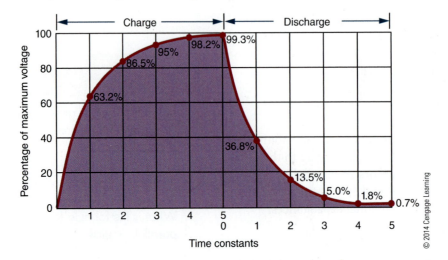

SUMMARY

- Capacitance is the ability to store electrical energy in an electrostatic field.
- A capacitor consists of two conductors separated by an insulator.
- The symbol for a fixed capacitor is:

© 2014 Cengage Learning

- The symbol for a variable capacitor is:

© 2014 Cengage Learning

- The unit of capacitance is the farad (F).
- Because the farad is large, microfarads (μF) and picofarads (pF) are more often used.
- The letter C represents capacitance.
- Capacitance is affected by
  1. Area of the capacitor plates
  2. Distance between the plates
  3. Types of dielectric materials
  4. Temperature
- Capacitor types include electrolytic, paper, plastic, ceramic, and variable.
- The formula for total capacitance in a series circuit is as follows:

$$\frac{1}{C_T} = \frac{1}{C_1} + \frac{1}{C_2} + \frac{1}{C_3} \cdots + \frac{1}{C_n}$$

- The formula for total capacitance in a parallel circuit is as follows:

$$C_T = C_1 + C_2 + C_3 \cdots + C_n$$

- The formula for the RC circuit time constant is as follows:

$$t = RC$$

- It takes five time constants to fully charge and discharge a capacitor.

CHAPTER 20 SELF-TEST

1. Where is the charge stored in a capacitor?
2. Describe the process of charging a capacitor with a DC voltage.
3. What happens to a capacitor when it is charged up and then removed from a circuit?
4. Describe how to discharge a charged capacitor.
5. How do the plate area and distance between plates affect the value of a capacitor?
6. What are the types and styles of capacitors?
7. Four capacitors are connected in series, 1.5 μF, 0.05 μF, 2000 pF, and 25 pF. What is the total capacitance of the circuit?
8. Four capacitors are connected in parallel, 1.5 μF, 0.05 μF, 2000 pF, and 25 pF. What is the total capacitance of the circuit?
9. Using a 100-volt source, create a chart showing the time constants, charge, and discharge voltage for each time constant.
10. Make a chart for the charge and discharge time constants for a 100-μF capacitor with a 10-kΩ resistor.

Section 2: DC Circuits

1. Design a voltage divider using series and parallel circuits. Before building, use Multisim to verify that the circuit works. Assemble the circuit using a printed circuit board, and test and compare results with Multisim

2. Design and build a simple DC motor using magnets and enamel-coated copper wire.

3. Design and build a tester that can measure the voltage of several different types of cells and batteries.

4. Design and build a circuit for determining unknown resistor values.

5. Design and build a circuit that takes advantage of a capacitor charging and/or discharging.

AC CIRCUITS

AC CIRCUITS

Hendrik Wade Bode (1905–1982)

In the 1930s, Bode devised a simple but accurate method for graphing gain and phase–shift plots.

Karl Ferdinand Braun (1850–1918)

In 1897, Braun built the first cathode ray tube (CRT) and cathode-ray tube oscilloscope.

Jacques-Arsène d'Arsonval (1851–1940)

In 1882, d'Arsonval was credited with inventing the moving–coil galvanometer.

Thomas Alva Edison (1847–1931)

In 1882, Edison created the concept and distribution of direct current (DC) electricity with a power station on Manhattan Island, New York and is credited with the first practical electric light bulb.

Donald Macadie

In 1923, Macadie invented the first multimeter, a meter that could measure amperes, volts, and ohms; this instrument was named the Avometer, and that was shortened to AVO.

Nikola Tesla (1856–1943)

Tesla was a major contributor to the birth of commercial alternating current electricity and is best known for his many revolutionary developments in the field of electromagnetism in the late nineteenth and early twentieth centuries.

George Westinghouse, Jr. (1846–1914)

In 1886, Westinghouse introduced and developed the first multi-voltage alternating current (AC) system in Great Barrington, Massachusetts, based on research by Nikola Tesla.

Alternating Current

OBJECTIVES

After completing this chapter, the student will be able to:

- Describe how an AC voltage is produced with an AC generator (alternator).
- Define alternation, cycle, hertz, sine wave, period, and frequency.
- Identify the parts of an AC generator (alternator).
- Define peak, peak-to-peak, effective, and rms.
- Explain the relationship between time and frequency.
- Identify and describe three basic nonsinusoidal waveforms.
- Describe how nonsinusoidal waveforms consist of the fundamental frequency and harmonics.
- Understand why AC is used in today's society.
- Describe how an AC distribution system works.

KEY TERMS

21-1	alternation
21-1	cycle
21-1	hertz (Hz)
21-1	sine wave
21-1	sinusoidal waveform
21-2	amplitude
21-2	degree of rotation
21-2	effective value
21-2	peak-to-peak value
21-2	peak value
21-2	period (t)
21-2	rms value

21-3	even harmonics
21-3	fundamental frequency
21-3	harmonics
21-3	nonsinusoidal waveform
21-3	odd harmonics
21-3	pulse width
21-3	sawtooth waveform
21-3	square waveform
21-3	triangular waveform

There are two types of electricity used to perform work, such as lighting bulbs or running motors. Direct current (DC), as its name implies, is a current that flows in one direction only. Batteries develop this type of energy. The second type of energy used today to perform useful work is alternating current (AC). Unlike its converse, direct current, the direction of alternating current flow reverses with regularity.

Today, alternating current is the most used method to transmit electrical energy from one location to another. This was not the case in the late 1800s. Thomas Edison utilized direct current. His vision of electrified cities incorporated the idea of many small, DC-generating stations scattered about in the community, because there was no way to change or alter direct current to recoup transmission losses. This limited the usefulness of DC to small, local areas and prohibited its usefulness in transmitting electrical energy over long distances.

In 1888, a young engineer discovered rotating magnetic fields and their potential for the practical generation and transmission of electrical energy that made the electrification of the United States and the rest of the world possible. The power lines that crisscross the cities and countryside are a testament to his genius. The father of alternating current and our electrified world was Nikola Tesla.

This chapter examines the reasons alternating current is important, the methods by which it is generated, and its important electrical characteristics.

21-1 GENERATING ALTERNATING CURRENT

An AC generator converts mechanical energy into electrical energy. The AC generator, also called an alternator, is capable of producing an alternating voltage by utilizing the principles of electromagnetic induction. Electromagnetic induction is the process of inducing a voltage in a conductor by passing it through a magnetic field.

As described in Chapter 18, the left-hand rule for generators can be used to determine the direction of current flow in a conductor that is being passed through a magnetic field: When the thumb is pointed in the direction of the conductor movement, the index finger (extended at right angles to the thumb) indicates the direction of the magnetic lines of flux from north to south, and the middle finger (extended at a

right angle to the index finger) indicates the direction of current flow in the conductor. Maximum voltage is induced when the conductor is moved perpendicular to the lines of flux. When the conductor is moved parallel to the lines of flux, no voltage is induced.

Figure 21-1A through 21-1F shows a wire loop being rotated through a magnetic field. In position B, the loop is parallel to the lines of force. As previously stated, no voltage is induced when the conductor is moved parallel to the lines of force. As the loop is rotated to position C, it passes through more lines of force, and the maximum voltage is induced when the loop is at right angles to the lines of force. As the loop continues to rotate to position D, fewer lines of force are cut and the induced voltage decreases. Movement from position B to position D represents a rotation of 180°. Rotating the loop to position E results in a reversal of current flow. Again, the maximum voltage is induced when the loop is at right angles to the lines of force. As the loop returns to its original position at F, the induced voltage returns to 0.

Each time an AC generator moves through one complete revolution, it is said to complete one **cycle**. Its output voltage is referred to as one cycle of output voltage. Similarly, the generator produces one cycle of

■ FIGURE 21-1A

Basic AC generator (alternator).

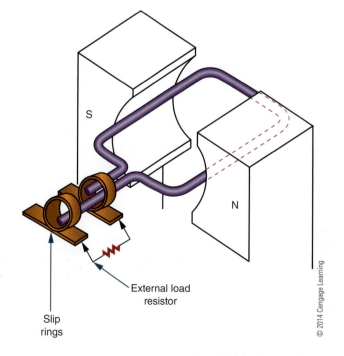

S

N

External load resistor

Slip rings

FIGURE 21-1B–F

AC generator inducing a voltage output.

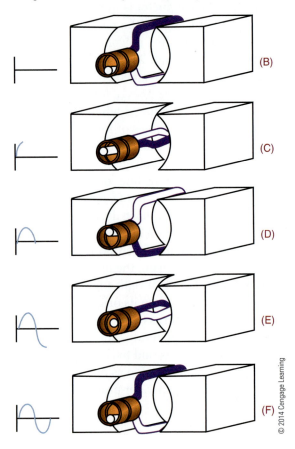

output current in a complete circuit. The two halves of a cycle are referred to as **alternations** (Figure 21-2). Each alternation is produced by a change in the polarity of the voltage. The voltage exhibits one polarity during half of the cycle (one alternation) and the opposite polarity during the other half of the cycle (second alternation). Generally, one alternation is referred to as the

FIGURE 21-2

Each cycle consists of a positive and negative alternation.

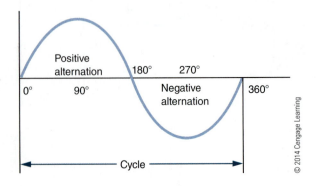

positive alternation and the other as the negative alternation. The positive alternation produces output with a positive polarity; the negative alternation produces output with a negative polarity. Two complete alternations of voltage with no reference to time make up a cycle. One cycle per second is defined as a **hertz (Hz)**.

The rotating loop of wire is referred to as the armature. The AC voltage that is induced in the rotating armature is removed from the ends of the loop by a sliding contact located at each end of the armature (Figure 21-3). Two metal rings, called slip rings, are attached to each end of the loop. Brushes slide against the slip rings to remove the AC voltage. The AC generator described generates a low voltage. To be practical, an AC generator must be made of many loops to increase the induced voltage.

The waveform produced by an AC generator is called a **sinusoidal waveform** or **sine wave** for short (Figure 21-4). The sine wave is the most basic

FIGURE 21-3

Voltage is removed from the armature of an AC generator through slip rings.

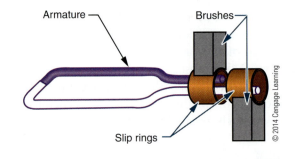

FIGURE 21-4

The sinusoidal waveform, the most basic of the AC waveforms.

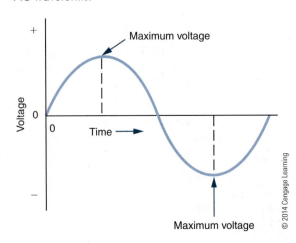

and widely used of all the AC waveforms and can be produced by both mechanical and electronic methods. The sine wave is identical to the trigonometric sine function. The values of a sine wave vary from 0 to maximum in the same manner as the sine value. Both voltage and current exist in sine-wave form.

21-1 QUESTIONS

1. What is the function of an AC generator?
2. Explain how an AC generator operates.
3. Define the following terms:
   a. Cycle
   b. Alternation
   c. Hertz
   d. Sine wave
4. Identify the major parts of an AC generator.
5. What is the difference between a positive and a negative alternation?
6. Would it be possible to design an AC generator whose armature generates the magnetic field and rotates, thus inducing an AC current in stationary coils placed around it?

21-2 AC VALUES

Each point on a sine wave has a pair of numbers associated with it. One value expresses the **degree of rotation** of the waveform. The degree of rotation is the angle to which the armature has turned. The second value indicates its **amplitude**. *Zero* is the maximum departure of the value of an alternating current or wave from the average value. There are several methods of expressing these values.

The **peak value** of a sine wave is the absolute value of the point on the waveform with the greatest amplitude (Figure 21-5). There are two peak values, one for the positive and one for the negative alternation. These peak values are equal.

The **peak-to-peak value** refers to the vertical distance between the two peaks (Figure 21-6). The peak-to-peak value can be determined by adding the absolute values of each peak.

The **effective value** of alternating current is the amount that produces the same degree of heat in a given resistance as an equal amount of direct current.

■ FIGURE 21-5

The peak value of a sine wave is the point on the AC waveform having the greatest amplitude. The peak value occurs during both the positive and the negative alternations of the waveform.

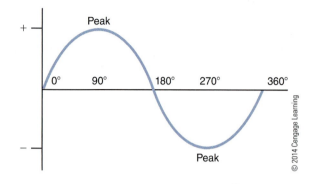

■ FIGURE 21-6

The peak-to-peak value can be determined by adding the peak values of the two alternations.

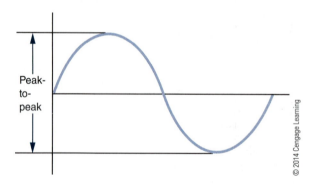

A mathematical process called the root-mean-square (rms) process can be used to determine the effective value . Therefore, the effective value is also referred to as the **rms value**. Using the rms process shows that the effective value of a sine wave is equal to 0.707 times the peak value. When an AC voltage or current is given or measured, it is assumed to be the effective value unless otherwise indicated. Most meters are calibrated to indicate the effective, or rms, value of voltage and current.

$$E_{rms} = 0.707\, E_p$$

where: E_{rms} = rms or effective voltage value
E_p = maximum voltage of one alternation

$$I_{rms} = 0.707\, I_p$$

where: I_{rms} = rms or effective current value
I_p = maximum current of one alternation

EXAMPLE: A current sine wave has a peak value of 10 amperes. What is the effective value?

Given	Solution
$I_{rms} = ?$	$I_{rms} = 0.707I_p$
$I_p = 10\,A$	$I_{rms} = (0.707)(10)$
	$I_{rms} = 7.07\,A$

EXAMPLE: A voltage sine wave has an effective value of 40 volts. What is the peak value of the sine wave?

Given	Solution
$E_p = ?$	$E_{rms} = 0.707\,E_p$
$E_{rms} = 40\,V$	$40 = 0.707\,E_p$
	$56.58\,V = E_p$

The time required to complete one cycle of a sine wave is called the **period (t)**. The period is usually measured in seconds. The letter t is used to represent the period.

The number of cycles that occurs in a specific period of time is called the frequency. The frequency of an AC sine wave is usually expressed in terms of cycles per second. The unit of frequency is the hertz. One hertz equals one cycle per second.

The period of a sine wave is inversely proportional to its frequency. The higher the frequency, the shorter the period. The relationship between the frequency and the period of a sine wave is expressed as

$$f = \frac{1}{t}$$

where: f = frequency
          t = period

EXAMPLE: What is the frequency of a sine wave with a period of 0.05 second?

Given	Solution
$f = ?$	$f = \dfrac{1}{t}$
$t = 0.05\,second$	$f = \dfrac{1}{0.05}$
	$f = 20\,Hz$

EXAMPLE: If a sine wave has a frequency of 60 hertz, what is its period?

Given	Solution
$f = 60\,Hz$	$t = \dfrac{1}{f}$
$t = ?$	$t = \dfrac{1}{60}$
	$t = 0.0167\,second,\ or\ 16.7\,msec$

21–2   QUESTIONS

1. Define the following values:
   a. peak
   b. peak-to-peak
   c. effective
   d. rms
2. A voltage sine wave has a peak value of 125 volts. What is its effective value?
3. What is the relationship between time and frequency?
4. A current sine wave has an effective value of 10 amperes. What is its peak value?
5. What is the period of a 400-hertz sine wave?

21–3   NONSINUSOIDAL WAVEFORMS

The sine wave is the most common AC waveform. However, it is not the only type of waveform used in electronics. Waveforms other than sinusoidal (sine wave) are classified as **nonsinusoidal waveforms**. Specially designed electronic circuits generate nonsinusoidal waveforms.

Figures 21-7 through 21-9 show three of the basic nonsinusoidal waveforms. These waveforms can represent either current or voltage. Figure 21-7 shows a **square wave**, so designated because its positive and negative alternations are square in shape. This indicates that the current or voltage immediately reaches its maximum value and remains there for the duration of the alternation. When the polarity changes, the current or voltage immediately reaches the opposite

■ **FIGURE 21-7**
Square waveform.

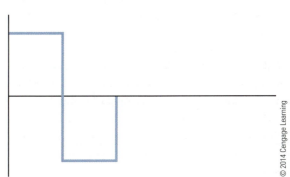

FIGURE 21-8

Triangular waveform.

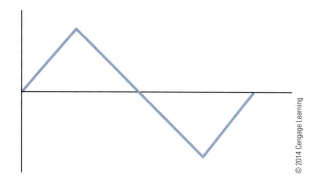

© 2014 Cengage Learning

FIGURE 21-9

Sawtooth waveform.

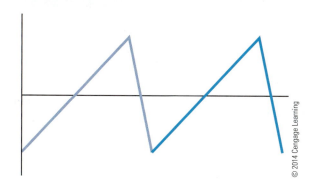

© 2014 Cengage Learning

peak value and remains there for the next duration. The **pulse width** is equal to one-half the period. The pulse width is the duration the voltage is at its maximum or peak value until it drops to its minimum value. The amplitude can be any value. The square wave is useful as an electronic signal because its characteristics can be changed easily.

Figure 21-8 shows a **triangular waveform**. It consists of positive and negative ramps of equal slope. During the positive alternation, the waveform rises at a linear rate from 0 to its peak value and then decreases back to 0. During the negative alternation, the waveform declines in the negative direction at a linear rate to peak value and then returns to 0. The period of this waveform is measured from peak to peak. Triangular waveforms are used primarily as electronic signals.

Figure 21-9 shows a **sawtooth waveform**, which is a special case of the triangular wave. First it rises at a linear rate and then rapidly declines to its negative peak value. The positive ramp is of relatively long duration and has a smaller slope than the short ramp. Sawtooth waves are used to trigger the operations of electronic circuits. In television sets and oscilloscopes, they are used to sweep the electron beam across the screen, creating the image.

Pulse waveforms and other Nonsinusoidal waveforms can be considered in two ways. One method considers the waveform as a momentary change in voltage followed, after a certain time interval, by another momentary change in voltage. The second method considers the waveform as the algebraic sum of many sine waves having different frequencies and amplitudes. This method is useful in the design of amplifiers. If the amplifier cannot pass all the sine-wave frequencies, distortion results.

Nonsinusoidal waves are composed of the fundamental frequency and harmonics. The **fundamental frequency** represents the repetition rate of the waveform. **Harmonics** are higher-frequency sine waves that are exact multiples of the fundamental frequency. **Odd harmonics** are frequencies that are odd multiples of the fundamental frequency. **Even harmonics** are frequencies that are even multiples of the fundamental frequency.

Square waves are composed of the fundamental frequency and all odd harmonics. Triangular waveforms are also composed of the fundamental frequency and all odd harmonics, but unlike the square wave, the odd harmonics are 180° out of phase with the fundamental frequency.

Sawtooth waveforms are composed of both even and odd harmonics. The even harmonics are 180° out of phase with the odd harmonics.

21-3 QUESTIONS

1. What are nonsinusoidal waveforms?
2. Draw two cycles of a
   a. Square wave
   b. Triangular wave
   c. Sawtooth wave
3. What are the two methods that nonsinusoidal waveforms are recognized in waveform development?
4. What are the applications of these nonsinusoidal waveforms?
5. Describe the fundamental frequency and harmonics of three different nonsinusoidal waveforms.

SUMMARY

- AC is the most commonly used type of electricity.
- AC consists of current flowing in one direction and then reversing and flowing in the opposite direction.
- One revolution of an AC generator is called a cycle.
- The two halves of a cycle are referred to as alternations.
- Two complete alternations with no reference to time make up a cycle.
- One cycle per second is defined as a hertz.
- The waveform produced by an AC generator is called a sinusoidal waveform or sine wave.
- The peak value of a sine wave is the absolute value of the point on the waveform with the greatest amplitude.
- The peak-to-peak value is the vertical distance from one peak to the other peak.
- The effective value of AC is the amount of current that produces the same degree of heat in a given resistance as an equal amount of direct current.

- The mathematical process called the root-mean square (rms) process can be used to determine the effective value.
- The rms value of a sine wave is equal to 0.707 times the peak value.

$$E_{rms} = 0.707E_p$$
$$I_{rms} = 0.707I_p$$

- The time required to complete one cycle of a sine wave is called the period (t).
- The number of cycles occurring in a specific period of time is called frequency (f).
- The relationship between frequency and period is:

$$f = \frac{1}{t}$$

- Square waves are composed of the fundamental frequency and all odd harmonics.
- Triangular waveforms are composed of the fundamental frequency and all odd harmonics 180° out of phase with the fundamental frequency.
- Sawtooth waveforms are composed of both even and odd harmonics, the even harmonics being 180° out of phase with the odd harmonics.

CHAPTER 21 SELF-TEST

1. What causes magnetic induction to occur?
2. Explain how the left-hand rule applies to AC generators.
3. Explain how the peak-to-peak value of a waveform is determined.
4. How is the effective value of alternating current determined?
5. What is the effective value of a voltage sine wave measuring 169 volts, peak to peak?
6. What is the peak value of a sine wave that has an effective value of 85 volts?
7. What is the frequency of a sine wave with a period of 0.02 second?
8. What is the period of a sine wave with a frequency of 400 Hz?
9. Draw examples of three nonsinusoidal waveforms that can represent current and voltage.
10. Why are harmonics important in the study of waveforms?

AC Measurements

OBJECTIVES

After completing this chapter, the student will be able to:

■ Identify the types of meters available for AC measurements.

■ Identify the types of meter movements used to make AC measurements.

■ Explain the function of an oscilloscope.

■ Identify the basic parts of an oscilloscope and explain their functions.

■ Demonstrate the proper setup of an oscilloscope.

■ Describe how to use an oscilloscope to take a measurement.

■ Explain how a counter works.

■ Identify the basic parts of a counter.

KEY TERMS

AC measurements of current, voltage, resistance, power, and frequency are necessary to operate and repair AC circuits and equipment.

This chapter covers some of the important test equipment used to take various AC measurements.

22–1 AC METERS

The **moving coil meter movement** is referred to as the d'Arsonval meter movement. The analog meter (Figure 22-1) is a moving coil meter. The easy-to-read digital meter (Figure 22-2) has recently replaced the analog meter.

Moving coil meters are designed to measure DC current. To measure AC current on this meter, the AC current must first be converted to DC current. This is accomplished with diodes called **rectifiers**. The process of converting AC to DC is called **rectification**. The rectifiers are placed between the meter input and

■ **FIGURE 22-1**

Analog meter used for AC measurements.

© 2014 Cengage Learning

■ **FIGURE 22-2**

Digital meter used for AC measurements.

© 2014 Cengage Learning

the meter movement and allow current to flow in only one direction (Figure 22-3). The rectifiers convert the sine wave into a pulsating DC current that is applied to the meter movement.

■ **FIGURE 22-3**

Rectifiers are used to convert the AC signal to a DC level before applying it to the meter movement.

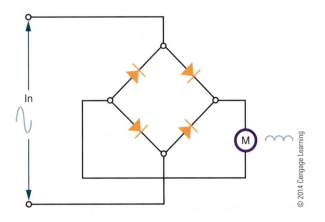

© 2014 Cengage Learning

■ **FIGURE 22-4**

Iron-vane meter movements do not require the conversion of AC to DC.

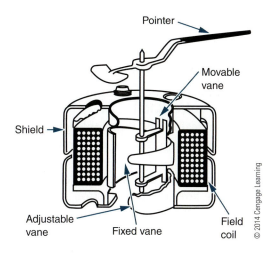

A second type of AC meter uses the **iron-vane meter movement** (Figure 22-4). This does not require the conversion of AC to DC. It consists of two iron vanes located within a coil: One vane is stationary and the other is movable. A pointer is attached to the movable vane and moves proportionally to the current passing through the coil. The magnetic field of the coil induces a north and south pole into the iron vanes. Because like poles repel and both vanes have the same polarity, they move away from each other. The iron-vane movement requires more current for a full-scale deflection than does the moving coil meter movement. For this reason, iron-vane movements find little application in low-power circuits. The iron-vane movement is also inaccurate for frequencies above 100 hertz. It is used primarily for 60-hertz applications.

The **clamp-on meter** is based on the principle that an AC current flowing in a conductor causes a magnetic field to expand and collapse as the current increases and decreases in value (Figure 22-5). Each time the AC current changes polarity, the magnetic field changes direction. The clamp-on AC meter uses a split-core transformer that allows the core to be opened and placed around the conductor. At the end of the core is a coil, which is cut by the magnetic lines of force. This induces a voltage into the coil, creating a flow of alternating current. This AC current must be rectified before being sent to the meter movement, generally a moving coil. This type of meter is used for measuring high values of AC current. The current in the conductor must be large in order to produce a magnetic field strong enough to induce a current flow in the core of the meter.

■ **FIGURE 22-5**

A clamp-on meter operates on the principle that current flowing through a wire generates a magnetic field around the wire.

The basic AC meter movement is a current measuring device. However, this type of meter movement can also be used to measure AC voltage and power. Because alternating current changes direction periodically, the meter leads can be hooked into an AC circuit in either direction.

CAUTION!

The meter must be connected in series with the circuit when measuring current. When measuring voltage, the meter must always be connected in parallel.

Always be sure the current or voltage being measured is within the range of the meter. For additional protection, always start at the highest scale and work down to the appropriate scale.

22-1 QUESTIONS

1. How is a d'Arsonval meter movement used to measure an AC voltage?
2. What makes an iron-vane meter movement desirable for measuring an AC signal?
3. Explain the principle behind a clamp-on AC meter.
4. Draw a circuit showing how to connect an AC ammeter.
5. Describe the proper procedure for using an AC voltmeter in a circuit. (Include switch settings where possible.)

22–2 OSCILLOSCOPES

An **oscilloscope** is the most versatile piece of test equipment available for working on electronic equipment and circuits. It provides a visual display of what is occurring in the circuit. Figure 22-6A shows an analog oscilloscope, and Figure 22-6B shows a digital oscilloscope. Both the analog and the digital oscilloscopes perform the same functions.

■ FIGURE 22-6

An example of an analog oscilloscope (A) and digital oscilloscope (B) available to a technician.

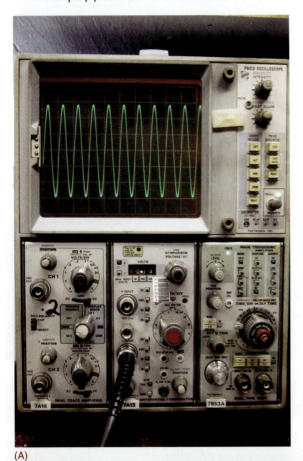

(A)

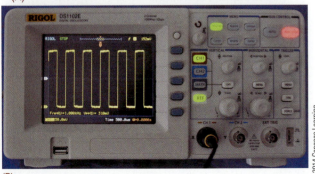

(B)

© 2014 Cengage Learning

An oscilloscope is capable of providing the following information about an electronic circuit:

1. The frequency of a signal
2. The duration of a signal
3. The phase relationship between signal waveforms
4. The shape of a signal's waveform
5. The amplitude of a signal

The basic parts of an analog oscilloscope are a **cathode-ray tube (CRT)**, a sweep generator, horizontal and vertical deflection amplifiers along with their synchronizing circuits, and power supplies (Figure 22-7).

The *sweep generator* provides a sawtooth waveform as input to the horizontal deflection amplifier. The horizontal and vertical deflection amplifiers increase the amplitude of the input voltage to the proper level for deflection of the electron beam in the cathode-ray tube. The power supply provides DC voltages to operate the amplifiers and cathode-ray tube.

The cathode-ray tube consists of three parts: a phosphor screen, deflection plates, and an electron gun (Figure 22-8). The **phosphor screen** emits light when struck by electrons. The electron gun generates the electron beam that strikes the screen. As the electron beam approaches the screen, the **deflection plates** change the direction of the electron beam as it approaches the screen. The horizontal deflection plate is attached to the sweep generator, which moves the

■ FIGURE 22-7

Block diagram of the basic parts of an oscilloscope.

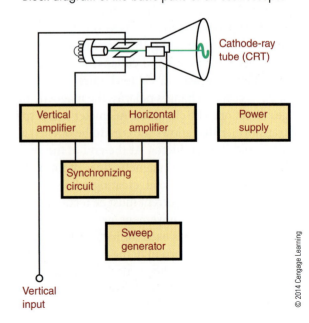

© 2014 Cengage Learning

FIGURE 22-8

Basic parts of a cathode-ray tube (CRT).

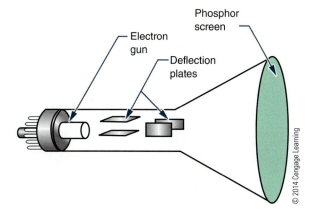

FIGURE 22-9

Oscilloscope faceplate.

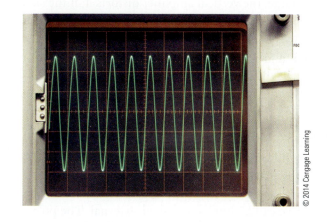

FIGURE 22-10

Oscilloscopes vary in the number of front panel controls.

electron beam back and forth across the screen. The vertical amplifier is connected to the input signal and controls its amplitude.

Using a **faceplate** that has been marked in centimeters along both a vertical and a horizontal axis (Figure 22-9), the oscilloscope can be calibrated with a known voltage before testing an unknown signal. Then, when the unknown signal is applied, its amplitude can be calculated. Rather than mark the actual faceplate of the CRT, oscilloscopes are provided with separate faceplates called **graticules** that are mounted in front of the CRT.

The power switch of an oscilloscope is usually located on the front panel (Figure 22-10). It may be a toggle, push-button, or rotary switch, and may be mounted separately or mounted with another switch. Its purpose is to apply line voltage to operate the oscilloscope.

The intensity (also called brightness) switch is used to control the electron beam within the CRT. It is a rotary control and allows the operator to adjust the electron beam for proper viewing.

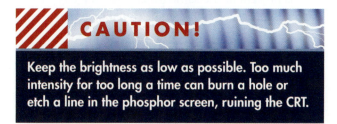

CAUTION!

Keep the brightness as low as possible. Too much intensity for too long a time can burn a hole or etch a line in the phosphor screen, ruining the CRT.

The focus and astigmatism controls are connected to the electron gun and are used to adjust the electron beam size and shape before it reaches the deflection plates. Both of these controls are rotary controls. To use them, the electron beam is slowly swept across the

screen of the CRT, and both controls are alternately adjusted until a perfectly round dot is obtained. On some newer oscilloscopes, the astigmatism control may be recessed below the front panel.

The horizontal and vertical position controls are also rotary controls, and they allow the electron beam to be positioned anywhere on the face of the CRT. Initially, they are set so that the electron beam sweeps across the center of the CRT. They are then adjusted to place the electron beam in a position to measure amplitude and time by its placement on a graticule.

The vertical block consists of a vertical input jack, an AC/DC switch, and a volts/cm rotary switch. The oscilloscope probe is connected to the input jack. The probe is then connected to the circuit to be tested. The AC/DC switch allows the signal to be sent directly to the vertical amplifier in the DC position or to a capacitor in the AC position. The capacitor in the AC position is used to block any DC component on the signal being checked. The volts/cm switch is used to adjust the amplitude of the input signal. If the signal is too large, the vertical amplifier distorts it; if the signal is too small, it is amplified. This control is calibrated to match the CRT faceplate. Where this control is set depends on the amplitude of the signal on the CRT.

The horizontal block, also referred to as the time base, consists of a time/cm rotary switch, a trigger-control switch, and a triggering level control. The time/cm switch sets the horizontal sweep rate that is to be represented by the horizontal graduations. The lower the setting, the fewer cycles per second are displayed. The trigger control switch selects the source and polarity of the trigger signal. The trigger source can be line, internal, or external. The polarity can be positive or negative. When line is selected as the trigger source, the power-line frequency of 60 hertz is used as the trigger frequency. When internal (INT) is selected as the trigger frequency, the input signal frequency is used as the trigger frequency. External (EXT) allows a trigger frequency from an external source to be used.

The level control sets the amplitude that the triggering signal must exceed before the sweep generator starts. If the level control is in the AUTO position, the oscilloscope is free running. Turning the level control results in a blank screen when no signal is present. The level control is turned to a point where the trace on the oscilloscope goes out, and then it is turned back to where the trace reappears. At this point, a stable presentation is on the screen. By using the triggering and level controls together to synchronize the sweep generator with the input signal, a stable trace can be obtained on the CRT.

Before use, the oscilloscope should be checked to determine that it is not faulty. Misadjusted controls can give false readings. Most oscilloscopes have a built-in test signal for setup. Initially the controls should be set in the following positions:

Intensity, Focus, Astigmatism, and Position controls (set to the center of their range):

Triggering:	INT +
Level:	AUTO
Time/cm:	1 msec
Volts/cm:	0.02
Power:	ON

The oscilloscope probe should be connected to the test jack of the voltage calibrator. A square wave should appear on the trace of the CRT. The display should be stable and show several cycles at the amplitude indicated by the voltage calibrator. The oscilloscope can now be used.

To use the oscilloscope, set the volts/cm selector switch to its highest setting. Connect the input signal to be observed, and rotate the volts/cm selector switch until the display is approximately two-thirds of the screen's height. Adjust the time-base controls to produce a stable trace and to select the desired number of cycles.

22–2 QUESTIONS

1. What can be learned about a waveform by using an oscilloscope?
2. What are the basic parts of an oscilloscope?
3. Describe a basic procedure for setting up an oscilloscope before its initial use.
4. What are applications of an oscilloscope when working with a circuit?
5. What does the graticule marked on the screen of the oscilloscope represent?

22–3 FREQUENCY COUNTERS

A **frequency counter** (Figure 22-11) measures frequency by comparing a known frequency against an input frequency. All frequency counters consist of the same basic sections: a time base, an input-signal conditioner, a gate-control circuit, a main gate, a decade counter, and a display (Figure 22-12). The signal conditioner converts the input signal to a wave shape and amplitude compatible with the circuitry in the counter. The main gate passes the conditioned input signal to the counter circuit if a signal from the time base is present at the same time. The time base drives the gate-control circuit, using a signal compatible with

■ **FIGURE 22-11**

A frequency counter is used widely in repair shops and industry.

© 2014 Cengage Learning

■ FIGURE 22-12

Block diagram of an electronic counter.

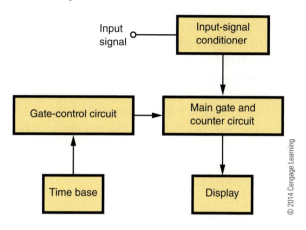

© 2014 Cengage Learning

the signal being measured. The gate control acts as the synchronization center of the counter. It controls the opening and closing of the main gate and also provides a signal to latch (lock onto) the count at the end of the counting period and reset the circuitry for the next count. The decade counter keeps a running tally of all the pulses that pass through the main gate. One decade counter is required for each digit displayed. The display, which provides a visual readout of the frequency being measured, can be of several types. The more common displays are gas-discharge tubes; light-emitting diode displays (LEDs), and liquid crystal displays (LCDs).

The electronic counter was once a piece of laboratory equipment but is now used in electronics repair shops, engineering departments, and ham-radio shacks, and on industrial production lines. The wide use of the counter can be attributed to the integrated circuit, which has reduced the size and price of the counter while increasing its accuracy, stability, reliability, and frequency range. (Integrated circuits are discussed in Section 4.)

22-4 BODE PLOTTERS

Bode plots were originated and named for H. W. Bode, who used the plots for studying amplifier feedback. The original plots required semilog graph paper. Two graphs were required: one for gain in decibels plotted against frequency on the log scale and the other for phase shift in degrees plotted against frequency on the log scale. The arrival of computer simulations such as National Instrument's Multisim has resulted in greater utilization of Bode plotters.

The **Bode plotter** in Multisim allows the user to produce a graph of a circuit's frequency response and is quite useful in analyzing filter circuits. The plotter measures a signal's voltage gain or phase shift. When the plotter is attached to a circuit, a frequency-amplitude and phase-angle analysis is performed.

The Bode plotter generates a range of frequencies over a specified spectrum. An AC source must be present somewhere in the circuit for it to work. The frequency of the AC source does not affect the plotter.

The Bode plotter measures the ratio of magnitudes (voltage gain in decibels) between two points (V+ and V−). It also measures the phase shift (in degrees) between two points. A logarithmic base is used when the compared values have a large range, which is typical when analyzing frequency response. The horizontal axis, or x–axis, always shows frequency in hertz.

When measuring the voltage gain, the vertical axis, or y–axis, shows the ratio of the circuit's output voltage to its input voltage. The units are in decibels. When measuring phase, the vertical axis always shows the phase angle in degrees.

22-3 QUESTIONS

1. What is the function of an electronic counter?
2. What are the basic sections of a counter?
3. Draw and label a block diagram of an electronic counter.
4. What is the function of the signal conditioner in a counter?
5. Why has the counter increased in popularity?

22-4 QUESTIONS

1. What is the function of a Bode plotter?
2. What are the two graphs that a Bode plotter generates?
3. What is required to use the Bode plotter?
4. What units must the input magnitude be in?
5. With both graphs, what does the horizontal axis represent?

SUMMARY

- To measure AC current or voltage on a moving coil meter, the current or voltage must first be converted to DC.
- Iron-vane meter movement does not require conversion to DC.
- A clamp-on AC meter is based on the principle that current flowing through a wire generates a magnetic field.
- An oscilloscope provides the following information about a circuit:
  - Frequency of the signal
  - Duration of the signal
  - Phase relationships between signal waveforms
  - Shape of the signal's waveform
  - Amplitude of the signal
- The basic parts of an analog oscilloscope are as follows:
  - Cathode-ray tube
  - Sweep generator
  - Horizontal deflection amplifier
  - Vertical deflection amplifier
  - Power supply
- A counter measures frequency by comparing an unknown frequency to a known frequency.
- The basic parts of a frequency counter are listed here:
  - Time base
  - Input-signal conditioner
  - Gate-control circuit
  - Main gate
  - Decade counter
  - Display
- A Bode plotter produces a graph of a circuit frequency response against gain and phase shift.

CHAPTER 22 SELF-TEST

1. Describe how a DC meter can be adapted to measure an AC signal.
2. Explain how a clamp-on ammeter is used for measuring current.
3. What type of information does an oscilloscope provide?
4. Describe the process for examining an analog oscilloscope to determine whether it is operating properly.
5. What is the last step before connecting an oscilloscope to a circuit?
6. Describe what needs to be done to the control settings before connecting an oscilloscope to an input signal.
7. Identify and describe the function of each of the major blocks of a counter.
8. What has been the major factor in bringing the counter to the workbench of repair shops?
9. Where would a digital counter be used, and what type of information would it provide?
10. Describe what signals a Bode plotter can measure.

Resistive AC Circuits

OBJECTIVES

After completing this chapter, the student will be able to:

- Describe the phase relationship between current and voltage in a resistive circuit.
- Apply Ohm's law to AC resistive circuits.
- Solve for unknown quantities in series AC resistive circuits.
- Solve for unknown quantities in parallel AC resistive circuits.
- Solve for power in AC resistive circuits.

KEY TERMS

23-1 AC circuit

23-1 resistive

23-2 series AC circuit

23-4 power curve

The relationship of current, voltage, and resistance is similar in DC and AC circuits. The simple AC circuit must be understood before moving on to more complex circuits containing capacitance and inductance.

23-1 BASIC AC RESISTIVE CIRCUITS

A basic **AC circuit** (Figure 23-1) consists of an AC source, conductors, and a resistive load. The AC source can be an AC generator or a circuit that generates an AC voltage. The **resistive** load can be a resistor, a heater, a lamp, or any similar device.

When an AC voltage is applied to the resistive load, the AC current's amplitude and direction vary in the same manner as those of the applied voltage. When the applied voltage changes polarity, the current also changes. They are said to be in phase. Figure 23-2 shows the in-phase relationship that exists between the current and the applied voltage in a pure resistive circuit. The current and voltage waveforms pass through 0 and maximum values at the same time. However, the two waveforms do not have the same peak amplitudes because they represent different quantities, measured in different units.

The AC current flowing through the resistor varies with the voltage and the resistance in the circuit. The current at any instant can be determined by applying Ohm's law.

Effective values are used in most measurements. As stated previously, the effective value is the amount of AC voltage that produces the same degree of heat as a DC voltage of the same value. The effective value can be considered the DC equivalent value. Ohm's law can be used with effective AC values, just as with DC values, in a pure resistive circuit.

EXAMPLE: If a circuit has an AC voltage source of 120 volts and a resistance of 1000 ohms, what is the effective current in the circuit? (Remember, an AC

■ FIGURE 23-1
A basic AC circuit consists of an AC source, conductors, and a resistive load.
© 2014 Cengage Learning

■ FIGURE 23-2
The voltage and current are in phase in a pure resistive circuit.

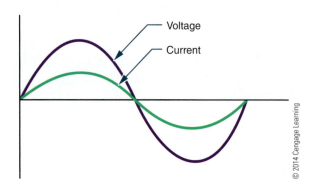

© 2014 Cengage Learning

voltage or current is assumed to be effective unless otherwise specified.)

Given	Solution
$I = ?$	$I = \dfrac{E}{R}$
$E = 120\,V$	
$R = 1000\,\Omega$	$I = \dfrac{120}{1000}$
	$I = 0.12\,A$

EXAMPLE: If an effective current of 1.7 amperes flows through a 68-ohm resistor, what is the effective value of the applied voltage?

Given	Solution
$I = 1.7\,A$	$I = \dfrac{E}{R}$
$E = ?$	
$R = 68\,\Omega$	$1.7 = \dfrac{E}{68}$
	$E = 115.6\,V$

23-1 QUESTIONS

1. What is the phase relationship between current and voltage in a pure resistive circuit?
2. What AC value is used for most measurements?
3. What is the effective current in a circuit of 10,000 Ω with 12 V applied?
4. What is the voltage (rms) of a circuit with 250 mA flowing through 100 Ω?
5. What is the resistance of a circuit when 350 μA are produced with an AC voltage of 12 V?

23-2 SERIES AC CIRCUITS

The current in a resistive circuit depends on the applied voltage. The current is always in phase with the voltage regardless of the number of resistors in the circuit. At any point in the circuit, the current has the same value.

Figure 23-3 shows a simple **series AC circuit**. The current flow is the same through both resistors. Using Ohm's law, the voltage drop across each resistor can be determined. The voltage drops added together equal the applied voltage. Figure 23-4 shows the phase relationships of the voltage drops, the applied voltage, and the current in the circuit. All the voltages and the current are in phase with one another in a pure resistive circuit.

EXAMPLE: If an AC circuit has an effective value of 120 volts applied across two resistors ($R_1 = 470$ ohms and $R_2 = 1000$ ohms), what is the voltage drop across each resistor?

First, find the total resistance (R_T).

Given	Solution
$R_1 = 470 \, \Omega$	$R_T = R_1 + R_2$
$R_2 = 1000 \, \Omega$	$R_T = 470 + 1000$
$E_{R_1} = ?$	$R_T = 1470 \, \Omega$
$E_{R_2} = ?$	

Using R_T, now find the total current (I_T).

$$I_T = \frac{E_T}{R_T}$$

$E_T = 120 \, V$

$$I_T = \frac{120}{1470}$$

$$I_T = 0.082 \, A$$

In a series circuit, $I_T = I_{R_1} = I_{R_2}$

$$\therefore I_{R_1} = 0.082 \, A$$
$$I_{R_2} = 0.082 \, A$$

■ FIGURE 23-3

Simple series AC circuit.

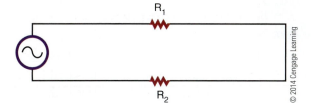

© 2014 Cengage Learning

■ FIGURE 23-4

The in-phase relationship of the voltage drops, applied voltage, and current in a series AC circuit.

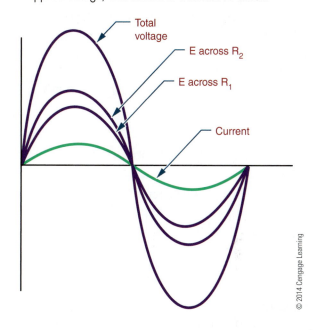

© 2014 Cengage Learning

Use I_{R_1} to find the voltage drop across resistor R_1.

$$I_{R_1} = \frac{E_{R_1}}{R_1}$$

$$0.082 = \frac{E_{R_1}}{470}$$

$$E_{R_1} = 38.54 \, V$$

Use I_2 to find the voltage drop across resistor R_2.

$$I_{R_2} = \frac{E_{R_2}}{R_2}$$

$$0.082 = \frac{E_{R_2}}{1000}$$

$$E_{R_2} = 82 \, V$$

The voltage drop across each resistor is the effective voltage drop.

23-2 QUESTIONS

1. What is the voltage drop across two resistors, 22 kilohms and 47 kilohms, connected in series with 24 volts applied?

2. What is the voltage drop across the following series resistors?
  a. $E_T = 100$ V, $R_1 = 680$ Ω, $R_2 = 1200$ Ω
  b. $E_T = 12$ V, $R_1 = 2.2$ kΩ, $R_2 = 4.7$ kΩ
  c. $I_T = 250$ mA, $R_1 = 100$ Ω, $R_2 = 500$ Ω
  d. $R_T = 10$ kΩ, $I_R = 1$ mA, $R_2 = 4.7$ kΩ
  e. $E_T = 120$ V, $R_1 = 720$ Ω, $I_{R_2} = 125$ mA

23-3 PARALLEL AC CIRCUITS

The voltage in a parallel circuit (Figure 23-5) remains constant across each individual branch. However, the total current divides among the individual branches. In a parallel AC circuit, the total current is in phase with the applied voltage (Figure 23-6). The individual currents are also in phase with the applied voltage.

■ FIGURE 23-5

A simple parallel AC circuit.

© 2014 Cengage Learning

■ FIGURE 23-6

The in-phase relationship of the applied voltage, total current, and individual branch currents in a parallel AC circuit.

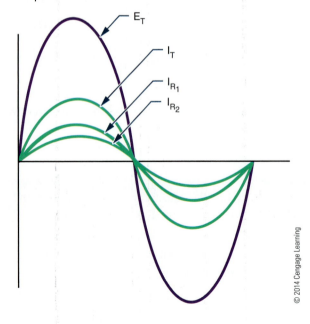

© 2014 Cengage Learning

All current and voltage values are the effective values. These values are used the same way DC values are used.

EXAMPLE: If an AC circuit has an effective voltage of 120 volts applied across two parallel resistors of 470 ohms and 1000 ohms, respectively, what is the current flowing through each of the resistors?

Given	Solution
$I_{R_1} = ?$	*In a parallel circuit* $E_T = E_{R_1} = E_{R_2}$
$I_{R_2} = ?$	$\therefore E_{R_1} = 120$ V
$E_T = 120$ V	$E_{R_2} = 120$ V
$R_1 = 470$ ohms	

Use E_1 to find the current through resistor R_1.

$$I_{R_1} = \frac{E_{R_1}}{R_1}$$

$R_2 = 1000$ Ω

$$I_{R_1} = \frac{120}{470}$$

$$I_{R_2} = 0.26 \text{ A or } 260 \text{ mA}$$

Use E_2 to find the current through resistor R_2.

$$I_{R_2} = \frac{E_{R_2}}{R_2}$$

$$I_{R_2} = \frac{120}{1000}$$

$$I_{R_2} = 0.12 \text{ A or } 120 \text{ mA}$$

23-3 QUESTIONS

1. What is the current flow through the following parallel AC resistive circuits?
2. $E_T = 100$ V, $R_1 = 470$ Ω, $R_2 = 1000$ Ω
3. $E_T = 24$ V, $R_1 = 22$ kΩ, $R_2 = 47$ kΩ
4. $E_T = 150$ V, $R_1 = 100$ Ω, $R_2 = 500$ Ω
5. $I_T = 0.0075$ A, $E_{R_1} = 10$ V, $R_2 = 4.7$ kΩ
6. $R_T = 4700$ Ω, $I_{R_1} = 11$ mA, $E_{R_2} = 120$ V

23-4 POWER IN AC CIRCUITS

Power is dissipated in AC resistive circuits the same way as in DC resistive circuits. Power is measured in watts and is equal to the current times the voltage in the circuit.

■ FIGURE 23-7

The relationship of power, current, and voltage in a resistive AC circuit.

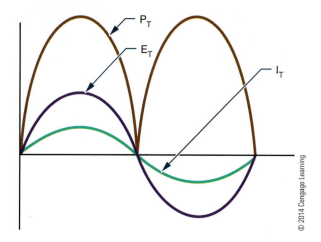

© 2014 Cengage Learning

The power consumed by the resistor in an AC circuit varies with the amount of current flowing through it and the voltage applied across it. Figure 23-7 shows the relationship of power, current, and voltage. The **power curve** does not fall below the reference line because the power is dissipated in the form of heat. It does not matter in which direction the current is flowing, as power is assumed to have a positive value.

Power varies between the peak value and 0. Midway between peak value and 0 is the average power consumed by the circuit. In an AC circuit, the average power is the power consumed. This can be determined by multiplying the effective voltage value by the effective current value. This is expressed as

$$P = IE$$

EXAMPLE: What is the power consumption in an AC circuit with 120 volts applied across 150 ohms of resistance? (Remember, when the voltage value is not specified for an AC source, it is assumed to be the effective value.)

Given	Solution
$I_T = ?$	*First, find the total current (I_T).*
$E_T = 120\,V$	$I_T = \dfrac{E_T}{R_T}$
$R_T = 150\,\Omega$	
$P_T = ?$	$I_T = \dfrac{120}{150}$
	$I_T = 0.80\,A$

Now, find the total power (P_T).

$$P_T = I_T E_T$$
$$P_T = (0.80)(120)$$
$$P_T = 96\,W$$

When the current is not given, the current must be determined before using the power formula. The resistance in this circuit consumes 96 watts of power.

23–4 QUESTIONS

What is the total power consumption of the following circuits?

Series:

1. $E_T = 100\,V$, $R_1 = 680\,\Omega$, $R_2 = 1200\,\Omega$
2. $I_T = 250\,mA$, $R_1 = 100\,\Omega$, $R_2 = 500\,\Omega$

Parallel:

3. $E_T = 100\,V$, $R_1 = 470\,\Omega$, $R_2 = 1000\,\Omega$
4. $I_T = 7.5\,mA$, $E_{R_1} = 10\,V$, $R_2 = 4.7\,k\Omega$
5. Find the power consumption of each individual component in the following circuit.

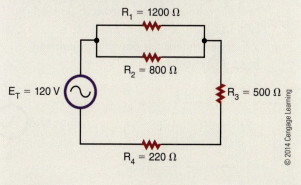

© 2014 Cengage Learning

SUMMARY

- A basic AC resistive circuit consists of a voltage source, conductors, and a resistive load.
- The current is in phase with the applied voltage in a resistive circuit.
- The effective value of AC current or voltage produces the same results as the equivalent DC voltage or current.
- The effective values are the most widely used measurement values.
- Ohm's law can be used with all effective values.
- AC voltage or current values are assumed to be the effective values if not otherwise specified.

CHAPTER 23 SELF-TEST

1. Explain the phase relationship between current and voltage in a pure resistive circuit.
2. What is the effective voltage of an AC circuit with 25 mA flowing through 4.7 kΩ?
3. What is the voltage drop across two resistors of 4.7 kΩ and 3.9 kΩ in series with an AC voltage of 12 V applied?
4. If two parallel resistors of 2.2 kΩ and 5.6 kΩ have an AC effective voltage of 120 V applied across their input, what is the current developed through each of the resistors?
5. When discussing in-phase relationships, why are voltage waveforms shown in series circuits and current waveforms shown in parallel circuits?
6. What determines the power consumption in an AC circuit?

7. What is the power consumption in an AC circuit with 120 V applied across a load of 1200 Ω?
8. What is the total power consumption in the following figure?

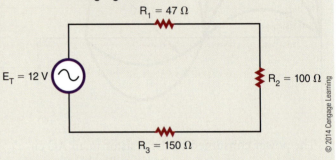

9. What is the power consumption of resistor R_1 in the following figure?

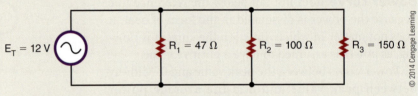

10. What is the power consumption for resistor R_3 in the following figure?

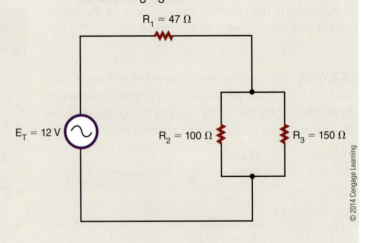

CHAPTER 24

Capacitive AC Circuits

Capacitors are key components of AC circuits. Capacitors combined with resistors and inductors form useful electronic networks.

24-1 CAPACITORS IN AC CIRCUITS

When an AC voltage is applied to a capacitor, it gives the appearance that electrons are flowing in the circuit. However, electrons do not pass through the dielectric of the capacitor. As the applied AC voltage increases and decreases in amplitude, the capacitor charges and discharges. The resulting movement of electrons from one plate of the capacitor to the other represents current flow.

The current and applied voltage in a capacitive AC circuit differs from those in a pure resistive circuit. In a pure resistive circuit, the current flows in phase with the applied voltage. Current and voltage in a capacitive AC circuit do not flow in phase with each other (Figure 24-1). When the voltage starts to increase, current is at maximum because the capacitor is discharged. As soon as the capacitor charges to the peak AC voltage, the charging current drops to 0. As the voltage begins to drop, the capacitor begins discharging. The current begins to increase in a negative direction. When the current is at maximum, the voltage is at 0. This relationship is described as 90° out of phase. The current leads the applied voltage in a capacitive circuit. The negative voltage peaks when the voltage equals 0 volts. The phase difference continues through each cycle. In a purely capacitive circuit, the current leads the voltage by an angle of 90°. This can be represented by the acronym **ICE**. Current (I) leads the voltage (E) in a capacitive (C) circuit.

FIGURE 24-1

Note the out-of-phase relationship between the current and the voltage in a capacitive AC circuit. The current leads the applied voltage.

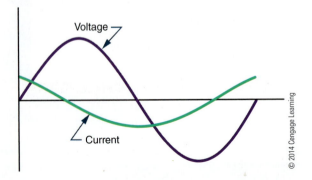

© 2014 Cengage Learning

In a capacitive AC circuit, the applied voltage is constantly changing, causing the capacitor to charge and discharge. After the capacitor is initially charged, the voltage stored on its plates opposes any change in the applied voltage. The opposition that the capacitor offers to the applied AC voltage is called **capacitive reactance**. Capacitive reactance is represented by X_C and is measured in ohms.

Capacitive reactance can be calculated by using the formula:

$$X_C = \frac{1}{2\pi fC}$$

where: π = **pi**, the constant 3.14
     f = frequency in hertz
     C = capacitance in farads

Capacitive reactance is a function of the frequency of the applied AC voltage and the capacitance. Increasing the frequency decreases the reactance, resulting in greater current flow. Decreasing the frequency increases the opposition and decreases current flow.

EXAMPLE: What is the capacitive reactance of a 1-microfarad capacitor at 60 hertz?

Given

$X_C = ?$

$\pi = 3.14$

$f = 60\,Hz$

$C = 1\,\mu F$
    $= 0.000001\,F$

Solution

$$X_C = \frac{1}{2\pi fC}$$

$$X_C = \frac{1}{(2)(3.14)(60)(0.000001)}$$

$$X_C = 2653.93\,\Omega$$

EXAMPLE: What is the capacitive reactance of a 1-microfarad capacitor at 400 hertz?

Given

$X_C = ?$

$\pi = 3.14$

$f = 4000\,Hz$

$C = 1\,\mu F$
    $= 0.000001\,F$

Solution

$$X_C = \frac{1}{2\pi fC}$$

$$X_C = \frac{1}{(2)(3.14)(400)(0.000001)}$$

$$X_C = 398.09\,\Omega$$

EXAMPLE: What is the capacitive reactance of a 0.1-microfarad capacitor at 60 hertz?

Given

$X_C = ?$

$\pi = 3.14$

$f = 60\,Hz$

$C = 0.1\,\mu F$
    $= 0.0000001\,F$

Solution

$$X_C = \frac{1}{2\pi fC}$$

$$X_C = \frac{1}{(2)(3.14)(60)(0.0000001)}$$

$$X_C = 26{,}539.28\,\Omega$$

EXAMPLE: What is the capacitive reactance of a 10-microfarad capacitor at 60 hertz?

Given
$X_C = ?$
$\pi = 3.14$
$f = 60\,Hz$
$C = 10\,\mu F$
$= 0.00001F$

Solution
$$X_C = \frac{1}{2\pi fC}$$
$$X_C = \frac{1}{(2)(3.14)(60)(0.00001)}$$
$$X_C = 265.39\,\Omega$$

Capacitive reactance is the opposition to changes in the applied AC voltage by a capacitor. In an AC circuit, a capacitor is thus an effective way of controlling current. Using Ohm's law, the current is directly proportional to the applied voltage and inversely proportional to the capacitive reactance. This is expressed as:

$$I = \frac{E}{X_C}$$

NOTE	X_C (capacitive reactance) has been substituted for R (resistance) in Ohm's law.

It is important to keep in mind that capacitive reactance depends on the frequency of the applied voltage and the capacitance in the circuit.

EXAMPLE: A 100-microfarad capacitor has 12 volts applied across it at 60 hertz. How much current is flowing through it?

Given
$I = ?$
$E = 12\,VAC$
$\pi = 3.14$
$f = 60\,Hz$
$C = 100\,\mu F$
$= 0.0001\,F$

Solution
First, find the capacitive reactance (X_C).
$$X_C = \frac{1}{2\pi fC}$$
$$X_C = \frac{1}{(2)(3.14)(60)(0.0001)}$$
$$X_C = 26.54\,\Omega$$

Using X_C, now find the current flow:
$$I = \frac{E}{X_C}$$
$$I = \frac{12}{26.54}$$
$$I = 0.452\,A,\ or\ 452\,mA$$

EXAMPLE: A 10-microfarad capacitor has 250 milliamperes of current flowing through it. How much voltage at 60 hertz is applied across the capacitor?

Given
$X_C = ?$
$\pi = 3.14$
$f = 60\,Hz$
$C = 10\,\mu F$
$= 0.00001F$
$I = 250\,mA$ or 0.25 amps
$E = ?$

Solution
First, find the capacitive reactance (X_C).
$$X_C = \frac{1}{2\pi fC}$$
$$X_C = \frac{1}{(2)(3.14)(60)(0.00001)}$$
$$X_C = 265.39\,\Omega$$

Now, find the voltage drop (E).
$$I = \frac{E}{X_C}$$
$$0.25 = \frac{E}{265.39}$$
$$E = 66.35V$$

When capacitors are connected in series, the capacitive reactance is equal to the sum of the individual capacitive reactance values.

$$X_{C_T} = X_{C_1} + X_{C_2} + X_{C_3} \ldots + X_{C_n}$$

When capacitors are connected in parallel, the reciprocal of the capacitive reactance is equal to the sum of the reciprocals of the individual capacitive reactance values.

$$\frac{1}{X_{C_T}} = \frac{1}{X_{C_1}} + \frac{1}{X_{C_2}} + \frac{1}{X_{C_3}} \ldots + \frac{1}{X_{C_n}}$$

24–1   QUESTIONS

1. Describe how an AC voltage appears to make current move through a capacitor.
2. What is the relationship between current and voltage in a capacitive circuit?
3. What is capacitive reactance?
4. What is the capacitive reactance of a 10-microfarad capacitor at 400 hertz?
5. How much current is flowing through a 100 μF capacitor with 120 volts at 60 hertz applied across it?

<table>
<tr><td>

24–2 APPLICATIONS OF CAPACITIVE CIRCUITS

</td></tr>
</table>

Capacitors can be used alone or combined with resistors to form RC (resistor-capacitor) networks. RC networks are used for filtering, decoupling, DC blocking, or coupling phase-shift circuits.

A **filter** is a circuit that discriminates among frequencies, attenuating (weakening) some while allowing others to pass. It works by establishing a cutoff point between frequencies that are passed and frequencies that are attenuated. The two most common types of filters are **low-pass filters** and **high-pass filters**. A low-pass filter allows low frequencies to pass with little opposition while attenuating the high frequencies. A high-pass filter permits frequencies above the cutoff point to pass while attenuating frequencies below the cutoff point.

A low-pass filter (Figure 24-2) consists of a capacitor and a resistor in series. The input voltage is applied across both capacitor and resistor. The output is taken from across the capacitor. At low frequencies, the capacitive reactance is higher than the resistance, so most of the voltage is dropped across the capacitor. Therefore, most of the voltage appears across the output. As the input frequency increases, the capacitive reactance decreases, and less voltage is dropped across the capacitor. Therefore, more voltage is dropped across the resistor, with a decrease in the output voltage. The cutoff is gradual. Frequencies above the cutoff point pass with a gradual attenuation in the output voltage. Figure 24-3 shows a frequency response curve for an RC low-pass filter.

A high-pass filter also consists of a resistor and capacitor in series (Figure 24-4). However, the output is taken across the resistor. At high frequencies, the capacitive reactance is low, and most of the voltage is dropped across the resistor. As the frequency decreases, the capacitive reactance increases, and more voltage is dropped across the capacitor. This results in a decrease in the output voltage across the resistor. Again, the drop in output voltage is gradual. Figure 24-5 shows a frequency response curve for an RC high-pass filter.

■ **FIGURE 24-2**

RC low-pass filter.

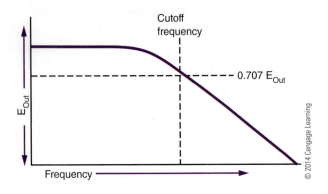

■ **FIGURE 24-4**

RC high-pass filter.

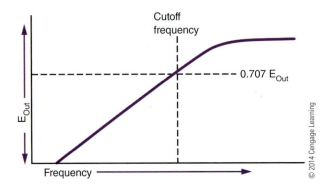

■ **FIGURE 24-3**

Frequency response of an RC low-pass filter.

■ **FIGURE 24-5**

Frequency response of an RC high-pass filter.

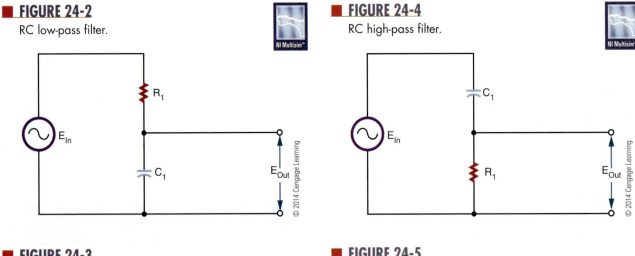

Most electronic circuits include both AC and DC voltages. This results in an AC signal superimposed on a DC signal. If the DC signal is used to operate equipment, it is advantageous to remove the AC signals. An RC low-pass filter can be used for this purpose. A **decoupling network** (Figure 24-6) allows the DC signal to pass while attenuating or eliminating the AC signal. The AC signal may be in the form of oscillations, noises, or transient spikes. By adjusting the cutoff frequency, most of the AC signal can be filtered out, leaving only the DC voltage across the capacitor.

In another application, it may be desirable to pass the AC signal while blocking the DC voltage. This type of circuit is called a **coupling network** (Figure 24-7). An RC high-pass filter can be used. Initially the capacitor charges to the DC voltage level. Once the capacitor is charged, no more DC current can flow in the circuit. The AC source causes the capacitor to charge and discharge at the AC rate, creating current flow through the resistor. The component values are chosen so that the AC signal is passed with a minimum of attenuation.

At times, it is necessary to shift the phase of the AC output signal with respect to the input signal. RC networks can be used for phase shifting applications. The **RC phase-shift networks** are used only where small amounts, less than 60°, of phase shift are desired.

Figure 24-8 shows an RC phase-shift network with the input applied across the resistor-capacitor combination and the output taken across the resistor. In the circuit, the current leads the voltage because of the capacitor. The voltage across the resistor is in phase with the resistor, which results in the output voltage leading the input voltage.

In Figure 24-9, the output is taken across the capacitor. The current in the circuit leads the applied voltage. However, the voltage across the capacitor lags behind the applied voltage.

To achieve greater phase shifts, several RC networks may be cascaded (connected) together (Figure 24-10). Cascaded networks reduce the output

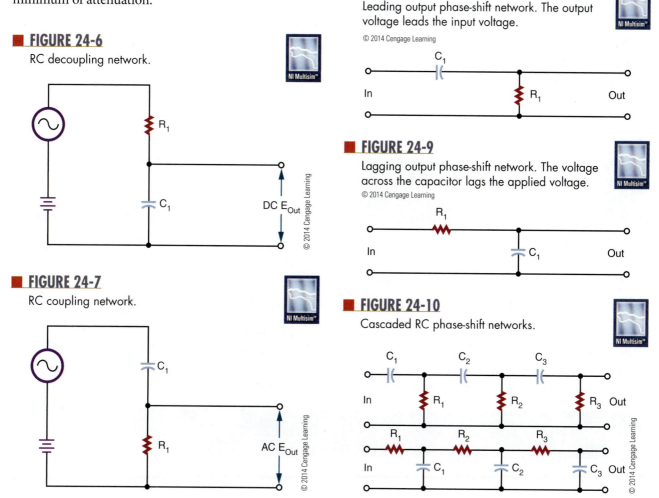

■ **FIGURE 24-6**

RC decoupling network.

■ **FIGURE 24-7**

RC coupling network.

■ **FIGURE 24-8**

Leading output phase-shift network. The output voltage leads the input voltage.

© 2014 Cengage Learning

■ **FIGURE 24-9**

Lagging output phase-shift network. The voltage across the capacitor lags the applied voltage.

© 2014 Cengage Learning

■ **FIGURE 24-10**

Cascaded RC phase-shift networks.

© 2014 Cengage Learning

voltage, however. An amplifier is needed to raise the output voltage to the proper operating level.

Phase-shift networks are valid for only one frequency, because the capacitive reactance varies with changes in frequency. Changing the reactance results in a different phase shift.

24–2 QUESTIONS

1. What are three uses of resistor-capacitor networks in electronic circuits?
2. Draw a diagram of a low-pass filter, and describe how it works.
3. Draw a diagram of a high-pass filter, and describe how it works.
4. What is the purpose of a decoupling network?
5. What do RC phase shift networks do?

SUMMARY

- When an AC voltage is applied to a capacitor, it gives the appearance of current flow.
- The capacitor charging and discharging represents current flow.
- The current leads the applied voltage by 90° in a capacitive circuit.
- Capacitive reactance is the opposition a capacitor offers to the applied voltage.
- X_C represents capacitive reactance.
- Capacitive reactance is measured in ohms.
- Capacitive reactance can be calculated by the formula:

$$X_C = \frac{1}{2\pi fC}$$

- RC networks are used for filtering, coupling, and phase shifting.
- A filter is a circuit that discriminates against certain frequencies.
- A low-pass filter passes frequencies below a cutoff frequency. It consists of a resistor and capacitor in series.
- A high-pass filter passes frequencies above a cutoff frequency. It consists of a capacitor and resistor in series.
- Coupling networks pass AC signals but block DC signals.

CHAPTER 24 SELF-TEST

1. What is the relationship between current and the applied voltage in a capacitive circuit?
2. What is capacitive reactance a function of?
3. What is the capacitive reactance of a 1000 μF capacitor at 60 Hz?
4. In question 3, what is the current flow through the capacitor with 12 V applied?
5. List three applications for capacitive circuits.
6. Explain how an RC low-pass filter functions.
7. How can an AC signal be removed from a DC source?
8. Why are capacitive coupling circuits important?
9. Describe how a phase-shift network shifts the phase of an AC signal.
10. How can the phase shift be changed for a different frequency in a cascade phase-shift network?

CHAPTER 25

Inductive AC Circuits

nductors, like capacitors, oppose current flow in AC circuits. They may also introduce a phase shift between the voltage and the current in AC circuits. A large number of electronic circuits are composed of inductors and resistors.

25–1 INDUCTORS IN AC CIRCUITS

Inductors in AC circuits offer opposition to current flow. When an AC voltage is placed across an inductor, it creates a magnetic field. As the AC voltage changes polarity, it causes the magnetic field to expand and collapse. It also induces a voltage in the inductor coil. This induced voltage is called a **counter electromotive force (cemf)**; the greater the inductance, the greater the cemf. The cemf is out of phase with the applied voltage by 180° (Figure 25-1) and opposes the applied voltage. This opposition is as effective in reducing current flow as a resistor.

The amount of voltage induced in the inductor depends on the rate of change of the magnetic field. The faster the magnetic field expands and collapses, the greater the induced voltage. The total effective voltage across the inductor is the difference between the applied voltage and the induced voltage. The induced voltage is always less than the applied voltage.

Figure 25-2 shows the relationship of the current to the applied voltage. In a purely inductive circuit, the current lags behind the applied voltage by 90°. Another way of stating this is that the applied voltage leads the current by 90° in a pure inductive current. This can be represented by the acronym **ELI**. Voltage (E) leads current (I) in an inductive (L) circuit.

■ FIGURE 25-1

The applied voltage and the induced voltage are 180° out of phase with each other in an inductive circuit.

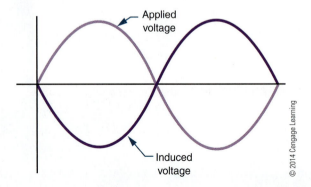

Applied voltage

Induced voltage

© 2014 Cengage Learning

■ FIGURE 25-2

The current lags the applied voltage in an AC inductive circuit.

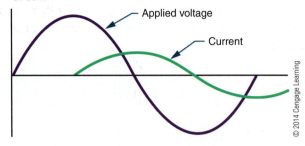

Applied voltage

Current

© 2014 Cengage Learning

The opposition offered to current flow by an inductor in an AC circuit is called **inductive reactance** and is measured in ohms. The amount of inductive reactance offered by an inductor depends on its inductance and the frequency of the applied voltage. The larger the inductance, the larger the magnetic field generated and the greater the opposition to current flow. Also, the higher the frequency, the greater the opposition the inductor has to current flow.

Inductive reactance is expressed by the symbol X_L and is determined by the following formula:

$$X_L = 2\pi fL$$

where: π = pi, or 3.14
$\quad$ f = frequency in hertz
$\quad$ L = inductance in henries

EXAMPLE: What is the inductive reactance of a 0.15-henry coil at 60 hertz?

Given
$X_L = ?$
$\pi = 3.14$
$f = 60\,Hz$
$L = 0.15\,H$

Solution
$X_L = 2\pi fL$
$X_L = (2)(3.14)(60)(0.15)$
$X_L = 56.52\,\Omega$

EXAMPLE: What is the inductive reactance of a 0.15-henry coil at 400 hertz?

Given
$X_L = ?$
$\pi = 3.14$
$f = 400\,Hz$
$L = 0.15\,H$

Solution
$X_L = 2\pi fL$
$X_L = (2)(3.14)(400)(0.15)$
$X_L = 376.80\,\Omega$

NOTE Notice that the inductive reactance increases with an increase in frequency.

Ohm's law applies to inductive reactance in AC circuits in the same manner that it applies to resistance. The inductive reactance in an AC circuit is directly proportional to the applied voltage and inversely proportional to the current. This relationship is expressed as

$$I = \frac{E}{X_L}$$

The current increases with an increase in voltage or a decrease in the inductive reactance. Likewise, decreasing the voltage or increasing the inductive reactance causes a decrease in current.

EXAMPLE: How much current flows through a 250-millihenry inductor when a signal of 12 volts and 60 hertz is applied across it?

Given

$X_L = ?$

$\pi = 3.14$

$f = 60\,Hz$

$L = 0.25\,H$

$I = ?$

$E = 12\,V$

Solution

First, find the inductive reactance (X_L).

$X_L = 2\pi fL$

$X_L = (2)(3.14)(60)(0.25)$

$X_L = 94.20\,\Omega$

Using X_L, now find the current flow (I).

$I = \frac{E}{X_L}$

$I = \frac{12}{94.20}$

$I = 0.127\,amp,\ or\ 127\,mA$

EXAMPLE: How much voltage is necessary to cause 10 milliamperes to flow through a 15-millihenry choke at 400 Hz?

Given

$X_L = ?$

$\pi = 3.14$

$f = 400\,Hz$

$L = 0.015H$

$I = 0.01A$

$E = ?$

Solution

First, find the inductive reactance (X_L).

$X_L = 2\pi fL$

$X_L = (2)(3.14)(400)(0.015)$

$X_L = 37.68\,\Omega$

Using X_L, now find the voltage (E).

$I = \frac{E}{X_L}$

$0.01 = \frac{E}{37.68}$

$E = 0.38\,V$

EXAMPLE: What is the inductive reactance of a coil that has 120 volts applied across it with 120 milliamperes of current flow?

Given

$I = 0.12\,A$

$E = 120\,V$

$X_L = ?$

Solution

$I = \frac{E}{X_L}$

$0.12 = \frac{120}{X_L}$

$X_L = 1000\,\Omega$

The impedance of a circuit containing both inductance and resistance is the total opposition to current flow by both the inductor and the resistor. Because of the phase shift caused by the inductor, the inductive reactance and the resistance cannot be added directly. The **impedance (Z)** is the vector sum of the inductive reactance and the resistance in the circuit. The impedance is expressed in ohms and is designated by the letter Z. Impedance can be defined in terms of Ohm's law as:

$$I = \frac{E}{Z}$$

The most common inductive circuit consists of a resistor and an inductor connected in series. This is referred to as an RL circuit. The impedance of a series RL circuit is the square root of the sum of the squares of the inductive reactance and the resistance. This can be expressed as:

$$Z = \sqrt{R^2 + X_L^2}$$

EXAMPLE: What is the impedance of a 100-millihenry choke in series with a 470-ohm resistor with 12 volts, 60 hertz applied across them?

Given

$X_L = ?$

$\pi = 3.14$

$f = 60\,Hz$

$L = 100\,mH = 0.1\,H$

$Z = ?$

$R = 470\,\Omega$

Solution

First, find the inductive reactance (X_L).

$X_L = 2\pi fL$

$X_L = (2)(3.14)(60)(0.1)$

$X_L = 37.68\,\Omega$

Using X_L, now find the impedance (Z).

$Z = \sqrt{R^2 + X_L^2}$

$Z = \sqrt{(470)^2 + (37.68)^2}$

$Z = 471.51\,\Omega$

When inductors are connected in series, the inductive reactance is equal to the sum of the individual inductive reactance values.

$$X_{L_T} = X_{L_1} + X_{L_2} + X_{L_3} \ldots + X_{L_n}$$

When inductors are connected in parallel, the reciprocal of the inductive reactance is equal to the sum of the reciprocals of the individual inductive reactance values.

$$\frac{1}{X_{L_T}} = \frac{1}{X_{L_1}} + \frac{1}{X_{L_2}} + \frac{1}{X_{L_3}} \ldots + \frac{1}{X_{L_n}}$$

25–1 QUESTIONS

1. What does an AC voltage do when applied across an inductor?
2. What is the relationship between current and voltage in an inductive circuit?
3. What is inductive reactance?
4. What is the inductive reactance of a 200-millihenry inductor at 10,000 hertz?
5. How is the impedance determined for an inductor-resistor network?

25–2 APPLICATIONS OF INDUCTIVE CIRCUITS

Inductive circuits are widely used in electronics. Inductors compete with capacitors for filtering and phase-shift applications. Because inductors are larger, heavier, and more expensive than capacitors, they have fewer applications than capacitors. However, inductors have the advantage of providing a reactive effect while still completing a DC circuit path. Capacitors can provide a reactive effect but block the DC elements.

Inductors are sometimes combined with capacitors to improve the performance of a circuit. The reactive effect of the capacitor then opposes the reactive effect of the inductor. The end result is that they complement each other in a circuit.

Series **RL networks** are used as low- and high-pass filters. Figure 25-3 shows the two basic types of filters. The circuits are essentially resistor-inductor voltage dividers. Figure 25-3A is a low-pass filter. The input is applied across the inductor and resistor. The output is taken from across the resistor. At low

■ **FIGURE 25-3**

RL filters.

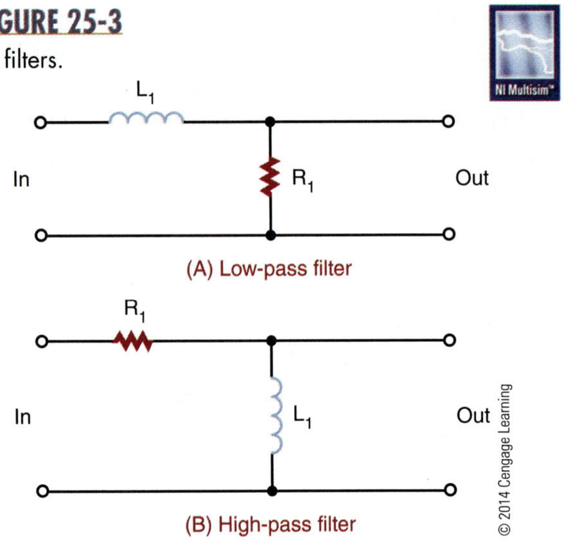

(A) Low-pass filter

(B) High-pass filter

© 2014 Cengage Learning

frequencies, the reactance of the coil is low. Therefore, it opposes little current, and most of the voltage is dropped across the resistor. As the input frequency increases, the inductive reactance increases and offers more opposition to current flow so that more voltage drops across the inductor. With more voltage dropped across the inductor, less voltage is dropped across the resistor. Increasing the input frequency decreases the output voltage. Low frequencies are passed with little reduction in amplitude, whereas high frequencies are greatly reduced in amplitude.

Figure 25-3B is a high-pass filter. The input is applied across the inductor and resistor, and the output is taken across the inductor. At high frequencies, the inductive reactance of the coil is high, causing most of the voltage to be dropped across the coil. As the frequency decreases, the inductive reactance decreases, offering less opposition to the current flow. This causes less voltage to be dropped across the coil and more voltage to be dropped across the resistor. The frequency above or below the frequencies passed or attenuated is called the **cutoff frequency**. The symbol for the cutoff frequency is **f$_{co}$**. The cutoff frequency can be determined by the formula:

$$f_{CO} = \frac{R}{2\pi L}.$$

where:

f$_{CO}$ = cutoff frequency in hertz
R = resistance in ohms
π = 3.14
L = inductance in henries

25–2 QUESTIONS

1. What are the disadvantages of using inductors in circuits?
2. What are the advantages of using inductors in circuits?
3. Draw a diagram of an RL low-pass filter, and explain how it operates.
4. Draw a diagram of an RL high-pass filter, and explain how it operates.
5. How can the cutoff frequency of an RL circuit be determined?

SUMMARY

- In a pure inductive circuit, the current lags the applied voltage by 90°.
- Inductive reactance is the opposition to current flow offered by an inductor in an AC circuit.
- X_L represents inductive reactance.
- Inductive reactance is measured in ohms.
- Inductive reactance can be calculated by the formula:

$$X_L = 2\pi fL$$

- Impedance is the vector sum of the inductive reactance and the resistance in the circuit.
- Series RL circuits are used for low- and high-pass filters.

CHAPTER 25 SELF-TEST

1. What is the relationship between current and the applied voltage in an inductive circuit?
2. What factor affects the inductive reactance of an inductive circuit?
3. What is the inductive reactance of a 100 mH coil at 60 Hz?
4. How much current would flow through the inductor in question 3 if 24 V were applied?
5. How are inductors used in circuits?
6. How much voltage is required for 86 mA to flow through a 100 mH inductor at 50 Hz?
7. Calculate the impedance of a 150 mH inductor in series with a 680 Ω resistor at 60 Hz.
8. What is the inductance of a choke in a series RL circuit with a resistor of 910 Ω with 32 V across it and an applied voltage of 45 V at 400 Hz?
9. What are the differences when connecting low-pass versus high-pass RL filters?
10. What is the cutoff frequency of an inductive circuit?

Reactance and Resonance Circuits

OBJECTIVES

After completing this chapter, the student will be able to:

- Identify the formulas for determining capacitive and inductive reactance.
- Identify how AC current and voltage react in capacitors and inductors.
- Determine the reactance of a series circuit, and identify whether it is capacitive or inductive.
- Define the term *impedance*.
- Solve problems for impedance that contain both resistance and capacitance or inductance.
- Discuss how Ohm's law must be modified before using it for AC circuits.
- Solve for X_C, X_L, X, Z, and I_T in RLC series circuits.
- Solve for I_C, I_L, I_X, I_R, and I_Z in RLC parallel circuits.

KEY TERMS

26-1 in phase
26-1 out of phase
26-1 phase angle
26-1 Pythagorean theorem
26-1 trigonometric function

26-1 vector
26-1 vector diagram
26-3 apparent power
26-3 power factor
26-4 resonance
26-4 resonant circuit

n previous chapters, resistance, capacitance, and inductance in AC circuits were looked at individually in a circuit. In this chapter, resistance, capacitance, and inductance are observed as connected together in an AC circuit. The concepts covered in this chapter do not present any new material but apply all of the principles presented so far.

When the reactance of the inductor equals the reactance of the capacitor, it forms a resonant circuit. Resonant circuits are used in a variety of circuits in electronics.

26–1   REACTANCE IN SERIES CIRCUITS

In a series circuit (Figure 26-1), the DC voltage is typically supplied by a battery. The value of the amount of current flowing in a circuit may be determined by Ohm's law, which uses the battery voltage and resistance of the circuit.

When an AC voltage source is applied to a circuit (Figure 26-2), the instantaneous value of current flowing through the resistor varies with the alternating output voltage applied. Ohm's law applies to the AC circuit just as it does with the DC circuit. Peak current may be calculated from the source's peak voltage, and rms current may be calculated from the rms voltage.

Figure 26-3 shows a graph of one cycle of AC. It shows that when the voltage reaches a peak, the

■ FIGURE 26-1

DC resistive circuit.

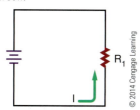

■ FIGURE 26-2

AC resistive circuit.

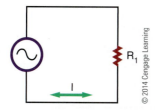

■ FIGURE 26-3

In a resistive AC circuit, current and voltage are in phase.

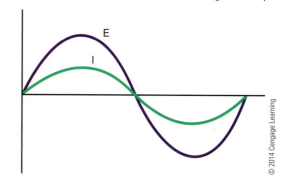

current also reaches a peak. Both the voltage and the current cross the zero line together. The two waveforms are said to be **in phase**. This is the condition that occurs when a circuit contains pure resistance and no reactive components.

The effects of pure inductance or capacitance cause the voltage or current of a circuit to be 90° **out of phase**. This situation becomes more complex when both reactive and resistance components are combined—a condition that is typical of an AC circuit.

In Figure 26-4, the components are shown connected in a series circuit, and the current flows equally through both components. Resistor R_1 has a value of 47 ohms, and the calculated value of inductive reactance for inductor L_1 is 25.25 ohms, for the frequency of 60 hertz. The current flow in the circuit is 2.25 amperes, and the voltage drops across resistor R_1 and inductor L_1 are 105.75 and 56.81 volts, respectively.

■ FIGURE 26-4

Voltage in either RL circuits such as this one or in RC circuits are not in phase and cannot be added directly.

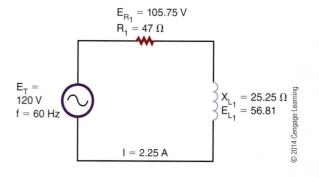

These voltages appear to be incorrect because they total more than the supply voltage of 120 volts. This occurs because the voltage across the resistor is out of phase with the voltage across the inductor.

Voltage and current are in phase in the resistive portion of the circuit. In the inductive portion of the circuit, voltage leads the current by 90°. Because the current at all points in a series circuit must be in phase, E_R and E_L are out of phase.

A way to represent voltage and current in circuits, such as the one in Figure 26-4, is to use **vectors**, arrows that start at the origin of a coordinate system and point in a particular direction (Figure 26-5). The length of the arrow indicates the magnitude: The longer the arrow, the larger the value. The angle the arrow makes with the x-axis indicates its phase in degrees. The positive x-axis is 0°, and the degrees increase as the arrow is moved in a counterclockwise direction.

In Figure 26-4, the current flows through both components with the same phase. The current vector is used as the zero-degree reference and lies on the x-axis. Voltage E_R is in-phase with the current and its vector is also placed on the x-axis.

The voltage across the inductor, E_L, is 90° ahead of the current. Therefore, its vector points straight up, or 90° from the x-axis. Each vector is drawn to scale. The source voltage vector (E_T) is started at the origin and ends at the maximum values of E_R and E_L, as shown in Figure 26-5. The angle of the vector $E_T(\theta)$ is the phase between the source voltage and current, and its length indicates the voltage's magnitude.

The base voltage diagram may be rearranged to form a right triangle, with the hypotenuse representing the longest vector (Figure 26-6). Scale drawings

FIGURE 26-6

Vectors can be rearranged to form a right triangle.

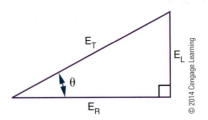

are not needed to determine magnitude because the **Pythagorean theorem** states the following:

$$E_T^2 = E_R^2 + E_L^2 \text{ or } E_T = \sqrt{E_R^2 + E_L^2}$$

This formula may be used to calculate any vector when the other two legs are known.

Vector representation also allows the use of **trigonometric functions** to determine voltage when only one voltage and the **phase angle** is known. It also determines the phase angle when the two voltages are known. These relationships are:

$$Sin\theta = E_L/E_T$$
$$Cos\theta = E_R/E_T$$
$$Tan\theta = E_L/E_R$$

Using the example in Figure 26-4 and any of the relationships shown, the phase difference may be determined between the supply voltage and current as 28.26°.

Experimenting with different values of E_L and E_R reveals several useful tips to remember:

- When a circuit is purely resistive, the phase angle is 0 because voltage and current are in phase.
- As the inductive reactance increases, the phase angle becomes greater until it reaches 45° when the resistance and the reactance are equal in value. As the inductive reactance increases further, the angle approaches 90°.
- When a circuit contains pure reactance with no resistance, the phase angle increases to 90°.

Current flow is the same through all components in a series circuit. The voltage drop across any component in the circuit is proportional to the resistance or reactance of that component. If vectors were drawn for the resistance and reactance of the circuit, they would be proportional to those drawn for the voltage (Figure 26-4). The resistance, R, would be drawn at 0° and the inductive reactance, X_L, would be drawn at 90°.

FIGURE 26-5

Vectors can be used to show the relationship between voltages in a reactive circuit.

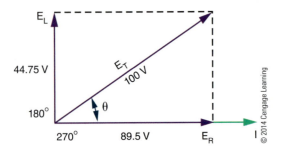

■ FIGURE 26-7

Vectors can also be used to describe impedance.

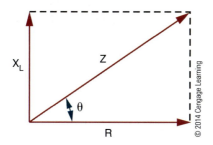

© 2014 Cengage Learning

The combined effect of resistance and reactance is called impedance and is represented by the symbol Z. Impedance must be used to calculate the current in a reactive circuit when the supply voltage is known. Dividing the source voltage by the resistance added to the reactance yields an incorrect answer because the voltages involved are not in phase with each other. Vectors may be used to describe the circuit impedance (Figure 26-7).

The series RC circuit in Figure 26-8A is described by the **vector diagrams** in Figures 26-8B and 26-8C. Again, the current is used as the zero-degree reference point, with E_R at 0° because it is in-phase with the current. Remember, in a capacitive circuit, the voltage lags 90° behind the current, so its voltage vector (E_C) is drawn downward. The phase angle of such a circuit is sometimes given a negative value, although it is just as acceptable to specify "leading" or "lagging" instead. All the same, trigonometric equations and the Pythagorean theorem may be applied to the vectors.

1. Draw a graph showing the relationship of current and voltage in a purely resistive series circuit.
2. Show the relationship between current and voltage for the circuit shown in Figure 26-9.
3. For the circuit in Figure 26-9, draw a graph showing the vector relationship. Include all arrows, and label each vector.
4. Using the Pythagorean theorem, determine the impedance for the circuit shown in Figure 26-9.
5. Determine the phase angle for the circuit shown in Figure 26-9.
6. Verify the impedance for the circuit shown in Figure 26-9.
7. Verify the current flow for the circuit shown in Figure 26-9.

■ FIGURE 26-9

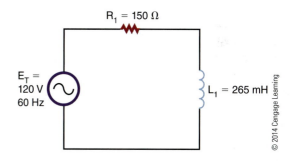

$R_1 = 150\ \Omega$

$E_T = 120\ V$ 60 Hz

$L_1 = 265\ mH$

© 2014 Cengage Learning

■ FIGURE 26-8

Vectors can be used to describe capacitive AC circuits, the same as for inductive circuits.

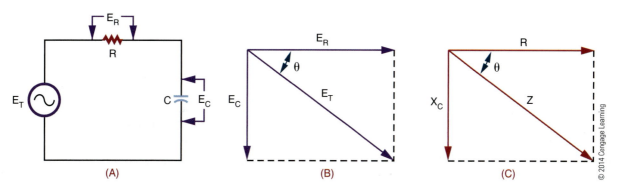

© 2014 Cengage Learning

26-2   REACTANCE IN PARALLEL CIRCUITS

Parallel circuits containing inductors and capacitors may also be analyzed with vector diagrams. However, the vectors used are current vectors because the voltage across each component must be equal and in phase with each other.

Figure 26-10 shows a parallel R_L circuit and its resulting vector diagram. The vector for the resistor current is placed at 0°. It is a simple matter to plot the vector for the inductor's current, IL. The current

through the inductor lags behind the voltage by 90°, so the vector is drawn downward.

Figure 26-11 shows the vector diagram for a parallel RC circuit. Notice that the capacitive and inductive vectors (Figure 26-10) for parallel circuits are drawn in the opposite direction to those of a series circuit. Remember, in parallel circuits current, not voltage, is examined.

Just as the resistance of a parallel circuit is always less than the value of the smallest resistor, the impedance of a parallel R_L or R_C circuit is smaller than both the individual resistance and reactance. Vectors may

■ **FIGURE 26-10**

Vectors can be used to analyze parallel inductive circuits. Current flow, not voltage, is used because the voltage across each component is equal and in phase.

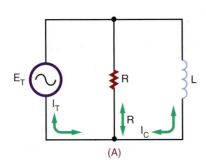
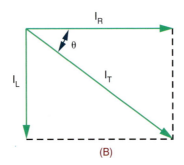
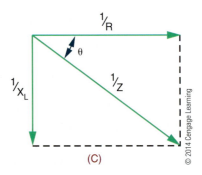

(A)   (B)   (C)

© 2014 Cengage Learning

■ **FIGURE 26-11**

Vectors can be used to analyze parallel capacitive circuits.

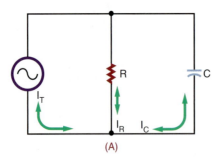
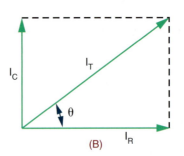

(A)   (B)

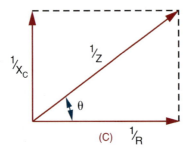

(C)

© 2014 Cengage Learning

be used to determine parallel circuit impedance; however, the reciprocal values of R, X, and Z must be used. This makes impedance calculations a little more complex for parallel circuits. All the trigonometric functions mentioned previously are applicable to parallel circuits provided that the vectors are drawn correctly. Mathematically, for Figures 26-10 and 26-11:

Inductive circuits:

$$I = \sqrt{I_R^2 + I_L^2}$$

$$1/Z = 1/R + 1/X_L \angle 90 \text{ or } \angle \arctan X_L/R$$

Capacitive circuits:

$$I = \sqrt{I_R^2 + I_C^2}$$

$$1/Z = 1/R + 1/X_C \angle -90 \text{ or } \angle \arctan X_C/R$$

26-2 QUESTIONS

1. Why is current used when analyzing parallel circuits using vector diagrams?
2. In an inductive circuit, does the current lead or lag behind the voltage?
3. In reactive circuits, is the impedance larger or smaller than the individual resistance or reactance?
4. Using the Pythagorean theorem, calculate the resultant current flow for the circuit shown in Figure 26-10. The total voltage is 120 V at 60 Hz, R = 150 Ω, and L = 265 mH.
5. Calculate the impedance of the circuit in Figure 26-12.

■ FIGURE 26-12

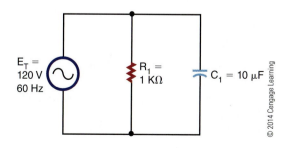

© 2014 Cengage Learning

26-3 POWER

The power consumption of a purely resistive AC circuit is easy to determine. Calculate the product of the rms current and rms voltage to obtain the average power. Figure 26-13A shows a graphical method in which instantaneous power consumption may be calculated by plotting current and voltage on the same axes and then performing successive multiplications to plot the power curve.

The same principle may be applied to the reactive circuit (b). Remember that a circuit containing just pure inductance shows current lagging voltage by 90°. Plotting the power curve for this circuit yields an alternating waveform that is centered on 0. The net power consumption of an inductive circuit is low.

During the positive portion of the waveform, the inductor takes energy and stores it in the form of a magnetic field. During the negative portion of the waveform, the field collapses and the coil returns energy to the circuit. A similar situation occurs with pure capacitance, except the capacitor stores energy as an electrostatic field and the voltage current phase relationship is reversed.

As resistance is introduced to a circuit, the phase angle becomes less than 90° and the power curve shifts to a more positive value, showing that the circuit is taking more energy than it is returning. However, the capacitive or inductive part of the circuit still stores and releases energy and consumes no power. The power loss is due entirely to the resistance. Remember, only the resistive part of the circuit consumes power.

Figure 26-14 shows a capacitive circuit. Using the Pythagorean theorem, the resistance of 100 ohms and the capacitance reactance of 50 ohms gives a combined impedance (Z) of approximately 112 ohms, as follows:

$$Z = \sqrt{R_1^2 + X_C^2}$$

$$Z = \sqrt{100^2 + 50^2}$$

$$Z = 111.8 \ \Omega$$

Using Ohm's law, this would allow a current of approximately 1 amp to flow in the circuit. The voltage drop across the resistor would be 100 volts, so the true power dissipated by the resistor would be approximately 100 watts.

FIGURE 26-13

(A) Power dissipation in a resistive circuit has a non-zero value. (B) In a reactive circuit, there is no average or net power loss.

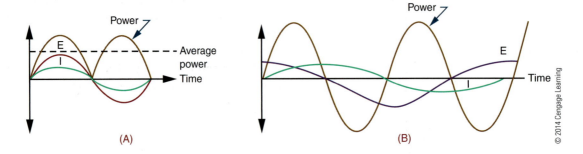

(A)   (B)

© 2014 Cengage Learning

FIGURE 26-14

In a reactive circuit, the true power dissipated with resistance and the reactive power supplied to its reactance vectorly sum to produce an apparent power vector.

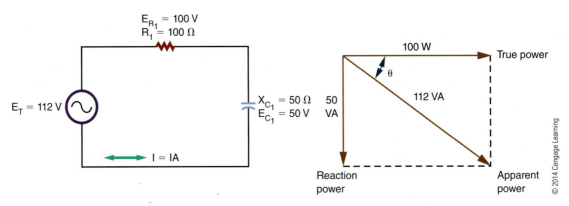

© 2014 Cengage Learning

The capacitive part of the circuit consumes no power, yet multiplying the source voltage by circuit current yields an answer of 112 watts. The difference is accounted for by the difference in phases between the various voltages. As opposed to true power, the figure obtained by multiplying the source voltage and current is known as **apparent power**, and it is specified as 112 VA, or 112 volt-amperes. The true power of the circuit can never be greater than the apparent power. The ratio of true power in watts to apparent power in volt-amps is called the **power factor**.

Vector diagrams may be used to analyze the power factor. If the phase angle of the circuit is known, the power factor may be calculated by taking the cosine of the angle. In the circuit shown in Figure 26-14, it is the same as the ratio of E_R to source voltage. Several methods are used to obtain the power factor based on the data available. In a pure resistive circuit, the true power is the same as the apparent power, so the power factor is 1. In a pure inductive or capacitive circuit, the true power is 0, no matter what the apparent power is, so the power factor is 0. Power factor is an important consideration in heavy industrial power distribution where the cables must be capable of handling the apparent power load.

26-3   QUESTIONS

1. What is the net power consumption of an inductive circuit? Why?
2. What is the relationship between true power and apparent power?
3. Define *power factor*.
4. Calculate the apparent power for the circuit shown in Figure 26-15.
5. Why is the power factor important to industry?

■ FIGURE 26-15

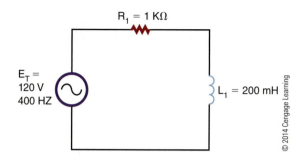

$R_1 = 1\ K\Omega$

$E_T = $ 120 V 400 HZ

$L_1 = 200\ mH$

© 2014 Cengage Learning

26-4 INTRODUCTION TO RESONANCE

Resonance is a phenomenon that occurs in many fields, including electronics. A resonance device produces a broadening and dampening effect. In electronics, resonant circuits pass desired frequencies and reject all others. The ability of a series or parallel combination of X_L or X_C to produce resonance provides a number of unique applications. Resonant circuits make it possible for a radio or television receiver to tune in and receive a station at a particular frequency. The tuning circuit consists of a coil of wire in parallel with a capacitor. Parallel tuned circuits have maximum impedance at resonant frequencies. Tuned circuits are vital to a variety of types of communication equipment, from radios to radar.

Resonance occurs when a circuit's inductive reactance and capacitive reactance are balanced. Previously, it was mentioned that inductive reactance increases with frequency, and capacitive reactance decreases with frequency. If both inductive and capacitive components are in an AC circuit, at one particular frequency their reactances will be equal but opposite. This condition is referred to as **resonance**, and a circuit that contains this characteristic is called a **resonant circuit**. Both inductance and capacitance must be present for this condition to occur.

In any resonant circuit, some resistance is usually present. Though the resistance does not have an effect on the resonant frequency, it does affect other resonant circuit parameters that are explained later in this text.

The value of inductance and capacitance determines the specific frequency of resonance of a circuit. Changing either or both results in a different resonant frequency. Typically, the larger the value of inductance and capacitance, the lower the resonant frequency.

Therefore, the smaller the value of inductance and capacitance, the higher the resonant frequency.

Above and below the resonant frequency of any LC circuit, the circuit behaves as any standard AC circuit. Resonance is desired with radio frequencies in tuning receivers and transmitters, certain industrial equipment, and test equipment. It is undesired in audio amplifiers and power supplies. Resonant circuits are not used in the audio bands of frequencies.

26-4 QUESTIONS

1. When does resonance occur?
2. Where are resonance circuits used?
3. What is the relationship between size of components and resonance frequencies?
4. What is a *resonant circuit*?
5. What happens to the frequencies above the resonant frequency of a resonant circuit?
6. Where is resonance desired in a circuit?

SUMMARY

- Ohm's law applies to AC circuits, just as it does to DC circuits.
- The AC current lags the voltage by 90° in an inductor (ELI).
- The AC current leads the voltage by 90° in a capacitor (ICE).
- Vector representation allows the use of trigonometric functions to determine voltage or current when the phase angle is known.
- The combined effect of resistance and inductance or capacitance is called impedance.
- Apparent power is obtained by multiplying the source voltage and current with units of volt-amperes.
- The ratio of true power to apparent power in volts-amperes is called the power factor.
- The power factor is very important in the consideration of heavy industrial power distribution.
- Resonance circuits make it possible for a circuit to tune a station to a particular frequency.
- Resonance is desired for radio frequency in tuning circuits.

CHAPTER 26 SELF-TEST

1. What are the values of X_C, X_L, X, Z, and I_T for the circuit shown in the following figure?

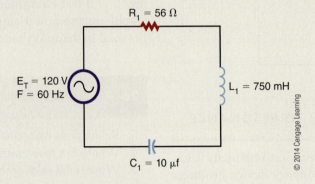

R$_1$ = 56 Ω

E$_T$ = 120 V
F = 60 Hz

L$_1$ = 750 mH

C$_1$ = 10 μf

© 2014 Cengage Learning

2. What are the values of I_C, I_L, I_X, I_R, and I_Z for the circuit shown in the following figure? What is the apparent power of this figure?

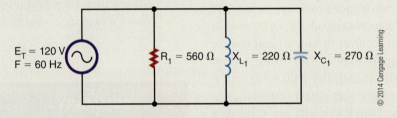

E$_T$ = 120 V
F = 60 Hz

R$_1$ = 560 Ω   X$_{L_1}$ = 220 Ω   X$_{C_1}$ = 270 Ω

© 2014 Cengage Learning

Transformers

OBJECTIVES

After completing this chapter, the student will be able to:

- Describe how a transformer operates.
- Explain how transformers are rated.
- Explain how transformers operate in a circuit.
- Describe the differences between step-up, step-down, and isolation transformers.
- Describe how the ratio of the voltage, current, and number of turns is related with a transformer.
- Describe applications of a transformer.
- Identify different types of transformers.

KEY TERMS

27-1 center-tapped secondary

27-1 coefficient of coupling

27-1 primary winding

27-1 secondary winding

27-1 transformer

27-1 volt-ampere (VA)

27-2 mutual inductance

27-3 impedance matching

27-3 step-down transformer

27-3 step-up transformer

27-3 turns ratio

27-4 isolation transformer

27-4 autotransformer

27-4 phase shift

Transformers allow the transfer of an AC signal from one circuit to another. The transfer may involve stepping up the voltage, stepping down the voltage, or passing the voltage unchanged.

27–1 ELECTROMAGNETIC INDUCTION

If two electrically isolated coils are placed next to each other and an AC voltage is put across one coil, a changing magnetic field results. This changing magnetic field induces a voltage into the second coil. This action is referred to as electromagnetic induction. The device is called a **transformer**.

In a transformer, the coil containing the AC voltage is referred to as the **primary winding**. The other coil, in which the voltage is induced, is referred to as the **secondary winding**. The amount of voltage induced depends on the amount of mutual induction between the two coils. The amount of mutual induction is determined by the **coefficient of coupling**. The coefficient of coupling is a number from 0 to 1, with 1 indicating that all the primary flux lines cut the secondary windings and 0 indicating that none of the primary flux lines cut the windings.

The design of a transformer is determined by the frequency at which it will be used, the power it must handle, and the voltage it must handle. For example, the application of the transformer determines the type of core material that the coils are wound on. For low-frequency applications, iron cores are used. For high-frequency applications, air cores with nonmetallic cores used to reduce losses associated with the higher frequencies.

Transformers are rated in **volt-amperes (VA)** rather than in power (watts). This is because of the loads that can be placed on the secondary winding. If the load is a pure capacitive load, the reactance could cause the current to be excessive. The power rating has little meaning where a volt-ampere rating can identify the maximum current the transformer can handle.

Figure 27-1 shows the schematic symbol of a transformer. The direction of the primary and secondary windings on the core determines the polarity of the induced voltage in the secondary winding. The AC voltage can either be in phase or 180° out of phase with the induced voltage. Dots are used on the schematic symbol of the transformer to indicate polarity.

■ FIGURE 27-1

Transformer schematic symbol showing phase indication.

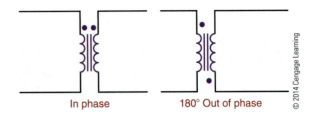

In phase   180° Out of phase

© 2014 Cengage Learning

■ FIGURE 27-2

(A) Transformers with a center-tapped secondary.
(B) Schematic symbol showing transformer with a center-tapped secondary.

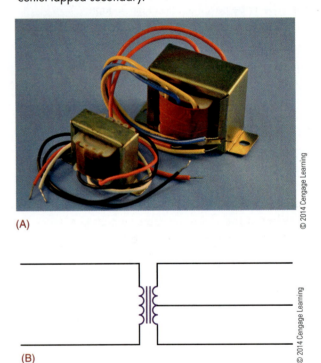

(A)

(B)

© 2014 Cengage Learning

Transformers are wound with tapped secondaries (Figure 27-2). A **center-tapped secondary** is the equivalent of two secondary windings, each with half of the total voltage across them. The center tap is used for power supply to convert AC voltages to DC voltages. A transformer may have taps on the primary to compensate for line voltages that are too high or too low.

1. How does a transformer operate?
2. What determines the design of a transformer?
3. Give an example of how the application of a transformer determines its design.
4. How are transformers rated?
5. Draw and label the schematic symbol for a transformer.

27-2 MUTUAL INDUCTANCE

When a transformer is operated without a load (Figure 27-3), there is no secondary current flow. There is a primary current flow because the transformer is connected across a voltage source. The amount of primary current depends on the size of the primary windings. The primary windings act like an inductor. Exciting current is the small amount of primary current that flows. The exciting current overcomes the AC resistance of the primary winding and supports the magnetic field of the core. Because of inductive reactance, the exciting current lags behind the applied voltage. These conditions change when a load is applied across the secondary.

When a load is connected across the secondary winding (Figure 27-4), a current is induced into the

■ FIGURE 27-3

Transformer without a load on the secondary.

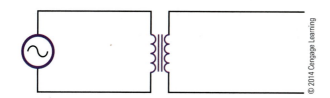

© 2014 Cengage Learning

■ FIGURE 27-4

Transformer with a loaded secondary.

NI Multisim™

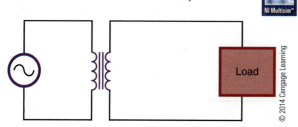

© 2014 Cengage Learning

secondary. Transformers are wound with the secondary on top of the primary. The magnetic field created by the primary current cuts the secondary windings. The current in the secondary establishes a magnetic field of its own. The expanding magnetic field in the secondary cuts the primary turns, inducing a voltage back into the primary. This magnetic field expands in the same direction as the current in the primary, aiding it and causing it to increase, with an effect called **mutual inductance**. The primary induces a voltage into the secondary, and the secondary induces a voltage back into the primary.

1. How does loading a transformer affect its operation?
2. What does a transformer act like with no load applied?
3. How is a transformer typically wound?
4. Define mutual inductance.
5. Describe how a transformer's secondary induces a voltage back into its primary.

27-3 TURNS RATIO

The turns ratio of a transformer determines whether the transformer is used to step up, step down, or pass voltage unchanged. The **turns ratio** is the number of turns in the secondary winding divided by the number of turns in the primary winding. This can be expressed as:

$$\text{Turns ratio} = \frac{N_S}{N_P}$$

where: N = number of turns
P = primary
S = secondary

A transformer with secondary voltage greater than its primary voltage is called a **step-up transformer**. The amount the voltage is stepped up depends on the turns ratio. The ratio of secondary to primary voltage is equal to the ratio of secondary to primary turns. This is expressed as

$$\frac{E_S}{E_P} = \frac{N_S}{N_P}$$

Thus the turns ratio of a step-up transformer is always greater than 1.

EXAMPLE: A transformer has 400 turns on the primary and 1200 turns on the secondary. If 120 volts of AC current are applied across the primary, what voltage is induced into the secondary?

Given	Solution
$E_S = ?$	$\dfrac{E_S}{E_P} = \dfrac{N_S}{N_P}$
$E_P = 120\,V$	
$N_S = 1200\,turns$	$\dfrac{E_S}{120} = \dfrac{1200}{400}$
$N_P = 400\,turns$	
	$E_S = 360\,V$

A transformer that produces a secondary voltage less than its primary voltage is called a **step-down transformer**. The amount the voltage is stepped down is determined by the turns ratio. In a step-down transformer, the turns ratio is always less than 1.

EXAMPLE: A transformer has 500 turns on the primary and 100 turns on the secondary. If 120 volts AC are applied across the primary, what is the voltage induced in the secondary?

Given	Solution
$E_S = ?$	$\dfrac{E_S}{E_P} = \dfrac{N_S}{N_P}$
$E_P = 120\,V$	
$N_S = 100\,turns$	$\dfrac{E_S}{120} = \dfrac{100}{500}$
$N_P = 500\,turns$	
	$E_S = 24\,V$

Assuming no transformer losses, the power in the secondary must equal the power in the primary. Although the transformer can step up voltage, it cannot step up power. The power removed from the secondary can never be more than the power supplied to the primary. Therefore, when a transformer steps up the voltage, it steps down the current, so the output power remains the same. This can be expressed as

$$P_P = P_S$$
$$(I_P)(E_P) = (I_S)(E_S)$$

The current is inversely proportional to the turns ratio. This can be expressed as

$$\frac{I_P}{I_S} = \frac{N_S}{N_P}$$

EXAMPLE: A transformer has a 10:1 turns ratio. If the primary has a current of 100 milliamperes, how much current flows in the secondary?

> **NOTE** The first number in the ratio refers to the primary, and the second number, to the secondary.

Given	Solution
$I_S = ?$	$\dfrac{I_P}{I_S} = \dfrac{N_S}{N_P}$
$N_P = 10$	
$N_S = 1$	$\dfrac{0.1}{I_S} = \dfrac{1}{10}$
$I_P = 100\,mA = 0.1\,A$	
	$I_S = 1\,A$

An important application of transformers is in **impedance matching**. Maximum power is transferred when the impedance of the load matches the impedance of the source. When the impedance does not match, power is wasted.

For example, if a transistor amplifier can efficiently drive a 100-ohm load, it will not efficiently drive a 4-ohm speaker. A transformer used between the transistor amplifier and speaker can make the impedance of the speaker appear to be in proportion. This is accomplished by choosing the proper turns ratio.

The impedance ratio is equal to the turns ratio squared. This is expressed as

$$\frac{Z_P}{Z_S} = \left(\frac{N_P}{N_S}\right)^2$$

EXAMPLE: What must the turns ratio of a transformer be to match a 4-ohm speaker to a 100-ohm source?

Given	Solution
$N_P = ?$	$\dfrac{Z_P}{Z_S} = \left(\dfrac{N_P}{N_S}\right)^2$
$N_S = ?$	
$Z_P = 100$	$\dfrac{100}{4} = \left(\dfrac{N_P}{N_S}\right)^2$
$Z_S = 4$	
	$\sqrt{25} = \dfrac{N_P}{N_S}$
	$\dfrac{5}{1} = \dfrac{N_P}{N_S}$

The turns ratio is 5:1.

27-3 QUESTIONS

1. What determines whether a transformer is a step-up or a step-down transformer?
2. Write the formula for determining the turns ratio of a transformer.
3. Write the formula for determining voltage based on the turns ratio of a transformer.
4. What is the secondary output of a transformer with 100 turns on the primary and 1800 turns on the secondary and 120 volts applied?
5. What is the turns ratio required to match a 10 kΩ source with a 600 Ω load.

27-4 APPLICATIONS

Transformers have many applications. Among them are stepping up and stepping down voltage and current, impedance matching, phase shifting, isolation, blocking DC while passing AC, and producing several signals at various voltage levels.

Transmitting electrical power to homes and industry requires the use of transformers. Power stations are located next to sources of energy, and electrical power must often be transmitted over great distances. The wires used to carry the power have resistance, which causes power loss during the transmission. The power is equal to the current times the voltage:

$$P = IE$$

Ohm's law states that current is directly proportional to voltage and inversely proportional to resistance:

$$I = \frac{E}{R}$$

The amount of power lost, then, is proportional to the amount of resistance in the line. The easiest way to reduce power losses is to keep the current low.

EXAMPLE: A power station produces 8500 volts at 10 amperes. The power lines have 100 ohms of resistance. What is the power loss of the lines?

Given
P = ?
I = 10 A
E = ?
R = 100 Ω

Solution
First, find the amount of voltage drop.

$$I = \frac{E}{R}$$

$$10 = \frac{E}{100}$$

$$E = 1000 \text{ V}$$

Using E, find the power loss.

$$P = IE$$
$$P = (10)(1000)$$
$$P = 10,000 \text{ W}$$

Using a transformer to step the voltage up to 85,000 volts at 1 ampere, what is the power loss?

Given
I = 1 A
E = ?
R = 100 Ω

Solution
First, find the amount of voltage drop.

$$I = \frac{E}{R}$$

$$1 = \frac{E}{100}$$

$$E = 10 \text{ V}$$

Using E, find the power loss.

$$P = IE$$
$$P = (1)(100)$$
$$P = 100 \text{ W}$$

How the transformer is wound determines whether it produces a **phase shift** or not. When both the primary and secondary windings are wound in the same direction, there is no phase shift of the output signal with respect to the input signal. If the secondary winding is wound in the opposite direction to the primary winding, the output waveform will be shifted 180° from the input waveform and represents a phase shift. The application determines how important the phase shift is (Figure 27-5).

NOTE Simply reversing the leads to the load can shift the phase.

■ **FIGURE 27-5**

A transformer can be used to generate a phase shift.

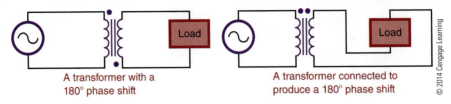

A transformer with a
180° phase shift

A transformer connected to
produce a 180° phase shift

If DC voltage is applied to a transformer, nothing occurs in the secondary once the magnetic field is established. A changing current in the primary winding is necessary to induce a voltage in the secondary winding. A transformer can be used to isolate the secondary from any DC voltage in the primary (Figure 27-6).

An **isolation transformer** is used to isolate electronic equipment from 120-volts AC, 60-hertz power while it is being tested (Figure 27-7). The reason for using a transformer is to prevent shocks. Without the transformer, one side of the power source is connected to the chassis. When the chassis is removed from the cabinet, the "hot" chassis presents a shock hazard. This condition is more likely to occur if the power cord can be plugged in either way. A transformer prevents

connecting either side of the equipment to ground. An isolation transformer does not step up or step down the voltage.

An **autotransformer** is a device used to step up or step down applied voltage. It is a special type of transformer in which the primary and secondary windings are both part of the same core. Figure 27-8A shows an autotransformer stepping down a voltage. Because the secondary consists of fewer turns, the voltage is stepped down. Figure 27-8B shows an autotransformer stepping up a voltage. Because the secondary has more turns than the primary, the voltage is stepped up. A disadvantage of the autotransformer is that the secondary is not isolated from the primary. The advantage is that the autotransformer is cheaper and easier to construct than a transformer.

A special type of autotransformer is a variable autotransformer, in which the load is connected to a movable arm and one side of the autotransformer (Figure 27-9). Moving the arm varies the turns ratio, producing a change in the voltage across the load. The output voltage can be varied from 0 to 130 VAC.

■ **FIGURE 27-6**

A transformer can be used to block DC voltage.

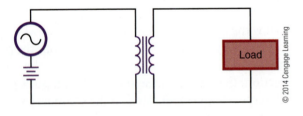

■ **FIGURE 27-7**

An isolation transformer prevents electrical shock by isolating the equipment from ground.

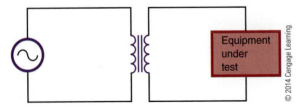

■ **FIGURE 27-8**

An autotransformer is a special type of transformer used to step-up or step-down the voltage.

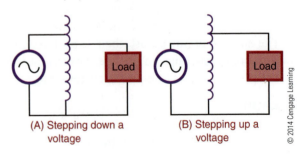

(A) Stepping down a
voltage

(B) Stepping up a
voltage

■ FIGURE 27-9

A variable autotransformer.

© 2014 Cengage Learning

27-4 QUESTIONS

1. What are the applications of transformers?
2. How are transformers used in transmitting electrical power to the home?
3. How can a transformer produce a phase shift of the input signal?
4. Why are isolation transformers important when working on electronic equipment?
5. What is an autotransformer used for?

SUMMARY

- A transformer consists of two coils: a primary winding and a secondary winding.
- An AC voltage is put across the primary winding, inducing a voltage in the secondary winding.
- Transformers allow an AC signal to be transferred from one circuit to another.
- Transformers allow stepping up, stepping down, or passing the signal unchanged.
- Transformers are designed to operate at certain frequencies.
- Transformers are rated in volt-amperes (VA).
- The schematic symbol used for iron-core transformers is as shown here:

© 2014 Cengage Learning

- The turns ratio determines whether a transformer is used to step up, step down, or pass voltage unchanged.

$$\text{Turns ratio} = \frac{N_S}{N_P}$$

- The ratio of secondary to primary voltage is equal to the ratio of secondary to primary turns.
- A transformer that produces a secondary voltage greater than its primary voltage is called a step-up transformer.
- The turns ratio of a step-up transformer is always greater than 1.
- A transformer that produces a secondary voltage less than its primary voltage is called a step-down transformer.
- The turns ratio of a step-down transformer is always less than 1.
- The amount the voltage is stepped up or down is determined by the turns ratio.
- Transformer applications include impedance matching, phase shifting, isolation, blocking DC while passing AC, and producing several signals at different voltage levels.
- An isolation transformer passes the signal unchanged.
- An isolation transformer is used to prevent electric shocks.
- An autotransformer is used to step up or step down voltage.
- An autotransformer is a special transformer that does not provide isolation.

CHAPTER 27 SELF-TEST

1. Explain how electromagnetic induction induces a voltage into the secondary of a transformer.
2. Why are transformers rated in volt-amperes rather than in watts?
3. What is the difference between two transformers, one that has voltage applied to the primary without a load on the secondary and one that has a load on the secondary?
4. How can a transformer generate a phase shift?
5. Describe how a transformer can block a DC signal.
6. What turns ratio is required on the secondary of a transformer if the primary has 400 turns?

The applied voltage is 120 VAC and the secondary voltage is 12 V.

7. What turns ratio is required for an impedance-matching transformer to match a 4-ohm speaker to a 16 Ω source?
8. Explain why transformers are important for transmitting electrical power to residential and industrial needs.
9. How does an isolation transformer prevent electrical shock?
10. Where would a variable autotransformer be used?

Section 3

1. Use various pieces of test equipment to check whether a circuit is working. Draw input and output waveforms.

2. Use Multisim to emulate various circuits and test equipment displays.

3. Design and build a tester that can provide a visual reading for a 120/240 V single-phase circuit.

4. Using fellow students, form a product development team. Assess each other's strengths. Identify who will fill the roles that will be needed to sustain a productive team (i.e., leadership, thinking, technical, bilingual, marketing, etc.), and then form the team.

5. Using available test instruments, develop a set of quality standards and criteria that apply to the electricity that your laboratory uses. Develop a plan that you can implement that monitors these quality factors and how they can be improved upon.

6. Using a development team, prepare a report or presentation electronically (i.e., PowerPoint) that outlines the current trends in the semiconductor manufacturing industry. Use graphics to clearly illustrate the process.

SECTION 4

SEMICONDUCTOR DEVICES

SEMICONDUCTOR DEVICES

John Bardeen (1908–1991)
In 1947, Bardeen co-invented the first transistor.

Walter Houser Brattain (1902–1987)
In 1947, Brattain co-invented the first transistor.

Karl Ferdinand Braun (1850–1918)
Around 1898, Braun invented a crystal diode rectifier of cat's whisker diode.

Frank William "Bill" Gutzwiller
In 1957, Gutzwiller commercialized the silicon-controlled rectifier (SCR) for General Electric.

Nick Holonyak Jr. (1928–)
In 1962, Holonyak developed the first practical visible-spectrum (red) LED and is referred to as the "father of the light-emitting diode."

Werner Jacobi (1907–1970)
1949, Jacobi filed a patent for an integrated circuit–like amplifying device with no commercial application.

Jack St. Clair Kilby (1923–2005)
In July 1958, Kilby successfully demonstrated the first working integrated circuit made of germanium.

Julius Edgar Lilienfeld (1882–1963)
In 1930, Lilienfeld was granted a patent for an FET-like transistor.

Jun-ichi Nishizawa (1926–)
In 1950, Nishizawa and colleagues invented the PIN photodiode.

Robert Noyce (1927–1990)
In 1959, Noyce came up with his own idea of an integrated circuit made of silicon a half-year later than Kilby and solved many of the practical problems that Kilby's chip had not. Noyce is referred to as the "Mayor of Silicon Valley."

Russell Shoemaker Ohl (1898–1987)
In 1939, Ohl discovered the P–N junction, which led to the development of the first silicon solar cells.

John Northrup Shive (1913–1984)
Shive of Bell Telephone laboratories invented the phototransistor in 1948, which was announced in 1950.

William Bradford Shockley Jr. (1910–1989)
In 1947, Shockley co-invented the first transistor and in 1950 proposed the silicon-controlled rectifier (SCR).

Clarence Melvin Zener (1905–1993)
Zener first described the electrical property of the zener diode in 1934, which is also named for him.

Semiconductor Fundamentals

OBJECTIVES

After completing this chapter, the student will be able to:

- Identify materials that act as semiconductors.
- Define covalent bonding.
- Describe the doping process for creating N- and P-type semiconductor materials.
- Explain how doping supports current flow in a semiconductor material.

KEY TERMS

28-1	active material
28-1	covalent bonding
28-1	germanium
28-1	intrinsic material
28-1	negative temperature coefficient
28-1	semiconductor
28-1	silicon
28-1	valence
28-1	valence shell
28-2	electron-hole pair
28-2	free electron
28-2	hole
28-3	acceptor atoms
28-3	donor atom
28-3	doping
28-3	majority carrier
28-3	minority carrier
28-3	N-type material
28-3	P-type material
28-3	pentavalent
28-3	trivalent

emiconductors are the basic components of electronic equipment. The more commonly used semiconductors are the diode (used to rectify), the transistor (used to amplify), and the integrated circuit (used to switch or amplify). The primary function of semiconductor devices is to control voltage or current for some desired result.

Advantages of semiconductors include the following:

- Small size and weight
- Low power consumption at low voltages
- High efficiency
- Great reliability
- Ability to operate in hazardous environments
- Instant operation when power is applied
- Economic mass production

Disadvantages of semiconductors include these:

- Great susceptibility to changes in temperature
- Extra components required for stabilization
- Easily damaged (by exceeding power limits or reversing polarity of operating voltage, and by excess heat when soldering into a circuit)

28-1 SEMICONDUCTION IN GERMANIUM AND SILICON

Semiconductor materials have characteristics that fall between those of insulators and conductors. Three pure semiconductor elements are carbon (C),

germanium (Ge), and silicon (Si). Those suitable for electronic applications are germanium and silicon.

Germanium is a brittle, grayish-white element discovered in 1886. A powder, germanium dioxide, is recovered from the ashes of certain types of coal. The powder is then reduced to the solid form of pure germanium.

Silicon was discovered in 1823. It is found extensively in the earth's crust as a white or sometimes colorless compound, silicon dioxide. Silicon dioxide (silica) can be found abundantly in sand, quartz, agate, and flint. It is then chemically reduced to pure silicon in a solid form. Silicon is the most commonly used semiconductor material.

Once the pure, or **intrinsic material** is available, it must be modified to produce the qualities necessary for semiconductor devices.

As described in Chapter 10, the center of the atom is the nucleus, which contains protons and neutrons. The protons have a positive charge and the neutrons have no charge. Electrons orbit around the nucleus and have a negative charge. Figure 28-1 shows the structure of the silicon atom. The first orbit contains two electrons; the second orbit contains eight electrons; and the outer orbit, or **valence shell**, contains four electrons. **Valence** is an indication of the atom's ability to gain or lose electrons and determines the electrical and chemical properties of the atom. Figure 28-2

■ FIGURE 28-1

Atomic structure of silicon.

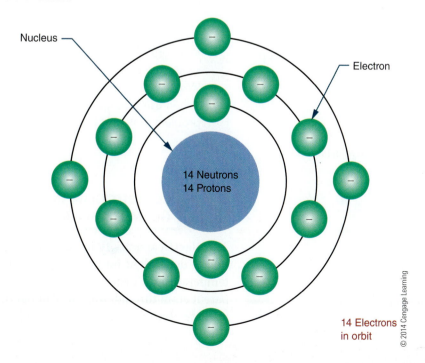

Nucleus

Electron

14 Neutrons
14 Protons

14 Electrons
in orbit

© 2014 Cengage Learning

■ FIGURE 28-2

Simplified silicon atom shown with only valence electrons.

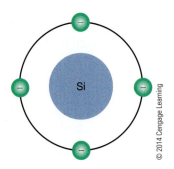

© 2014 Cengage Learning

shows a simplified drawing of the silicon atom, with only four electrons in the valence shell.

Materials that need electrons to complete their valence shell are not stable and are referred to as **active materials**. To gain stability, an active material must acquire electrons in its valence shell. Silicon atoms are able to share their valence electrons with other silicon atoms in a process called **covalent bonding** (Figure 28-3). Covalent bonding is the process of sharing valence electrons, resulting in the formation of crystals.

Each atom in such a crystalline structure has four of its own electrons and four shared electrons from four other atoms—a total of eight valence electrons. This covalent bond cannot support electrical activity because of its stability.

At room temperature, pure silicon crystals are poor conductors and behave like insulators. If heat energy is applied to the crystals, however, some of the electrons absorb the energy and move to a higher orbit, breaking the covalent bond and allowing the crystals to support current flow.

■ FIGURE 28-3

Crystalline structure of silicon with covalent bonding.

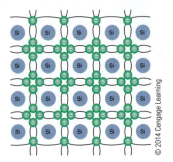

© 2014 Cengage Learning

Silicon, like other semiconductor materials, is said to have a **negative temperature coefficient**, because as the temperature increases, its resistance decreases. The resistance is cut in half for every 6 degrees Celsius of rise in temperature.

Like silicon, germanium has four electrons in its valence shell and can form a crystalline structure. Germanium's resistance is cut in half for every 10 degrees Celsius of temperature rise. Thus germanium appears to be more stable with respect to temperature change than silicon. However, germanium requires less heat energy to dislodge its electrons than does silicon. Silicon has a thousand times more resistance than germanium at room temperature.

Heat is a potential source of trouble for semiconductors and is not easy to control. Good circuit design minimizes heat changes. Its resistance is what makes silicon preferable to germanium in most circuits. In some applications, heat-sensitive devices are necessary. In these applications, the germanium temperature coefficient can be an advantage; therefore, germanium is used.

All early transistors were made of germanium. The first silicon transistor was not made until 1954. Today, silicon is used for most solid-state applications.

28-1 **QUESTIONS**

1. What is a semiconductor material?
2. Define the following terms:
   a. Covalent bonding
   b. Negative temperature coefficient
3. Why are silicon and germanium considered semiconductor materials?
4. Why is silicon preferred over germanium?
5. When using semiconductors, what is the major problem that must be controlled?

28-2 **CONDUCTION IN PURE GERMANIUM AND SILICON**

The electrical activity in semiconductor material is highly dependent on temperature. At extremely low temperatures, valence electrons are held tightly to the parent atom through the covalent bond. Because these valence electrons cannot drift, the material cannot support current flow. Germanium and silicon crystals function as insulators at low temperatures.

As the temperature increases, the valence electrons become agitated. Some of the electrons break the covalent bonds and drift randomly from one atom to the next. These **free electrons** are able to carry a small amount of electrical current if an electrical voltage is applied. At room temperature, enough heat energy is available to produce a small number of free electrons and to support a small amount of current. As the temperature increases, the material begins to acquire the characteristics of a conductor. Only at extremely high temperatures does silicon conduct current as ordinary conductors do. Typically, such high temperatures are not encountered under normal usage.

When an electron breaks away from its covalent bond, the space previously occupied by the electron is referred to as a **hole** (Figure 28-4). As described in Chapter 11, a hole simply represents the absence of an electron. Because an electron has a negative charge, its absence represents the loss of a negative charge. A hole thus has the characteristic of a positively charged particle. As an electron jumps from one valence shell to another valence shell with a hole, it leaves a hole behind it. If this action continues, the hole appears to move in the opposite direction to the electron.

■ **FIGURE 28-4**

A hole is created when an electron breaks its covalent bond.

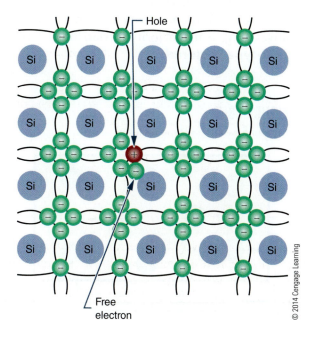

© 2014 Cengage Learning

■ **FIGURE 28-5**

Current flow in pure semiconductor material.

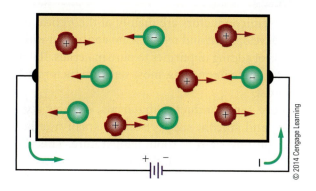

© 2014 Cengage Learning

Each corresponding electron and hole are referred to as an **electron-hole pair**. The number of electron hole pairs increases with an increase in temperature. At room temperature, a small number of electron-hole pairs exists.

When pure semiconductor material is subjected to a voltage, the free electrons are attracted to the positive terminal of the voltage source (Figure 28-5). The holes created by movement of the free electrons drift toward the negative terminal. As the free electrons flow into the positive terminal, an equal number leave the negative terminal. As the holes and electrons recombine, both holes and free electrons cease to exist.

In review, holes constantly drift toward the negative terminal of the voltage source. Electrons always flow toward the positive terminal. Current flow in a semiconductor consists of the movement of both electrons and holes. The amount of current flow is determined by the number of electron-hole pairs in the material. The ability to support current flow increases with the temperature of the material.

28-2 **QUESTIONS**

1. How can pure germanium support a current flow?
2. Describe the process of electrons moving through semiconductor material.
3. When a potential is applied to pure germanium, in what direction do the electrons and holes move?
4. What happens when holes and electrons recombine?
5. What determines the amount of current flow in a pure semiconductor material?

CONDUCTION IN DOPED GERMANIUM AND SILICON

Pure semiconductors are mainly of theoretical interest. Development and research are concerned with the effects of adding impurities to pure materials. If it were not for these impurities, most semiconductors would not exist.

Pure semiconductor materials, such as germanium and silicon, support only a small number of electron-hole pairs at room temperature. This allows for conduction of very little current. To increase their conductivity, a process called doping is used.

Doping is the process of adding impurities to a semiconductor material. Two types of impurities are used. The first, called **pentavalent**, is made of atoms with five valence electrons. Examples are arsenic and antimony. The other, called **trivalent**, is made of atoms with three valence electrons. Examples are indium and gallium.

When pure semiconductor material is doped with a pentavalent material such as arsenic (As), some of the existing atoms are displaced with arsenic atoms (Figure 28-6). The arsenic atom shares four of its valence electrons with adjacent silicon atoms in a covalent bond. Its fifth electron is loosely attached to the nucleus and is easily set free.

The arsenic atom is referred to as a **donor atom** because it gives its extra electron away. There are many donor atoms in a semiconductor material that has been doped. This means that many free electrons are available to support current flow.

At room temperature the number of donated free electrons exceeds the number of electron-hole pairs. This means that there are more electrons than holes. The electrons are therefore called the **majority carrier**. The holes are **minority carriers**. Because the negative charge is the majority carrier, the material is called **N-type material**.

If voltage is applied to N-type material (Figure 28-7), the free electrons contributed by the donor atoms flow toward the positive terminal. Additional electrons break away from their covalent bonds and flow toward the positive terminal. These free electrons, in breaking their covalent bonds, create electron-hole pairs. The corresponding holes move toward the negative terminal.

When semiconductor materials are doped with trivalent materials such as indium (I), the indium atom shares its three valence electrons with three adjacent atoms (Figure 28-8). This creates a hole in the covalent bond.

The presence of additional holes allows the electrons to drift easily from one covalent bond to the next. Because holes easily accept electrons, atoms that contribute extra holes are called **acceptor atoms**.

Under normal conditions, the number of holes greatly exceeds the number of electrons in such material. Therefore, the holes are the majority carrier and the electrons are the minority carrier. Because the positive charge is the majority carrier, the material is called **P-type material**.

■ **FIGURE 28-6**

Silicon semiconductor material doped with an arsenic atom.

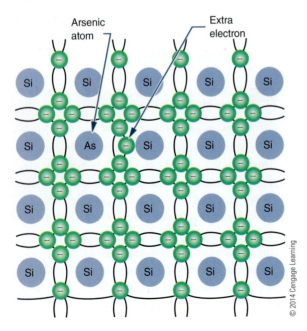

■ **FIGURE 28-7**

Current flowing in N-type material.

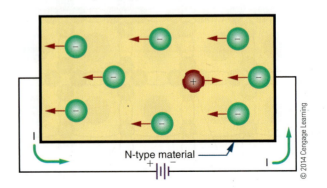

■ FIGURE 28-8

Silicon semiconductor material doped with an indium atom.

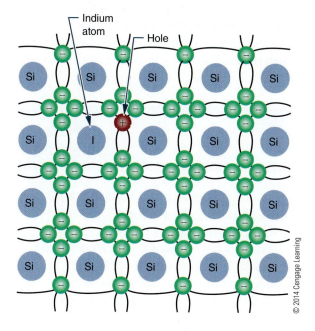

© 2014 Cengage Learning

If voltage is applied to P-type material, it causes the holes to move toward the negative terminal and the electrons to move toward the positive terminal (Figure 28-9). In addition to the holes provided by the acceptor atom, holes are produced as electrons break away from their covalent bonds, creating electron-hole pairs.

N- and P-type semiconductor materials have much higher conductivity than pure semiconductor materials. This conductivity can be increased or decreased by the addition or deletion of impurities. The more heavily a semiconductor material is doped, the lower its electrical resistance.

■ FIGURE 28-9

Current flow in P-type material.

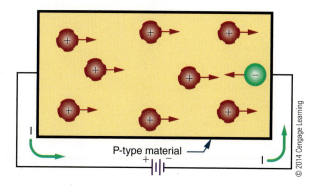

P-type material

© 2014 Cengage Learning

28–3 **QUESTIONS**

1. Describe the process of doping a semiconductor material.
2. What are the two types of impurities used for doping?
3. What determines whether a material, when doped, is an N-type or P-type semiconductor material?
4. How does doping support current flow in a semiconductor material?
5. What determines the conductivity of a semiconductor material?

SUMMARY

- Semiconductor materials are any materials with characteristics that fall between those of insulators and conductors.
- Pure semiconductor materials are germanium (Ge), silicon (Si), and carbon (C).
- Silicon is used for most semiconductor devices.
- Valence is an indication of an atom's ability to gain or lose electrons.
- Semiconductor materials have valence shells that are half full.
- Crystals are formed by atoms sharing their valence electrons through covalent bonding.
- Semiconductor materials have a negative temperature coefficient: As the temperature rises, their resistance decreases.
- Heat creates problems with semiconductor materials by allowing electrons to break their covalent bonds.
- As the temperature increases in a semiconductor material, electrons drift from one atom to another.
- A hole represents the absence of an electron in the valence shell.
- A difference of potential, applied to pure semiconductor material, creates a current flow toward the positive terminal and a hole flow toward the negative terminal.
- Current flow in semiconductor materials consists of both electron flow and hole movement.

- Doping is the process of adding impurities to a semiconductor material.
- Pentavalent materials have atoms with five valence electrons and are used to make N-type material.
- Trivalent materials have atoms with three valence electrons and are used to make P-type material.

- In N-type material, electrons are the majority carrier, and holes are the minority carrier.
- In P-type material, holes are the majority carrier, and electrons are the minority carrier.
- N- and P-type semiconductor materials have a higher conductivity than pure semiconductor materials.

CHAPTER 28 SELF-TEST

1. What makes silicon more desirable to use than germanium?
2. What happens with a negative temperature coefficient?
3. Why is covalent bonding important in the formation of semiconductor materials?
4. Describe how an electron travels through a block of pure silicon at room temperature.
5. When can germanium's temperature coefficient be an advantage?
6. How can current flow be supported in semiconductor material?
7. Describe the process of converting a block of pure silicon to N-type material.
8. How is N-type material defined?
9. Describe what happens to a block of N-type material when a voltage is applied.
10. Does doped semiconductor material have a higher or lower conductivity than pure semiconductor material?
11. How can the conductivity of semiconductor material be increased?

P–N Junction Diodes

OBJECTIVES

After completing this chapter, the student will be able to:

- Describe what a junction diode is and how it is made.
- Define depletion region and barrier voltage.
- Explain the difference between forward bias and reverse bias of a diode.
- Draw and label the schematic symbol for a diode.
- Describe three diode construction techniques.
- Identify the most common diode packages.
- Test diodes using an ohmmeter.

KEY TERMS

29-1 barrier voltage
29-1 depletion region
29-1 diode
29-1 junction diode
29-1 mobile charge
29-2 bias voltage
29-2 forward bias
29-2 forward voltage drop (E_F)
29-2 reverse bias

29-2 reverse current (I_R)
29-3 anode
29-3 cathode
29-3 maximum forward current (I_F max)
29-3 peak inverse voltage (PIV)
29-4 seed

iodes are the simplest type of semiconductor. They allow current to flow in only one direction. The knowledge of semiconductors that is acquired by studying diodes is also applicable to other types of semiconductor devices.

29-1 P–N JUNCTIONS

When pure or intrinsic semiconductor material is doped with a pentavalent or trivalent material, the doped material is called N- or P-type based on the majority carrier. The electrical charge of each type is neutral because each atom contributes an equal number of protons and electrons.

Independent electrical charges exist in each type of semiconductor material, because electrons are free to drift. The electrons and holes that drift are referred to as **mobile charges**. In addition to the mobile charges, each atom that gains an electron has more electrons than protons and assumes a negative charge. Similarly, each atom that loses an electron has more protons than electrons and therefore assumes a positive charge. As described in Chapter 10, these individual charged atoms are called negative and positive ions. There is always an equal number of mobile and ionic charges within N-type and P-type semiconductor materials.

A **diode** is created by joining N- and P-type materials together (Figure 29-1). Where the materials come in contact with each other, a junction is formed. This device is referred to as a **junction diode**.

When the junction is formed, the mobile charges in the vicinity of the junction are strongly attracted to their opposites and drift toward the junction. As the charges accumulate, the action increases. Some electrons move across the junction and fill some of the holes near the junction in the P-type material. In the N-type material, the electrons become depleted near the junction. This region near the junction where the electrons and holes are depleted is called the **depletion region**, which extends only a short distance on either side of the junction.

There are no majority carriers in the depletion region, and the N-type and P-type materials are no longer electrically neutral. The N-type material takes on a positive charge near the junction, and the P-type material takes on a negative charge.

The depletion region does not get larger. The combining action tapers off quickly, and the region remains small. The size is limited by the opposite charges that build up on each side of the junction. As the negative charge builds up, it repels further electrons and keeps them from crossing the junction. The positive charge absorbs free electrons and aids in holding them back.

These opposite charges that build up on each side of the junction create a voltage, referred to as the **barrier voltage**. It can be represented as an external voltage source, even though it exists across the P–N junction (Figure 29-2).

The barrier voltage is quite small, measuring only several tenths of a volt. Typically, the barrier voltage is 0.3 V for a germanium P–N junction and 0.7 V for silicon P–N junction. This voltage becomes apparent when an external voltage source is applied.

■ FIGURE 29-1

Diode formed by joining P- and N-type material to form a PN junction.

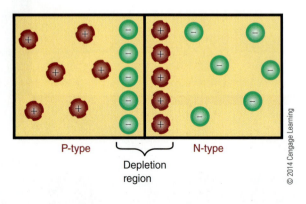

P-type   Depletion region   N-type

© 2014 Cengage Learning

■ FIGURE 29-2

Barrier voltage as it exists across a PN junction.

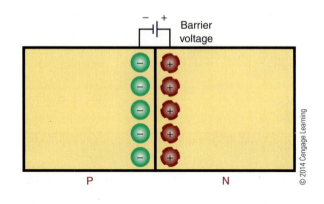

Barrier voltage

P   N

© 2014 Cengage Learning

QUESTIONS

1. Define the following terms:
   a. Donor atom
   b. Acceptor atom
   c. Diode
2. What occurs when an N-type material is joined with a P-type material?
3. How is the depletion region formed?
4. What is the barrier voltage?
5. What are typical barrier voltages for germanium and silicon diodes?

29-2 DIODE BIASING

When a voltage is applied to a diode, it is referred to as a **bias voltage**. Figure 29-3 shows a P–N junction diode connected to a voltage source. A resistor is added for limiting current to a safe value.

In the circuit shown, the negative terminal of the voltage source is connected to the N-type material. This forces electrons away from the terminal, toward the P–N junction. The free electrons that accumulate on the P side of the junction are attracted by the positive terminal. This action cancels the negative charge on the P side; the barrier voltage is eliminated, and a current is able to flow. Current flow occurs only if the external voltage is greater than the barrier voltage.

The voltage source supplies a constant flow of electrons, which drift through the N-type material along

with the free electrons contained in it. Holes in the P material also drift toward the junction. The holes and electrons combine at the junction and appear to cancel each other. However, as the electrons and holes combine, new electrons and holes appear at the terminals of the voltage source. The majority carriers continue to move toward the P–N junction as long as the voltage source is applied.

Electrons flow through the P side of the diode, attracted by the positive terminal of the voltage source. As electrons leave the P material, holes are created that drift toward the P–N junction, where they combine with other electrons. When the current flows from the N-type toward the P-type material, the diode is said to have a **forward bias**.

The resistance of the P and N materials and the external resistance of the circuit limits the current that flows when a diode is forward biased. The diode resistance is small. Therefore, connecting a voltage directly to a forward-biased diode creates a large current flow. This can generate enough heat to destroy the diode. To limit the forward current flow, an external resistor must be connected in series with the diode.

A diode conducts current in the forward direction only if the external voltage is larger than the barrier voltage and is connected properly. A germanium diode requires a minimum forward bias of 0.3 V; a silicon diode requires a minimum forward bias of 0.7 V.

Once a diode starts conducting, a voltage drop occurs. This voltage drop is equal to the barrier voltage and is referred to as the **forward voltage drop (E_F)**. The voltage drop is 0.3 V for a germanium diode and 0.7 V for a silicon diode. The amount of forward current (I_F) is a function of the external voltage (E), the forward voltage drop (E_F), and the external resistance (R). The relationship can be shown using Ohm's law:

$$I = \frac{E}{R}$$

$$I_F = \frac{E - E_F}{R}$$

EXAMPLE: A silicon diode has an external bias voltage of 12 V with an external resistor of 150 Ω. What is the total forward current?

■ FIGURE 29-3

PN junction diode with forward bias.

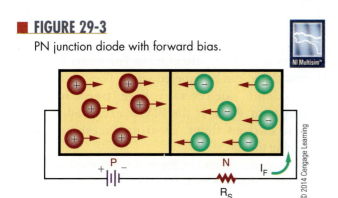

Given

$I_F = ?$

$E = 12 \text{ V}$

$R = 150 \ \Omega$

$E_F = 0.7 \text{ V}$

Solution

$$I_F = \frac{E - E_F}{R}$$

$$I_F = \frac{12 - 0.7}{150}$$

$$I_F = 0.075 \text{ A, or } 75 \text{ mA}$$

In a diode that is forward biased, the negative terminal of the external voltage source is connected to the N material, and the positive terminal is connected to the P material. If these terminals are reversed, the diode does not conduct and is said to be connected in **reverse bias** (Figure 29-4). In this configuration, the free electrons in the N material are attracted toward the positive terminal of the external voltage source. This increases the number of positive ions in the area of the P–N junction, which increases the width of the depletion region on the N side of the junction. Electrons also leave the negative terminal of the voltage source and enter the P material. These electrons fill holes near the P–N junction, causing the holes to move toward the negative terminal, which increases the width of the depletion region on the P side of the junction. The overall effect is that the depletion region is wider than in an unbiased or forward-biased diode.

The reverse-biased voltage increases the barrier voltage. If the barrier voltage is equal to the external voltage source, holes and electrons cannot support current flow. A small current flows with a reverse bias applied. This leakage current is referred to as **reverse current (I_R)** and exists because of minority carriers. At room temperature, the minority carriers are few in number. As the temperature increases, more electron-hole pairs are created. This increases the number of majority carriers and the leakage current.

All P–N junction diodes produce a small leakage current. In germanium diodes, it is measured in microamperes; in silicon diodes, it is measured in nanoamperes. Germanium has more leakage current because it is more sensitive to temperature. This disadvantage of germanium is offset by its smaller barrier voltage.

In summary, a P–N junction diode is a one-directional device. When it is forward biased, a current flows. When it is reverse biased, only a small leakage current flows. It is this characteristic that allows the diode to be used as a rectifier. A rectifier converts an AC voltage to a DC voltage.

29–2   QUESTIONS

1. What is a bias voltage?
2. What is the minimum amount of voltage needed to produce current flow across a P–N junction diode?
3. What is the total forward current for a germanium diode with a 9 V external bias voltage and a 180 Ω external resistor?
4. What is the difference between forward and reverse biasing?
5. What is leakage current in a P–N junction diode?

29–3   DIODE CHARACTERISTICS

Both germanium and silicon diodes can be damaged by excessive heat and excessive reverse voltage. Manufacturers specify the **maximum forward current (I_F max)** that can be handled safely. They also specify the maximum safe reverse voltage (**peak inverse voltage**, or **PIV**). If the PIV is exceeded, a large reverse current flows, creating excess heat and damaging the diode.

At room temperature, the reverse current is small. As the temperature increases, the reverse current increases, interfering with proper operation of the diode.

■ FIGURE 29-4

PN junction diode with reverse bias.

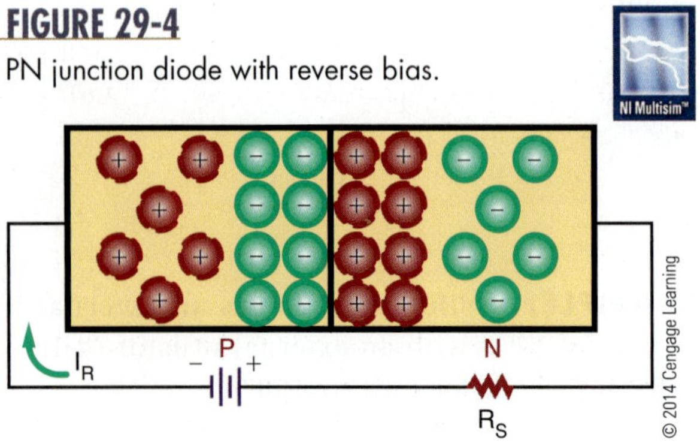

© 2014 Cengage Learning

■ FIGURE 29-5

Diode schematic symbol.

© 2014 Cengage Learning

Anode ▶ Cathode

In germanium diodes, the reverse current is higher than in silicon diodes, doubling with approximately every 10°C of increased temperature.

The diode symbol is shown in Figure 29-5. The P section is represented by an arrow, and the N section by a bar. Forward current flows from the N section to the P section (against the arrow). The N section is called the **cathode**, and the P section is called the **anode**. The cathode supplies and the anode collects the electrons.

Figure 29-6 shows a properly connected forward-biased diode. The negative terminal is connected to the cathode, and the positive terminal is connected to the anode. This setup conducts a forward current. A resistor (R_S) is added in series to limit the forward current to a safe value.

Figure 29-7 shows a diode connected in reverse bias. The negative terminal is connected to the anode, and the positive terminal is connected to the cathode. In reverse bias, a small reverse current (I_R) flows.

■ FIGURE 29-6

Diode connected with forward bias.

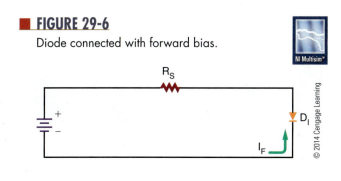

© 2014 Cengage Learning

■ FIGURE 29-7

Diode connected with reverse bias.

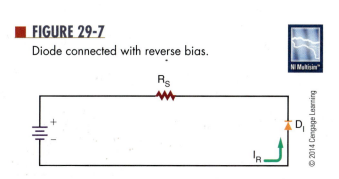

© 2014 Cengage Learning

29-3 QUESTIONS

1. What problem can a reverse current create in a germanium or silicon diode?
2. Draw and label the schematic symbol for a diode.
3. Draw a circuit that includes a forward-biased diode.
4. Draw a circuit that includes a reverse-biased diode.
5. Why should a resistor be connected in series with a forward-biased diode?

29-4 DIODE CONSTRUCTION TECHNIQUES

The P-N junction of a diode may be one of three types: a grown junction, an alloyed junction, or a diffused junction. Each involves a different construction technique.

In a grown junction method (the earliest technique used), intrinsic semiconductor material and P-type impurities are placed in a quartz container and heated until they melt. A small semiconductor crystal, called a **seed**, is then lowered into the molten mixture. The seed crystal is slowly rotated and withdrawn from the molten mixture slowly enough to allow the molten mixture to cling to the seed. The molten mixture, clinging to the seed crystal, cools and rehardens, assuming the same crystalline characteristics as the seed. As the seed crystal is withdrawn, it is alternately doped with N- and P-type impurities. Doping is the process of adding impurities to pure semiconductor crystals to increase the number of free electrons or the number of holes. This creates N and P layers in the crystal as it is grown. The resulting crystal is then sliced into many P-N sections.

The alloyed junction method of forming a semiconductor is extremely simple. A small pellet of trivalent material, such as indium, is placed on an N-type semiconductor crystal. The pellet and crystal are heated until the pellet melts and partially fuses with the semiconductor crystal. The area where the two materials combine forms the P-type material. When the heat is removed, the material recrystallizes and a solid P-N junction is formed.

The diffused junction method is the method most in use today. A mask with openings is placed on a thin section of N- or P-type semiconductor material called a wafer. The wafer is then placed in an oven and exposed to an impurity in a gaseous state. At an extremely high temperature, the impure atoms penetrate or diffuse through the exposed surfaces of the wafer. The depth of diffusion is controlled by the length of the exposure and the temperature.

Once the P–N junction is formed, the diode must be packaged to protect it from both environmental and mechanical stresses. The package must also provide a means of connecting the diode to a circuit. Package style is determined by the purpose or application of the diode (Figure 29-8). If large currents are to flow through the diode, the package must be designed to keep the junction from overheating. Figure 29-9 shows the package for diodes rated at 3 A or less. A black, white, or silver band on the end identifies the cathode.

■ FIGURE 29-8

Common diode packages.

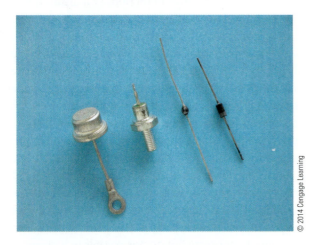

© 2014 Cengage Learning

■ FIGURE 29-9

Packages for diodes.

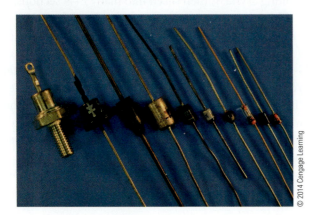

© 2014 Cengage Learning

29-4 QUESTIONS

1. Describe three methods of diode fabrication.
2. Which method of diode fabrication is favored over the others?
3. Draw four common diode packages.
4. How is the cathode identified on diode packages rated at less than 3 A?
5. Why are the diodes after fabrication put in a package?

29-5 TESTING P-N JUNCTION DIODES

A diode can be tested by checking the forward-to-reverse-resistance ratio with an ohmmeter. The resistance ratio indicates the ability of the diode to pass current in one direction and block current in the other direction.

A germanium diode has a low forward resistance of several hundred Ω. The reverse resistance is high, greater than 100,000 Ω. Silicon diodes have a higher forward and reverse resistance than germanium. An ohmmeter test of a diode should show a low forward resistance and a high reverse resistance.

CAUTION!

Some ohmmeters use high-voltage batteries that can destroy a P–N junction.

The polarity of the terminals in an ohmmeter appears at the leads of the ohmmeter: Red is positive and black is negative or common. If the positive ohmmeter lead is connected to the anode of the diode and the negative lead to the cathode, the diode is forward biased. Current should then flow through the diode, and the meter should indicate a low resistance. If the meter leads are reversed, the diode is reverse biased. Little current should flow, and the meter should measure a high resistance.

If a diode shows both low forward and low reverse resistance, it is probably shorted. If the diode measures both high forward and high reverse resistance, then it is probably opened. An accurate diode test can be made with most types of ohmmeters.

NOTE Some ohmmeters used for trouble-shooting have an open-circuit lead voltage of less than 0.3 V. This type of meter cannot be used to measure the forward resistance of a diode.

The forward resistance voltage must be larger than the barrier voltage of the diode (0.7 V for silicon and 0.3 V for germanium) for conduction to take place.

An ohmmeter can also be used to determine the cathode and anode of an unmarked diode. When there is a low reading, the positive lead is connected to the anode, and the negative lead is connected to the cathode.

29-5 QUESTIONS

1. How is a diode tested with an ohmmeter?
2. What precaution(s) must be taken when testing diodes with an ohmmeter?
3. How does the ohmmeter indicate that a diode is shorted?
4. How does the ohmmeter indicate that a diode is opened?
5. How can an ohmmeter be used to determine the cathode end of an unmarked diode?

SUMMARY

- A junction diode is created by joining N-type and P-type materials together.
- The region near the junction is referred to as the depletion region. Electrons cross the junction from the N-type to the P-type material, and thus both the holes and the electrons near the junction are depleted.
- The size of the depletion region is limited by the charge on each side of the junction.
- The charge at the junction creates a voltage called the barrier voltage.
- The barrier voltage is 0.3 V for germanium and 0.7 V for silicon.
- A current flows through a diode only when the external voltage is greater than the barrier voltage.
- A diode that is forward biased conducts current. The P-type material is connected to the positive terminal, and the N-type material is connected to the negative terminal.
- A diode that is reverse biased conducts only a small leakage current.
- A diode is a one-directional device.
- The manufacturer specifies a diode's maximum forward current and reverse voltages.
- The schematic symbol for a diode is

© 2014 Cengage Learning

- In a diode, the cathode is the N-type material, and the anode is the P-type material.
- Diodes can be constructed by the grown junction, alloyed junction, or diffused junction method.
- The diffused junction method is the one most often used.
- Packages for diodes of less than 3 A identify the cathode end of the diode with a black, white, or silver band.
- A diode is tested by comparing the forward to the reverse resistance with an ohmmeter.
- When a diode is forward biased, the resistance is low.
- When a diode is reverse biased, the resistance is high.

CHAPTER 29 SELF-TEST

1. What does a P–N junction diode accomplish?
2. Under what conditions will a silicon P–N junction diode turn on?
3. Draw examples of a P–N junction diode in forward and reverse bias. (Use schematic symbols.)
4. Describe the process for forming a diode.
5. What are the majority carriers in the depletion region?
6. What is represented as an external voltage in a diode?
7. Why must an external resistor be connected to a diode when connecting to a voltage source?
8. How much voltage must be applied to a silicon diode before it starts to conduct?
9. If a silicon diode is connected to a 12-volt source with 100 mA of current flowing through it, how big is the external resistor?
10. Can a diode be made to conduct in either direction?
11. How can the leakage current in a diode be made to increase?
12. Describe the process for testing a diode and identifying the cathode.

Zener Diodes

OBJECTIVES

After completing this chapter, the student will be able to:

- Describe the function and characteristics of a zener diode.
- Draw and label the schematic symbol for a zener diode.
- Explain how a zener diode operates as a voltage regulator.
- Describe the procedure for testing zener diodes.

KEY TERMS

30-1 breakdown voltage (E_Z)

30-1 derating factor

30-1 peak reverse voltage

30-1 power dissipation rating

30-1 zener diode

30-1 zener region

30-1 zener test current (I_{ZT})

30-2 maximum zener current (I_{ZM})

30-2 reverse current (I_R)

30-2 reverse voltage (E_R)

30-3 load current (I_L)

30-3 zener voltage rating (E_Z)

Zener diodes are closely related to P–N junction diodes. They are constructed to take advantage of reverse current. Zener diodes find wide application for controlling voltage in all types of circuits.

30–1 ZENER DIODE CHARACTERISTICS

As previously stated, a high reverse-bias voltage applied to a diode may create a high reverse current, which can generate excessive heat and cause a diode to break down. The applied reverse voltage at which the breakdown occurs is called the **breakdown voltage (E_Z)** or **peak reverse voltage**. A special diode called a **zener diode** is connected to operate in the reverse-bias mode. It is designed to operate at those voltages that exceed the breakdown voltage. This breakdown region is called the **zener region**.

When a reverse-bias voltage is applied that is high enough to cause breakdown in a zener diode, a high reverse current (I_Z) flows. The reverse current is low until breakdown occurs. After breakdown, the reverse current increases rapidly. This occurs because the resistance of the zener diode decreases as the reverse voltage increases.

The breakdown voltage, or zener voltage (E_Z), is determined by the resistivity of the diode. This, in turn, is controlled by the doping technique used during manufacturing. The rated breakdown voltage represents the reverse voltage at the **zener test current (I_{ZT})**. The zener test current is somewhat less than the maximum reverse current the diode can handle. The breakdown voltage is typically rated with 1% to 20% tolerance.

The ability of a zener diode to dissipate power decreases as the temperature increases. Therefore, **power dissipation ratings** are given for specific temperatures. Power ratings are also based on lead lengths: shorter leads dissipate more power. A **derating factor** is given by the diode manufacturer to determine the power rating at different temperatures from the ones specified in their tables. For example, a derating factor of 6 mW per degree Celsius means that the diode power rating decreases 6 mW for each degree of change in temperature.

Zener diodes are packaged like P–N junction diodes (Figure 30-1). Low-power zener diodes are mounted in either glass or epoxy. High-power zener diodes are stud mounted with a metal case. The schematic symbol for the zener diode is similar to the P–N junction diode except for the diagonal lines on the cathode bar (Figure 30-2).

■ FIGURE 30-1

Zener diode packages.

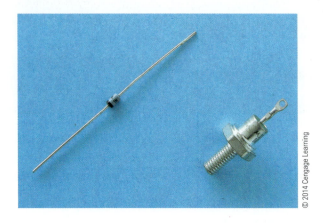

© 2014 Cengage Learning

■ FIGURE 30-2

Schematic symbol for a zener diode.

© 2014 Cengage Learning

Anode   Cathode

30–1 QUESTIONS

1. What is the unique feature of a zener diode?
2. How is a zener diode connected into a circuit?
3. What determines the voltage at which a zener diode breaks down?
4. What considerations go into determining the power dissipation rating of a zener diode?
5. Draw and label the schematic symbol used to represent a zener diode.

30–2 ZENER DIODE RATINGS

The **maximum zener current (I_{ZM})** is the maximum reverse current that can flow in a zener diode without exceeding the power dissipation rating specified by the manufacturer. The **reverse current (I_R)** represents the leakage current before breakdown and is specified at a certain **reverse voltage (E_R)**. The reverse voltage is approximately 80% of the zener voltage (E_Z).

Zener diodes that have a breakdown voltage of 5 V or more have a positive zener voltage temperature coefficient, which means that the breakdown voltage increases as the temperature increases. Zener diodes that have a breakdown voltage of less than 4 V have a negative zener voltage-temperature coefficient, which means that the breakdown voltage decreases with an increase in temperature. Zener diodes with a breakdown voltage between 4 and 5 volts may have a positive or negative voltage-temperature coefficient.

Connecting a zener diode in series with a P–N junction diode, with the P–N junction diode forward biased and the zener diode reverse biased, forms a temperature-compensated zener diode. By careful selection of the diodes, temperature coefficients can be selected that are equal and opposite. More than one P–N junction diode may be needed for proper compensation.

30-2 QUESTIONS

1. What determines the maximum zener current of a zener diode?
2. What is the difference between the maximum zener current and the reverse current for a zener diode?
3. What does a positive zener voltage temperature coefficient signify?
4. What does a negative zener voltage temperature coefficient signify?
5. How can a zener diode be temperature compensated?

30-3 VOLTAGE REGULATION WITH ZENER DIODES

A zener diode can be used to stabilize or regulate voltage. For example, it can be used to compensate for power-line voltage changes or load-resistance changes while maintaining a constant DC output.

Figure 30-3 shows a typical zener diode regulator circuit. The zener diode is connected in series with resistor RS. The resistor allows enough current to flow for the zener diode to operate in the zener breakdown region. The DC input voltage must be higher than the zener diode breakdown voltage. The voltage drop across the zener diode is equal to the zener diode's voltage rating. Zener diodes are manufactured to

■ **FIGURE 30-3**

Typical zener diode regulator circuit.

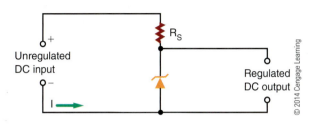

have a specific breakdown voltage rating that is often referred to as the diode's **zener voltage rating (E_Z)**. The voltage drop across the resistor is equal to the difference between the zener (breakdown) voltage and the input voltage.

The input voltage may increase or decrease. This causes the current through the zener diode to increase or decrease accordingly. When the zener diode is operating in the zener voltage, or breakdown region, a large current flows through the zener, with an increase in input voltage. However, the zener voltage remains the same. The zener diode opposes an increase in input voltage, because when the current increases, the resistance drops. This allows the zener output voltage of the zener diode to remain constant as the input voltage changes. The change in the input voltage appears across the series resistor. The resistor is in series with the zener diode, and the sum of the voltage drop must equal the input voltage. The output voltage is taken across the zener diode.

The output voltage can be increased or decreased by changing the zener diode and the series resistor. The circuit just described supplies a constant voltage. When a circuit is designed, the current in the circuit must be considered as well as the voltage. The external load requires a specific **load current (I_L)** determined by the load resistance and output voltage (Figure 30-4). The load current and the zener current

■ **FIGURE 30-4**

Zener diode voltage regulator with load.

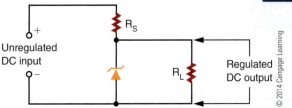

■ FIGURE 30-5

Setup for testing zener diode regulation.

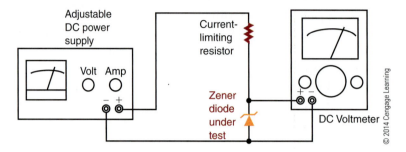

flow through the series resistor. The series resistor must be chosen so that the zener current is adequate to keep the zener diode in the breakdown zone and allow the current to flow.

When the load resistance increases, the load current decreases, which should increase the voltage across the load resistance. But the zener diode opposes any change by conducting more current. The sum of the zener current and the load current through the series resistor remains constant. This action maintains the same voltage across the series resistor.

Similarly, when the load current increases, the zener current decreases, maintaining a constant voltage. This action allows the circuit to regulate for change in output current as well as input voltage.

do not provide information on whether the zener diode is regulating at the rated value. For that information, a regulation test must be performed with a metered power supply that can indicate both voltage and current.

Figure 30-5 shows the proper setup for a zener diode regulation test. The output of the power supply is connected with a limiting resistor in series with a zener diode to be tested. A voltmeter is connected across the zener diode under test to monitor the zener voltage. The output voltage is slowly increased until the specified current is flowing through the zener diode. The current is then varied on either side of the specified zener current (I_Z). If the voltage remains constant, the zener diode is operating properly.

30-3   QUESTIONS

1. What is a practical function for a zener diode?
2. Draw a schematic diagram of a zener diode regulator circuit.
3. How can the voltage of a zener diode voltage regulator circuit be changed?
4. What must be considered in designing a zener diode voltage regulator?
5. Describe how a zener diode voltage regulator maintains a constant output voltage.

30-4   TESTING ZENER DIODES

Zener diodes can be quickly tested for opens, shorts, or leakage with an ohmmeter. The ohmmeter is connected in forward and reverse bias in the same manner as with P–N junction diodes. However, these tests

30-4   QUESTIONS

1. Describe the process for testing a zener diode with an ohmmeter.
2. What parameters are not tested when using an ohmmeter to test a zener diode?
3. Draw a schematic diagram showing how to connect a zener diode to check its breakdown voltage.
4. Describe how the circuit in question 3 operates to determine whether the zener diode is operating properly.
5. How can the cathode end of a zener diode be determined using an ohmmeter?

SUMMARY

- Zener diodes are designed to operate at voltages greater than the breakdown voltage (peak reverse voltage).
- The breakdown voltage of a zener diode is determined by the resistivity of the diode.
- Zener diodes are manufactured with a specific breakdown (zener) voltage.
- Power dissipation of a zener diode is based on temperature and lead lengths.
- The schematic symbol for a zener diode is

© 2014 Cengage Learning

- Zener diodes are packaged the same as P–N junction diodes.

- Zener diodes with a breakdown voltage greater than 5 V have a positive zener voltage- temperature coefficient.
- Zener diodes with a breakdown voltage less than 4 V have a negative zener voltage temperature coefficient.
- Zener diodes are used to stabilize or regulate voltage.
- Zener diode regulators provide a constant output voltage despite changes in the input voltage or output current.
- Zener diodes can be tested for opens, shorts, or leakage with an ohmmeter.
- To determine whether a zener diode is regulating at the proper voltage, a regulation test must be performed.

CHAPTER 30 SELF-TEST

1. What happens to a zener diode after the breakdown voltage is exceeded?
2. How is the breakdown voltage of a zener diode determined?
3. What happens to the zener diode's power-handling capabilities as its temperature is increased?
4. List the specifications that determine a zener diode's rating.
5. Draw a circuit using schematic symbols for regulating voltage on a load with a zener diode.
6. Explain how a zener diode functions in a voltage regulator circuit.
7. How does a zener diode maintain a constant voltage in a voltage regulator?
8. Draw a setup for testing a zener diode, using schematic symbols.
9. Describe the process for testing the voltage rating of a zener diode.
10. How can a zener diode test determine whether the zener diode is operating properly?

Bipolar Transistors

OBJECTIVES

After completing this chapter, the student will be able to:

- Describe how a transistor is constructed and its two different configurations.
- Draw and label the schematic symbol for an NPN transistor and a PNP transistor.
- Identify the ways of classifying transistors.
- Identify the function of a transistor using a reference manual and the identification number (2NXXXX).
- Identify commonly used transistor packages.
- Describe how to bias a transistor for operation.
- Explain how to test a transistor with both a transistor tester and an ohmmeter.
- Describe the process used for substituting a transistor.

KEY TERMS

31-1 base (B)
31-1 bipolar transistor
31-1 collector (C)
31-1 emitter (E)
31-1 junction transistor

31-1 NPN transistor
31-1 PNP transistor
31-1 transistor
31-2 TO (transistor outline)

n 1948, Bell Laboratories developed the first working junction transistor. A transistor is a three-element, two-junction device used to control electron flow. By varying the amount of voltage applied to the three elements, the amount of current can be controlled for purposes of amplification, oscillation, and switching. These applications are covered in Chapters 38, 39, and 40.

31-1 TRANSISTOR CONSTRUCTION

When a third layer is added to a semiconductor diode, a device is produced that can amplify power, current, or voltage. The device is called a **bipolar transistor**, also referred to as a **junction transistor** or **transistor**. The term *transistor* will be used here.

A transistor, like a junction diode, can be constructed of germanium or silicon, but silicon is more popular. A transistor consists of three alternately doped regions (as compared to two in a diode). The three regions are arranged in one of two ways.

In the first method, the P-type material is sandwiched between two N-type materials, forming an **NPN transistor** (Figure 31-1). In the second method, a layer of N-type material is sandwiched between two layers of P-type material, forming a **PNP transistor** (Figure 31-2).

■ FIGURE 31-1
NPN transistor.

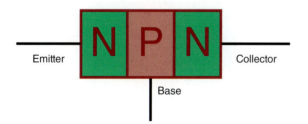

(A) Block diagram of an NPN transistor

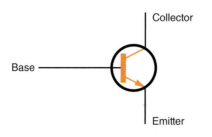

(B) Schematic symbol for an NPN transistor

© 2014 Cengage Learning

■ FIGURE 31-2
PNP transistor.

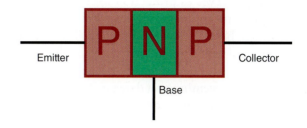

(A) Block diagram of a PNP transistor

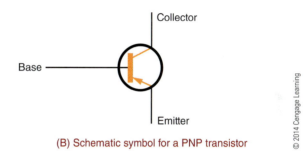

(B) Schematic symbol for a PNP transistor

© 2014 Cengage Learning

In both types of transistor, the middle region is called the **base** and the outer regions are called the **emitter** and **collector**. The emitter, base, and collector are identified by the letters **E**, **B**, and **C**, respectively.

31-1 QUESTIONS

1. How does the construction of a transistor differ from the construction of a PN junction diode?
2. What are the two types of transistors?
3. What are the three parts of a transistor called?
4. Draw and label the schematic symbols for an NPN transistor and a PNP transistor.
5. What are transistors used for?

31-2 TRANSISTOR TYPES AND PACKAGING

Transistors are classified by the following methods:

1. According to type (either NPN or PNP)
2. According to the material used (germanium or silicon)
3. According to major use (high or low power, switching, or high frequency)

Most transistors are identified by a number. This number begins with a 2 and the letter N and has up to four more digits. These symbols identify the device as a transistor and indicate that it has two junctions.

A package serves as protection for the transistor and provides a means of making electrical connections to the emitter, base, and collector regions. The package also serves as a heat sink, or an area from which heat can be extracted, removing excess heat from the transistor and preventing heat damage. Many different packages are available, covering a wide range of applications (Figure 31-3).

Transistor packages are designated by size and configuration. The most common package identifier consists of the letters **TO (transistor outline)** followed by a number. Some common transistor packages are shown in Figure 31-4.

Because of the large assortment of transistor packages available, it is difficult to develop rules for identifying the emitter, base, and collector leads of each device. It is best to refer to the manufacturer's specification sheet to identify the leads of each device.

■ FIGURE 31-3

Various transistor packages.

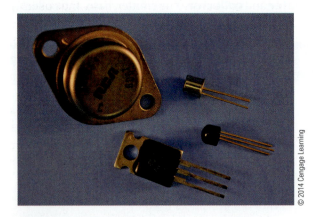

■ FIGURE 31-4

Typical transistor packages.

© 2014 Cengage Learning

31-2   QUESTIONS

1. How are transistors classified?
2. What symbols are used to identify transistors?
3. What purposes does the packaging of a transistor serve?
4. How are transistor packages labeled?
5. What is the best method for determining which leads of a transistor are the base, emitter, and collector?

31-3   BASIC TRANSISTOR OPERATION

A diode is a rectifier, and a transistor is an amplifier. A transistor may be used in a variety of ways, but its basic functions are to provide current amplification of a signal or to switch the signal.

A transistor must be properly biased by external voltages so that the emitter, base, and collector regions interact in the desired manner. In a properly biased transistor, the emitter junction is forward biased and the collector junction is reverse biased. A properly biased NPN transistor is shown in Figure 31-5.

■ FIGURE 31-5

Properly biased NPN transistor.

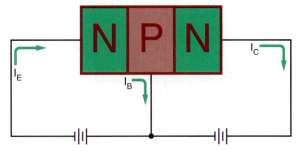

(A) Block diagram of a biased NPN transistor

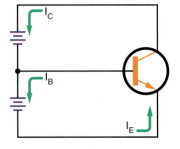

(B) Schematic diagram of a biased NPN transistor

© 2014 Cengage Learning

Electrons are caused to flow from an NPN transistor emitter by a forward bias. Forward bias is a positive voltage on the base terminal with respect to the emitter terminal. A positive potential attracts electrons, creating an electron flow from the emitter. The electrons that are attracted into the base are now influenced by the positive potential applied to the collector. The majority of electrons are attracted to the collector and into the positive side of the reverse-biased voltage source. A few electrons are absorbed into the base region and support a small electron flow from it. For this action to occur, the base region must be extremely thin. In a properly biased PNP transistor, the batteries must be reversed (Figure 31-6).

The difference between the NPN and PNP transistors is twofold: The batteries have opposite polarities, and the direction of the electron flow is reversed.

As with the diode, a barrier voltage exists within the transistor. In a transistor, the barrier voltage is produced across the emitter-base junction. This voltage must be exceeded before electrons can flow through the junction. The internal barrier voltage is determined by the type of semiconductor material used. As in diodes, the internal barrier voltage is 0.3 V for germanium transistors and 0.7 V for silicon transistors.

The collector-base junction of a transistor must also be subjected to a positive potential that is high enough to attract most of the electrons supplied by the emitter. The reverse-bias voltage applied to the collector-base junction is usually much higher than the forward-bias voltage across the emitter-base junction, supplying this higher voltage.

31-3 QUESTIONS

1. What are the basic functions of a transistor?
2. What is the proper method for biasing a transistor?
3. What is the difference between biasing an NPN transistor and biasing a PNP transistor?
4. What are the barrier voltages for a germanium transistor and a silicon transistor?
5. What is the difference between the collector-base and emitter-base junction bias voltages?

31-4 TRANSISTOR TESTING

Transistors are semiconductor devices that usually operate for long periods of time without failure. If a transistor does fail, excessively high temperature, current, or voltage generally causes the failure. Failure can also be caused by extreme mechanical stress. As a result of this electrical or mechanical abuse, a transistor may open or short internally, or its characteristics may alter enough to affect its operation. There are two methods for checking a transistor to determine whether it is functioning properly: with an ohmmeter and with a transistor tester.

A conventional ohmmeter can help detect a defective transistor in an out-of-circuit test. Resistance tests are made between the two junctions of a transistor in the following way: emitter to base, collector to base, and collector to emitter. In testing the transistor, the resistance is measured between any two terminals with the meter leads connected one way. The meter leads are then reversed. In one-meter connection, the resistance should be high, 10,000 ohms or more. In the other meter connection, the resistance should be lower, less than 10,000 ohms.

Each junction of a transistor exhibits a low resistance when it is forward biased and a high resistance when reverse biased. The battery in the ohmmeter is the source of the forward- and reverse-bias voltage. The exact resistance measured varies with different types of transistors, but there is always a change

■ **FIGURE 31-6**

Properly biased PNP transistor.

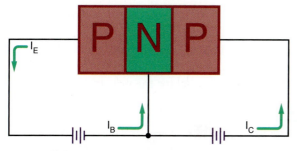

(A) Block diagram of a biased PNP transistor

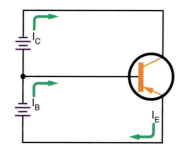

(B) Schematic diagram of a biased PNP transistor

■ **FIGURE 31-7**

Resistance measurements of transistor junctions.

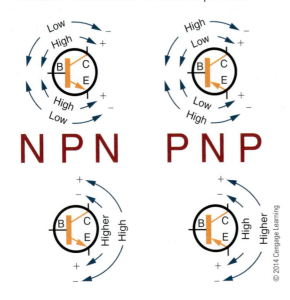

N P N   P N P

when the ohmmeter leads are reversed. This method of checking works for either NPN or PNP transistors (Figure 31-7).

If a transistor fails this test, it is defective. If it passes, it may still be defective. A more reliable means of testing a transistor is by using a transistor tester.

Transistor testers are designed specifically for testing transistors and diodes. There are two types: an in-circuit tester and an out-of-circuit tester. Both may be housed in the same package (Figure 31-8).

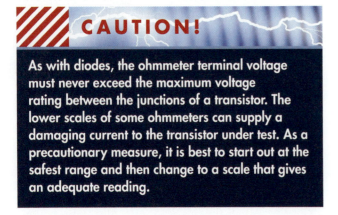

CAUTION!

As with diodes, the ohmmeter terminal voltage must never exceed the maximum voltage rating between the junctions of a transistor. The lower scales of some ohmmeters can supply a damaging current to the transistor under test. As a precautionary measure, it is best to start out at the safest range and then change to a scale that gives an adequate reading.

The transistor's ability to amplify is taken as a rough measure of its performance. There is an advantage with an in-circuit tester because the transistor does not have to be removed from the circuit. Not only can an out-of-circuit transistor tester determine

■ **FIGURE 31-8**

Transistor tester.

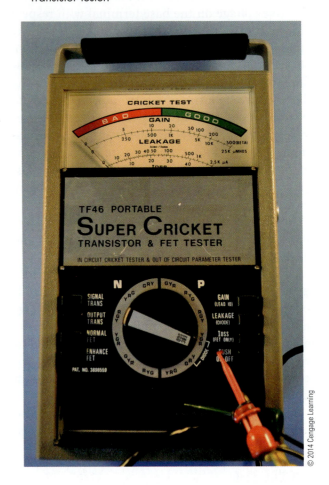

whether the transistor is good or defective, it can also determine the leakage current.

Transistor testers contain controls for adjusting voltage, current, and signal. Refer to the manufacturer's instruction manual for the proper settings.

31-4   QUESTIONS

1. What may cause a transistor to fail?
2. What are two methods of testing a transistor?
3. When using an ohmmeter, what should the results be for an NPN transistor?
4. What caution must be observed when checking a transistor with an ohmmeter?
5. What are the two types of commercial transistor testers?

31–5 TRANSISTOR SUBSTITUTION

Numerous guides have been prepared by manufacturers to provide cross-references for transistor substitution, with many being posted on the Internet. Most substitutions can be made with confidence.

If the transistor is unlisted or the number of the transistor is missing, the following procedure can be used to make an accurate replacement selection:

1. NPN or PNP? The first source of information is the symbol on the schematic diagram. If a schematic is not available, the polarity of the voltage source between the emitter and collector must be determined. If the collector voltage is positive with respect to the emitter voltage, then it is an NPN device. If the collector voltage is negative with respect to the emitter voltage, then it is a PNP device. An easy way to remember the polarity of the collector voltage for each type of transistor is shown in Figure 31-9.
2. Germanium or silicon? Measure the voltage from the emitter to the base. If the voltage is approximately 0.3 V, the transistor is germanium. If the voltage is approximately 0.7 V, the transistor is silicon.
3. Operating frequency range? Identify the type of circuit and determine whether it is working in the audio range, the kilohertz range, or the megahertz range.
4. Operating voltage? Voltages from collector to emitter, collector to base, and emitter to base should be noted either from the schematic diagram or by actual voltage measurement. The transistor selected for replacement should have voltage ratings that are at least three to four times the actual operating voltage. This helps to protect against voltage spikes, transients, and surges that are inherent in most circuits.
5. Collector current requirements? The easiest way to determine the actual current is to measure the

current in the collector circuit with an ammeter. This measurement should be taken under maximum power conditions. Again, a safety factor of three to four times the measured current should be allowed.
6. Maximum power dissipation? Use maximum voltage and collector current requirements to determine maximum power requirements (P = IE). The transistor is a major factor in determining power dissipation in the following types of circuits:
   * Input stages, AF, or RF (50 to 200 mW)
   * IF stages and driver stages (200 mW to 1 W)
   * High-power output stages (1 W and higher)
7. Current gain? The common emitter small signal DC current gain referred to as hfe or Beta (ß) should be considered. Here are some typical gain categories:
   * RF mixers, IF and AF (80 to 150)
   * RF and AF drivers (25 to 80)
   * RF and AF output (4 to 40)
   * High-gain preamps and sync separators (150 to 500)
8. Case style? Frequently, there is no difference between the case styles of original parts and recommended replacements. Case types and sizes need only be considered where an exact mechanical fit is required. Silicon grease should always be used with power devices, to promote heat transfer.
9. Lead configuration? This is not a prime consideration for replacement transistors, although it may be desirable for ease of insertion and appearance.

FIGURE 31-9
How to remember the polarity of the collector voltage.

31-5 QUESTIONS

1. Where can suggestions for transistor replacements be found?
2. Why does it matter whether a transistor is germanium or silicon?
3. Why are the operating frequency, voltage, current, and power ratings important when replacing a transistor?
4. What does the transistor's Beta refer to?
5. Are the transistor's case and lead configuration important when substituting a transistor?

SUMMARY

- A transistor is a three-layer device used to amplify and switch power and voltage.
- A bipolar transistor is also called a junction transistor or simply a transistor.
- Transistors can be configured as NPN or PNP.
- The middle region of the transistor is called the base, and the two outer regions are called the emitter and collector.
- The schematic symbols used for NPN and PNP transistors are

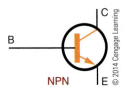
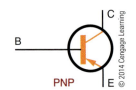

- A transistor is classified according to whether it is NPN or PNP, silicon or germanium, high or low power, and switching or high frequency.

- Transistors are identified with a prefix of 2N followed by up to four digits.
- The transistor package provides protection, a heat sink, and a support for the leads.
- Transistor packages are identified with the letters TO (transistor outline).
- In a properly biased transistor, the emitter-base junction is forward biased and the collector-base junction is reverse biased.
- PNP transistor bias sources are the reverse of NPN bias sources.
- The internal barrier voltage for germanium transistors is 0.3 V and for silicon transistors is 0.7 V.
- The reverse-bias voltage applied to the collector-base junction is higher than the forward-bias voltage applied to the emitter-base junction.
- When a transistor is tested with an ohmmeter, each junction exhibits a low resistance when it is forward biased and a high resistance when it is reverse biased.
- Transistor testers are available for testing transistors in and out of circuit.

CHAPTER 31 SELF-TEST

1. Describe the construction of both NPN and PNP transistors.
2. How is a transistor identified after it is packaged?
3. What functions does a package offer a transistor?
4. How can the leads of a transistor package be determined?
5. What is the difference between a diode and a transistor?
6. The junction of a transistor can be forward biased, reverse biased, or unbiased. What are the normal conditions of bias across the emitter-base and collector-base junctions of a transistor?
7. When checking a good transistor with an ohmmeter, what kind of resistance should exist across each junction?
8. Using an ohmmeter, what difficulty, if any, would be experienced in identifying the transistor material type and emitter, base, and collector leads of an unknown transistor?
9. When connecting a transistor in a circuit, why must the technician know whether a transistor is NPN or PNP?
10. How does the testing of a transistor with an ohmmeter compare to testing a transistor with a transistor tester?

Field Effect Transistors (FETs)

OBJECTIVES

After completing this chapter, the student will be able to:

- Describe the differences among transistors, JFETs, and MOSFETs.
- Draw schematic symbols for both P-channel and N-channel JFETs, depletion MOSFETs, and enhancement MOSFETs.
- Describe how a JFET, a depletion MOSFET, and an enhancement MOSFET operate.
- Identify the parts of JFETs and MOSFETs.
- Describe the safety precautions that must be observed when handling MOSFETs.
- Describe the procedure for testing JFETs and MOSFETs with an ohmmeter.

KEY TERMS

32-1 channel

32-1 drain (D)

32-1 drain current (I_D)

32-1 drain-to-source voltage (E_{DS})

32-1 gate (G)

32-1 gate-to-source cutoff voltage ($E_{GS\ (OFF)}$)

32-1 gate-to-source voltage (E_{GS})

32-1 junction field effect transistor (JFET)

32-1 N-channel JFET

32-1 P-channel JFET

32-1 pinch off

32-1 pinch-off voltage (E_P)

32-1 source (S)

32-1 substrate

32-2 depletion mode

32-2 enhancement mode

32-2 metal-oxide semiconductor field effect transistor (MOSFET or MOST)

32-2 N-channel depletion MOSFET

32-2 P-channel depletion MOSFET

32-3 N-channel enhancement MOSFET

32-3 P-channel enhancement MOSFET

32-4 transient

The history of field effect transistors (FETs) dates back to 1925, when Julius Lillenfield invented both the junction FET and the insulated gate FET. Both of these devices currently dominate today's electronics technology. This chapter is an introduction to the theory of junction and insulated gate FETs.

32–1 JUNCTION FETs

The **junction field effect transistor (JFET)** is a unipolar transistor that functions using only majority carriers. The JFET is a voltage-operated device. JFETs are constructed from N-type and P-type semiconductor materials and are capable of amplifying electronic signals, but they are constructed differently from bipolar transistors and operate on different principles. Knowing how a JFET is constructed helps to understand how it operates.

Construction of a JFET begins with a substrate, or base, of lightly doped semiconductor material. The **substrate** can be either P- or N-type material. The P–N junction in the substrate is made using both the growth and diffusion methods (see Chapter 29). The shape of the P–N junction is important. Figure 32-1 shows a cross section of the embedded region within the substrate. The U-shaped region is called the **channel** and is flush with the upper surface of the substrate. When the channel is made of N-type material in a P-type substrate, an **N-channel JFET** is formed. When the channel is made of P-type material in an N-type substrate, a **P-channel JFET** is formed.

Three electrical connections are made to a JFET (Figure 32-2). One lead is connected to the substrate to form the **gate (G)**. One lead is connected to each end of the channel to form the **source (S)** and the **drain (D)**. It does not matter which lead is attached to the source or drain, because the channel is symmetrical.

■ FIGURE 32-2

Lead connections for an N-channel JFET.

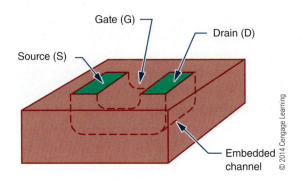

The operation of a JFET requires two external bias voltages. One of the voltage sources (E_{DS}) is connected between the source and the drain, forcing the current to flow through the channel. The other voltage source (E_{GS}) is connected between the gate and the source. It controls the amount of current flowing through the channel. Figure 32-3 shows a properly biased N-channel JFET.

Voltage source EDS is connected so that the source is made negative with respect to the drain. This causes a current to flow, because the majority carriers are electrons in the N-type material. The source-to-drain current is called the FET's drain current (I_D). The channel serves as resistance to the supply voltage (E_{DS}).

The **gate-to-source voltage (E_{GS})** is connected so that the gate is negative with respect to the source. This causes the P–N junction formed by the gate and channel to be reverse biased. This creates a depletion region in the vicinity of the P–N junction, which spreads inward along the length of the channel. The depletion region is wider at the drain end because the E_{DS} voltage adds to the E_{GS} voltage, creating a higher-reverse bias voltage than that appearing across the source end.

■ FIGURE 32-1

Cross section of an N-channel JFET.

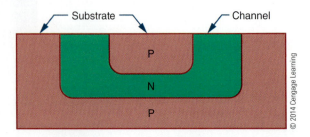

■ FIGURE 32-3

Properly biased N-channel JFET.

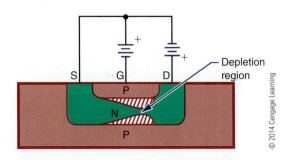

The size of the depletion region is controlled by E_{GS}. As E_{GS} increases, so does the depletion region. A decrease in E_{GS} causes a decrease in the depletion region. When the depletion region increases, it effectively reduces the size of the channel, which in turn reduces the amount of current that is able to flow through it. E_{GS} can thus be used to control the drain current (I_D) that flows through the channel. An increase in E_{GS} causes a decrease in I_D.

In normal operation, the input voltage is applied between the gate and the source. The resulting output current is the **drain current (I_D)**. In a JFET, the input voltage is used to control the output current. In a transistor, it is the input current, not the voltage, which is used to control the output current.

Because the gate-to-source voltage is reverse biased, the JFET has an extremely high input resistance. If the gate-to-source voltage were forward biased, a large current would flow through the channel, causing the input resistance to drop and reducing the gain of the device. The amount of gate-to-source voltage required to reduce I_D to zero is called the **gate-to-source cutoff voltage ($E_{GS(off)}$)**. The manufacturer of the device specifies this value.

The **drain-to-source voltage (E_{DS})** has control over the depletion region within the JFET. As E_{DS} increases, I_D also increases. A point is then reached where I_D levels off, increasing only slightly as E_{DS} continues to rise. This occurs because the size of the depletion region has increased also, to the point where the channel is depleted of minority carriers and cannot allow I_D to increase proportionally with E_{DS}. The resistance of the channel also increases with an increase of E_{DS}, with the result that I_D increases at a slower rate. However, I_D levels off because the depletion region expands and reduces the channel's width. When this occurs, I_D is said to pinch off. The value of E_{DS} required to **pinch off**, or limit, I_D is called the **pinch-off voltage (E_P)**. The manufacturer usually gives the E_P for an E_{GS} of 0. E_P is always close to $E_{GS(off)}$ when E_{GS} is equal to 0. When E_P is equal to E_{GS}, the drain current is pinched off.

P-channel and N-channel JFETs have the same characteristics. The main difference between them is the direction of the drain current (I_D) through the channel. In a P-channel JFET, the polarity of the bias voltages (E_{GS}, E_{DS}) is opposite to that in an N-channel JFET.

The schematic symbols used for N-channel and P-channel JFETs are shown in Figure 32-4. The polarities required to bias an N-channel JFET are shown in Figure 32-5 and for a P-channel JFET in Figure 32-6.

■ **FIGURE 32-4**

Schematic symbols for JFETs.

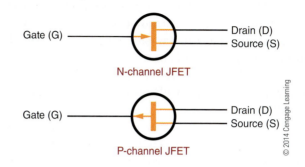

N-channel JFET

P-channel JFET

© 2014 Cengage Learning

■ **FIGURE 32-5**

The polarities required to bias an N-channel JFET.

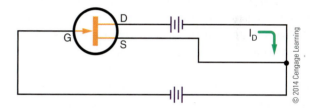

© 2014 Cengage Learning

■ **FIGURE 32-6**

The polarities required to bias a P-channel JFET.

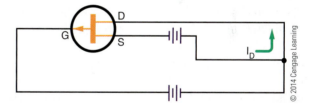

© 2014 Cengage Learning

32–1   QUESTIONS

1. Describe how a JFET differs in construction from a bipolar transistor.
2. Identify the three electrical connections to the JFET.
3. How is the current shut off in a JFET?
4. Define the following with reference to JFETs:
   a. Depletion region
   b. Pinch-off voltage
   c. Source
   d. Drain
5. Draw and label schematic diagrams of P-channel and N-channel JFETs.

32-2 DEPLETION INSULATED GATE FETs (MOSFETs)

Insulated gate FETs do not use a P–N junction. Instead, they use a metal gate, which is electrically isolated from the semiconductor channel by a thin layer of oxide. This device is known as a **metal-oxide semiconductor field effect transistor (MOSFET or MOST)**.

There are two important types of MOSFETs: N-type units with N channels and P-type units with P channels. N-type units with N channels are called **depletion mode** devices because they conduct when zero bias is applied to the gate. In the *depletion mode,* the electrons are conducting until the gate bias voltage depletes them. The drain current is depleted as a negative bias is applied to the gate. P-type units with P channels are **enhancement mode** devices. In the *enhancement mode,* the electron flow is normally cut off until it is aided or enhanced by the bias voltage on the gate. Although P-channel depletion MOSFETs and N-channel enhancement MOSFETs exist, they are not commonly used.

Figure 32-7 shows a cross section of a **N-channel depletion MOSFET**. It is formed by implanting an N channel in a P substrate. A thin insulating layer of silicon dioxide is then deposited on the channel, leaving the ends of the channel exposed to be attached to wires and act as the source and the drain. A thin metallic layer is attached to the insulating layer over the N channel. The metallic layer serves as the gate. An additional lead is attached to the substrate. The metal gate is insulated from the semiconductor channel so that the gate and the channel do not form a P–N junction. The metal gate is used to control the conductivity of the channel, as in the JFET.

The MOSFET shown in Figure 32-8 has an N channel. The drain is always made positive with respect to the source, as in the JFET. The majority carriers are electrons in the N channel, which allow drain current (I_D) to flow from the source to the drain. As with the JFET, the drain current is controlled by the gate-to-source bias voltage (E_{GS}). When the source voltage is 0, a substantial drain current flows through the device because a large number of majority carriers (electrons) are present in the channel. When the gate is made negative with respect to the source, the drain current decreases as the majority carriers are depleted. If the negative gate voltage is increased sufficiently, the drain current drops to 0. One difference between MOSFETs and JFETs is that the gate of the N-channel depletion MOSFET can also be made positive with respect to the source. This cannot be done with a JFET because it would cause the gate and channel P–N junction to be forward biased.

When the gate voltage of a depletion MOSFET is made positive, the silicon-dioxide insulating layer prevents any current from flowing through the gate lead. The input resistance remains high, and more majority carriers (electrons) are drawn into the channel, enhancing the conductivity of the channel. A positive gate voltage can be used to increase the MOSFET's drain current, and a negative gate voltage can be used to decrease the drain current. Because a negative gate voltage is required to deplete the N-channel MOSFET, it is called a depletion mode device. A large amount of drain current flows when the gate voltage is equal to 0. All depletion mode devices are considered to be turned on when the gate voltage is equal to 0.

An N-channel depletion MOSFET is represented by the schematic symbol shown in Figure 32-9. Note that the gate lead is separated from the source and drain

■ **FIGURE 32-7**

N-channel depletion MOSFET.

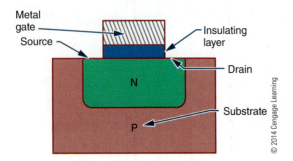

© 2014 Cengage Learning

■ **FIGURE 32-8**

N-channel depletion MOSFET with bias supply.

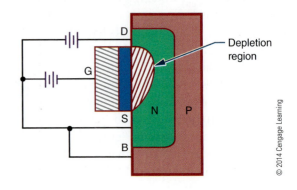

© 2014 Cengage Learning

■ FIGURE 32-9
Schematic symbol for an N-channel depletion MOSFET.

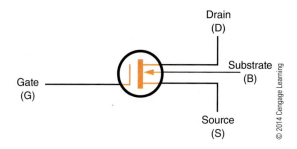

■ FIGURE 32-11
Schematic symbol for a P-channel depletion MOSFET.

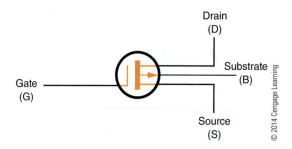

leads. The arrow on the substrate lead points inward to represent an N-channel device. Some MOSFETs are constructed with the substrate connected internally to the source lead, and the separate substrate lead is not used.

A properly biased N-channel depletion MOSFET is shown in Figure 32-10. Note that it is biased the same way as an N-channel JFET. The drain-to-source voltage (E_{DS}) must always be applied so that the drain is positive with respect to the source. The gate-to-source voltage (E_{DS}) can be applied with the polarity reversed. The substrate is usually connected to the source, either internally or externally. In special applications, the substrate may be connected to the gate or another point within the FET circuit.

A depletion MOSFET may also be constructed as a P-channel device. P-channel devices operate in the same manner as N-channel devices. The difference is that the majority carriers are holes. The drain lead is made negative with respect to the source, and the drain current flows in the opposite direction. The gate may be positive or negative with respect to the source.

The schematic symbol for a **P-channel depletion MOSFET** is shown in Figure 32-11. The only difference between the N-channel and P-channel symbols is the direction of the arrow on the substrate lead.

Both N-channel and P-channel depletion MOSFETs are symmetrical. The source and drain leads may be interchanged. In special applications, the gate may be offset from the drain region to reduce capacitance between the gate and drain. When the gate is offset, the source and drain leads cannot be interchanged.

32-2 QUESTIONS

1. How does a MOSFET differ in construction from a JFET?
2. Describe how a MOSFET conducts a current.
3. What is the major difference in operation between a MOSFET and a JFET?
4. Draw and label schematic diagrams of P-channel and N-channel MOSFETs.
5. What leads can be interchanged on JFETs and MOSFETs?

32-3 ENHANCEMENT INSULATED GATE FETs (MOSFETs)

Depletion MOSFETs are devices that are normally on. That is, they conduct a substantial amount of drain current when the gate-to-source voltage is 0. This is useful in many applications. It is also useful to have a device that is normally off—that is, a device that conducts only when a suitable value of E_{GS} is applied. Figure 32-12 shows a MOSFET that functions as a normally off device. It is similar to a depletion MOSFET, but it does not have a conducting channel. Instead, the source and drain regions are diffused separately into the substrate. The figure shows an N-type substrate and P-type source and drain regions. The opposite arrangement could also be used. The lead arrangements are the same as with a depletion MOSFET.

A **P-channel enhancement MOSFET** must be biased so that the drain is negative with respect to

■ FIGURE 32-10
Properly biased N-channel depletion MOSFET.

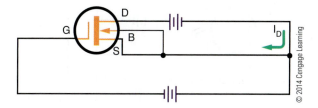

■ **FIGURE 32-12**

P-channel enhancement MOSFET.

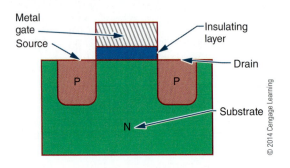

■ **FIGURE 32-14**

Properly biased P-channel enhancement MOSFET.

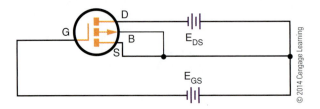

the source. When only the drain-to-source voltage (E_{DS}) is applied, a drain current does not flow. This is because there is no conducting channel between the source and drain. When the gate is made negative with respect to the source, holes are drawn toward the gate, where they gather to create a P-type channel that allows current to flow from the drain to the source. When the negative gate voltage is increased, the size of the channel increases, allowing even more current to flow. An increase in gate voltage tends to enhance the drain current.

The gate of a P-channel enhancement MOSFET can be made positive with respect to the source without affecting the operation. The MOSFET's drain current is 0 and cannot be reduced with the application of a positive gate voltage.

The schematic symbol for a P-channel enhancement MOSFET is shown in Figure 32-13. It is the same as that for a P-channel depletion MOSFET except that a broken line is used to interconnect the source, drain, and substrate region. This indicates the normally off condition. The arrow points outward to indicate a P channel.

A properly biased P-channel enhancement MOSFET is shown in Figure 32-14. Notice that E_{DS} makes

the MOSFET's drain negative with respect to the source. E_{GS} also makes the gate negative with respect to the source. Only when E_{GS} increases from 0 volts and applies a negative voltage to the gate does a substantial amount of drain current flow. The substrate is normally connected to the source, but in special applications the substrate and source may be at different potentials.

N-channel enhancement MOSFETs may also be constructed. These devices operate with a positive gate voltage so that electrons are attracted toward the gate to form an N-type channel. Otherwise, these devices function like P-channel devices.

Figure 32-15 shows the schematic symbol for an N-channel enhancement MOSFET. It is similar to the P-channel device except that the arrow points inward to identify the N channel. Figure 32-16 shows a properly biased N-channel enhancement MOSFET.

■ **FIGURE 32-15**

Schematic symbol for an N-channel enhancement MOSFET.

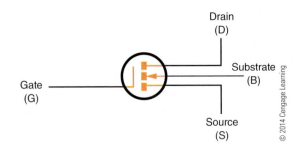

■ **FIGURE 32-13**

Schematic symbol for a P-channel enhancement MOSFET.

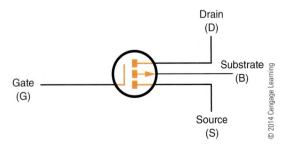

■ **FIGURE 32-16**

Properly biased N-channel enhancement MOSFET.

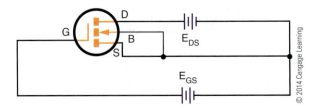

MOSFETs are usually symmetrical, like JFETs. Therefore, the source and drain can usually be reversed or interchanged.

32-3 QUESTIONS

1. How do depletion and enhancement MOSFETs differ from each other?
2. Describe how an enhancement insulated gate FET operates.
3. Draw and label the schematic symbols for P-channel and N-channel enhancement MOSFETs.
4. Why are there four leads on a MOSFET?
5. What leads on an enhancement MOSFET can be reversed?

32-4 MOSFET SAFETY PRECAUTIONS

Certain safety precautions must be observed when handling and using MOSFETs. It is important to check the manufacturer's specification sheet for maximum rating of E_{GS}.

CAUTION!

If E_{GS} is increased too much, the thin insulating layer ruptures, ruining the device. The insulating layer is so sensitive that it can be damaged by a static charge that has built up on the leads of the device. Electrostatic charges on fingers can be transferred to the MOSFET's leads when handling or mounting the device.

To avoid damage to the device, MOSFETs are usually shipped with the leads shorted together. Shorting techniques include wrapping leads with a shorting wire, inserting the device into a shorting ring, pressing the device into conductive foam, taping several devices together, shipping in antistatic tubes, and wrapping the device in metal foil.

Newer MOSFETs are protected with zener diodes electrically connected between the gate and source internally. The diodes protect against static discharges

and in-circuit transients and eliminate the need for external shorting devices. In electronics, a **transient** is a temporary component of current existing in a circuit during adjustment to a load change, voltage source difference, or line impulse.

If the following procedures are followed, unprotected MOSFETs can be handled safely:

1. Prior to installation into a circuit, the leads should be kept shorted together.
2. The hand used to handle the device should be grounded with a metallic wristband.
3. The soldering iron tip should be grounded.
4. A MOSFET should never be inserted or removed from its circuit when the power is on.

32-4 QUESTIONS

1. What is the reason that MOSFETs have to be handled very carefully?
2. What voltage source, if exceeded, can ruin a MOSFET?
3. What methods are used to protect MOSFETs during shipping?
4. What precautions have been taken to protect newer MOSFETs?
5. Describe the procedures that must be observed when handling unprotected MOSFETs.

32-5 TESTING FETs

Testing a field effect transistor is more complicated than testing a normal transistor. The following points must be considered before actually testing an FET:

1. Is the device a JFET or a MOSFET?
2. Is the FET an N-channel device or a P-channel device?
3. With MOSFETs, is the device an enhancement mode device or a depletion mode device?

Before removing an FET from a circuit or handling it, check to see whether it is a JFET or a MOSFET. MOSFETs can be damaged easily unless certain precautions in handling are followed.

1. Keep all of the leads of the MOSFET shorted until ready to use.

2. Make sure the hand used to handle the MOSFET is grounded.
3. Ensure that the power to the circuit is removed before insertion or removal of the MOSFET.

Both JFETs and MOSFETs can be tested using commercial transistor test equipment or an ohmmeter. If using commercial transistor test equipment, refer to the operations manual for proper switch settings.

Testing JFETs with an Ohmmeter

1. Use a low-voltage ohmmeter in the R X 100 range.
2. Determine the polarity of the test leads. Red is positive and black is negative.
3. Determine the forward resistance as follows:
   a. N-channel JFETs: Connect the positive lead to the gate and the negative lead to the source or drain. Because a channel connects the source and drain, only one side needs to be tested. The forward resistance should be a low reading.
   b. P-channel JFETs: Connect the negative lead to the gate and the positive lead to the source or drain.
4. Determine the reverse resistance as follows:
   a. N-channel JFETs: Connect the negative test lead of the ohmmeter to the gate and the positive test lead to the source or drain. The JFET should indicate an infinite resistance. A lower reading indicates a short or leakage.
   b. P-channel JFETs: Connect the positive test lead of the ohmmeter to the gate and the negative test lead to the source or drain.

Testing MOSFETs with an Ohmmeter

The forward and reverse resistance should be checked with a low-voltage ohmmeter on its highest scale. MOSFETs have extremely high input resistance because of the insulated gate. The meter should register an infinite resistance in both the forward- and reverse-resistance tests between the gate and source or drain. A lower reading indicates a breakdown of the insulation between the gate and source or drain.

32–5 QUESTIONS

1. What questions must be answered before actually testing an FET?
2. Why is it important to know whether a device is a JFET or a MOSFET before removing it from a circuit?
3. Describe how to test a JFET using an ohmmeter.
4. Describe how to test a MOSFET using an ohmmeter.
5. What procedure is used to test either a JFET or MOSFET with a commercial transistor tester?

SUMMARY

- A JFET uses a channel instead of junctions (as in transistors) for controlling a signal.
- The three leads of a JFET are attached to the gate, source, and drain.
- The input signal is applied between the gate and the source for controlling a JFET.
- JFETs have extremely high input resistance.
- The schematic symbols for JFETs are as follows:

N-channel
© 2014 Cengage Learning

P-channel
© 2014 Cengage Learning

- MOSFETs (insulated gate FETs) isolate the metal gate from the channel with a thin oxide layer.
- Depletion mode MOSFETs are usually N-channel devices and are classified as normally on.
- Enhancement mode MOSFETs are usually P-channel devices and are normally off.
- One difference between JFETs and MOSFETs is that the gate can be made positive or negative on MOSFETs.
- The schematic symbols for a depletion MOSFET are as follows;

N-channel
© 2014 Cengage Learning

P-channel
© 2014 Cengage Learning

- The source and drain leads can be interchanged on most JFETs and MOSFETs because the devices are symmetrical.
- The schematic symbols for an enhancement MOSFET are as follows:

N-channel
© 2014 Cengage Learning

P-channel
© 2014 Cengage Learning

- MOSFETs must be handled carefully to avoid rupture of the thin oxide layer separating the metal gate from the channel.

- Electrostatic charges from fingers can damage a MOSFET.
- Before using, keep the leads of a MOSFET shorted together.
- Wear a grounded metallic wrist strap when handling MOSFETs.
- Use a grounded soldering iron when soldering MOSFETs into a circuit, and make sure the power to the circuit is off.
- JFETs and MOSFETs can be tested using a commercial transistor tester or an ohmmeter.

CHAPTER 32 SELF-TEST

1. Draw schematic symbols for N- and P-channel JFETs and MOSFETs.
2. What are the two external bias voltages for a JFET called?
3. Explain what is meant by pinch-off voltage for an FET.
4. How is the pinch-off voltage for a JFET determined?
5. Describe what is meant by a depletion-type MOSFET.
6. What is the difference between a depletion MOSFET and an enhancement MOSFET?
7. In what mode of operation is an enhancement MOSFET likely to be cut off?
8. Develop a list of safety precautions that must be observed when handling MOSFETs.
9. Describe how a JFET can be tested with an ohmmeter.
10. Describe how a MOSFET can be tested with an ohmmeter.

Thyristors

OBJECTIVES

After completing this chapter, the student will be able to:

- Identify common types of thyristors.
- Describe how an SCR, TRIAC, or DIAC operates in a circuit.
- Draw and label schematic symbols for an SCR, TRIAC, and DIAC.
- Identify circuit applications of the different types of thyristors.
- Identify the packaging used with the different types of thyristors.
- Test thyristors using an ohmmeter.

KEY TERMS

33   bistable action

33   regenerative feedback

33   thyristor

33-1   silicon-controlled rectifier (SCR)

33-2   main terminal 1 (MT1)

33-2   main terminal 2 (MT2)

33-2   TRIAC

33-3   DIAC

Thyristors are a broad range of semiconductor components used for electronically controlled switches. They are semiconductor devices with a bistable action that depends on a PNPN regenerative feedback to stay turned on or off. **Bistable action** refers to locking onto one of two stable states. **Regenerative feedback** is a method of obtaining an increased output by feeding part of the output back to the input.

Thyristors are widely used for applications where DC and AC power must be controlled. They are used to apply power to a load or remove power from a load. In addition, they can also be used to regulate power or adjust the amount of power applied to a load—for example, a dimmer control for a light or a motor speed control.

33–1 SILICON-CONTROLLED RECTIFIERS

Silicon-controlled rectifiers are the best known of the thyristors and are generally referred to as **SCRs**. They have three terminals (anode, cathode, and gate) and are used primarily as switches. An SCR is basically a rectifier because it controls current in only one direction. The advantage of the SCR over a power transistor is that it can control a large current, dependent on an external circuit, with a small trigger signal. An SCR requires a current flowing to stay turned on after the gate signal is removed. If the current flow drops to 0, the SCR shuts off and a gate signal must be reapplied to turn the SCR back on. A power transistor would require 10 times the trigger signal of an SCR to control the same amount of current.

An SCR is a solid-state device consisting of four alternately doped semiconductor layers. It is made from silicon by the diffusion or diffusion-alloy method. Figure 33-1 shows a simplified diagram of an SCR. The four layers are sandwiched together to form three junctions. Leads are attached to only three of the layers to form anode, cathode, and gate.

Figure 33-2 shows the four layers divided into two 3-layer devices. They are a PNP and an NPN transistor interconnected to form a regenerative feedback pair. Figure 33-3 shows the schematic symbols for these transistors. The figure shows that the anode is positive with respect to the cathode, and the gate is open. The NPN transistor does not conduct because its emitter junction is not subject to a forward-bias voltage (provided by the PNP transistor's collector or gate signal). Because the NPN transistor's collector is not conducting, the PNP transistor is not conducting (the NPN transistor's collector provides the base drive for the PNP transistor). The circuit does not allow current to flow from the cathode to the anode under these conditions.

If the gate is made positive with respect to the cathode, the emitter junction of the NPN transistor

■ **FIGURE 33-2**

Equivalent SCR.

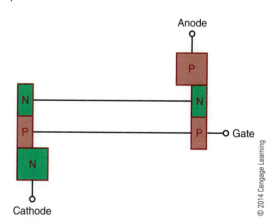

■ **FIGURE 33-3**

Schematic representation of an equivalent SCR.

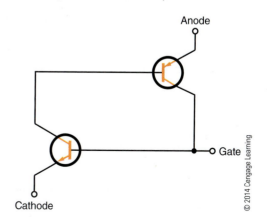

■ **FIGURE 33-1**

Simplified SCR.

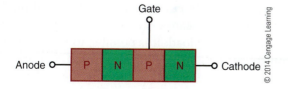

becomes forward biased and the NPN transistor conducts. This causes base current to flow through the PNP transistor, which in turn allows the PNP transistor to conduct. The collector current flowing through the PNP transistor causes the base current to flow through the NPN transistor. The two transistors hold each other in the conducting state, allowing current to flow continuously from the cathode to the anode. This action takes place even though the gate voltage is applied only momentarily. The momentary gate voltage causes the circuit to switch to the conducting state and the circuit to continue conducting even though the gate voltage is removed. The anode current is limited only by the external circuit. To switch the SCR off, it is necessary to reduce the anode-to-cathode voltage below the hold-on value, typically close to 0. This causes both transistors to turn off and remain off until a gate voltage is again applied.

The SCR is turned on by a positive input gate voltage and turned off by reducing the anode-to-cathode voltage to 0. When the SCR is turned on and is conducting a high cathode-to-anode current, it is conducting in the forward direction. If the polarity of the cathode-to-anode bias voltage is reversed, only a small leakage current flows in the reverse direction.

Figure 33-4 shows the schematic symbol for an SCR. This is a diode symbol with a gate lead attached. The leads are typically identified by the letters K (cathode), A (anode), and G (gate). Figure 33-5 shows several SCR packages.

A properly biased SCR is shown in Figure 33-6. The switch is used to apply and remove gate voltage. The resistor R_G is used to limit current to the specified gate current. The anode-to-cathode voltage is provided by the AC voltage source. The series resistor (R_L) is used to limit the anode-to-cathode current to the specified gate current when the device is turned on. Without resistor R_L the SCR would conduct an anode-to-cathode current high enough to damage the SCR.

SCRs are used primarily to control the application of DC and AC power to various types of loads. They can be used as switches to open or close circuits. They

FIGURE 33-5

Common SCR packages.

© 2014 Cengage Learning

FIGURE 33-6

Properly biased SCR.

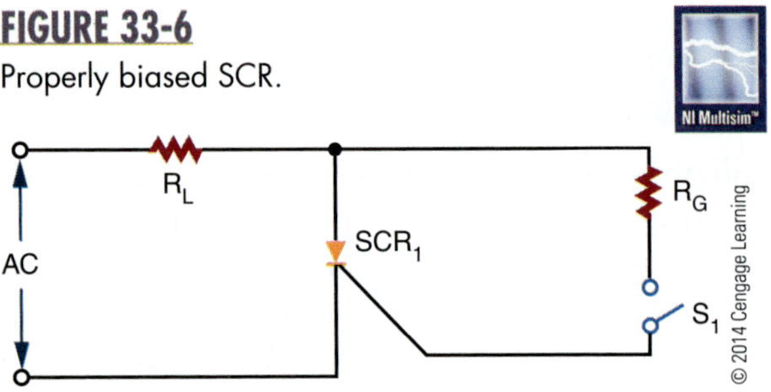

NI Multisim™

© 2014 Cengage Learning

can also be used to vary the amount of power applied to a load. In using an SCR, a small gate current can control a large load current.

When an SCR is used in a DC circuit, there is no inexpensive method of turning off the SCR without removing power from the load. This problem can be solved by connecting a switch across the SCR (Figure 33-7). When switch S_2 is closed, it shorts out the SCR. This reduces the anode-to-cathode voltage to 0, reducing the forward current to below the holding value and turning off the SCR.

When an SCR is used in an AC circuit, it is capable of conducting only one of the alternations of each AC input cycle, the alternation that makes the anode positive with respect to the cathode. When the gate current is applied continuously, the SCR conducts continuously. When the gate current is removed, the SCR turns off within one-half of the AC signal and remains off until the gate current is reapplied. It should be noted that this means only half of the available power is applied to the load. It is possible to use the SCR to control current during both alternations of each cycle. This is accomplished by rectifying the AC signal so

FIGURE 33-4

Schematic symbol for an SCR.

© 2014 Cengage Learning

■ FIGURE 33-7

Removing power in a DC circuit.

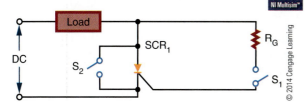

■ FIGURE 33-8

Variable half-wave circuit.

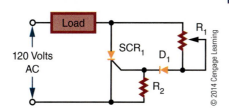

■ FIGURE 33-9

Equivalent TRIAC.

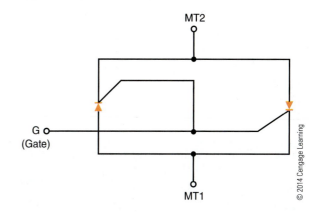

that both alternations of each cycle are made to flow in the same direction before being applied to the SCR.

Figure 33-8 shows a simple variable half-wave circuit. The circuit provides a phase shift from 0 to 90 electrical degrees of the anode voltage signal. Diode D_1 blocks the reverse gate voltage on the negative half-cycle of the anode supply voltage.

1. Why is an SCR considered better for switching than a transistor?
2. Describe how an SCR is constructed.
3. Explain how an SCR operates.
4. Draw and label the schematic symbol for an SCR.
5. What are some applications of an SCR?

33–2 TRIACs

TRIAC is an acronym for **Tri**ode **AC** semiconductor. TRIACs have the same switching characteristics as SCRs but conduct both directions of AC current flow. A TRIAC is equivalent to two SCRs connected in parallel, back to back (Figure 33-9).

Because a TRIAC can control current flowing in either direction, it is widely used to control application of AC power to various types of loads. It can be turned on by applying a gate current and turned off by reducing the operating current to below the holding level. It is designed to conduct forward and reverse current through the terminals.

Figure 33-10 shows a simplified diagram of a TRIAC. A TRIAC is a four-layer NPNP device in parallel with a PNPN device. It is designed to respond to a gating current through a single gate. The input and output terminals are identified as **main terminal 1 (MT1)** and **main terminal 2 (MT2)**. The terminals are connected across the PN junctions at the opposite ends of the device. Terminal MT1 is the reference point for measurement of voltage and current at the gate terminal. The gate (G) is connected to the P–N junction at the same end as MT1. From MT1 to MT2, the signal must pass through an NPNP series of layers or a PNPN series of layers.

■ FIGURE 33-10

Simplified TRIAC.

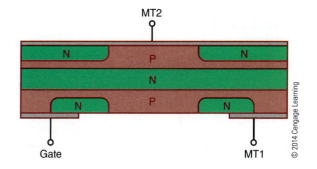

■ **FIGURE 33-11**

Schematic symbol for a TRIAC.

© 2014 Cengage Learning

■ **FIGURE 33-12**

Common TRIAC packages.

© 2014 Cengage Learning

The schematic symbol for a TRIAC is shown in Figure 33-11. It consists of two parallel diodes connected in opposite directions with a single gate lead. The terminals are identified as MT1, MT2, and G (gate). Several TRIAC packages are shown in Figure 33-12.

A TRIAC can be used as an AC switch (Figure 33-13). It can also be used to control the amount of AC power applied to a load (Figure 33-14). TRIACs apply all

■ **FIGURE 33-13**

TRIAC AC switch circuit.

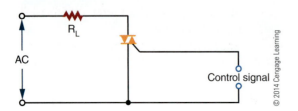

© 2014 Cengage Learning

■ **FIGURE 33-14**

TRIAC AC control circuit.

available power to the load. When a TRIAC is used to vary the amount of AC power applied to the load, a special triggering device is needed to ensure that the TRIAC functions at the proper time. The triggering device is necessary because the TRIAC is not equally sensitive to the gate current flowing in opposite directions.

TRIACs have disadvantages when compared to SCRs. TRIACs have current ratings as high as 25 amperes, but SCRs are available with current ratings as high as 1400 amperes. A TRIAC's maximum voltage rating is 500 volts, compared to an SCR's 2600 volts. TRIACs are designed for low frequency (50 to 400 hertz), whereas SCRs can handle up to 30,000 hertz. TRIACs also have difficulty switching power to inductive loads.

33–2 QUESTIONS

1. What is the difference between a TRIAC and an SCR?
2. Describe how a TRIAC is constructed.
3. Draw and label the schematic symbol for a TRIAC.
4. What are some applications of a TRIAC?
5. Compare the advantages and disadvantages of TRIACs and SCRs.

33–3 BIDIRECTIONAL TRIGGER DIODES

Bidirectional (two-directional) trigger diodes are used in TRIAC circuits because TRIACs have nonsymmetrical triggering characteristics; that is, they are not equally sensitive to gate current flowing in opposite directions. The most frequently used triggering device is the **DIAC** (**Di**ode **AC**).

The DIAC is constructed in the same manner as the transistor. It has three alternately doped layers (Figure 33-15). The only difference in the construction

■ **FIGURE 33-15**

Simplified DIAC.

© 2014 Cengage Learning

is that the doping concentration around both junctions in the DIAC is equal. Leads are only attached to the outer layers. Because there are only two leads, the device is packaged like a P–N junction diode.

Both junctions are equally doped, so a DIAC has the same effect on current regardless of the direction of flow. One of the junctions is forward biased and the other is reverse biased. The reverse-biased junction controls the current flowing through the DIAC. The DIAC performs as if it contains two P–N junction diodes connected in series back to back (Figure 33-16). The DIAC remains in the off state until an applied voltage in either direction is high enough to cause its reverse-biased junction to break over and start conducting much like a zener diode. Break over is the point at which conduction starts to occur. This causes the DIAC to turn on and the current to rise to a value limited by a series resistor in the circuit.

The schematic symbol for a DIAC is shown in Figure 33-17. It is similar to the symbol for a TRIAC. The difference is that a DIAC does not have a gate lead.

DIACs are most commonly used as a triggering device for TRIACs. Each time the DIAC turns on, it allows current to flow through the TRIAC gate, thus turning the TRIAC on. The DIAC is used in conjunction with the TRIAC to provide full-wave control of AC signals. Figure 33-18 shows a variable full-wave phase-control circuit. Variable resistor R_1 and capacitor C_1 form a phase-shift network. When the voltage across C_1 reaches the breakover voltage of the DIAC, C_1 partially discharges through the DIAC into the gate of the TRIAC. This discharge creates a pulse that triggers the TRIAC into conduction. This circuit is useful for controlling lamps, heaters, and speeds of small electrical motors.

■ **FIGURE 33-16**

Equivalent DIAC.

© 2014 Cengage Learning

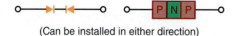

(Can be installed in either direction)

■ **FIGURE 33-17**

Schematic symbol for a DIAC.

© 2014 Cengage Learning

■ **FIGURE 33-18**

Variable full-wave phase-control circuit.

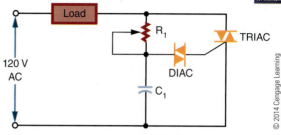

© 2014 Cengage Learning

33-3 QUESTIONS

1. Where are DIACs used in circuits?
2. Describe how a DIAC is constructed.
3. Explain how a DIAC works in a circuit.
4. Draw the schematic symbol for a DIAC.
5. Draw the schematic for a full-wave phase-control circuit using DIACs and TRIACs.

33-4 TESTING THYRISTORS

Like other semiconductor devices, thyristors may fail. They can be tested with commercial test equipment or with an ohmmeter.

To use commercial test equipment for testing thyristors, refer to the operator's manual for proper switch settings and readings.

An ohmmeter can detect the majority of defective thyristors. It cannot detect marginal or voltage-sensitive devices. It can, however, give a good indication of the condition of the thyristor.

Testing SCRs with an Ohmmeter

1. Determine the polarity of the ohmmeter leads. The red lead is positive and the black lead is negative.
2. Connect the ohmmeter leads, positive to the cathode and negative to the anode. The resistance should exceed 1 megohm.
3. Reverse the leads, negative to the cathode and positive to the anode. The resistance should again exceed 1 megohm.

4. With the ohmmeter leads connected as in step 3, short the gate to the anode (touch the gate lead to the anode lead). The resistance should drop to less than 1 megohm.

5. Remove the short between the gate and the anode. If a low-resistance range of the ohmmeter is used, the resistance should stay low. If a high-resistance range is used, the resistance should return to above 1 megohm. In the higher resistance ranges, the ohmmeter does not supply enough current to keep the gate latched (turned on) when the short is removed.

6. Remove the ohmmeter leads from the SCR and repeat the test. Because some ohmmeters do not give significant results on step 5, step 4 is sufficient.

Testing TRIACs with an Ohmmeter

1. Determine the polarity of the ohmmeter leads.
2. Connect the positive lead to MT1 and the negative lead to MT2. The resistance should be high.
3. With the leads still connected as in step 2, short the gate to MT1. The resistance should drop.
4. Remove the short. The low resistance should remain. The ohmmeter may not supply enough current to keep the TRIAC latched if a large gate current is required.
5. Remove the leads and reconnect as specified in step 2. The resistance should again be high.
6. Short the gate to MT2. The resistance should drop.
7. Remove the short. The low resistance should remain.
8. Remove and reverse the leads, the negative lead to MT1 and the positive lead to MT2. The resistance should read high.
9. Short the gate to MT1. The resistance should drop.
10. Remove the short. The low resistance should remain.
11. Remove the leads and reconnect in the same configuration. The resistance should again be high.
12. Short the gate to MT2. The resistance should drop.
13. Remove the short. The low resistance should remain.
14. Remove and reconnect the leads. The resistance should be high.

■ FIGURE 33-19
Dynamic DIAC test.

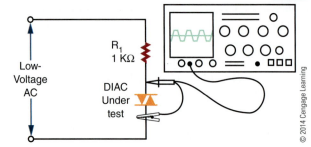

© 2014 Cengage Learning

Testing DIACs with an Ohmmeter

In testing DIACs with an ohmmeter, a low resistance in either direction indicates that the device is not opened (defective). This does not indicate a shorted device. Further testing of the DIAC requires a special circuit setup to check the voltage at the terminals (Figure 33-19).

33–4   QUESTIONS

1. Describe the switch settings and indications for using a transistor tester for testing an SCR. (Refer to the manual.)
2. Describe the switch settings and indications for using a transistor tester for testing a TRIAC. (Refer to the manual.)
3. Describe the procedure for testing an SCR with an ohmmeter.
4. Describe the procedure for testing a TRIAC with an ohmmeter.
5. Describe the procedure for testing a DIAC with an ohmmeter.

SUMMARY

- Thyristors include SCRs (silicon-controlled rectifiers), TRIACs, and DIACs.
- An SCR controls current in one direction by a positive gate signal.
- Reducing the anode-to-cathode voltage to 0 turns off an SCR.

- SCRs can be used to control current in both AC and DC circuits.
- The schematic symbol for an SCR is

© 2014 Cengage Learning

- TRIACs are bidirectional triode thyristors.
- TRIACs control current in either direction by either a positive or negative gate signal.
- The schematic symbol for a TRIAC is

© 2014 Cengage Learning

- SCRs can handle up to 1400 A compared to 25 A for TRIACs.

- SCRs have voltage ratings up to 2600 V compared to 500 V for TRIACs.
- SCRs can handle frequencies of up to 30,000 hertz compared to 400 hertz for TRIACs.
- Because TRIACs have nonsymmetrical triggering characteristics, they require the use of a DIAC.
- DIACs are bidirectional trigger diodes.
- The schematic symbol for a DIAC is

© 2014 Cengage Learning

- DIACs are mostly used as triggering devices for TRIACs.
- Thyristors can be tested using commercial transistor testers or ohmmeters.

CHAPTER 33 SELF-TEST

1. What is the difference between a P–N junction diode and an SCR?
2. After the SCR has been turned on, what effect does the anode supply voltage have on the anode current that flows through the SCR?
3. Draw a sine wave and identify on it where an SCR would be shut off.
4. What effect does the load resistance have on the current flowing through an SCR circuit?
5. What are the advantages of a TRIAC over an SCR?
6. What are the disadvantages of a TRIAC over an SCR?
7. Why is a DIAC used in the gate circuit of a TRIAC?
8. Describe the process for testing an SCR.
9. When testing thyristors, what cannot be tested with an ohmmeter?
10. Draw a schematic of the setup for testing a DIAC.

Integrated Circuits

OBJECTIVES

After completing this chapter, the student will be able to:

- Explain the importance of integrated circuits.
- Identify advantages and disadvantages of integrated circuits.
- Identify the major components of an integrated circuit.
- Describe the four processes used to construct integrated circuits.
- Identify the major integrated circuit packages.
- List the families of integrated circuits.

KEY TERMS

34-1 integrated circuit (IC)

34-2 hybrid integrated circuit

34-2 monolithic integrated circuit

34-2 thick-film integrated circuit

34-2 thin-film integrated circuit

34-2 yield

34-3 dual in-line package (DIP)

34-3 large-scale integration (LSI)

34-3 leadless chip carrier (LCC)

34-3 medium-scale integration (MSI)

34-3 pin grid array (PGA)

34-3 plastic quad flat package (PQFP)

34-3 small-scale integration (SSI)

34-3 thin small outline package (TSOP)

34-3 ultra-large-scale integration (ULSI)

34-3 very large-scale integration (VLSI)

Transistors and other semiconductor devices have made it possible to reduce the size of electronic circuits because of their small size and low power consumption. It is now possible to extend the principles behind semiconductors to complete circuits as well as individual components. The goal of the integrated circuit is to develop a single device to perform a specific function, such as amplification or switching, thus eliminating the separation between components and circuits.

Several factors have made the integrated circuit popular:

- It is reliable with complex circuits.
- It meets the need for low power consumption.
- It offers small size and weight.
- It is economical to produce.
- It offers new and better solutions to system problems.

34-1 INTRODUCTION TO INTEGRATED CIRCUITS

The first integrated circuit was conceived by Geoffrey W. A. Dummer, a radar scientist for the Royal Radar Establishment of the British Ministry of Defense. He published his idea in Washington, D.C., on May 7, 1952. But by 1956, he had still failed to build a successful circuit based on his model.

At Texas Instruments, Jack Kilby was assigned in the summer of 1958 to develop a smaller electrical circuit. His idea evolved to fabricating all the components and the chip out of the same block (monolith) of semiconductor material. In September 1958, he constructed the first integrated circuit, and it worked perfectly. Making all the parts out of the same block of material and adding the metal needed to connect them as a top layer resulted in not needing individual discrete components and wires. The circuits could be made smaller, and the manufacturing process could be automated. Jack Kilby is probably most famous for his invention of the integrated circuit, for which he received the Nobel Prize in Physics in the year 2000.

Robert Noyce of Fairchild Semiconductor independently developed his own idea for integrated circuit construction a half a year later than Jack Kilby. Noyce's circuit solved Kilby's problem of interconnecting all the components on the chip. He added the metal as a final layer and then removed some of it so that the wires needed to connect the components were formed. This made the integrated circuit suitable for mass production.

Today, an **integrated circuit (IC)** is a complete electronic circuit in a package no larger than that of a conventional low-power transistor. The circuit consists of diodes, transistors, resistors, and capacitors. Integrated circuits are produced with the same technology and materials used in making transistors and other semiconductor devices.

The most obvious advantage of the integrated circuit is its small size. An integrated circuit is constructed of a chip of semiconductor material approximately one-eighth of an inch square. It is because of the integrated circuit's small size that it finds extensive use in military and aerospace programs. The integrated circuit has also transformed the calculator from a desktop instrument to a handheld instrument. Computer systems, once the size of rooms, are now available in portable models because of integrated circuits.

This small, integrated circuit consumes less power and operates at higher speeds than a conventional transistor circuit. The electron travel time is reduced by direct connection of the internal components.

Integrated circuits are more reliable than directly connected transistor circuits. In the integrated circuit, internal components are connected permanently. All the components are formed at the same time, reducing the chance for error. After the integrated circuit is formed, it is pretested before final assembly.

Many integrated circuits are produced at the same time. This results in substantial cost savings. Manufacturers offer a complete and standard line of integrated circuits. Special-purpose integrated circuits can still be produced to specification, but this results in higher costs if quantities are small.

Integrated circuits reduce the number of parts needed to construct electronic equipment. This reduces inventory, resulting in less overhead for the manufacturer and further reducing the cost of electronic equipment.

Integrated circuits do have some disadvantages. They cannot handle large amounts of current or voltage. High current generates excessive heat, damaging the device. High voltage breaks down the insulation between the various internal components. Most integrated circuits are low-power devices, consuming from 5 to 15 volts in the milliamp range. This results in power consumption of less than 1 watt.

Only four types of components are included in integrated circuits: diodes, transistors, resistors, and capacitors. Diodes and transistors are the easiest to construct. Resistors increase in size as they increase in resistance. Capacitors require more space than resistors and also increase in size as their capacity increases.

Integrated circuits cannot be repaired. This is because the internal components cannot be separated. Therefore, problems are identified by individual circuit instead of by individual component. The advantage of this disadvantage is that it greatly simplifies maintaining highly complex systems. It also reduces the amount of time required for maintenance personnel to service equipment.

When all factors are considered, the advantages outweigh the disadvantages. Integrated circuits reduce the size, weight, and cost of electronic equipment while increasing its reliability. As integrated circuits become more sophisticated, they become capable of performing a wider range of tasks.

34–1 QUESTIONS

1. Define integrated circuit.
2. What are the advantages of integrated circuits?
3. What are the disadvantages of integrated circuits?
4. What components can be included in integrated circuits?
5. What is the procedure for repairing a faulty integrated circuit?

34–2 INTEGRATED CIRCUIT CONSTRUCTION TECHNIQUES

Integrated circuits are classified according to their construction techniques. The more common construction techniques are monolithic, thin film, thick film, and hybrid.

Monolithic integrated circuits are constructed in the same manner as transistors but include a few additional steps. The integrated circuit begins with a circular silicon wafer 3 to 4 inches in diameter and about 0.010 inch thick. This serves as a substrate (base) on which the integrated circuits are formed. Many integrated circuits are formed at the same time on this wafer—up to several hundred, depending on the size of the wafer. The integrated circuits on the wafer are generally all the same size and type and contain the same number and type of components (Figure 34-1).

■ FIGURE 34-1

Integrated circuits on a wafer.

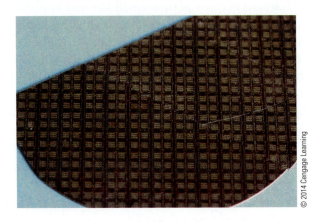

© 2014 Cengage Learning

After fabrication, the integrated circuits are tested on the wafer. After testing, the wafer is sliced into individual chips. Each chip represents one complete integrated circuit containing all the components and connections associated with that circuit. Each chip that passes quality control tests is then mounted into a package. Even though a large number of integrated circuits are fabricated at the same time, only a certain number are usable. This is referred to as the **yield**, which is the maximum number of usable integrated circuits compared with the rejects.

Thin-film integrated circuits are formed on the surface of an insulating substrate of glass or ceramic, usually less than an inch square. The components (resistors and capacitors) are formed by an extremely thin film of metal and oxides deposited on the substrate. Thin strips of metal are then deposited to connect the components. Diodes and transistors are formed as separate semiconductor devices and attached in the proper location. Resistors are formed by depositing tantalum or nichrome on the surface of the substrate in a thin film 0.0001 inch thick. The value of the resistor is determined by the length, width, and thickness of each strip. Conductors are formed of low-resistance metal such as gold, platinum, or aluminum. This process can produce a resistor accurate to within $\pm$ 0.1%. It can also obtain a ratio between resistors accurate to $\pm 0.01\%$. Accurate ratios are important for the proper operation of a particular circuit.

Thin-film capacitors consist of two thin layers of metal separated by an extremely thin dielectric. A metal layer is deposited on the substrate. An oxide coating is then formed over the metal for a dielectric. The dielectric is formed from an insulating material such as tantalum oxide, silicon oxide, or aluminum oxide. The top plate, formed from gold, tantalum, or platinum, is deposited on the dielectric. The capacitor value is obtained by adjusting the area of the plate and by varying the thickness and type of dielectric.

Diode and transistor chips are formed using the monolithic technique. The diode and transistor chips are permanently mounted on the substrate. They are then electrically connected to the thin-film circuit using extremely thin wires.

The materials used for components and conductors are deposited on the substrate using an evaporation or sputtering process. The evaporation process requires that the material deposited on the substrate be placed in a vacuum and heated until it evaporates. The vapors are then allowed to condense on the substrate, forming a thin film.

The sputtering process subjects a gas-filled chamber to a high voltage. The high voltage ionizes the gas, and the material to be deposited is bombarded with ions. This dislodges the atoms within the material, which drift toward the substrate, where they are deposited as a thin film. A mask is used to ensure proper location of the film deposit. Another method is to coat the substrate completely and cut or etch away the undesired portions.

In **thick-film integrated circuits** construction, the resistors, capacitors, and conductors are formed on the substrate using a screen-printing process: A fine wire screen is placed over the substrate, and metallized ink is forced through the screen with a squeegee. The screen acts as a mask. The substrate and ink are then heated to over 1112°F (600°C) to harden the ink.

Thick-film capacitors have low values (in picofarads). When a higher value is required, discrete capacitors are used. The thick-film components are 0.001 inch thick; they have no discernable thickness. The thick-film components resemble conventional discrete components.

Hybrid integrated circuits are formed using monolithic, thin-film, thick-film, and discrete components. This allows for a high degree of circuit complexity by using monolithic circuits while taking advantage of extremely accurate component values and tolerances with the film techniques. Discrete components are used because they can handle relatively large amounts of power.

If only a few circuits are to be made, it is cheaper to use hybrid integrated circuits. The major expense in hybrid construction is the wiring and assembly of components and the final packaging of the device.

Because hybrid circuits use discrete components, they are larger and heavier than monolithic integrated circuits. The use of discrete components tends to make hybrid circuits less reliable than monolithic circuits.

34-2 QUESTIONS

1. What methods are used to construct integrated circuits?
2. Describe the monolithic process of integrated circuit fabrication.
3. What are the differences between thin-film and thick-film fabrication processes?
4. How is a hybrid integrated circuit constructed?
5. What determines which process will be used when fabricating an integrated circuit?

34-3 INTEGRATED CIRCUIT PACKAGING

An integrated circuit is a thin chip consisting of hundreds to millions of interconnected semiconductor devices, such as transistors, resistors, and capacitors. In 2004, a typical chip size of 0.15 in.^{2} (1 cm^2) or smaller was common, although larger sizes were also produced.

Integrated circuits are mounted in packages designed to protect them from moisture, dust, and other contaminants. The most popular package design is the **dual in-line package (DIP)** (Figure 34-2). DIPs are manufactured in several sizes to accommodate the

■ **FIGURE 34-2**

Integrated circuit dual in-line packages.

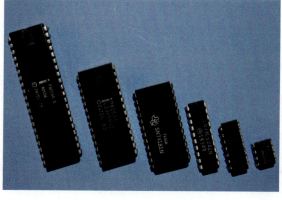

© 2014 Cengage Learning

■ **FIGURE 34-3**
Families of integrated circuits.

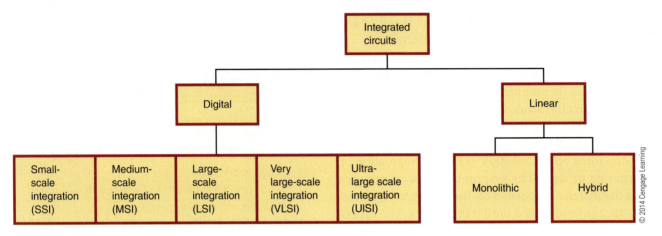

various sizes of integrated circuits: **small-scale integration (SSI)** with up to 100 electrical components per chip (early 1960s), **medium-scale integration (MSI)** with 100 to 3000 electrical components per chip (late 1960s), **large-scale integration (LSI)** with 3000 to 100,000 electrical components per chip (mid-1970s), and **very large-scale integration (VLSI)** with 100,000 to 1,000,000 electrical components per chip (1980s). **ULSI** stands for **ultra-large-scale integration** and provides no qualitative leap from VLSI but is reserved to emphasize chip complexity in marketing. Integrated circuit families are shown in Figure 34-3.

Wafer-scale integration (WSI), which uses uncut wafers containing entire computer systems including memory, was developed in the 1980s. It failed because of problems with creating a defect-free manufacturing process. WSI was followed by *system-on-chip (SOC)* design. This process takes components that were typically manufactured on separate chips and wired together on a printed circuit board and designs them to occupy a single chip that contains the whole electronic system: microprocessor, memory, peripheral interface, input/output logic control, a/d converters, and so on.

The pin count of the VLSI circuits in the 1980s exceeded the practical limits for DIP packaging. This led to **pin grid array (PGA)** and **leadless chip carrier (LCC)** packages. Surface-mount packaging, which uses a fine lead pitch, became popular in the late 1980s. In the late 1990s, high pin count devices used **plastic quad flat package (PQFP)** and **thin small outline package (TSOP)** packaging.

The package is made in either ceramic or plastic. Plastic is less expensive and is suitable for most applications where the operating temperature falls between 32°F and 158°F (0°C and 70°C). Ceramic devices are more expensive but offer better protection against moisture and contamination. They also operate in a wider range of temperatures of −67°F to −193°F (−55°C to +125°C). Ceramic devices are recommended for military, aerospace, and severe industrial applications.

A small 8-pin DIP, called a mini-DIP, is used for devices with a minimum number of inputs and outputs. It is used mostly with monolithic integrated circuits.

A flat-pack is smaller and thinner than a DIP and is used where space is limited. It is made from metal or ceramic and operates in a temperature range of −67°F to −193°F (−55°C to +125°C).

After the integrated circuit is packaged, it is tested over a wide range of temperatures to ensure that it meets all electrical specifications.

34-3   QUESTIONS

1. What are the functions of an integrated circuit package?
2. What package is most often used for integrated circuits?
3. What materials are used for integrated circuit packages?
4. What are the advantages of ceramic packages?
5. What is the advantage of the flat-pack integrated circuit package?

34-4 HANDLING OF INTEGRATED CIRCUITS

Integrated circuits are easily damaged with static electricity that is referred to as electrostatic discharge (ESD). ESD is a static discharge and occurs when two positively charged substances are rubbed together or separated, causing a transfer or electrons so that one substance becomes negatively charged. When either substance comes in contact with a conductor, an electrical current flows until it is at the same electrical potential as ground. ESD is commonly experienced more during the winter months when the environment is dry.

Most static discharges that do damage to an integrated circuit will be too weak to feel as they are less than 3000 volts. Most ICs are packaged with warning labels (Figure 34-4) to identify that they are sensitive to damage by static electricity. Others may be shielded in antistatic tubes or carriers. Keep in mind that just because an IC does not have a symbol or label does not mean it is not ESD sensitive.

Inside an integrated circuit, a static discharge could destroy oxide layers or junctions. These static discharges can create an ESD latent defect, causing the device to appear to operate properly even though it is damaged. It has been weakened and failure will result at a later time. There is no method for testing for an ESD latent defect.

The following list will help to minimize damage to an IC through electrostatic discharge:

1. Always check manuals and the IC package materials for ESD warnings and instructions.
2. Wear a grounded wrist strap to discharge any static charge built up on the body.
3. Do not permit the IC to come in contact with clothing or other ungrounded materials that might have a static charge.
4. When handling an IC, minimize physical movement such as scuffing feet.
5. Always discharge the package of the IC before removing it. Keep the package grounded until the device is placed in the circuit.
6. Before touching an IC, always touch the surface on which it rests, in order to provide a discharge path.
7. Handle the IC only when it is ready to place in the circuit.
8. When removing and replacing an IC, avoid touching the leads.
9. When working on a circuit containing ICs, do not touch any material that will create a static charge.

■ **FIGURE 34-4**

Warning labels printed on packing for device that can be easily damaged by ESD.

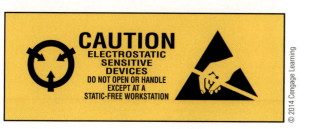

© 2014 Cengage Learning

34-4 QUESTIONS

1. What is electrostatic discharge?
2. How does ESD occur?
3. How are integrated circuits identified as sensitive to static electricity?
4. What does a static discharge cause in an integrated circuit?
5. Describe the best way to handle an IC that is identified as sensitive to ESD.

SUMMARY

- Integrated circuits are popular because they have the following advantages:
  - Are reliable with complex circuits
  - Consume little power
  - Are small and light
  - Are economical to produce
  - Provide new and better solutions to problems
- Integrated circuits cannot handle large amounts of current or voltage.
- Only diodes, transistors, resistors, and capacitors are available as integrated circuits.
- Integrated circuits cannot be repaired, only replaced.
- Integrated circuits are constructed by monolithic, thin-film, thick-film, or hybrid techniques.
- The most popular integrated circuit package is the DIP (dual-inline package).
- Integrated circuit packages are made of ceramic or plastic, with plastic most often used.
- Electrostatic discharge (ESD) can damage integrated circuits.
- Integrated circuits easily affected by ESD damage are packaged with warning labels.
- ESD destroys oxide layers and/or junctions, which can cause hidden defects.
- There is no known method for testing an integrated circuit for hidden ESD defects.
- Wear a grounded wrist strap at all times when handling integrated circuits.
- Always avoid touching the lead of an integrated circuit.
- Do not touch any materials that will create a static charge when working on an integrated circuit.

CHAPTER 34 SELF-TEST

1. What does an integrated circuit consist of?
2. How would a problem be isolated in a circuit containing integrated circuits?
3. What are three construction techniques used for fabricating an integrated circuit?
4. What does the word *chip* identify?
5. What components does a hybrid integrated circuit contain?
6. When is it beneficial to use hybrid integrated circuits?
7. What are the popular DIP package sizes?
8. What does SOC reference with respect to WSI?
9. What types of packages are best for use by the aerospace industry and the military?
10. What is the difference between a DIP and a flat pack?

CHAPTER 35

Optoelectric Devices

OBJECTIVES

After completing this chapter, the student will be able to:

- Identify the three categories of semiconductor devices that react to light.
- Classify the major frequency ranges of light.
- Identify major light-sensitive devices and describe their operation and applications.
- Identify major light-emitting devices and describe their operation and applications.
- Draw and label the schematic symbols associated with optoelectric devices.
- Identify packages used for optoelectric devices.

KEY TERMS

35-1 light

35-2 photoconductive cell (photocell)

35-2 photodiode

35-2 phototransistor

35-2 photovoltaic cell (solar cell)

35-3 light-emitting diode (LED)

35-3 optical coupler

35-3 PIN photodiode

Semiconductors, in general, and semiconductor diodes, in particular, have important uses in optoelectronics. Here, a device is designed to interact with electromagnetic radiation (light energy) in the visible, infrared, and ultraviolet ranges.

Three types of devices interact with light:

• Light-detection devices
• Light-conversion devices
• Light-emitting devices

The semiconductor material and the doping techniques used determine the relevant light wavelength of a particular device.

35-1   BASIC PRINCIPLES OF LIGHT

Light is electromagnetic radiation that is visible to the human eye. Light is thought to travel in a form similar to radio waves. Like radio waves, light is measured in wavelengths.

Light travels at 186,000 miles per second, or 30,000,000,000 centimeters per second, through a vacuum. The velocity is reduced as it passes through various types of mediums. The frequency range of light is 300 to 300,000,000 gigahertz (giga = 1,000,000,000). Of this frequency range, only a small band is visible to the human eye. The visible region spans from approximately 400,000 to 750,000 gigahertz. Infrared light falls below 400,000 gigahertz, and ultraviolet light lies above 750,000 gigahertz. Light waves at the upper end of the frequency range have more energy than light waves at the lower end.

35-1   QUESTIONS

1. What is light?
2. What frequency range of light is visible to the human eye?
3. What is infrared light?
4. What is ultraviolet light?
5. What type of light wave has the most energy?

35-2   LIGHT-SENSITIVE DEVICES

The **photoconductive cell (photocell)** is the oldest optoelectric device. It is a light-sensitive device in which the internal resistance changes with a change

■ FIGURE 35-1

Photo cell.

© 2014 Cengage Learning

in light intensity. The resistance change is not proportional to the light striking it. The photocell is made from light-sensitive material such as cadmium sulfide (CdS) or cadmium selenide (CdSe).

Figure 35-1 shows some typical photocells. The light-sensitive material is deposited on an insulating substrate of glass or ceramic in an S-shape to allow greater contact length. The photocell is more sensitive to light than any other device. The resistance can vary from several hundred ohms to several hundred million ohms. The photocell is useful for low-light applications. It can withstand high operating voltages of 200 to 300 volts, with a low-power consumption of up to 300 milliwatts. A disadvantage of the photocell is its slow response to light change.

Figure 35-2 shows the schematic symbols used to represent a photocell. The arrows indicate a light-sensitive device. Sometimes the Greek letter lambda (λ) is used to represent a light-sensitive device.

Photocells are used in light meters for photographic equipment, instrusion detectors, automatic door openers, and various kinds of test equipment to measure light intensity.

■ FIGURE 35-2

Schematic symbol for a photocell.
© 2014 Cengage Learning

■ **FIGURE 35-3**

Construction of a solar cell.

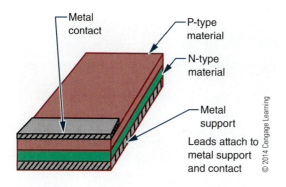

The **photovoltaic cell (solar cell)** converts light energy directly into electrical energy. The solar cell finds most of its applications in converting solar energy into electrical energy.

The solar cell is basically a P–N junction device made from semiconductor materials, most commonly from silicon. Figure 35-3 shows the construction technique. The P and N layers form a P–N junction. The metal support and the metal contact act as the electrical contacts. It is designed with a large surface area.

The light striking the surface of the solar cell imparts much of its energy to the atoms in the semiconductor material. The light energy knocks valence electrons from their orbits, creating free electrons. The electrons near the depletion region are drawn to the N-type material, producing a small voltage across the P–N junction. The voltage increases with an increase in light intensity. All the light energy striking the solar cell does not create free electrons, however. In fact, the solar cell is a highly inefficient device, with a top efficiency of 15% to 20%. The cell is inefficient when expressed in terms of electrical power output compared to the total power contained in the input light energy.

Solar cells have a low voltage output of about 0.45 volt at 50 milliamperes. They must be connected in a series and parallel network to obtain the desired voltage and current output.

Applications include light meters for photographic equipment, motion-picture projector soundtrack decoders, and battery chargers on satellites.

Schematic symbols for the solar cell are shown in Figure 35-4. The positive terminal is identified with a plus (+) sign.

The **photodiode** also uses a P–N junction and has a construction similar to that of the solar cell. It is

■ **FIGURE 35-4**

Schematic symbol for a solar cell.

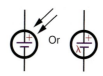

used the same way as the photocell, as a light-variable resistor. Photodiodes are semiconductor devices that are made primarily from silicon. They are constructed in two ways. One method is as a simple PN junction (Figure 35-5). The other method inserts an intrinsic (undoped) layer between the P and N regions (Figure 35-6), forming a PIN photodiode.

A P–N junction photodiode operates on the same principles as a photovoltaic cell except that it is used to control current flow, not generate it. A reverse-bias voltage is applied across the photodiode, forming a wide depletion region. When light energy strikes the photodiode, it enters the depletion region, creating free electrons. The electrons are drawn toward the positive side of the bias source. A small current flows through the photodiode in the reverse direction. As the light energy increases, more free electrons are generated, creating more current flow.

■ **FIGURE 35-5**

PN junction photodiode.

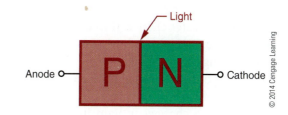

■ **FIGURE 35-6**

PIN junction photodiode.

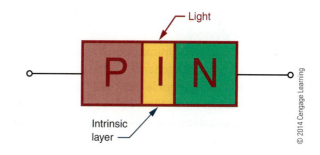

■ **FIGURE 35-7**

Photodiode package.

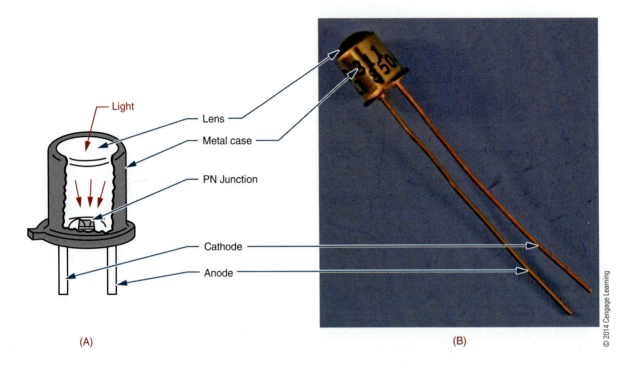

■ **FIGURE 35-7**

Photodiode package.

(A)

(B)

© 2014 Cengage Learning

A **PIN photodiode** has an intrinsic layer between the P and N regions. This effectively extends the depletion region. The wider depletion region allows the PIN photodiode to respond to lower light frequencies, which have less energy and so must penetrate deeper into the depletion region before generating free electrons. The wider depletion region offers a greater chance of free electrons being generated. The PIN photodiode is more efficient over a wide range.

The PIN photodiode has a lower internal capacitance because of the intrinsic layer. This allows for a faster response to changes in light intensity. Also, a more linear change in reverse current with light intensity is produced.

The advantage of the photodiode is fast response to light changes, the fastest of any photosensitive device. The disadvantage is a low output compared to other photosensitive devices.

Figure 35-7 shows a typical photodiode package. A glass window allows light energy to strike the photodiode. The schematic symbol is shown in Figure 35-8. A typical circuit is shown in Figure 35-9.

A **phototransistor** is constructed like other transistors with two P–N junctions. It resembles a standard NPN transistor. It is used like a photodiode and packaged like a photodiode, except that it has three leads (emitter, base, and collector). Figure 35-10 shows the

■ **FIGURE 35-8**

Schematic symbol for a photodiode.

© 2014 Cengage Learning

■ **FIGURE 35-9**

Voltage divider using a photodiode.

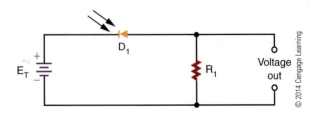

© 2014 Cengage Learning

■ FIGURE 35-10

Equivalent circuit for a phototransistor.

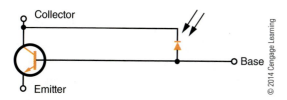

equivalent circuit. The transistor conduction depends on the conduction of the photodiode. The base lead is seldom used. When it is used, its purpose is to adjust the turn-on point of the transistor.

Phototransistors can produce higher output current than photodiodes. Their response to light changes is not as fast as that of the photodiode. There is a sacrifice of speed for a higher output current.

Applications include photo tachometers, photographic exposure controls, flame detectors, object counters, and mechanical positioners.

Figure 35-11 shows the schematic symbol for a phototransistor. Figure 35-12 shows a typical circuit application.

■ FIGURE 35-11

Schematic symbol for a phototransistor.

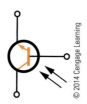

■ FIGURE 35-12

Darkness-on DC switch.

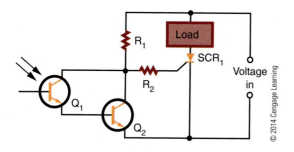

QUESTIONS

1. Explain how a photoconductive cell operates.
2. Explain how a solar cell functions.
3. What is the difference between the two types of photodiodes?
4. How is the phototransistor an improvement over the photodiode?
5. Draw and label the schematic symbols for a photoconductive cell, solar cell, photodiode, and phototransistor.

35-3 **LIGHT-EMITTING DEVICES**

Light-emitting devices produce light when subject to a current flow, converting electrical energy to light energy. A **light-emitting diode (LED)** is the most common semiconductor light-emitting device. Being a semiconductor component, it has an unlimited life span due to the absence of a filament, a cause of trouble in conventional lamps.

Any P–N junction diode can emit light when subject to an electrical current flow. The light is produced when the free electrons combine with holes, and the extra energy is released in the form of light. The frequency of the light emitted is determined by the type of semiconductor material used in construction of the diode. Regular diodes do not emit light because of the opaque material used in their packaging.

An LED is simply a P–N junction diode that emits light when a current flows through its junction. The light is visible because the LED is packaged in semitransparent material. The frequency of the light emitted depends on the material used in the construction of the LED. Gallium arsenide (GaAs) produces light in the infrared region, which is invisible to the human eye. Gallium arsenide phosphide (GaAsP) emits a visible red light. By changing the phosphide content, different frequencies of light can be produced. LEDs are available in the following colors: infrared, red, orange, yellow, green, blue, ultraviolet, pink, purple, and white. The color of an LED is not determined by the color of the plastic lens but by the semiconductor material used to make it.

■ **FIGURE 35-13**

LED construction.

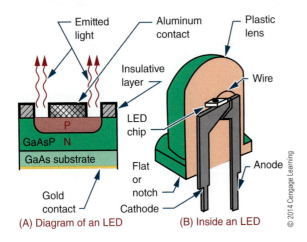

(A) Diagram of an LED   (B) Inside an LED

Figure 35-13 shows the construction of an LED. The P layer is made thin so that light energy near the P–N junction only needs to travel a short distance through it.

After the LED is formed, it is mounted in a package designed for optimum emission of light. Figure 35-14 shows some of the more common LED packages and colors. Most LEDs contain a lens that gathers and intensifies the light. The case may act as a colored filter to enhance the natural light emitted.

> **NOTE** The cathode lead of an LED is identified as the shorter lead, a flat or notch on the plastic lens base or the lead with a flag-like structure when held up to a light source.

■ **FIGURE 35-14**

Common LED packages.

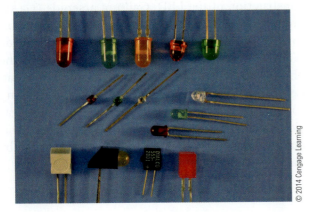

■ **FIGURE 35-15**

Forward-biased LED.

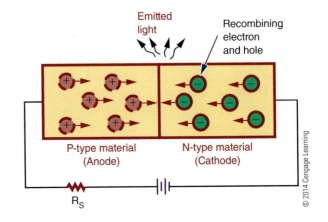

P-type material (Anode)   N-type material (Cathode)

CAUTION!

Never connect an LED directly to a battery or power supply.

In a circuit, an LED is connected with forward bias to emit light (Figure 35-15). The forward bias must exceed 1.2 volts before a forward current can flow. Because LEDs can be easily damaged by an excessive amount of voltage or current, a series resistor is connected to limit current flow.

The schematic symbol for an LED is shown in Figure 35-16. Figure 35-17 shows a properly biased circuit. The series resistor (R_S) is used to limit the forward current (I_F) based on the applied voltage.

■ **FIGURE 35-16**

Schematic symbol for an LED.

© 2014 Cengage Learning

■ **FIGURE 35-17**

Properly biased LED circuit.

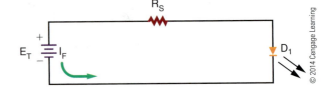

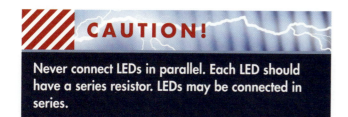

CAUTION!

Never connect LEDs in parallel. Each LED should have a series resistor. LEDs may be connected in series.

Figure 35-18 shows LEDs arranged to form a seven-segment display used for digital readouts. Figure 35-19 shows an LED used in conjunction with a phototransistor to form an optical coupler. Both devices are housed in the same package. An **optical coupler** consists of an LED and a phototransistor. The output of the optical coupler is coupled to the input using light produced by the LED.

NOTE Super bright LEDs are now used in common applications such as automobile and truck lights, replacement lights in the home and office, traffic lights, and so on. They are used where low energy consumption and long life are required in a light source.

The signal to the LED can vary, which in turn varies the amount of light available. The phototransistor converts the varying light back into electrical energy. An optical coupler allows one circuit to pass a signal to another circuit while providing a high degree of electrical insulation.

■ **FIGURE 35-18**

Seven-segment display of LEDS for digital readout.

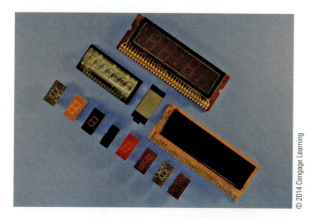

© 2014 Cengage Learning

■ **FIGURE 35-19**

Commercial optical coupler.

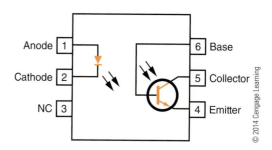

© 2014 Cengage Learning

35-3 QUESTIONS

1. Explain how an LED differs from a conventional diode.
2. How does an LED emit different colors of light?
3. How may the LED package enhance the light emitted?
4. Draw and label the schematic symbol for an LED.
5. What is the function of an optical coupler?

SUMMARY

- Semiconductor devices that interact with light can be classified as light-detection devices, light-conversion devices, or light-emitting devices.
- Light is electromagnetic radiation that is visible to the human eye.
- The frequency range of light is as follows:
  - Infrared light—less than 400,000 gigahertz
  - Visible light—400,000 to 750,000 gigahertz
  - Ultraviolet light—greater than 750,000 gigahertz
- Light-sensitive devices include photocells, solar cells, photodiodes, and phototransistors.
- The LED (light-emitting diode) is a light-emitting device.
- LEDs are available in many colors.
- An optical coupler combines a light-sensitive device with a light-emitting device.
- The schematic symbols for light-sensitive devices are as shown here:
  - Photocell

© 2014 Cengage Learning

- Solar cell

© 2014 Cengage Learning

- Photodiode cell

© 2014 Cengage Learning

- Phototransistor cell

© 2014 Cengage Learning

- The schematic symbol for an LED is

© 2014 Cengage Learning

CHAPTER 35 SELF-TEST

1. Make a table showing light spectrum and frequencies.
2. Identify why the light waves at the upper frequency range have more energy.
3. Which light-sensitive device has the fastest response time for changes in light intensity?
4. Why are photocells not used where a quick response is necessary?
5. Describe how a solar cell functions.
6. How are solar cells connected to create a practical application?
7. Which light-sensitive device would lend itself to a wider range of applications: a photodiode or a phototransistor? Why?
8. How does an LED produce light?
9. How does the amount of current flowing in an LED affect the intensity of the light emitted?
10. What is the function of an optical coupler?

PRACTICAL APPLICATIONS

Section 4

1. Using an assortment of semiconductor devices, identify how to determine whether the device is working or defective.

2. Using an assortment of semiconductor devices, look up each item in a catalog and/or on the Internet to determine specifications and price.

3. Build a device that can test different types of semiconductor devices. Use the device with an oscilloscope to determine what the device is and whether it is functional.

4. Make a Power Point presentation on how to test various semiconductor devices.

5. Make a video presentation on how a particular semiconductor device functions.

LINEAR ELECTRONIC CIRCUITS

LINEAR ELECTRONIC CIRCUITS

Harold Stephen Black (1898–1983)

In 1927, Black invented negative feedback (NFB), which allowed an amplifier to trade gain for reduced distortion and also resulted in reduced output impedance.

James Kilton Clapp (1897–1965)

In 1948, Clapp invented the Clapp oscillator, one of several types of electronic oscillators constructed from a transistor and a positive feedback network.

Edwin Henry Colpitts (1872–1949)

In 1927, Colpitts invented the Colpitts oscillator, in wich the feedback signal is taken from a voltage divider made by two capacitors in series.

Sidney Darlington (1906–1997)

In 1953, Darlington invented the Darlington configuration.

Lee De Forest (1873–1961)

In 1908, De Forest invented the audion (renamed the triode in 1919), a vacuum tube that takes a relatively weak electrical signal and amplifies it. De Forest was named one of the fathers of the "electronic age."

Thomas Alva Edison (1847–1931)

In the early 1880s, Edison was credited as being the first inventor of the electric fuse. He also developed the idea for a circuit breaker in 1879.

Heinrich Greinacher (1880–1974)

Greinacher invented the half-wave multiplier, also called the Villard cascade.

Ralph Vinton Lyon Hartley (1888–1970)

In 1915, Hartley invented the Hartley oscillator, in which the feedback needed for oscillation is taken from a tap on the tank coil.

Hung Chang Lin (1919–2009)

In 1955, Lin invented the quasi-complementary amplifier, which is used in many commercial audio amplifiers.

George Washington Pierce (1872–1956)

Pierce invented the Pierce oscillator, a derivative of the Colpitts oscillator, which is well suited for implementing crystal oscillator circuits.

Otto Herbert Schmitt (1913–1998)

Schmitt invented the differential amplifier, an electronic amplifier that multiplies the difference between two inputs by some constant factor referred to as the differential gain.

Power Supplies

OBJECTIVES

After completing this chapter, the student will be able to:

- Explain the purpose of a power supply.
- Draw a block diagram of the circuits and parts of a power supply.
- Describe the three different rectifier configurations.
- Explain the function of a filter.
- Describe the two basic types of voltage regulators and how they operate.
- Explain the function of a voltage multiplier.
- Describe the function of a voltage divider.
- Identify over-voltage and overcurrent protection devices.

KEY TERMS

36-1 polarized plug

36-2 bridge rectifier

36-2 full-wave rectifier

36-2 half-wave rectifier

36-2 rectifier circuit

36-2 ripple frequency

36-3 capacitor filter

36-3 filter

36-4 error voltage

36-4 sampling circuit

36-4 series voltage regulator

36-4 shunt voltage regulator

36-4 voltage regulator

36-5 voltage doubler

36-5 voltage multiplier

36-5 voltage tripler

36-6 voltage divider

36-7 circuit breaker

36-7 crowbar

36-7 fuse

36-7 metal-oxide varistor (MOV)

36-7 normal blow fuse

36-7 over-voltage protection circuit

36-7 slow blow fuse

36-7 varistor

Power supplies are used to supply voltage to a variety of circuits. Their basic principles are the same.

The primary function of the power supply is to convert alternating current (AC) to direct current (DC). The power supply may increase or decrease the incoming AC voltage by means of a transformer.

Once the voltage is at the desired level, it is converted to a DC voltage through a process called rectification. The rectified voltage still contains an AC signal, referred to as a ripple frequency. The ripple frequency is removed with a filter.

To ensure that the output voltage remains at a constant level, a voltage regulator is used. The voltage regulator holds the output voltage at a constant level.

36-1 TRANSFORMERS

The circuit shown in Figure 36-1 represents a basic circuit that converts AC to DC. It converts the sine wave from the AC power source to direct current, without a transformer. One disadvantage with this type of circuit occurs when the load requires a higher or lower voltage.

Another disadvantage of this basic circuit is that one side of the AC power source is connected to the rectifier circuit and the other side is connected directly to the chassis of the device. This can present a serious shock hazard when the chassis is not isolated from the user.

When a device uses a two-prong plug, it provides two ways of plugging the device to the AC power source. One way could connect the chassis directly to the hot side of the AC source. If the user touches ground and the chassis, it could result in a lethal shock.

When using test equipment with two-prong plugs, if one piece is plugged in with a reversed connection, a voltage could exist between two pieces of test

■ **FIGURE 36-2**

The white plug is polarized. One prong is wider than the other. The wide prong is plugged into the neutral side of the AC power source, reducing the risk of lethal shock.

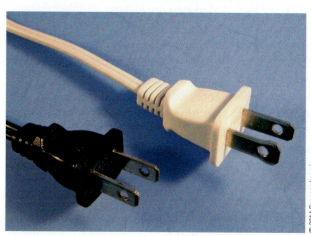

© 2014 Cengage Learning

equipment that could be lethal. Figure 36-2 shows two 2-prong plugs. In one, the prongs are the same width; the other shows that one prong is wider than the other and is referred to as a **polarized plug**. The polarized plug eliminates the chance of plugging in a device and making the chassis hot. The wide prong is plugged into the neutral side of the AC power source.

Devices that are built without transformers tend to be inexpensive consumer products. These devices use plastic cabinets, knobs, and switches to eliminate potential shock hazards. When servicing these devices without transformers, an isolation transformer must be used (Figure 36-3). The isolation transformer

■ **FIGURE 36-3**

An isolation transformer isolates the chassis of a device from the AC power source, preventing accidental electric shock.

© 2014 Cengage Learning

■ **FIGURE 36-1**

Basic circuit for converting AC to DC.

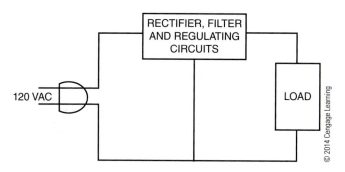

© 2014 Cengage Learning

isolates the chassis from the AC power source, preventing accidental electric shock.

Transformers are therefore used in power supplies to isolate the power supply from the AC voltage source. They are also used to step up voltages if higher voltages are required or to step down voltages if lower voltages are required. If transformers are used in power supplies, the AC power source is connected only to the primary of the transformer. This isolates the electrical circuits from the power source. When selecting a transformer, its primary power rating is the first concern. The most common primary ratings are 110 to 120 volts and 220 to 240 volts. The next concern is the transformer's frequency. Some frequencies are 50 to 60 hertz, 400 hertz, and 10,000 hertz. The third concern is the secondary voltage and current ratings of the transformer. The final concern is its power-handling capability, or volt-ampere rating, which is essentially the amount of power that can be delivered to the secondary of the transformer. It is given as volt-amperes because of the loads that can be placed on the secondary.

36–1 QUESTIONS

1. Why are transformers used in power supplies?
2. How is a transformer connected in a power supply?
3. What safety consideration is given to transformers used in a power supply?
4. What are the important considerations when selecting a transformer for a power supply?
5. How are transformers rated?

36–2 RECTIFIER CIRCUITS

The **rectifier circuit** is the heart of the power supply. Its function is to convert the incoming AC voltage to a DC voltage. There are three basic types of rectifier circuits used with power supplies: **half-wave rectifiers**, **full-wave rectifiers**, and **bridge rectifiers**.

Figure 36-4 shows a basic half-wave rectifier. The diode is located in series with the load. The current in the circuit flows in only one direction because of the diode.

Figure 36-5 shows a half-wave rectifier during the positive alternation of the sine wave. The diode is

■ FIGURE 36-4

Basic half-wave rectifier.

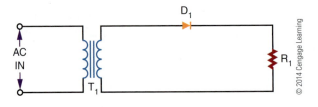

■ FIGURE 36-5

Half-wave rectifier during positive alternation.

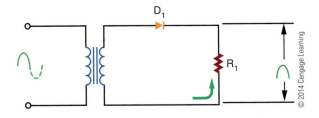

forward biased, allowing a current to flow through the load. This allows the positive alternation of the cycle to develop across the load.

Figure 36-6 shows the circuit during the negative alternation of the sine wave. The diode is now reverse biased and does not conduct. Because no current flows through the load, no voltage is dropped across the load.

The half-wave rectifier operates during only one-half of the input cycle. The output is a series of positive or negative pulses, depending on how the diode is connected in the circuit. The frequency of the pulses is the same as the input frequency and is called the **ripple frequency**.

The polarity of the output depends on which way the diode is connected in the circuit (Figure 36-7). The current flows through the diode from cathode to anode. When current flows through a diode, a

■ FIGURE 36-6

Half-wave rectifier during negative alternation.

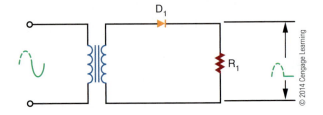

FIGURE 36-7

The diode determines the direction of current flow.

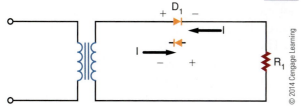

deficiency of electrons exists at the anode end, making it the positive end of the diode. Reversing the diode can reverse the polarity of the power supply.

There is a serious disadvantage with the half-wave rectifier because current flows during only half of each cycle. To overcome this disadvantage, a full-wave rectifier can be used.

Figure 36-8 shows a basic full-wave rectifier circuit. It requires two diodes and a center-tapped transformer. The center tap is grounded. The voltages at each end of the transformer are 180° out of phase with each other.

Figure 36-9 shows a full-wave rectifier during the positive alternation of the input signal. The anode of

diode D_1 is positive, and the anode of diode D_2 is negative. Diode D_1 is forward biased and conducts current, whereas diode D_2 is reverse biased and does not conduct current. The current flows from the center tap of the transformer, through the load and diode D_1, to the top of the secondary of the transformer. This permits the positive half of the cycle to be felt across the load.

Figure 36-10 shows the full-wave rectifier during the negative half of the cycle. The anode of diode D_2 becomes positive, and the anode of diode D_1 becomes negative. Diode D_2 is now forward biased and conducts. Diode D_1 is reverse-biased and does not conduct. The current flows from the center tap of the transformer, through the load and diode D_2, to the bottom of the secondary of the transformer.

With a full-wave rectifier, the current flows during both half-cycles. This means that the ripple frequency is twice the input frequency. There is a disadvantage with the full-wave rectifier because the output voltage is half that of a half-wave rectifier for the same transformer. This disadvantage can be overcome by the use of a bridge rectifier circuit.

Figure 36-11 shows a bridge rectifier circuit. The four diodes are arranged so that the current flows in only one direction through the load.

FIGURE 36-8

Basic full-wave rectifier circuit.

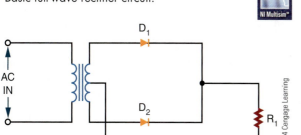

FIGURE 36-9

Full-wave rectifier during positive alternation.

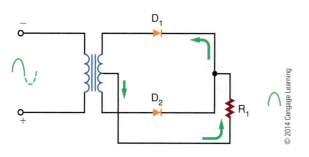

FIGURE 36-10

Full-wave rectifier during negative alternation.

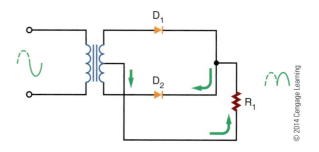

FIGURE 36-11

Bridge rectifier circuit.

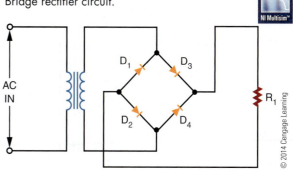

Figure 36-12 shows the current flow during the positive alternation of the input signal. Current flows from the bottom of the secondary side of the transformer, up through diode D_4, through the load, through diode D_2, to the top of the secondary of the transformer. The entire voltage is dropped across the load.

Figure 36-13 shows the current flow during the negative alternation of the input signal. The top of the secondary is negative, and the bottom is positive. The current flows from the top of the secondary, down through diode D_1, through the load and diode D_3, to the bottom of the secondary. Note that the current flows in the same direction through the load as during the positive alternation. Again the entire voltage is dropped across the load.

A bridge rectifier is a type of full-wave rectifier because it operates on both half-cycles of the input sine wave. The advantage of the bridge rectifier is that the circuit does not require a center-tapped secondary. This circuit does not require a transformer to operate. A transformer is used only to step up or step down the voltage or provide isolation.

■ FIGURE 36-12

Bridge rectifier during positive alternation.

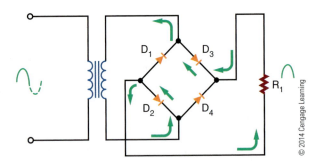

© 2014 Cengage Learning

■ FIGURE 36-13

Bridge rectifier during negative alternation.

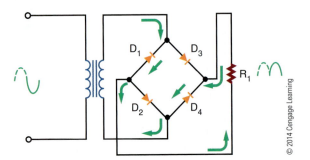

© 2014 Cengage Learning

To summarize the differences in rectifiers: The advantages of the half-wave rectifier are its simplicity and low cost. It requires one diode and a transformer. It is not very efficient, because only half of the input signal is used. It is also restricted to low-current applications.

The full-wave rectifier is more efficient than the half-wave rectifier. It operates on both alternations of the sine wave. The higher ripple frequency of the full-wave rectifier is easier to filter. A disadvantage is that it requires a center-tapped transformer. Its output voltage is lower than that of a half-wave rectifier for the same transformer because of the center tap.

The bridge rectifier can operate without a transformer. However, a transformer is needed to step up or step down the voltage. The output of the bridge rectifier is higher than that of either the full-wave or half-wave rectifier. A disadvantage is that the bridge rectifier requires four diodes. However, the diodes are inexpensive compared to a center-tapped transformer.

36-2 QUESTIONS

1. What is the function of the rectifier in a power supply?
2. What are three configurations for connecting rectifiers to power supplies?
3. What are the differences in operation of the three configurations?
4. What are the advantages of one rectifier configuration over another?
5. Which rectifier configuration represents the best selection? Why?

36-3 FILTER CIRCUITS

The output of the rectifier circuit is a pulsating DC voltage. This is not suitable for most electronic circuits, so a **filter** follows the rectifier in most power supplies. The filter converts the pulsating DC voltage to a smooth DC voltage. The simplest filter is a **capacitor filter** connected across the output of the rectifier (Figure 36-14). Figure 36-15 compares the outputs of a rectifier without and with the addition of a filter capacitor.

■ **FIGURE 36-14**

Half-wave rectifier with filter capacitor.

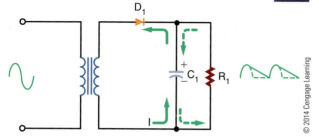

■ **FIGURE 36-15**

Output of a half-wave rectifier without and with a filter capacitor.

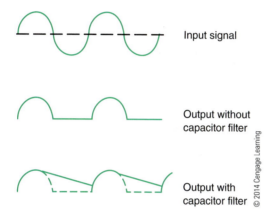

A capacitor affects the circuit in the following manner. When the anode of the diode is positive, current flows in the circuit. At the same time, the filter capacitor charges to the polarity indicated in Figure 36-14. After 90° of the input signal, the capacitor is fully charged to the peak potential of the circuit.

When the input signal starts to drop in the negative direction, the capacitor discharges through the load. The resistance of the load controls the rate the capacitor discharges by the RC time constant. The discharge time constant is long compared to the cycle time. Therefore the cycle is completed before the capacitor can discharge. Thus, after the first quarter cycle, the current through the load is supplied by the discharging capacitor. As the capacitor discharges, the voltage stored in the capacitor decreases. However, before the capacitor can completely discharge, the next cycle of the sine wave occurs. This causes the anode of the diode to become positive, allowing the diode to

■ **FIGURE 36-16**

Effects of different filter capacitors on output of half-wave rectifier.

conduct. The capacitor recharges and the cycle repeats. The end result is that the pulses are smoothed out and the output voltage is actually raised (Figure 36-16).

A larger capacitor has a longer RC time constant, resulting in a slower discharge of the capacitor and thus raising the output voltage. The presence of the capacitor allows the diode in the circuit to conduct for a shorter period of time. When the diode is not conducting, the capacitor is supplying the current to the load. If the current required by the load is large, a very large capacitor must be used.

A capacitor filter across a full-wave or bridge rectifier behaves much like the capacitor filter in the half-wave rectifier just described. Figure 36-17 shows the output of a full-wave or bridge rectifier. The ripple frequency is twice that of the half-wave rectifier. When

■ **FIGURE 36-17**

Effects of different filter capacitors on output of full-wave or bridge rectifier.

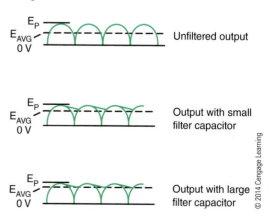

■ **FIGURE 36-18**

Half-wave rectifier (A) and full-wave rectifier (B) each followed by a filter capacitor.

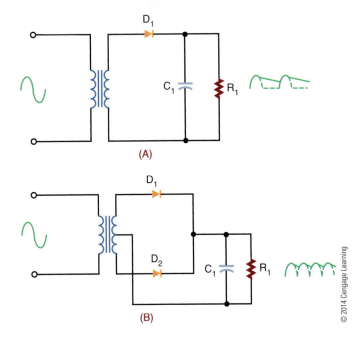

© 2014 Cengage Learning

the capacitor filter is added to the output of the rectifier, the capacitor does not discharge very far before the next pulse occurs. The output voltage is high. If a large capacitor is used, the output equals the peak voltage of the input signal. Thus, the capacitor does a better job of filtering in a full-wave circuit than in a half-wave circuit.

The purpose of the filter capacitor is to smooth out the pulsating DC voltage from the rectifier. The ripple remaining on the DC voltage determines the filter's performance. The ripple can be made lower by using a large capacitor or by increasing the load resistance. Typically, the load resistance is determined by the circuit design. Therefore, the size of the filter capacitor is determined by the amount of the ripple.

It should be realized that the filter capacitor places additional stress on the diodes used in the rectifier circuit. A half-wave rectifier and a full-wave rectifier, each followed by a filter capacitor, are shown in Figure 36-18. The capacitor charges to the peak value of the secondary voltage and holds this value throughout the input cycle. When the diode becomes reverse biased, it shuts off, and a maximum negative voltage is felt on the anode of the diode. The filter capacitor holds a maximum positive voltage on the cathode of the diode. The difference of potential across the diode is twice the peak value of the secondary. The diode must be selected to withstand this voltage.

The maximum voltage a diode can withstand when reverse biased is called the peak inverse voltage (PIV). A diode should be selected that has a PIV higher than twice the peak value. Ideally, the diode should be operated at 80% of its rated value to allow for changes in the input voltage. This holds true for both the half-wave and the full-wave rectifier but does not hold true for the bridge rectifier.

The diodes in a bridge rectifier are never exposed to more than the peak value of the secondary. In Figure 36-19, none of the diodes is exposed to more than the peak value of the input signal. The use of diodes with lower PIV ratings represents another advantage of the bridge rectifier.

■ **FIGURE 36-19**

Bridge rectifier with filter capacitor.

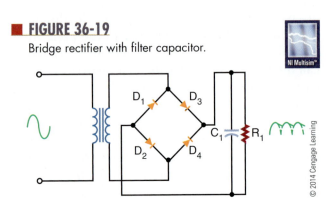

© 2014 Cengage Learning

1. What is the purpose of a filter in a power supply?
2. What is the simplest filter configuration?
3. What is the ripple frequency?
4. How is a filter capacitor selected?
5. What adverse effects result from the addition of a filter?

36-4 VOLTAGE REGULATORS

Two factors can cause the output voltage of a power supply to vary. First, the input voltage to the power supply can vary, resulting in increases and decreases in the output voltage. Second, the load resistance can vary, resulting in a change in current demand.

Many circuits are designed to operate at a certain voltage. If the voltage varies, the operation of the circuit is affected. Therefore, the power supply must produce the same output regardless of load and input voltage changes. To accomplish this, a **voltage regulator** is added after the filter.

There are two basic types of voltage regulators: **shunt voltage regulators** and **series voltage regulators**. They are named for the method by which they are connected with the load. The shunt regulator is connected in parallel with the load, whereas the series regulator is connected in series with the load. Series regulators are more popular than shunt regulators because they are more efficient and dissipate less power. The shunt regulator also acts as a control device, protecting the regulator from a short in the load.

Figure 36-20 shows a basic zener diode regulator circuit. This is a shunt regulator. The zener diode is connected in series with a resistor. The input voltage, unregulated DC voltage, is applied across both the zener diode and the resistor so as to make the zener

diode reverse biased. The resistor allows a small current to flow to keep the zener diode in its zener breakdown region. The input voltage must be higher than the zener breakdown voltage of the diode. The voltage across the zener diode is equal to the zener diode's voltage rating. The voltage dropped across the resistor is equal to the difference between the zener diode's voltage and the input voltage.

The circuit shown in Figure 36-20 provides a constant output voltage for a changing input voltage. Any change in the voltage appears across the resistor. The sum of the voltage drops must equal the input voltage. By changing the output zener diode and the series resistor, the output voltage can be increased or decreased.

The load resistance and the output voltage determine the current through the load. The load current and the zener current flow through the series resistor. The series resistor must be chosen carefully so that the zener current keeps the zener diode in the breakdown zone and allows the current to flow.

When the load current increases, the zener current decreases, and the load current and zener current together maintain a constant voltage. This allows the circuit to regulate for changes in output current as well as input voltage.

Figure 36-21 shows a shunt regulator circuit using a transistor. Note that transistor Q_1 is in parallel with the load. This protects the regulator in case a short develops with the load. More complex shunt regulators are available that use more than one transistor.

The series regulator is more popular than the shunt regulator. The simplest series regulator is a variable resistor in series with the load (Figure 36-22). The resistance is adjusted continuously to maintain a constant voltage across the load. When the DC voltage increases, the resistance increases, dropping more voltage. This maintains the voltage drop across the load by dropping the extra voltage across the series resistor.

■ **FIGURE 36-21**

Shunt regulator using a transistor.

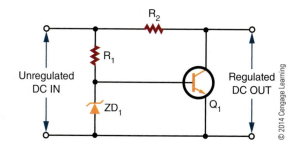

■ **FIGURE 36-20**

Basic zener diode regulator circuit.

© 2014 Cengage Learning

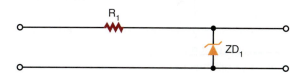

■ **FIGURE 36-22**

Series regulator using a variable resistor.

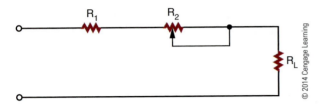

The variable resistor can also compensate for changes in load current. If the load current increases, more voltage drops across the variable resistor, which results in less voltage dropped across the load. If the resistance can be made to decrease at the same instant that the current increases, the voltage dropped across the variable resistor can remain constant, resulting in a constant output voltage despite changes in the load current.

In reality, it is too difficult to vary the resistance manually to compensate for voltage and current changes. It is more efficient to replace the variable resistor with a transistor (Figure 36-23) connected so that the load's current flows through it. By changing the current on the base of the transistor, the transistor can be biased to conduct more or less current. Additional components are required to make the circuit self-adjusting (Figure 36-24). These components allow the

■ **FIGURE 36-23**

Transistor series regulator using a manually adjusted variable resistor.

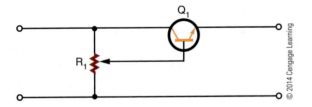

■ **FIGURE 36-24**

Self-adjusting series regulator.

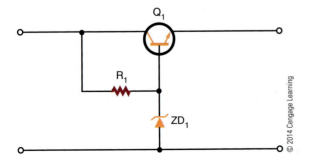

■ **FIGURE 36-25**

A simple series regulator.

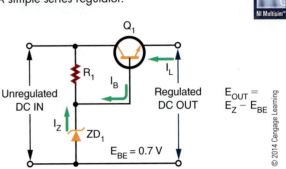

transistor to compensate automatically for changes in the input voltage or load current.

Figure 36-25 shows a simple series regulator. The input is unregulated DC voltage, and the output is a lower, regulated DC voltage. The transistor is connected as an emitter follower, meaning that there is a lack of phase inversion between the base and the emitter. The emitter voltage follows the base voltage. The load is connected between the emitter of the transistor and ground. The voltage at the base of the transistor is set by the zener diode. Therefore, the output voltage is equal to the zener voltage minus the 0.7-volt dropped across the base-emitter junction of the transistor.

When the input voltage increases through the transistor, the voltage out also tries to increase. The base voltage is set by the zener diode. If the emitter becomes more positive than the base, the conductance of the transistor decreases. When the transistor conducts less, it reacts the same way a large resistor would react if placed between the input and output. Most of the increase in input voltage is dropped across the transistor, and there is only a small increase in the output voltage.

There is a disadvantage with the emitter-follower regulator because the zener diode must have a large power rating. Zener diodes with large power-handling capabilities are expensive.

A more popular type of series regulator is the feedback regulator. A feedback regulator consists of a feedback circuit to monitor the output voltage. If the output voltage changes, a control signal is developed that controls the transistor's conductance. Figure 36-26 shows a block diagram of the feedback regulator. An unregulated DC voltage is applied to the input of the regulator. A lower, regulated DC output voltage appears at the output terminals of the regulator.

A **sampling circuit** appears across the output terminals. A sampling circuit is a voltage divider that sends

■ FIGURE 36-26

Block diagram of a feedback series regulator.

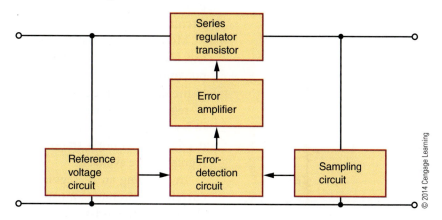

a sample of the output voltage to the error-detection circuit. The voltage changes if the output voltage changes.

The error-detection circuit compares the sampled voltage with a reference voltage. In order to produce the reference voltage, a zener diode is used. The difference between the sample and reference voltages is the **error voltage**. An error amplifier amplifies the error voltage. The error amplifier controls the conduction of the series transistor. The transistor conducts more or less to compensate for changes in the output voltage.

Figure 36-27 shows a feedback voltage regulator circuit. Resistors R_3, R_4, and R_5 form the sampling circuit. Transistor Q_2 acts as both the error-detection and error-amplification circuits. Zener diode D_1 and resistor R_1 produce the reference voltage. Transistor Q_1 is the series regulator transistor. Resistor R_2 is the collector load resistor for transistor Q_2 and the biasing resistor for transistor Q_1.

If the output voltage tries to increase, the sample voltage also increases. This increases the bias voltage on the base of transistor Q_2. The emitter voltage of transistor Q_2 is held constant by zener diode D_1. This results in transistor Q_2 conducting more and increasing the current through resistor R_2. In turn, the voltage on the collector of transistor Q_2 and the base of transistor Q_1 decreases. This decreases the forward bias of transistor Q_1, causing it to conduct less. When transistor Q_1 conducts less, less current flows. This causes a smaller voltage drop across the load and cancels the increase in voltage.

The output voltage can be adjusted accurately by varying potentiometer R_4. To increase the output voltage of the regulator, the wiper of potentiometer R_4 is moved in the more negative direction. This decreases the sample voltage on the base of transistor Q_2, decreasing the forward bias. This results in transistor Q_2 conducting less, causing the collector voltage of transistor Q_2 and the base of transistor Q_1 to increase. This increases the forward bias on transistor Q_1, causing it to conduct more. More current flows through the load, which increases the voltage out.

A serious disadvantage with the series regulator is that the transistor is in series with the load. A short in the load would result in a large current flowing through the transistor, which could destroy it. A circuit is needed to keep the current passing through the transistor at a safe level.

Figure 36-28 shows a circuit that limits current through the series regulator transistor. It is shown added to the feedback series voltage regulator. Transistor Q_3 and resistor R_6 form the current-limiting circuit. In order for transistor Q_3 to conduct, the base-to-emitter junction must be forward biased by a minimum of 0.7 volt. When

■ FIGURE 36-27

Basic feedback series regulator.

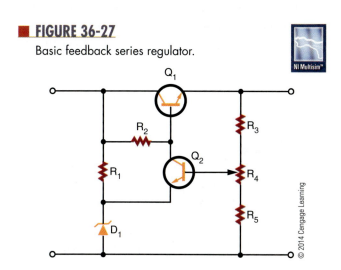

FIGURE 36-28

Feedback series regulator with current-limiting circuit.

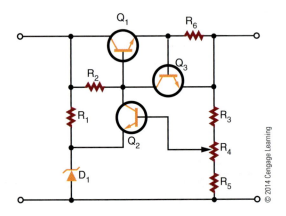

© 2014 Cengage Learning

0.7 volt is applied between the base and emitter, the transistor conducts. If R_6 is 1 ohm, the current necessary to develop 0.7 volt to the base of transistor Q_3 is

$$I = \frac{E}{R}$$

$$I = \frac{0.7}{1}$$

$$I = 0.7 \text{ A, or } 700 \text{ mA}$$

When less than 700 milliamperes of current flows, transistor Q_3's base-to-emitter voltage is less than 0.7 volt, keeping it shut off. When transistor Q_3 is shut off, the circuit acts as if it does not exist. When the current exceeds 700 milliamperes, the voltage dropped across resistor R_6 increases above 0.7 volt. This results in transistor Q_3 conducting through resistor R_2, thus decreasing the voltage on the base of transistor Q_1 and causing it to conduct less. The current cannot increase above 700 milliamperes. Changing the value of resistor R_6 can vary the amount of current that can be limited. Increasing the value of resistor R_6 lowers the limit on current value.

The feedback series regulator has another disadvantage: the number of components required. This problem can be overcome by the use of an integrated circuit (IC) regulator.

Modern IC regulators are low in cost and easy to use. Most IC regulators have only three terminals (input, output, and ground) and can be connected directly to the filtered output of a rectifier (Figure 36-29). The IC regulator provides a wide variety of output voltages of both positive and negative polarities. If a voltage is needed that is not a standard voltage, adjustable IC regulators are available.

In selecting an IC regulator, the voltage and load current requirements must be known along with the electrical characteristics of the unregulated power supply. Their output voltage classifies IC regulators. Fixed voltage regulators have three terminals and provide only one output voltage. They are available in both positive and negative voltages. Dual-polarity voltage regulators can supply both a positive voltage and a negative voltage. Both fixed- and dual-polarity voltage regulators are available as adjustable voltage regulators. If using any of the IC voltage regulators, refer to the manufacturer's specification sheet.

36–4   QUESTIONS

1. What is the purpose of a voltage regulator in a power supply?
2. What are the two basic types of voltage regulators?
3. Which type of voltage regulator is used more?
4. Draw a schematic of a simple zener diode voltage regulator, and explain how it operates.
5. Draw a block diagram of a series feedback regulator, and explain how it operates.

FIGURE 36-29

Three-terminal IC regulator.

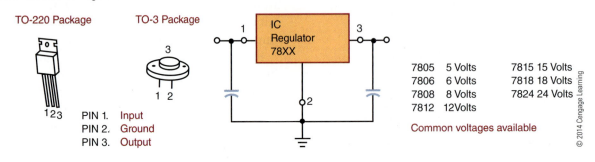

TO-220 Package

TO-3 Package

PIN 1. Input
PIN 2. Ground
PIN 3. Output

7805	5 Volts	7815	15 Volts
7806	6 Volts	7818	18 Volts
7808	8 Volts	7824	24 Volts
7812	12 Volts		

Common voltages available

© 2014 Cengage Learning

36–5 VOLTAGE MULTIPLIERS

In all previous cases, the DC voltage is limited to the peak value of the input sine wave. When higher DC voltages are required, a step-up transformer is used. However, higher DC voltages can be produced without a step-up transformer. Circuits that are capable of producing higher DC voltages without the benefit of a transformer are called voltage multipliers. Two **voltage multipliers** are the **voltage doubler** and the **voltage tripler**.

Figure 36-30 shows a half-wave voltage doubler. It produces a DC output voltage that is twice the peak value of the input signal. Figure 36-31 shows the circuit during the negative alternation of the input signal. Diode D_1 conducts, and the current flows along the path shown. Capacitor C_1 charges to the peak value of the input signal. Because there is no discharge path, capacitor C_1 remains charged. Figure 36-32 shows the positive alternation of the input signal. At this point, capacitor C_1 is charged to the negative peak value. This keeps diode D_1 reverse biased and makes diode D_2 forward biased. This allows diode D_2 to conduct, charging capacitor C_2. Because capacitor C_1 is charged to the maximum negative value, capacitor C_2 charges to twice the peak value of the input signal.

■ FIGURE 36-30

Half-wave voltage doubler.

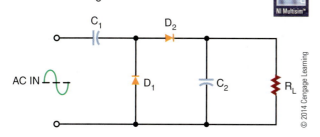

■ FIGURE 36-31

Half-wave voltage doubler during negative alternation of input signal.

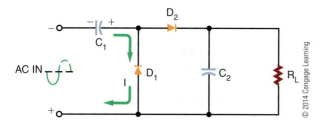

■ FIGURE 36-32

Half-wave voltage doubler during positive alternation of input signal.

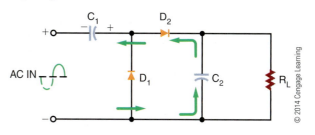

As the sine wave changes from the positive half cycle to the negative half cycle, diode D_2 is cut off because capacitor C_2 holds diode D_2 reverse biased. Capacitor C_2 discharges through the load, holding the voltage across the load constant. Therefore, it also acts as a filter capacitor. Capacitor C_2 recharges only during the positive cycle of the input signal, resulting in a ripple frequency of 60 hertz (hence the name "half-wave voltage doubler"). One disadvantage of the half-wave voltage doubler is that it is difficult to filter because of the 60-hertz ripple frequency. Another disadvantage is that capacitor C_2 must have a voltage rating of at least twice the peak value of the AC input signal.

A full-wave voltage doubler overcomes some of the disadvantages of the half-wave voltage doubler. Figure 36-33 is a schematic of a circuit that operates as a full-wave voltage doubler. Figure 36-34 shows that, on the positive alternation of the input signal, capacitor C_1 charges through diode D_1 to the peak value of the AC input signal. Figure 36-35 shows that on the negative alternation, capacitor C_2 charges through diode D_2 to the peak value of the input signal.

When the AC input signal is changing between the peaks of the alternations, capacitors C_1 and C_2 discharge in series through the load. Because each capacitor is charged to the peak value of the input signal,

■ FIGURE 36-33

Full-wave voltage doubler.

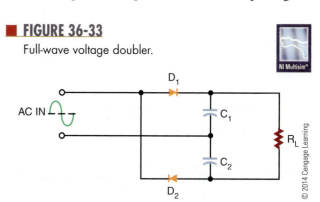

■ **FIGURE 36-34**

Full-wave voltage doubler during positive alternation of input signal.

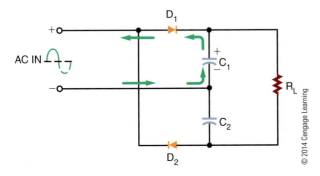

■ **FIGURE 36-35**

Full-wave voltage doubler during negative alternation of input signal.

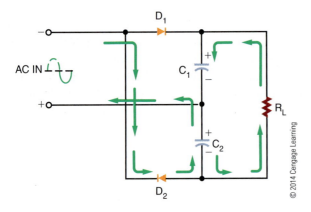

■ **FIGURE 36-36**

Voltage tripler.

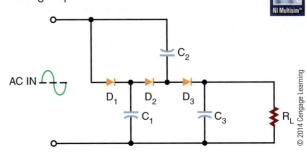

■ **FIGURE 36-37**

Voltage tripler during first positive alternation of input signal.

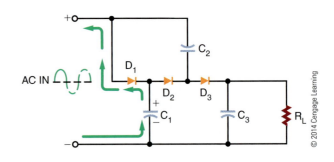

■ **FIGURE 36-38**

Voltage tripler during negative alternation of input signal.

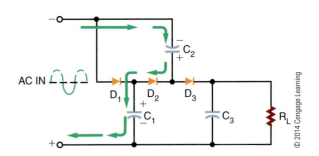

■ **FIGURE 36-39**

Voltage tripler during second positive alternation of input signal.

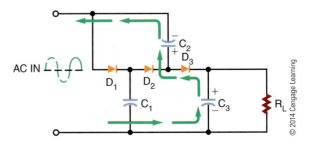

the total voltage across the load is two times the peak value of the input signal.

Capacitors C_1 and C_2 are charged during the peaks of the input signal. The ripple frequency is 120 hertz because both capacitors C_1 and C_2 are charged during each cycle. Capacitors C_1 and C_1 split the output voltage to the load, so each capacitor is subject to the peak value of the input signal.

Figure 36-36 shows the circuit of a voltage tripler. In Figure 36-37, the positive alternation of the input signal biases diode D_1 so that it conducts. This charges capacitor C_1 to the peak value of the input signal. Capacitor C_1 places a positive potential across diode D_2.

Figure 36-38 shows the negative alternation of the input signal. Because diode D_2 is now forward biased, current flows through it to capacitor C_1 via capacitor C_2. This charges capacitor C_2 to twice the peak value because of the voltage stored in capacitor C_1.

Figure 36-39 shows the occurrence of the next positive alternation. It places a difference of potential across capacitor C_2 that is three times the peak value. The top plate of capacitor C_2 has a positive peak value of two times the peak value. The anode of diode D_3 has a positive value of three times the peak value with respect to ground. This charges capacitor C_3 to three times the peak value. This is the voltage that is applied to the load.

36-5 QUESTIONS

1. What is the function of a voltage multiplier circuit?
2. Draw a schematic of a half-wave voltage doubler, and explain how it operates.
3. Draw a schematic of a full-wave voltage doubler.
4. Draw a schematic of a voltage tripler.
5. What requirement must be placed on capacitors used in voltage doubler and tripler circuits?

36-6 VOLTAGE DIVIDERS

Voltage dividers were discussed in Section 17–4 and will not be discussed extensively in this section. A voltage divider can be used to convert a higher voltage to a lower voltage for application requiring less voltage. Ohm's law (presented in Chapter 14) is one of the most important concepts to understand. It states that the current through a circuit is directly proportional to the voltage across the circuit and inversely proportional to the resistance.

$$Current = \frac{Voltage}{Resistance}$$

$$I = \frac{E}{R}$$

The example (Figure 36-40) is a practical example of a real-world voltage divider. The voltage source is the standard AC power source of 120V AC that has been converted to 120V DC (assuming no loss for this discussion). What is desired is a voltage-divider network that will divide this 120V DC source into a lower-voltage source specified by the manufacturer of a particular device (radio, cassette player, CD player, etc.). Assume the desired voltage is 6 volts, and the load will draw 200 mA. The remaining voltage will be dissipated as heat. Keep in mind that it is necessary to know the load current for the voltage required for the device in advance.

Referring to Figure 36-41, select the amount of current that is desired to flow through resistor R_2 starts the process. In this example, an arbitrary value of 10 mA is chosen. This means that there is 10 mA of current flowing through R_2 and the voltage at Junction A is 6 volts. Therefore R_2 can be determined as follows:

$$I_2 = \frac{E_2}{R_2}$$

$$0.01 = \frac{6}{R_2}$$

$$600\ \Omega = R_2$$

Next, inspect junction A between resistors R_1 and R_2. At this point, there is current summing. There is 10 mA of current flowing through R_2, and it is joined

■ FIGURE 36-40

Example of a real-world problem.

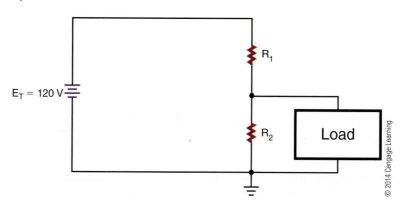

■ **FIGURE 36-41**

Determining current and resistor values.

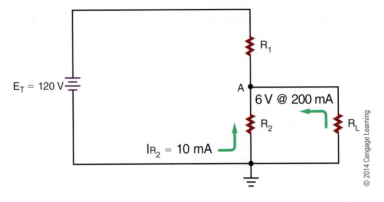

© 2014 Cengage Learning

or summed with 200 mA of current through the 6-volt load. This means that a total of 210 mA of current is flowing through resistor R_1.

The final step in the process is calculating the values of R_1. The final calculation to determine the value of resistor R_1 is based on the total current flowing through junction A, 210 mA. Also the desired voltage at junction A is 6 volts. The source voltage is 120 volts; therefore, resistor R_1 must drop 120 V − 6 V, or 114 V.

$$I_1 = \frac{E_1}{R_1}$$

$$0.210 = \frac{114}{R_1}$$

$$543\Omega = R_1$$

As in the real world, the resistor values required in this circuit are not exact values, but are rounded. The closest 1% resistor value is 549 Ω (Figure 36-42) and the closest 2%, 5%, and 10% resistor is 560 Ω (Figure 13-5).

Resistor R_1 consumes power and dissipates it as heat if it has too small a power rating. The amount of power that will be dissipated by the resistor can be determined using the power formula, P = IE (Chapter 17). The voltage across R_1 is 114 volts, and the current through it is 210 mA.

$$P_{R1} = I_{R1}E_{R1}$$

$$P_{R1} = (0.210)(114)$$

$$P_{R1} = 23.94 \text{ W}$$

Resistor R_1 must dissipate 23.94 watts. A safety factor would require selecting a power rating a minimum of two times the amount of power dissipated. A 560 Ω, 50 W resistor does exist and would work for this application.

■ **FIGURE 36-42**

One-percent resistor value chart.

1% Resistor Values			
100	178	316	562
102	182	324	576
105	187	332	590
107	191	340	604
110	196	348	619
113	200	357	634
115	205	365	649
118	210	374	665
121	215	383	681
124	221	392	698
127	226	402	715
130	232	412	732
133	237	422	750
137	243	432	768
140	249	442	787
143	255	453	806
147	261	464	825
150	267	475	845
154	274	487	866
158	280	499	887
162	287	511	909
165	294	523	931
169	301	536	953
174	309	549	976

© 2014 Cengage Learning

1. How does a voltage divider convert a higher voltage to a lower voltage?
2. How can the load current be determined when it is not listed with a device?
3. Why was 10 mA determined as the current through R_2 in Figure 36-41?
4. What is the voltage through R_2 in Figure 36-41 and why?
5. What could be added to the circuit in Figure 36-41 to insure the voltage is always 6 volts to the load?

36-7 CIRCUIT-PROTECTION DEVICES

To protect the load from failure of the power supply, an **over-voltage protection circuit** t is used.

Figure 36-43 shows an over-voltage protection circuit called a **crowbar**. The SCR, placed in parallel with the load, is normally cut off (not conducting). If the output voltage rises above a predetermined level, the SCR turns on and places a short circuit across the load. With the short circuit across the load, very little current flows through the load. This fully protects the load. The short circuit across the load does not protect the power supply. The output of the power supply is shorted, thereby blowing the fuse of the power supply.

The zener diode establishes the voltage level at which the SCR turns on. It protects the load from voltage above the zener voltage. As long as the supply voltage is less than the zener diode voltage rating, the diode does not conduct. This keeps the SCR turned off.

If the supply voltage rises above the zener voltage due to a malfunction, the zener diode conducts. This creates a gate current to the SCR, turning it on and shorting out

FIGURE 36-43

Crowbar overprotection circuit.

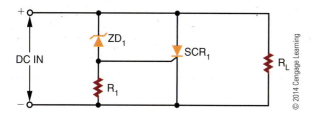

the load. It should be noted that the SCR must be large enough to handle the large short-circuit current.

To protect circuits against excessive transient voltage, a **varistor** is used. A varistor is an electronic component; the most common type is a **metal-oxide varistor (MOV)**. A MOV protects various types of electronic devices and semiconductor elements from switching and induced lightning surges. When exposed to high transient voltage, the MOV clamps the voltage to a safe level. An MOV absorbs potentially destructive energy and dissipates it as heat, protecting the circuit components.

MOV devices are made primarily of zinc oxide with small amounts of bismuth, cobalt, manganese, and other metal oxides. They can be connected in parallel for increased energy-handling capabilities. They can also be connected in series to provide higher voltage ratings or to provide voltage ratings between the standard increments.

MOVs are available with peak current ratings ranging from 40 A to 70,000 A.

Another protection device is a **fuse** (Figure 36-44). A fuse is a device that fails when an overload occurs. A fuse is essentially a small piece of wire between two metal terminals. A hollow glass cylinder holds the metal terminals apart and protects the wire. Typically, a fuse is placed in series with the primary of the power supply transformer. If a large current flows in the power supply, it causes the fuse wire to overheat and melt. This opens the circuit so that no more current can flow. The glass housing of the fuse allows a visual check to see whether the fuse is blown.

Fuses are classified as **normal** or **slow blow**. A normal fuse opens as soon as its current is exceeded. In some circuits this is an advantage because it removes the overload very quickly. A slow-blow fuse

FIGURE 36-44

Fuses used for protection of electronic circuits.

FIGURE 36-45

Circuit breakers used for protection of electronic circuits.

© 2014 Cengage Learning

can withstand a brief period of overloading before it blows. This brief period of overloading occurs because the fuse wire heats more slowly. If the overload is present for longer than a few seconds, it blows the fuse. A slow-blow fuse may contain a spring for pulling the fuse wire apart once it melts. Some circuits can withstand a surge in current. In these, a slow-blow fuse is preferable to a normal fuse.

The fuse is always installed after the switch on the "hot" (live and energized) lead of the AC power source. This disconnects the transformer from the AC power source when the fuse blows. By installing it after the switch, power can be removed from the fuse holder for added safety in replacing a blown fuse.

A disadvantage of the fuse is that it must be replaced each time it blows. A **circuit breaker** performs the same job but does not have to be replaced each time an overload occurs. Instead, the circuit breaker can be manually reset after the overload occurs (Figure 36-45). Circuit breakers are connected into the circuit in the same manner as fuses.

36-7   QUESTIONS

1. How does a crowbar over-voltage protection circuit operate?
2. How does a fuse operate when used in a circuit?
3. What are the different types of fuses?
4. Where is the fuse of any circuit-protection device located in a circuit?
5. What is the advantage of a circuit breaker over a fuse?

SUMMARY

- The primary purpose of a power supply is to convert AC to DC.
- Power supplies built with transformers are not isolated from the power source.
- Transformers are used in power supplies for isolation and to step up or step down the voltage.
- A rectifier circuit converts incoming AC voltage to pulsating DC voltage.
- The basic rectifier circuits are half-wave, full-wave, and bridge.
- Half-wave rectifiers are simpler and less expensive than either full-wave or bridge rectifiers.
- The full-wave rectifier is more efficient than the half-wave rectifier.
- The bridge rectifier can operate without a transformer.
- To convert pulsating DC voltage to a smooth DC voltage, a filter must follow the rectifier in the circuit.
- A capacitor in parallel with the load is an effective filter.
- A voltage regulator provides constant output regardless of load and input voltage changes.
- The voltage regulator is located after the filter in the circuit.
- The two basic types of voltage regulators are the shunt regulator and the series regulator.
- The series regulator is more efficient and therefore more popular than the shunt regulator.
- Voltage multipliers are circuits capable of providing higher DC voltages than the input voltage without the aid of a transformer.
- Voltage doublers and voltage triplers are voltage multipliers.
- Voltage dividers can reduce high voltages to lower voltages.
- A crowbar is a circuit designed for overvoltage protection.
- A fuse protects a circuit from a current overload.
- Fuses are classified as either normal or slow-blow.
- Circuit breakers perform the same job as fuses but do not have to be replaced each time there is an overload.

CHAPTER 36 SELF-TEST

1. What are the four areas of concern when selecting a transformer for a power supply?
2. What is the purpose of the transformer in a power supply?
3. What purpose does a rectifier serve with a power supply?
4. What are the advantages and disadvantages between the full-wave rectifier circuit and the bridge rectifier circuit?
5. Describe the process of how a filter capacitor converts a pulsating DC voltage to a smooth DC voltage.
6. On what basis is the size of the filter capacitor selected?
7. How does a series regulator maintain the output voltage at a constant level?
8. What circuit characteristics must be known when selecting a regulator circuit?
9. What practical use do voltage multipliers serve?
10. What advantages does a full-wave voltage doubler have over a half-wave voltage doubler?
11. What type of device is used for over-voltage protection?
12. What types of devices are used for over-current protection?

Amplifier Basics

OBJECTIVES

After completing this chapter, the student will be able to:

- Describe the purpose of an amplifier.
- Identify the three basic configurations of transistor amplifier circuits.
- Identify the classes of amplifiers.
- Describe the operation of direct coupled amplifiers, audio amplifiers, video amplifiers, RF amplifiers, IF amplifiers, and operational amplifiers.
- Draw and label schematic diagrams for the different types of amplifier circuits.

KEY TERMS

37-1 amplification

37-1 common-base circuit

37-1 common-collector circuit

37-1 common-emitter circuit

37-1 forward biased

37-1 V_{CC}

37-2 bypass capacitor

37-2 class A amplifier

37-2 class AB amplifier

37-2 class B amplifier

37-2 class C amplifier

37-2 degenerative (negative) feedback

37-2 thermal instability

37-3 coupling

37-3 direct coupling

37-3 impedance coupling

37-3 resistance-capacitance coupling

37-3 transformer coupling

mplifiers are electronic circuits that are used to increase the amplitude of an electronic signal. A circuit designed to convert a low voltage to a higher voltage is called a voltage amplifier. A circuit designed to convert a low current to a higher current is called a current amplifier.

37-1 AMPLIFIER CONFIGURATIONS

In order for a transistor to provide **amplification**, it must be able to accept an input signal and produce an output signal that is greater than the input.

The input signal controls current flow in the transistor. This in turn controls the voltage through the load. The transistor circuit is designed to take voltage from an external power source (**V_{CC}**) and apply it to a load resistor (R_L) in the form of an output voltage. The output voltage is controlled by a small input voltage.

The transistor is used primarily as an amplifying device. However, there is more than one way of achieving this amplification. The transistor can be connected in three different circuit configurations.

The three circuit configurations are the **common-base circuit**, the **common-emitter circuit**, and the **common-collector circuit**. In each configuration, one of the transistor's leads serves as a common reference point, and the other two leads serve as input and output connections. Each configuration can be constructed using either NPN or PNP transistors. In each, the transistor's emitter-base junction is **forward biased**, whereas the collector-base junction is reverse biased. Each configuration offers advantages and disadvantages.

In the common-base circuit (Figure 37-1), the input signal enters the emitter-base circuit, and the output leaves from the collector-base circuit. The base is the element common to both the input and output circuits.

In the common-emitter circuit (Figure 37-2), the input signal enters the base-emitter circuit, and the output signal leaves from the collector-emitter circuit. The emitter is common to both the input and output circuits. This method of connecting a transistor is the most widely used.

The third type of connection (Figure 37-3) is the common-collector circuit. In this configuration, the input signal enters the base-collector circuit, and the output signal leaves from the emitter-collector circuit. Here, the collector is common to both the

■ **FIGURE 37-1**

Common-base amplifier circuit.

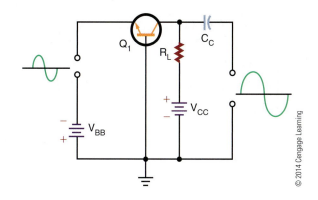

© 2014 Cengage Learning

■ **FIGURE 37-2**

Common-emitter amplifier circuit.

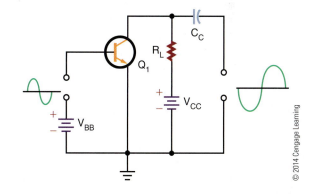

© 2014 Cengage Learning

■ **FIGURE 37-3**

Common-collector amplifier circuit.

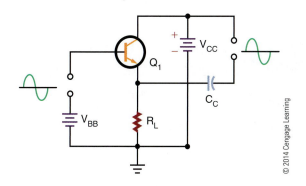

© 2014 Cengage Learning

input and output circuits. This circuit is used as an impedance-matching circuit.

Figure 37-4 charts the input–output resistance and the voltage, current, and power gains for the three circuit configurations. Figure 37-5 shows the phase relationship of input and output waveforms for the three configurations.

■ **FIGURE 37-4**

Amplifier circuit characteristics.

Circuit Type	Input Resistance	Output Resistance	Voltage Gain	Current Gain	Power Gain
Common base	Low	High	High	Less than 1	Medium
Common emitter	Medium	Medium	Medium	Medium	High
Common collector	High	Low	Less than 1	Medium	Medium

© 2014 Cengage Learning

■ **FIGURE 37-5**

Amplifier circuit input–output phase relationships.

Amplifier Type	Input Waveform	Output Waveform
Common base	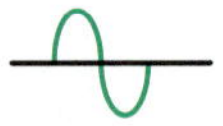	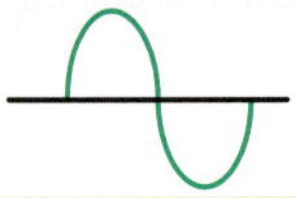
Common emitter	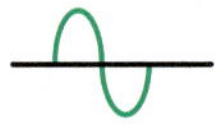	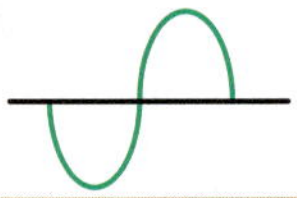
Common collector	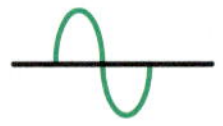	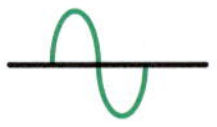

© 2014 Cengage Learning

NOTE **The common-emitter configuration provides a phase reversal of the input–output signal.**

37–1 QUESTIONS

1. Draw schematic diagrams of the three basic configurations of transistor amplifier circuits.
2. List the characteristics of the following circuit configurations:
   a. Common-base circuit
   b. Common-emitter circuit
   c. Common-collector circuit
3. Make a chart showing the input–output phase relationship of the three configurations.
4. Make a chart showing the input–output resistances of the three configurations.
5. Make a chart showing the voltage, current, and power gains of the three configurations.

37–2 AMPLIFIER BIASING

The basic configurations of transistor amplifier circuits are the common base, the common emitter, and the common collector. All require two voltages for proper biasing. The base-emitter junction must be forward biased and the base-collector junction must be reverse biased. However, both bias voltages can be provided from a single source.

Because the common-emitter circuit configuration is the most often used, it is described in detail here. The same principles apply to the common-base and common-collector circuits.

Figure 37-6 shows a common-emitter transistor amplifier using a single voltage source. The circuit is schematically diagrammed in Figure 37-7. The voltage source is identified as $+V_{CC}$. The ground symbol is the negative side of the voltage source V_{CC}. The single voltage source provides proper biasing for the base-emitter and base-collector junctions. The two resistors (R_B and R_L) are used to distribute the voltage for proper operation. Resistor R_L, the collector load

■ **FIGURE 37-6**

Common-emitter amplifier with single voltage source.

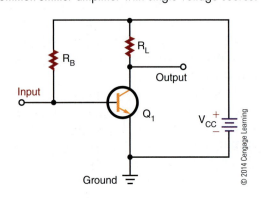

© 2014 Cengage Learning

■ **FIGURE 37-7**

Schematic representation of common-emitter amplifier with single voltage source.

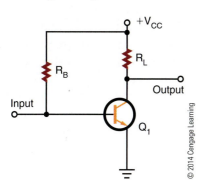

■ **FIGURE 37-8**

Common-emitter amplifier with collector feedback.

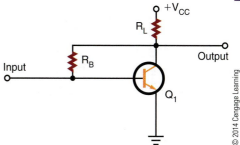

resistor, is in series with the collector. When a collector current flows, a voltage develops across resistor R_L. The voltage dropped across resistor R_L and the voltage dropped across the transistor's collector-to-emitter junction must add up to the total voltage applied.

Resistor R_B, connected between the base and the voltage source, controls the amount of current flowing out of the base. The base current flowing through resistor R_B creates a voltage across it. Most of the voltage from the source is dropped across it. A small amount of voltage drops across the transistor's base-to-emitter junction, providing the proper forward bias.

The single voltage source can provide the necessary forward-bias and reverse-bias voltages. For an NPN transistor, the transistor's base and collector must be positive with respect to the emitter. Therefore, the voltage source can be connected to the base and the collector through resistors R_B and R_L. This circuit is often called a base-biased circuit, because resistor R_B and the voltage source control the base current.

The input signal is applied between the transistor's base and emitter or between the input terminal and ground. The input signal either aids or opposes the forward bias across the emitter junction. This causes the collector current to vary, which then causes the voltage across R_L to vary. The output signal is developed between the output terminal and ground.

The circuit shown in Figure 37-7 is unstable, because it cannot compensate for changes in the bias current with no signal applied. Temperature changes cause the transistor's internal resistance to vary, which causes the bias currents to change. This shifts the transistor operating point, reducing the gain of the transistor. This process is referred to as **thermal instability**.

It is possible to compensate for temperature changes in a transistor amplifier circuit. If a portion of the unwanted output signal is fed back to the circuit input, the signal opposes the change. This is referred to as **degenerative** or **negative feedback** (Figure 37-8). In a circuit using degenerative feedback, the base resistor R_B is connected directly to the collector of the transistor. The voltage at the collector determines the current flowing through resistor R_B. If the temperature increases, the collector current increases, and the voltage across R_L increases. The collector-to-emitter voltage decreases, reducing the voltage applied to R_B. This reduces the base current, which causes the collector current to decrease. This is referred to as a *collector feedback circuit*.

Figure 37-9 shows another type of feedback. The circuit is similar to the one shown in Figure 37-7, except that a resistor R_E is connected in series with the emitter lead. Resistors R_B and R_E and the transistor's emitter-base junction are connected in series with the source voltage V_{CC}.

An increase in temperature causes the collector current to increase. The emitter current then also increases, causing the voltage across resistor R_E to increase and the voltage across resistor R_B to decrease.

■ **FIGURE 37-9**

Common-emitter amplifier with emitter feedback.

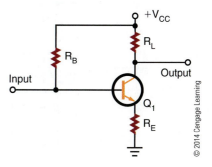

■ **FIGURE 37-10**

Emitter feedback with bypass capacitor.

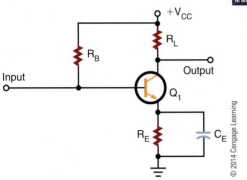

© 2014 Cengage Learning

The base current then decreases, which reduces both the collector current and the emitter current. Because the feedback is generated at the transistor's emitter, the circuit is called an *emitter feedback circuit*.

There is a problem with this type of circuit because an AC input signal develops across resistor R_E as well as across the load resistor R_L and the transistor. This reduces the overall gain of the circuit. With the addition of a capacitor across the emitter resistor R_E (Figure 37-10), the AC signal bypasses resistor R_E. The capacitor is often referred to as a **bypass capacitor**.

The bypass capacitor prevents any sudden voltage changes from appearing across resistor R_E by offering a lower impedance to the AC signal. The bypass capacitor holds the voltage across resistor R_E steady while not interfering with the feedback action provided by resistor R_E.

A voltage-divider feedback circuit (Figure 37-11) offers even more stability. This circuit is the one most

widely used. Resistor R_B is replaced with two resistors, R_1 and R_2, connected in series across the source voltage V_{CC}. The resistors divide the source voltage into two voltages, forming a voltage divider.

Resistor R_2 drops less voltage than resistor R_1. The voltage at the base, with respect to ground, is equal to the voltage developed across resistor R_2. The purpose of the voltage divider is to establish a constant voltage from the base of the transistor to ground. The current flow through resistor R_2 is toward the base. Therefore, the end of resistor R_2 attached to the base is positive with respect to ground.

Because the emitter current flows up through resistor R_E, the voltage dropped across resistor R_E is more positive at the end that is attached to the emitter. The voltage developed across the emitter-base junction is the difference between the two positive voltages developed across resistor R_2 and resistor R_E. For the proper forward bias to occur, the base must be slightly more positive than the emitter.

When the temperature increases, the collector and emitter currents also increase. The increase in the emitter current causes an increase in the voltage drop across the emitter resistor R_E. This results in the emitter becoming more positive with respect to ground. The forward bias across the emitter-base junction is then reduced, causing the base current to decrease. The reduction in the base current reduces the collector and emitter currents. The opposite action takes place if the temperature decreases: The base current increases, causing the collector and emitter currents to increase.

The amplifier circuits discussed so far have been biased so that all of the applied AC input signal appears at the output. Except for a higher voltage, the output signal is the same as the input signal. An amplifier that is biased so that the current flows throughout the entire cycle is operating as a **class A amplifier** (Figure 37-12).

■ **FIGURE 37-11**

Common-emitter amplifier with voltage-divider feedback.

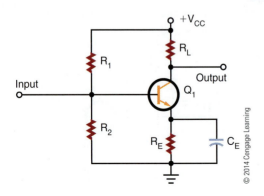

© 2014 Cengage Learning

■ **FIGURE 37-12**

Class A amplifier output.

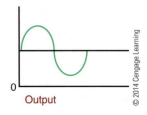

Input   Output

© 2014 Cengage Learning

An amplifier that is biased so that the output current flows for less than a full cycle but more than a half cycle is operating as a **class AB amplifier**. More than half but less than the full AC input signal is amplified in the class AB mode (Figure 37-13).

■ FIGURE 37-13

Class AB amplifier output.

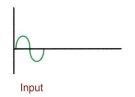
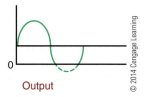

Input   Output

■ FIGURE 37-14

Class B amplifier output.

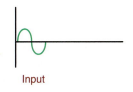
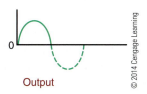

Input   Output

■ FIGURE 37-15

Class C amplifier output.

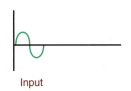
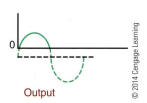

Input   Output

An amplifier that is biased so that the output current flows for only half of the input cycle is operating as a **class B amplifier**. Only one-half of the AC input signal is amplified in the class B mode (Figure 37-14).

An amplifier that is biased so that the output current flows for less than half of the AC input cycle is operating as a **class C amplifier**. Less than one alternation is amplified in the class C mode (Figure 37-15).

Class A amplifiers are the most linear of the types mentioned. They produce the least amount of distortion. They also have lower output ratings and are the least efficient. They find wide application where the full signal must be maintained, as in the amplification of audio signals in radios and televisions. However, because of the high power-handling capabilities required for class A operation, transistors are usually operated in the class AB or class B mode.

Class AB, B, and C amplifiers produce a substantial amount of distortion. This is because they amplify only a portion of the input signal. To amplify the full AC input signal, two transistors are needed, connected in a push–pull configuration (Figure 37-16). Class B amplifiers are used as output stages of stereo systems and public address amplifiers, and in many industrial controls. Class C amplifiers are used for high-power amplifiers and transmitters where only a single frequency is amplified, such as the RF (radio frequency) carrier used for radio and television transmission.

■ FIGURE 37-16

Push–pull amplifier configuration.

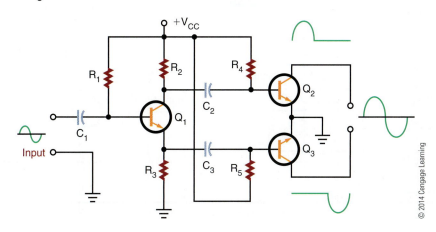

1. Draw a schematic diagram of a common-emitter transistor amplifier using a single voltage source.
2. How are temperature changes compensated for in a transistor amplifier?
3. Draw a schematic diagram of a voltage-divider feedback circuit.
4. List the classes of amplifiers and identify their outputs.
5. List the applications of each class of amplifier.

37–3   AMPLIFIER COUPLING

To obtain greater amplification, transistor amplifiers may be connected together. However, to prevent one amplifier's bias voltage from affecting the operation of the second amplifier, a **coupling** technique must be used. The coupling method used must not disrupt the operation of either circuit. Coupling methods used include **resistance-capacitance coupling, impedance coupling, transformer coupling**, and **direct coupling**.

Resistive-capacitive coupling, or RC coupling, consists of two resistors and a capacitor connected as shown in Figure 37-17. Resistor R_3 is the collector load resistor of the first stage. Capacitor C_1 is the

DC-blocking and AC-coupling capacitor. Resistors R_5 and R_6 are the input load resistor and DC return resistor for the base-emitter junction of the second stage. Resistance-capacitive coupling is used primarily in audio amplifiers.

Coupling capacitor C_1 must have a low reactance to minimize low-frequency attenuation. Typically, a high-capacitance value of 10 to 100 microfarads is used. The coupling capacitor is generally an electrolytic type.

The reactance of the coupling capacitor increases as the frequency decreases. The low-frequency limit is determined by the size of the coupling capacitor. The type of transistor used determines the high-frequency limit.

The impedance-coupling method is similar to the RC-coupling method, but an inductor is used in place of the collector load resistor for the first stage of amplification (Figure 37-18).

Impedance coupling works just like RC coupling. The advantage is that the inductor has a very low DC resistance across its windings. The AC output signal is developed across the inductor just as across the load resistor. However, the inductor consumes less power than the resistor, increasing the overall efficiency of the circuit.

The disadvantage of impedance coupling is that the inductance reactance increases with the frequency. The voltage gain varies with the frequency. This type of coupling is ideal for single-frequency amplification when a very narrow band of frequencies must be amplified.

■ **FIGURE 37-17**

RC coupling.

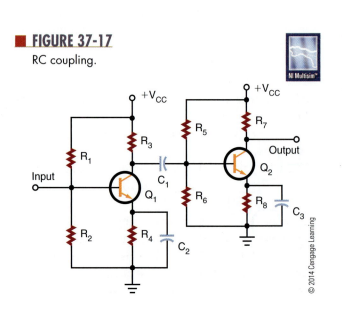

© 2014 Cengage Learning

■ **FIGURE 37-18**

Impedance coupling.

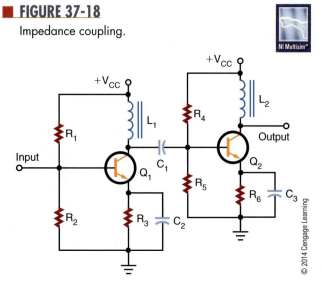

© 2014 Cengage Learning

In a transformer-coupled circuit, the two amplifier stages are coupled through the transformer (Figure 37-19). The transformer can effectively match a high-impedance source to a low-impedance load. The disadvantage is that transformers are large and expensive. Also, like the inductor-coupled amplifier, the transformer-coupled amplifier is useful only for a narrow band of frequencies.

When very low frequencies or a DC signal must be amplified, the direct-coupling technique must be used (Figure 37-20). Direct-coupled amplifiers provide a uniform current or voltage gain over a wide range of frequencies. This type of amplifier can amplify frequencies from 0 (DC) hertz to many thousands of hertz. However, direct-coupled amplifiers find their best applications with low frequencies.

A drawback of direct-coupled amplifiers is that they are not stable. Any change in the output current of the first stage is amplified by the second stage. This occurs because the second stage is essentially biased by the first stage. To improve the stability requires the use of expensive precision components.

■ FIGURE 37-19

Transformer coupling.

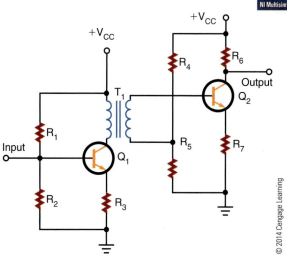

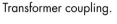

© 2014 Cengage Learning

■ FIGURE 37-20

Direct coupling.

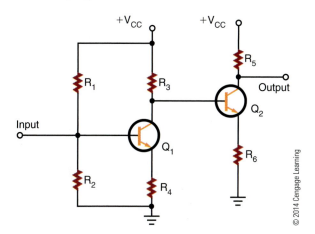

© 2014 Cengage Learning

37–3 QUESTIONS

1. What are the four methods for coupling amplifiers?
2. Where would RC coupling be used?
3. What determines low-frequency and high-frequency coupling?
4. Why would impedance coupling be used over RC coupling?
5. What is the drawback of using direct-coupled amplifiers?

SUMMARY

- Amplifiers are electronic circuits used to increase the amplitude of an electronic signal.
- The transistor is used primarily as an amplifying device.
- The three transistor amplifier configurations are common base, common collector, and common emitter.
- Common-collector amplifiers are used for impedance matching.
- Common-emitter amplifiers provide phase reversal of the input–output signal.
- All transistor amplifiers require two voltages for proper biasing.
- A single voltage source can provide the necessary forward-bias and reverse-bias voltages using a voltage-divider arrangement.
- A voltage-divider feedback arrangement is the most commonly used biasing arrangement.
- A transistor amplifier can be biased so that all or part of the input signal is present at the output.

- Class A amplifiers are biased so that the output current flows throughout the cycle.
- Class AB amplifiers are biased so that the output current flows for less than the full but more than half of the input cycle.
- Class B amplifiers are biased so that the output current flows for only half of the input cycle.
- Class C amplifiers are biased so that the output current flows for less than half of the input cycle.

- Coupling methods used to connect one transistor to another include resistance-capacitance coupling, impedance coupling, transformer coupling, and direct coupling.
- Direct-coupled amplifiers are used for high gain at low frequencies or amplification of a DC signal.

CHAPTER 37 SELF-TEST

1. Briefly describe how a transistor is used to provide amplification.
2. Why is the common-emitter amplifier the most widely used transistor amplifier configuration?
3. Describe thermal instability with transistors and methods used to compensate for it.
4. What class amplifier would be best for an audio sound system, and how would it be configured for minimal distortion of the input signal?
5. When using RC coupling to couple two amplifiers together, what function does the capacitor serve?
6. What factor affects the gain of the transistor, and what must be done to compensate for it?
7. How does an amplifier's class of operation affect the biasing of the amplifier?
8. What factor must be taken into consideration when connecting one amplifier to another?
9. How does the frequency of operation of an amplifier affect the coupling method used in connecting amplifiers together?
10. How can the drawbacks of direct-coupled amplifiers be overcome?

CHAPTER 38

Amplifier Applications

OBJECTIVES

After completing this chapter, the student will be able to:

- Describe the operation of the following amplifiers:
  - Direct-coupled amplifiers
  - Audio amplifiers
  - Video amplifiers
  - RF amplifiers
  - IF amplifiers
  - Operational amplifiers
- Identify schematic diagrams for the different types of amplifier circuits.

KEY TERMS

An amplifier may be defined as an electronic circuit designed to increase the amplitude of an electronic signal. Amplifier circuits are one of the most basic circuits in electronics. Amplifier circuits make signal level greater, sound louder, and provide the circuit with gain. Gain is the ability of an electronic circuit to increase the amplitude of an electronic signal. This chapter looks at some unique types of amplifier circuits.

38–1   DIRECT-COUPLED AMPLIFIERS

Direct-coupled, or **DC**, **amplifiers** are used for high gain at low frequencies or for amplification of a direct-current signal. The DC amplifier is also used to eliminate frequency loss through a coupling network. Applications of the DC amplifier include computers, measuring and test instruments, and industrial control equipment.

A simple DC amplifier is shown in Figure 38-1. The common-emitter amplifier is the one most frequently used. The circuit is shown with voltage-divider bias and emitter feedback. This type of circuit does not use a coupling capacitor. The input is applied directly to the base of the transistor. The output is taken from the collector.

The DC amplifier can provide both voltage and current gain. However, it is used primarily as a voltage amplifier. The voltage gain is uniform for both AC and DC signals.

In most applications, one stage of amplification is not enough; two or more stages are required to obtain a higher gain. Two or more stages connected together are referred to as a **multistage amplifier**. Figure 38-2 shows a two-stage amplifier. The input signal is amplified by the first stage. The amplified signal is then applied to the base of the transistor in

■ FIGURE 38-2
Two-stage DC amplifier.

the second stage. The overall gain of the circuit is the product of the voltage gains of the two stages. For example, if both the first and second stages have a voltage gain of 10, the overall gain of the circuit is 100.

Figure 38-3 shows another type of two-stage DC amplifier in which both an NPN transistor and a PNP transistor are used. This type of circuit is called a **complementary amplifier**. The circuit functions the same way as the circuit shown in Figure 38-2. The difference is that the second-stage transistor is a PNP transistor, which is reversed so that the emitter and collector are biased properly.

Figure 38-4 shows two transistors connected together to function as a single unit. This circuit arrangement is called a **Darlington arrangement**. Transistor Q_1 is used to control the conduction of transistor Q_2. The input signal applied to the base of transistor Q_1 controls the base of transistor Q_2. The Darlington arrangement may be a single package with three leads: emitter (E), base (B), and collector (C). It is used as a simple DC amplifier but offers a very high voltage gain.

■ FIGURE 38-1
Simple DC amplifier.

■ FIGURE 38-3
Complementary DC amplifier.

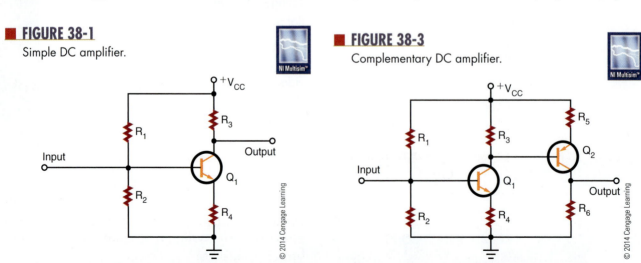

© 2014 Cengage Learning

■ FIGURE 38-4

Darlington arrangement.

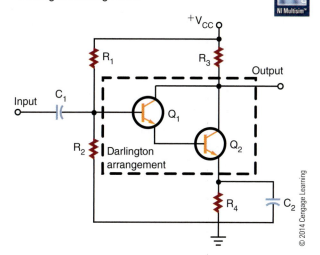

© 2014 Cengage Learning

The main disadvantage of multistage amplifiers is their high thermal instability. In circuits requiring three or four stages of DC amplification, the final output stage may not amplify the original DC or AC signal, but one that has been greatly distorted. The same problem exists with the Darlington arrangement.

In applications where both high gain and temperature stability are required, another type of amplifier is necessary. This type is called a **differential amplifier** (Figure 38-5). It is unique in that it has two separate inputs and can provide either one or two outputs. If a signal is applied to the input of transistor Q_1, an amplified signal is produced between output A and ground as in an ordinary amplifier. However, a small signal is also developed across resistor R_4 and is applied to the emitter of transistor Q_2. Transistor Q_2 functions as a common-base amplifier and amplifies the signal on its base. An amplified output signal is produced between output B and ground. The output produced at B

■ FIGURE 38-5

Differential amplifier.

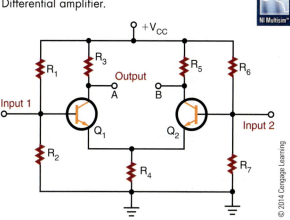

© 2014 Cengage Learning

is 180° out of phase with output A. This makes the differential amplifier much more versatile than a conventional amplifier.

The differential amplifier is generally not used to obtain an output between output and ground. The output signal is usually obtained between output A and output B. Because the two outputs are 180° out of phase, a substantial output voltage is developed between the two points. The input signal can be applied to either input.

The differential amplifier has a high degree of temperature stability because transistors Q_1 and Q_2 are mounted close together and therefore are equally affected by temperature changes. Also, the collector currents of transistors Q_1 and Q_2 tend to increase and decrease by the same amount so that the output voltage remains constant.

The differential amplifier is used extensively in integrated circuits and electronic equipment in general. It is used to amplify and/or compare the amplitudes of both DC and AC signals. It is possible to connect one or more differential amplifiers together to obtain a higher overall gain. In some cases, the differential amplifier is used as a first stage, with conventional amplifiers used in succeeding circuits. Because of their versatility and temperature stability, differential amplifiers are the most important type of direct-coupled amplifier.

38-1	**QUESTIONS**

1. When are direct-coupled amplifiers used in a circuit?
2. What kind of amplifiers are direct-coupled amplifiers primarily used for?
3. Draw schematic diagrams of the following circuits:
   a. Complementary amplifier
   b. Darlington arrangement
   c. Differential amplifier
4. How does a differential amplifier differ from a conventional amplifier?
5. Where are differential amplifiers primarily used?

38-2 AUDIO AMPLIFIERS

Audio amplifiers amplify AC signals in the frequency range of approximately 20 to 20,000 hertz. They may amplify the whole audio range or only a small portion of it.

Audio amplifiers are divided into two categories: **voltage amplifiers** and **power amplifiers**. Voltage amplifiers are primarily used to produce a high voltage gain. Power amplifiers are used to deliver a large amount of power to a load. For example, a voltage amplifier is typically used to increase the voltage level of a signal sufficiently to drive a power amplifier. The power amplifier then supplies a high output to drive a load such as a loudspeaker or other high-power device. Typically, voltage amplifiers are biased to operate as class A amplifiers, and power amplifiers are biased to operate as class B amplifiers.

Figure 38-6 shows a simple voltage amplifier. The circuit shown is a common-emitter circuit. It is biased to operate as a class A amplifier to provide a minimum amount of distortion. The amplifier can provide a substantial voltage gain over a wide frequency range. Because of the coupling capacitors, the circuit cannot amplify a DC signal.

Two or more voltage amplifiers can be connected together to provide higher amplification. The stages may be RC coupled or transformer coupled;

transformer coupling is more efficient. The transformer is used to match the input and output impedance of the two stages. This keeps the second stage from loading down the first stage. **Loading down** is the condition when a device creates too large a load and severely affects the output by drawing too much current. The transformer used to link the two stages together is called an **interstage transformer**.

Once a sufficient voltage level is available, a power amplifier is used to drive the load. Power amplifiers are designed to drive specific loads and are rated in watts. Typically a load may vary from 4 to 16 ohms.

Figure 38-7 shows a two-transistor power amplifier circuit, called a **push–pull amplifier**. The top half of the circuit is a mirror image of the bottom half. Each half is a single transistor amplifier. The output voltage is developed across the primary of the transformer during alternate half-cycles of the input signal. Both transistors are biased either class AB or class B. The input to a push–pull amplifier requires complementary signals. That is, one signal must be inverted compared to the other. However, both signals must have the same amplitude and the same frequency. The circuit that produces the complementary signal is called a **phase splitter**. A single-transistor phase splitter is shown in Figure 38-8. The complementary outputs are taken from the collector and the emitter of the transistor. The phase splitter is operated as a class A amplifier to provide minimum distortion. The coupling capacitors are necessary to offset the differences between the DC collector and emitter voltages.

A push–pull amplifier that does not require a phase splitter is called a **complementary push–pull amplifier**. It uses an NPN transistor and a PNP transistor to accomplish the push–pull action (Figure 38-9). The two transistors are connected in series with their emitters together. When each transistor is properly biased, there is 0.7 volt between the

■ **FIGURE 38-6**

Voltage amplifier.

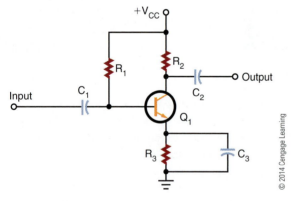

■ **FIGURE 38-7**

Push–pull power amplifier.

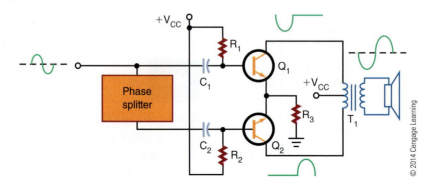

© 2014 Cengage Learning

■ FIGURE 38-8
Phase splitter.

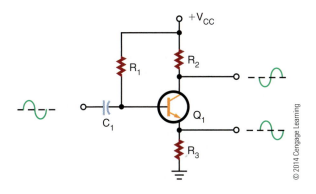

■ FIGURE 38-9
Complementary push–pull power amplifier.

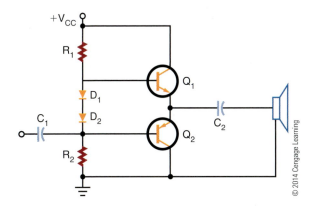

■ FIGURE 38-10
Quasi-complementary power amplifier.

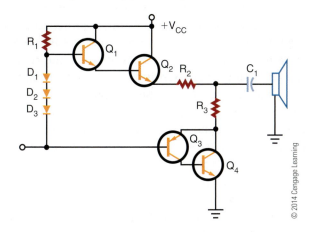

■ FIGURE 38-11
Different types of heat sinks available.

base and emitter or 1.4 volts between the two bases. The two diodes help to keep the 1.4-volt difference constant. The output is taken from between the two emitters through a coupling capacitor.

For amplifiers greater than 10 watts, it is difficult and expensive to match NPN and PNP transistors to ensure that they have the same characteristics. Figure 38-10 shows a circuit that uses two NPN transistors for the output-power transistors. The power transistors are driven by lower-power NPN and PNP transistors, and the upper set of transistors is connected in a Darlington configuration. The lower set of transistors uses a PNP transistor and an NPN transistor. Operating as a single unit, they respond like a PNP transistor. This type of amplifier is called a **quasi-complementary amplifier**. It operates like a complementary amplifier but does not require high-power complementary output transistors.

Because of the large amounts of power generated by power amplifiers, some components get hot. To assist in the removal of this heat buildup, a **heat sink** is

used. A heat sink is a device that provides a large area from which the heat can radiate. Figure 38-11 shows different types of heat sinks used with transistors.

38–2 QUESTIONS

1. For what frequency range are audio amplifiers used?
2. What are the two types of audio amplifiers?
3. What is an interstage transformer?
4. Draw schematic diagrams of the following:
   a. Push–pull amplifier
   b. Complementary push–pull amplifier
   c. Quasi-complementary amplifier
5. How can excess heat generated by an amplifier be removed?

38-3 VIDEO AMPLIFIERS

Video amplifiers are wideband amplifiers used to amplify video (picture) information. The frequency range of the video amplifier is greater than that of the audio amplifier, extending from a few hertz to 5 or 6 megahertz. For example, a television requires a uniform bandwidth of 60 hertz to 4 megahertz. Radar requires a bandwidth of 30 hertz to 2 megahertz. In circuits that use sawtooth or pulse waveforms, it is necessary to cover a range of frequencies from $\frac{1}{10}$ of the lowest frequency to 10 times the highest frequency. The extended range is necessary because nonsinusoidal waveforms contain many harmonics and they must all be amplified equally.

Because video amplifiers require good uniformity, in frequency response only direct or RC coupling is used. Direct coupling provides the best frequency response, whereas RC coupling has economic advantages. The RC-coupled amplifier also has a flat response in the middle frequency range that is suitable for video amplifiers. **Flat response** is the term used to indicate that the gain of an amplifier varies only slightly within a stated frequency range. The response curve plotted for such an amplifier is almost a straight line; hence the term *flat response*.

A factor that limits the high-frequency response in a transistor amplifier is the shunt capacitance of the circuit. A small capacitance exists between the junctions of the transistor. The capacitance is determined by the size of the junction and the spacing between the transistor's leads. The capacitance is further affected by the junction bias. A forward-biased base-emitter junction has a greater capacitance than a reverse-biased collector-base junction.

To reduce the effects of shunt capacitance and increase the frequency response in transistor video amplifiers, peaking coils are used. Figure 38-12 shows the **shunt-peaking** method. A small inductor is placed in series with the load resistor. At the low- and mid-frequency ranges, the peaking coil has little effect on the amplifier response. At the higher frequencies, the inductor resonates with the circuit's capacitance, which results in an increase in the output impedance and boosts the gain.

Another method is to insert a small inductor in series with the interstage coupling capacitor. This

■ FIGURE 38-12
Shunt peaking.

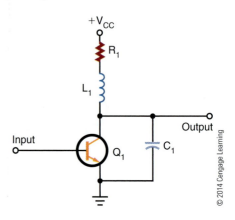

method is called **series peaking** (Figure 38-13). The peaking coil effectively isolates the input and output capacitance of the two stages.

Often series and shunt peaking are combined to gain the advantages of both (Figure 38-14). This combination can extend the bandwidth to over 5 megahertz.

The most common use of video amplifiers is in television receivers and computer monitors. (Figure 38-15). Transistor Q_1 is connected as an emitter-follower. Input to transistor Q_1 is from the video detector. The video detector recovers the video signal from the intermediate frequency. In the collector circuit of transistor Q_2 is a shunt-peaking coil (L_1). In the signal-output path is a series-peaking coil (L_2). The video signal is then coupled to the picture tube through coupling capacitor C_4.

38-3 QUESTIONS

1. What is a video amplifier?
2. What is the frequency range of a video amplifier?
3. What coupling techniques are used for video amplifiers?
4. Define the following:
   a. Shunt peaking
   b. Series peaking
5. Where are video amplifiers used?

■ **FIGURE 38-13**
Series peaking.

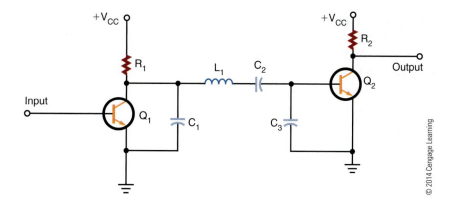

■ **FIGURE 38-14**
Series-shunt peaking.

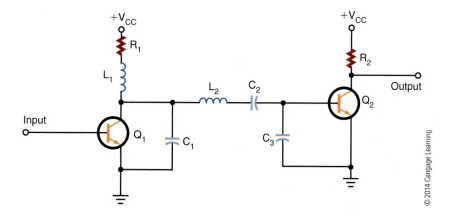

■ **FIGURE 38-15**
Video amplifier in a television receiver.

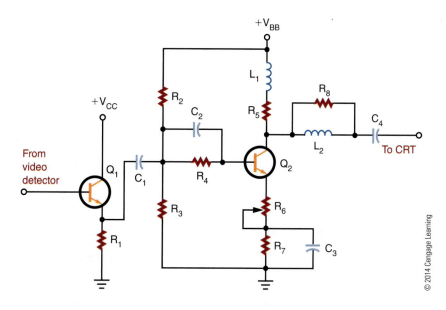

38-4 RF AND IF AMPLIFIERS

RF (radio-frequency) amplifiers usually are the first stage in AM, FM, or TV receivers and are similar to other amplifiers. They differ primarily in the frequency spectrum over which they operate, which is 10,000 hertz to 30,000 megahertz. There are two classes of RF amplifiers: untuned and tuned. In an **untuned amplifier**, a response is desired over a large RF range, and the main function is amplification. In a **tuned amplifier**, high amplification is desired over a small range of frequencies or a single frequency. Normally, when RF amplifiers are mentioned, they are assumed to be tuned unless otherwise specified.

In receiving equipment, the RF amplifier serves to amplify the signal and select the proper frequency. In transmitters, the RF amplifier serves to amplify a single frequency for application to the antenna. Basically, the receiver RF amplifier is a voltage amplifier, and the transmitter RF amplifier is a power amplifier.

In a receiver circuit, the RF amplifier must provide sufficient gain, produce low internal noise, provide good selectivity, and respond well to the selected frequencies.

Figure 38-16 shows an RF amplifier used for an AM radio. Capacitors C_1 and C_4 tune the antenna and the output transformer T_1 to the same frequency. The input signal is magnetically coupled to the base of transistor Q_1. Transistor Q_1 operates as a class A amplifier. Capacitor C_4 and transformer T_1 provide a high voltage gain at the resonant frequency for the collector load circuit. Transformer T_1 is tapped to provide a good impedance match for the transistor.

Figure 38-17 shows an RF amplifier used in a television VHF tuner. The circuit is tuned by coils L_{1A}, L_{1B}, and L_{1C}. When the channel selector is turned, a new set of coils is switched into the circuit. This provides the necessary bandwidth response for each channel. The input signal is developed across the tuned circuit consisting of L_{1A}, C_1, and C_2. Transistor Q_1 operates as a class A amplifier. The collector-output circuit is a double-tuned transformer. Coil L_{1B} is tuned by capacitor C_4, and coil L_{1C} is tuned by capacitor C_7. Resistor R_2 and capacitor C_6 form a decoupling filter to prevent any RF from entering the power supply to interact with other circuits.

In an AM radio, the incoming RF signal is converted to a constant **IF (intermediate frequency)** signal. A fixed-tuned IF amplifier is then used to increase the signal to a usable level. The **IF amplifier** is a single-frequency amplifier. Typically, two or more IF amplifiers are used to increase the signal to the proper level. The sensitivity of a receiver is determined by its signal-to-noise (S/N) ratio. The higher the gain, the better the sensitivity. Figure 38-18 shows a typical IF amplifier in an AM radio. The IF frequency is 455,000 hertz. Figure 38-19 shows an IF amplifier in a television receiver. Figure 38-20 compares the frequencies of radio and television receivers.

■ **FIGURE 38-16**
RF amplifier in an AM radio.

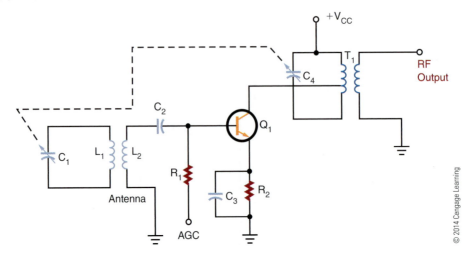

■ **FIGURE 38-17**

RF amplifier in a television VHF tuner.

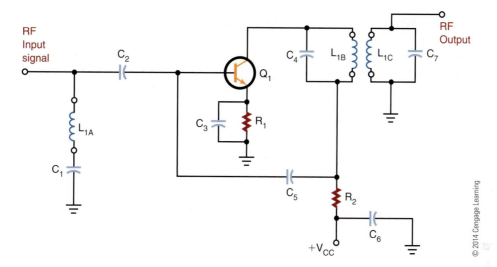

■ **FIGURE 38-18**

IF amplifier in an AM radio.

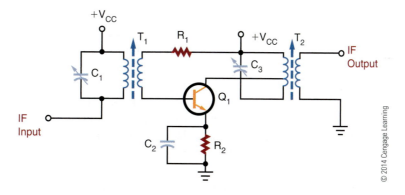

■ **FIGURE 38-19**

IF amplifier in a television receiver.

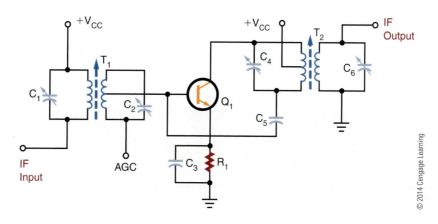

■ **FIGURE 38-20**

Comparison of radio and television frequencies.

Type	Received RF	Common IF	Bandwidth
AM Radio	535–1605 kHz	455 kHz	10 kHz
FM Radio	88–108 MHz	10.7 MHz	150 kHz
Television Channels 2–6 Channels 7–13 Channels 14–83	 54–88 MHz 174–216 MHz 470–890 MHz	 41–47 MHz	 6 MHz

© 2014 Cengage Learning

38-4 QUESTIONS

1. How do RF amplifiers differ from other amplifiers?
2. What are the two types of RF amplifiers?
3. Where are RF amplifiers used?
4. What is an IF amplifier?
5. What is significant about an IF amplifier?

38-5 OPERATIONAL AMPLIFIERS

Operational amplifiers are usually called op-amps. An **op-amp** is a very high gain amplifier used to amplify both AC and DC signals. Typically, op-amps have an output gain in the range of 20,000 to 1,000,000 times the input. Figure 38-21 shows the schematic symbol used for op-amps. The negative (−) input is called the **inverting input** and the positive (+) is called the **noninverting input**.

■ **FIGURE 38-21**

Schematic symbol for an op-amp.

© 2014 Cengage Learning

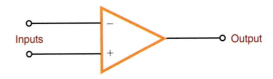

Figure 38-22 is a block diagram of an op-amp. The op-amp consists of three stages. Each stage is an amplifier with some unique characteristic.

The input stage is a differential amplifier. It allows the op-amp to respond only to the differences between input signals. Also, the differential amplifier amplifies only the differential input voltage and is unaffected by signals common to both inputs. This is referred to as **common-mode rejection**. Common-mode rejection is useful when measuring a small signal in the presence of 60-hertz noise. The 60-hertz noise common to both inputs is rejected, and the op-amp amplifies only the small difference between the two inputs. The differential amplifier has a low-frequency response that extends down to a DC level. This means

■ **FIGURE 38-22**

Block diagram of an op-amp.

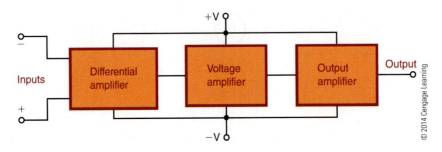

© 2014 Cengage Learning

that the differential amplifier can respond not only to low-frequency AC signals but to DC signals as well.

The second stage is a high-gain voltage amplifier. This stage is composed of several Darlington-pair transistors. This stage provides a voltage gain of 200,000 or more and supplies most of the op-amp's gain.

The last stage is an output amplifier. Typically, this is a complementary emitter-follower amplifier. It is used to give the op-amp a low output impedance. The op-amp can deliver several milliamperes of current to a load.

Generally, op-amps are designed to be powered by a dual-voltage power supply in the range of ±5 to ±15 volts. The positive power source delivers +5 to +15 volts with respect to ground. The negative power source delivers −5 to −15 volts with respect to ground. This allows the output voltage to swing from positive to negative with respect to ground. However, in certain cases, op-amps may be operated from a single voltage source.

A schematic diagram of a typical op-amp is shown in Figure 38-23. The op-amp shown is called a 741. This op-amp does not require frequency compensation, is short-circuit protected, and has no latch-up problems. It provides good performance for a low price and is the most commonly used op-amp. A device that contains two 741 op-amps in a single package is called a 747 op-amp. Because no coupling capacitors are used, the circuit can amplify DC signals as well as AC signals.

The op-amp's normal mode of operation is a closed loop. The **closed-loop mode** uses feedback, as compared to the **open-loop mode**, which does not use feedback. A lot of degenerative feedback is used with the closed-loop mode. This reduces the overall gain of the op-amp but provides better stability.

In closed-loop operation, the output signal is applied back to one of the input terminals as a feedback signal. This feedback signal opposes the input signal. There are two basic closed-loop circuits: inverting and noninverting. The inverting configuration is more popular.

Figure 38-24 shows the op-amp connected as an **inverting amplifier**. The input signal is applied to the op-amp's inverting (−) input through resistor R_1. The feedback is provided through resistor R_2. The signal at the inverting input is determined by both the input and output voltages.

A minus sign indicates a negative output signal when the input signal is positive. A plus sign indicates a positive output signal when the input signal is negative. The output is 180° out of phase with the input. Depending on the ratio of resistor R_2 to R_1, the gain of the inverting amplifier can be less than, equal to, or

■ **FIGURE 38-23**
Schematic diagram of an op-amp.

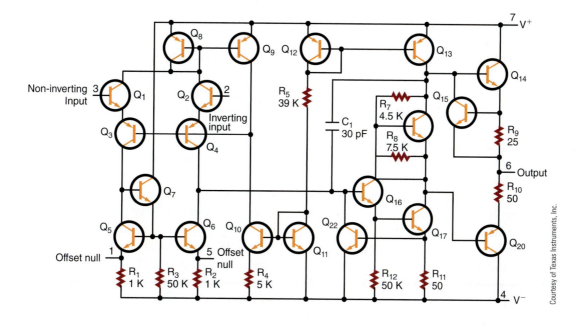

Courtesy of Texas Instruments, Inc.

■ **FIGURE 38-24**

Op-amp connected as an inverting amplifier.

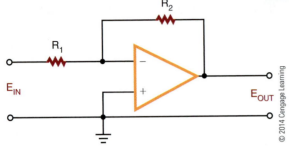

■ **FIGURE 38-25**

Op-amp connected as a noninverting amplifier.

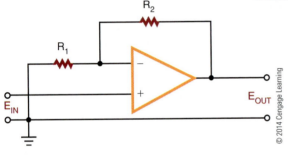

greater than 1. When the gain is equal to 1, it is called a **unity-gain amplifier** and is used to invert the polarity of the input signal.

Figure 38-25 shows an op-amp connected as a **noninverting amplifier**. The output signal is in phase with the input. The input signal is applied to the op-amp's noninverting input. The output voltage is divided between resistors R_1 and R_2 to produce a voltage that is fed back to the inverting (−) input. The voltage gain of a noninverting input amplifier is always greater than 1.

The gain of an op-amp varies with frequency. Typically, the gain given by a specification sheet is the DC gain. As the frequency increases, the gain decreases. Without some means to increase the bandwidth, the op-amp is only good for amplifying a DC signal. To increase the bandwidth, feedback is used to reduce the gain. By reducing the gain, the bandwidth increases by the same amount. In this way, the 741 op-amp bandwidth can be increased to 1 megahertz.

Besides its use for comparing, inverting, or noninverting a signal, the op-amp has several other applications. It can be used to add several signals together, as shown in Figure 38-26. This is referred to as a

■ **FIGURE 38-26**

Op-amp connected as a summing amplifier.

summing amplifier. The negative feedback holds the inverting input of the op-amp very close to ground potential. Therefore, all the input signals are electrically isolated from each other. The output of the amplifier is the inverted sum of the input signal.

In a summing amplifier, the resistor chosen for the noninverting input-to-ground is equal to the total parallel resistance of the input and feedback resistances. If the feedback resistance is increased, the circuit can provide gain. If different input resistances are used, the input signals can be added together with different gains.

Summing amplifiers are used when mixing audio signals together. Potentiometers are used for the input resistances to adjust the strength of each of the input signals.

Op-amps can also be used as active filters. Filters that use resistors, inductors, and capacitors are called **passive filters**. **Active filters** are inductorless filters using integrated circuits. The advantage of active filters is the absence of inductors, which is handy at low frequencies because of the size of inductors.

There are some disadvantages when using op-amps as active filters because they require a power supply, can generate noise, and can oscillate due to thermal drift or aging components.

Figure 38-27 shows an op-amp used as a high-pass filter. A **high-pass filter** rejects low frequencies while passing frequencies above a certain cut-off frequency. Figure 38-28 shows an op-amp used as a low-pass filter. A **low-pass filter** passes low frequencies while blocking frequencies above the cut-off frequency. Figure 38-29 shows an op-amp used as a band-pass filter. A **band-pass filter** passes frequencies around some central frequency while attenuating both higher and lower frequencies.

■ **FIGURE 38-27**

Op-amp connected as a high-pass filter.

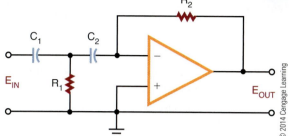

■ **FIGURE 38-28**

Op-amp connected as a low-pass filter.

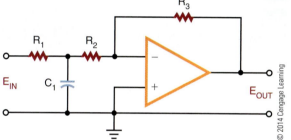

■ **FIGURE 38-29**

Op-amp connected as a band-pass filter.

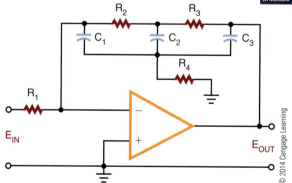

■ **FIGURE 38-30**

Op-amp connected as a difference amplifier.

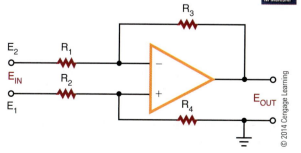

A **difference amplifier** subtracts one signal from another signal. Figure 38-30 shows a basic difference amplifier. This circuit is called a **subtractor** because it subtracts the value of E_2 from the value of E_1.

38–5 QUESTIONS

1. What is an op-amp?
2. Draw a block diagram of an op-amp.
3. Briefly explain how an op-amp works.
4. What is the normal mode of operation for an op-amp?
5. What type of gain can be obtained with an op-amp?
6. Draw schematic diagrams of the following:
   a. Inverting amplifier
   b. Summing amplifier
   c. High-pass filter
   d. Band-pass filter
   e. Difference amplifier

SUMMARY

- Direct-coupled amplifiers are used primarily as voltage amplifiers.
- A differential amplifier has two separate inputs and may provide either one or two outputs.
- Audio amplifiers amplify AC signals in the audio range of 20 to 20,000 hertz.
- The two types of audio amplifiers are voltage amplifiers and power amplifiers.
- Video amplifiers are wideband amplifiers used to amplify video information.
- Video frequencies extend from a few hertz to 5 or 6 megahertz.
- RF amplifiers operate from 10,000 hertz to 30,000 megahertz.
- The two types of RF amplifiers are tuned and untuned.
- Op-amps may provide output gains of 20,000 to 1,000,000 times the input.
- The two basic closed-loop modes are the inverting configuration and the noninverting configuration.

CHAPTER 38 SELF-TEST

1. Under what conditions would a DC amplifier be used?
2. How is the problem of temperature instability resolved with high-gain DC amplifiers?
3. What are the main differences between audio voltage amplifiers and audio power amplifiers?
4. What is the practical advantage of using a quasi-complementary power amplifier over a complementary push–pull amplifier?
5. How does a video amplifier differ from an audio amplifier?
6. What factor is involved in limiting the output of high-frequency video amplifiers?
7. What is the purpose of an RF amplifier?
8. How are IF amplifiers used in a circuit?
9. Identify the three stages of an op-amp, and describe its function.
10. For what type of application is the op-amp used?

Oscillators

An oscillator is a nonrotating device for producing alternating current. Oscillators are used extensively in the field of electronics: in radios and televisions, communication systems, computers, industrial controls, and timekeeping devices. Without the oscillator, very few electronic circuits would be possible.

39–1　FUNDAMENTALS OF OSCILLATORS

An **oscillator** is a circuit that generates a repetitive AC signal. The frequency of the AC signal may vary from a few hertz to many millions of hertz. The oscillator is an alternative to the mechanical generator used to produce electrical power. The advantages of the oscillator are the absence of moving parts and the range over which the AC signal can be produced. The output of an oscillator may be a sinusoidal, rectangular, or sawtooth waveform, depending on the type of oscillator used. The main requirement of an oscillator is that the output be uniform; that is, the output must not vary in either frequency or amplitude.

When an inductor and a capacitor are connected in parallel, they form what is called a **tank circuit**. When a tank circuit is excited by an external DC source, it oscillates; that is, it produces a back-and-forth current flow. If it were not for the resistance of the circuit, the tank circuit would oscillate forever. However, the resistance of the tank circuit absorbs energy from the current, and the oscillations of the circuit are dampened.

For the tank circuit to maintain its oscillation, the energy that is dissipated must be replaced. The energy that is replaced is referred to as **positive feedback**. Positive feedback is the feeding back of a portion of the output signal into the tank circuit to sustain oscillation. The feedback must be in phase with the signal in the tank circuit. Figure 39-1 shows a block diagram of an oscillator. The basic oscillator can be broken down into three sections. The frequency-determining oscillator circuit is usually an LC tank circuit. An amplifier

■ FIGURE 39-1

Block diagram of an oscillator.

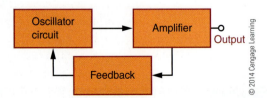

increases the output signal of the tank circuit. A feedback circuit delivers the proper amount of energy to the tank circuit to sustain oscillation. The oscillator circuit is essentially a closed loop that uses DC power to maintain AC oscillations.

39–1　QUESTIONS

1. What is an oscillator?
2. How does a tank circuit operate?
3. What makes a tank circuit continue to oscillate?
4. Draw and label a block diagram of an oscillator.
5. What are the functions of the basic parts of an oscillator?

39–2　SINUSOIDAL OSCILLATORS

Sinusoidal oscillators are oscillators that produce a sine-wave output. They are classified according to their frequency-determining components. The three basic types of sinusoidal oscillators are **LC oscillators**, **crystal oscillators**, and **RC oscillators**.

LC oscillators use a tank circuit of capacitors and inductors, connected either in series or parallel, to determine the frequency. Crystal oscillators are like LC oscillators except that crystal oscillators maintain a higher degree of stability. LC and crystal oscillators are used in the radio frequency (RF) range. They are not suitable for low-frequency applications. For low-frequency applications, RC oscillators are used. RC oscillators use a resistance-capacitance network to determine the oscillator frequency.

Three basic types of LC oscillators are the **Hartley oscillator**, the **Colpitts oscillator**, and the **Clapp oscillator**. Figure 39-2 and Figure 39-3 show the two basic types of Hartley oscillator. The tapped inductor in the tank circuit identifies these circuits as Hartley oscillators. The disadvantage of the **series-fed Hartley oscillator** (Figure 39-2) is that DC current flows through a portion of the tank circuit. The **shunt-fed Hartley oscillator** (Figure 39-3) overcomes the problem of DC current in the tank circuit by using a coupling capacitor in the feedback line.

The Colpitts oscillator (Figure 39-4) is like the shunt-fed Hartley except that two capacitors are

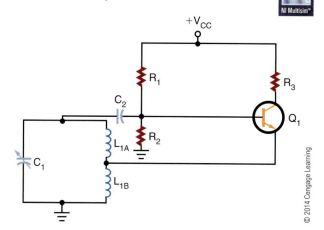

■ FIGURE 39-2
Series-fed Hartley oscillator.

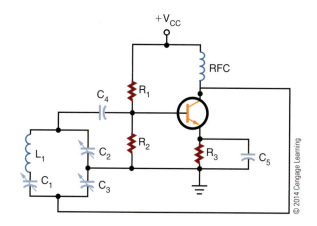

■ FIGURE 39-5
Clapp oscillator.

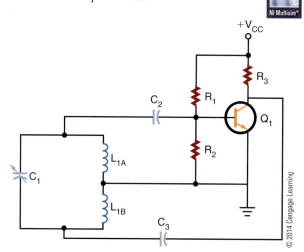

■ FIGURE 39-3
Shunt-fed Hartley oscillator.

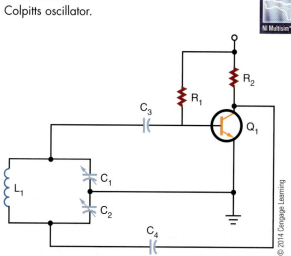

■ FIGURE 39-4
Colpitts oscillator.

substituted for the tapped inductor. The Colpitts is more stable than the Hartley and is more often used.

The Clapp oscillator (Figure 39-5) is a variation of the Colpitts oscillator. The main difference is the addition of a capacitor in series with the inductor in the tank circuit, which allows tuning of the oscillator frequency.

Temperature changes, aging of components, and varying load requirements cause oscillators to become unstable. When stability is a requirement, crystal oscillators are used.

Crystals are materials that can convert mechanical energy to electrical energy when pressure is applied or can convert electrical energy to mechanical energy when a voltage is applied. When an AC voltage is applied to a crystal, the crystal stretches and compresses, creating mechanical vibrations that correspond to the frequency of the AC signal.

Crystals, because of their structure, have a natural frequency of vibration. If the AC signal applied matches the natural frequency, the crystal vibrates more. If the AC signal is different from the crystal's natural frequency, little vibration is produced. The crystal's mechanical frequency of vibration is constant, making it ideal for oscillator circuits.

The most common materials used for crystals are Rochelle salt, tourmaline, and quartz. Rochelle salt has the most electrical activity, but it fractures easily. Tourmaline has the least electrical activity, but it is the strongest. Quartz is a compromise: It has good electrical activity and is strong. Quartz is the most commonly used crystal in oscillator circuits.

The crystal material is mounted between two metal plates, with pressure applied by a spring so that the metal

■ **FIGURE 39-6**

Schematic symbol for a crystal.

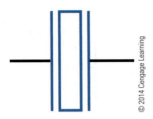

■ **FIGURE 39-7**

Crystal shunt-fed Hartley oscillator.

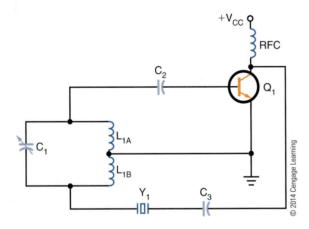

■ **FIGURE 39-8**

Colpitts crystal oscillator.

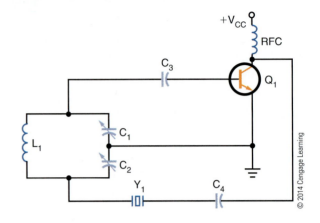

■ **FIGURE 39-9**

Pierce oscillator.

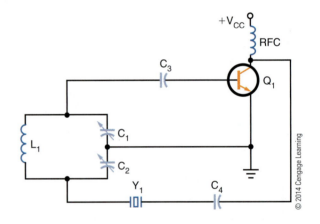

■ **FIGURE 39-10**

Butler oscillator.

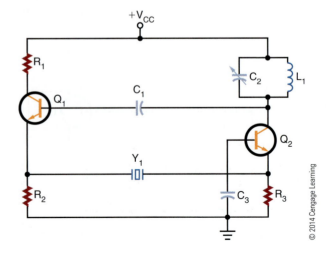

plates make electrical contact with the crystal. The crystal is then placed in a metal package. Figure 39-6 shows the schematic symbol used to represent the crystal. The letters Y or XTAL identify crystals in schematics.

Figure 39-7 shows a shunt-fed Hartley oscillator with the addition of a crystal. The crystal is connected in series with the feedback circuit. If the frequency of the tank circuit drifts from the crystal frequency, the impedance of the crystal increases, reducing feedback to the tank circuit. This allows the tank circuit to return to the frequency of the crystal.

Figure 39-8 shows a Colpitts oscillator connected the same way as the Hartley crystal oscillator. The crystal controls the feedback to the tank circuit. The LC tank circuit is tuned to the crystal frequency.

Figure 39-9 shows a **Pierce oscillator**. This circuit is similar to the Colpitts oscillator except that the tank-circuit inductor is replaced with a crystal. The crystal controls the tank-circuit impedance, which determines the feedback and stabilizes the oscillator.

Figure 39-10 shows a **Butler oscillator**. This is a two-transistor circuit. It uses a tank circuit, and the crystal determines the frequency. The tank circuit must be tuned to the crystal frequency or the oscillator will not work. The advantage of the Butler oscillator is that a small voltage exists across the crystal, reducing stress on the crystal. By replacing the tank-circuit components, the oscillator can be tuned to operate on one of the crystal's overtone frequencies.

RC oscillators use resistance-capacitance networks to determine the oscillator frequency. There are two basic types of RC oscillators that produce sinusoidal waveforms: the phase-shift oscillator and the Wien-bridge oscillator.

A **phase-shift oscillator** is a conventional amplifier with a phase-shifting RC feedback network (Figure 39-11). The feedback must shift the signal 180°. Because the capacitive reactance changes with a change in frequency, it is frequency sensitive. Stability is improved by reducing the amount of phase shift across each RC network. However, there is a power loss across the combined RC network. The transistor must have enough gain to offset these losses.

A **Wien-bridge oscillator** is a two-stage amplifier with a lead-lag network and voltage divider (Figure 39-12). The lead-lag network consists of a series RC network (R_1C_1) and a parallel network. It is called a lead-lag network because the output phase angle leads for some frequencies and lags for other frequencies. At the resonant frequency, the phase shift is 0°, and the output voltage is maximum. Resistors R_2 and R4 form the voltage divider network, which is used to develop the degenerative feedback. Regenerative feedback is applied to the base, and degenerative feedback is applied to the emitter of oscillator transistor Q_1. The output of transistor Q_1 is capacitively coupled to the base of transistor Q_2, where it is amplified and shifted and rotated 180°. The output is coupled by capacitor C_4 to the bridge network.

■ **FIGURE 39-11**

Phase-shift oscillator.

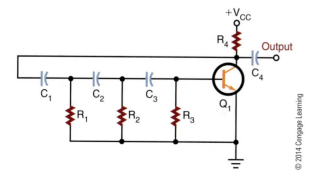

■ **FIGURE 39-12**

Wien-bridge oscillator.

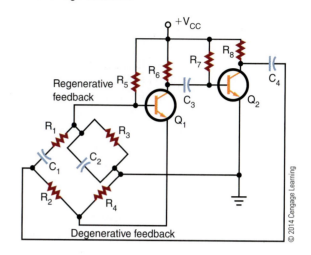

■ **FIGURE 39-13**

IC Wien-bridge oscillator.

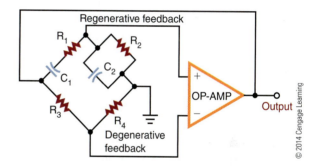

Figure 39-13 shows an integrated circuit Wien-bridge oscillator. The inverting and noninverting inputs of the op-amp are ideal for use as a Wien-bridge oscillator. The gain of the op-amp is high, which offsets the circuit losses.

39–2 QUESTIONS

1. What are the three types of sinusoidal oscillators?
2. Draw and label schematic diagrams of the three types of LC oscillators.
3. What is the difference between a Colpitts oscillator and a Hartley oscillator?
4. How can the stability of an LC oscillator be improved?
5. What are the two types of RC oscillators used for producing sinusoidal waves?

39-3 NONSINUSOIDAL OSCILLATORS

Nonsinusoidal oscillators are oscillators that do not produce a sine-wave output. There is no specific nonsinusoidal waveshape. The nonsinusoidal oscillator output may be a square, sawtooth, rectangular, or triangular waveform or a combination of two waveforms. A common characteristic of all nonsinusoidal oscillators is that they are a form of **relaxation oscillator**. A relaxation oscillator stores energy in a reactive component during one phase of the oscillation cycle and gradually releases the energy during the relaxation phase of the cycle.

Blocking oscillators and **multivibrators** are relaxation oscillators. Figure 39-14 shows a blocking oscillator circuit. The reason for the name is that the transistor is easily driven into the blocking (cutoff) mode. The blocking condition is determined by the discharge from capacitor C_1. Capacitor C_1 is charged through the emitter-base junction of transistor Q_1. However, once capacitor C_1 is charged, the only discharge path is through resistor R_1. The RC time constant of resistor R_1 and capacitor C_1 determines how long the transistor is blocked, or cut off, and also determines the oscillator frequency. A long time constant produces a low frequency; a short time constant produces a high frequency.

If the output is taken from an RC network in the emitter circuit of the transistor, the output is a sawtooth waveshape (Figure 39-15). The RC network determines the frequency of oscillation and produces the sawtooth output. Transistor Q_1 is forward biased by resistor R_1. As transistor Q_1 conducts, capacitor C_1 charges quickly. The positive potential on the top plate of capacitor C_1 reverse biases the emitter junction, turning off transistor Q_1. Capacitor C_1 discharges through resistor R_2, producing the trailing portion of the sawtooth output. When capacitor C_1 discharges,

FIGURE 39-15

Sawtooth waveform generated by a blocking oscillator.

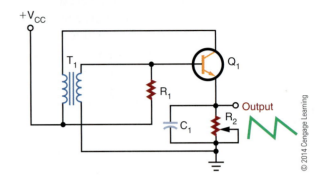

© 2014 Cengage Learning

transistor Q_1 is again forward biased and conducts, repeating the action.

Capacitor C_1 and resistor R_2 determine the frequency of oscillation. By making resistor R_2 variable, the frequency can be adjusted. If resistor R_2 offers high resistance, a long RC time constant results, producing a low frequency of oscillation. If resistor R_2 offers low resistance, a short RC time constant results, producing a high frequency of oscillation.

A multivibrator is a relaxation oscillator that can function in either of two temporarily stable conditions and is capable of rapidly switching from one temporary state to the other.

Figure 39-16 shows a basic free-running multivibrator circuit. It is basically an oscillator consisting of two stages coupled so that the input signal to each stage is taken from the output of the other. One stage conducts while the other stage is cut off, until a point is reached where the stages reverse their conditions. The circuit is free-running because of regenerative feedback.

FIGURE 39-14

Blocking oscillator.

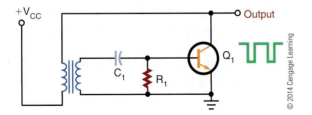

© 2014 Cengage Learning

FIGURE 39-16

Free-running multivibrator.

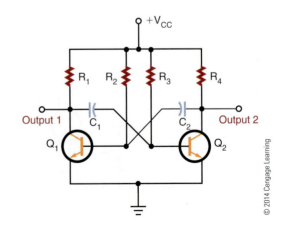

© 2014 Cengage Learning

■ FIGURE 39-17

Diagram of a 555 timer integrated circuit.

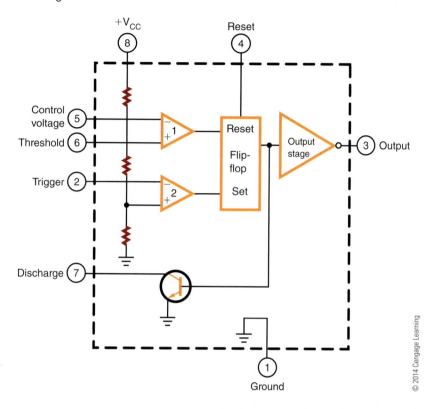

© 2014 Cengage Learning

The coupling circuit determines the frequency of oscillation.

An **astable multivibrator** is one type of free-running multivibrator. The output of an astable multivibrator is rectangular. By varying the RC time constants of the coupling circuits, rectangular pulses of any desired width can be obtained. By changing the values of the resistor and capacitor, the operating frequency can be changed. The frequency stability of the multivibrator is better than that of the typical blocking oscillator.

An integrated circuit that can be used as an astable multivibrator is the 555 timer (Figure 39-17). This integrated circuit can perform many functions. It consists of two comparators, a flip-flop, an output stage, and a discharge transistor. Figure 39-18 shows a schematic diagram in which the 555 timer is used as an astable multivibrator. The output frequency is determined by resistors R_A and R_B and capacitor C_1. This circuit finds wide application in industry.

■ FIGURE 39-18

Astable multivibrator using a 555 timer.

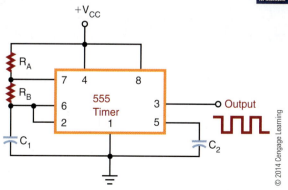

© 2014 Cengage Learning

39-3 QUESTIONS

1. Draw the more commonly used nonsinusoidal waveforms.
2. What is a relaxation oscillator?
3. Give two examples of relaxation oscillators.
4. Draw a schematic diagram of a blocking oscillator.
5. Draw a schematic diagram of a 555 timer used as an astable multivibrator.

SUMMARY

- An oscillator is a nonrotating device for producing alternating current.
- The output of an oscillator can be sinusoidal, rectangular, or sawtooth.
- The main requirement of an oscillator is that the output be uniform and not vary in frequency or amplitude.
- A tank circuit is formed when a capacitor is connected in parallel with an inductor.
- A tank circuit oscillates when an external voltage source is applied.
- The oscillations of a tank circuit are dampened by the resistance of the circuit.
- For a tank circuit to maintain oscillation, positive feedback is required.
- An oscillator has three basic parts: a frequency-determining device, an amplifier, and a feedback circuit.
- The three basic types of sinusoidal oscillators are LC oscillators, crystal oscillators, and RC oscillators.
- The three basic types of LC oscillators are the Hartley, the Colpitts, and the Clapp.
- Crystal oscillators provide more stability than LC oscillators.
- RC oscillators use resistance-capacitance networks to determine the oscillator frequency.
- Nonsinusoidal oscillators do not produce a sine-wave output.
- Nonsinusoidal oscillator outputs include square, sawtooth, rectangular, and triangular waveforms and combinations of two waveforms.
- A relaxation oscillator is the basis of all nonsinusoidal oscillators.
- A relaxation oscillator stores energy in a reactive component during one part of the oscillation cycle.
- Examples of relaxation oscillators are blocking oscillators and multivibrators.

CHAPTER 39 SELF-TEST

1. Identify the parts of an oscillator, and explain what each part does in making the oscillator function.
2. What is the function of feedback in an oscillator?
3. What is the difference between the series-fed and the shunt-fed Hartley oscillators?
4. What function does the crystal perform in an oscillator?
5. Explain how a tank circuit can sustain an oscillation.
6. What are the major types of sinusoidal oscillators?
7. How are crystals used in oscillator circuits?
8. How does a nonsinusoidal oscillator differ from a sinusoidal oscillator?
9. What types of components make up nonsinusoidal oscillators?
10. Draw three types of waveforms produced by nonsinusoidal oscillators.

Waveshaping Circuits

OBJECTIVES

After completing this chapter, the student will be able to:

■ Identify ways in which waveform shapes can be changed.

■ Explain the frequency-domain concept in waveform construction.

■ Define pulse width, duty cycle, rise time, fall time, undershoot, overshoot, and ringing as they relate to waveforms.

■ Explain how differentiators and integrators work.

■ Describe clipper and clamper circuits.

■ Describe the differences between monostable and bistable multivibrators.

■ Draw schematic diagrams of waveshaping circuits.

KEY TERMS

n electronics, it is sometimes necessary to change the shape of a waveform. Sine waves may have to be changed to square waves, rectangular waveforms may have to be changed to pulse waveforms, and pulse waveforms may have to be changed to square or rectangular waveforms. Waveforms can be analyzed by two methods. Analyzing a waveform by its amplitude per unit of time is called a *time-domain analysis*. Analyzing a waveform by the sine waves that make it up is called a *frequency-domain analysis*. This concept assumes that all periodic waveforms are composed of sine waves.

40-1   NONSINUSOIDAL WAVEFORMS

Figure 40-1 shows three basic waveforms that are represented by the **time domain** concept. The three waveforms shown are sine wave, square wave, and sawtooth wave. Although the three waveforms are different, they all have the same period of frequency. By using various electronic circuits, these waveforms can be changed from one shape to another.

A **periodic waveform** is one with the same waveform for all cycles. According to the **frequency domain** concept, all periodic waveforms are made up of sine waves. In other words, superimposing a number of sine waves having different amplitudes, phases, and frequencies can form any periodic wave. Sine waves are important because they are the only waveform that cannot be distorted by RC, RL, or LC circuits.

■ FIGURE 40-1

Three basic waveforms: (A) sine wave, (B) square wave, (C) sawtooth wave.

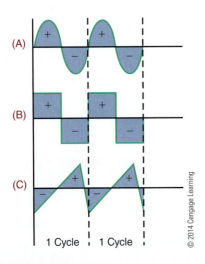

© 2014 Cengage Learning

■ FIGURE 40-2

Chart of fundamental frequency of 1000 hertz and some of its harmonics.

(Fundamental)	1st Harmonic	1000 Hz
	2nd Harmonic	2000 Hz
	3rd Harmonic	3000 Hz
	4th Harmonic	4000 Hz
	5th Harmonic	5000 Hz

© 2014 Cengage Learning

The sine wave with the same frequency as the periodic waveform is called the fundamental frequency. The fundamental frequency is also called the **first harmonic**. Harmonics are multiples of the fundamental frequency. The second harmonic is twice the fundamental, the third harmonic is three times the fundamental, and so on. Figure 40-2 shows the fundamental frequency of 1000 hertz and some of its harmonics. Harmonics can be combined in an infinite number of ways to produce any periodic waveform. The type and number of harmonics included in the periodic waveform depend on the shape of the waveform.

For example, Figure 40-3 shows a square wave. Figure 40-4 shows how a square wave is formed by the combination of the fundamental frequency with an infinite number of odd harmonics that cross the zero reference line in phase with the fundamental.

Figure 40-5 shows the formation of a sawtooth waveform. It consists of the fundamental frequency plus odd harmonics in-phase and even harmonics crossing the zero reference line 180° out of phase with the fundamental.

An oscilloscope displays waveforms in the time domain. A spectrum analyzer (Figure 40-6) displays waveforms in the frequency domain. Frequency domain analysis can be used to determine how a circuit affects a waveform.

■ FIGURE 40-3

Square wave.

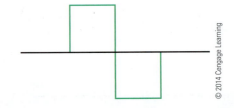

© 2014 Cengage Learning

■ **FIGURE 40-4**

Formation of a square wave by the frequency domain method.

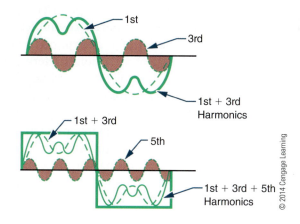

■ **FIGURE 40-5**

Formation of a sawtooth wave by the frequency domain method.

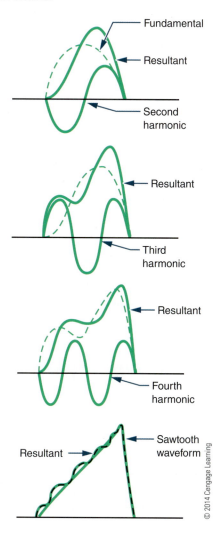

■ **FIGURE 40-6**

Spectrum analyzer.

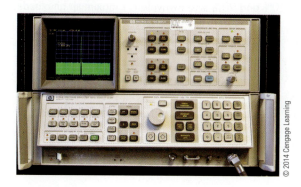

■ **FIGURE 40-7**

Period of a waveform.

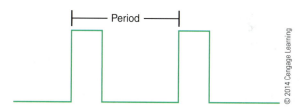

Periodic waveforms are waveforms that occur at regular intervals. The period of a waveform is measured from any point on one cycle to the same point on the next cycle (Figure 40-7).

The pulse width is the length of the pulse (Figure 40-8). The **duty cycle** is the ratio of the pulse width to the period. It can be represented as a percentage indicating the amount of time that the pulse exists during each cycle.

$$\text{duty cycle} = \frac{\text{pulse width}}{\text{period}}$$

All pulses have rise and fall times. The **rise time** is the time it takes for the pulse to rise from 10% to 90% of its maximum amplitude. The **fall time** is the time it takes for a pulse to fall from 90% to 10% of its maximum amplitude (Figure 40-9). Overshoot,

■ **FIGURE 40-8**

Pulse width of a waveform.

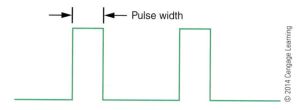

■ **FIGURE 40-9**

The rise and fall times of a waveform are measured at 10% and 90% of the waveform's maximum amplitude.

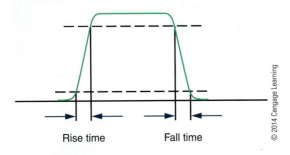

Rise time   Fall time

© 2014 Cengage Learning

■ **FIGURE 40-10**

Overshoot, undershoot, and ringing.

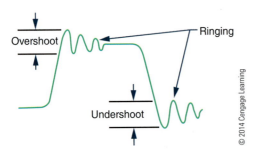

© 2014 Cengage Learning

undershoot, and ringing are conditions common to high-frequency pulses (Figure 40-10).

Overshoot occurs when the leading edge of a waveform exceeds its normal maximum value. **Undershoot** occurs when the trailing edge exceeds its normal minimum value. (The **leading edge** is the front edge of the waveform; the **trailing edge** is the back edge of the waveform.) Both conditions are followed by damped oscillations known as **ringing**. These conditions are undesirable but exist because of imperfect circuits.

40-1   QUESTIONS

1. Define the frequency domain concept.
2. How are the following waveforms constructed according to the frequency domain concept?
   a. Square wave
   b. Sawtooth wave
3. What is a periodic waveform?
4. What is a duty cycle?
5. Draw examples of overshoot, undershoot, and ringing as they apply to a waveform.

40-2   WAVESHAPING CIRCUITS

An RC network can change the shape of complex waveforms so that the output barely resembles the input. The amount of distortion is determined by the RC time constant. The type of distortion is determined by the component the output is taken across. If the output is taken across the resistor, the circuit is called a **differentiator**. A differentiator is used to produce a pip or peaked waveform from square or rectangular waveforms for timing or synchronizing circuits. It is also used to produce trigger or marker pulses. If the output is taken across the capacitor, the circuit is called an **integrator**. An integrator is used for waveshaping in radio, television, radar, and computers.

Figure 40-11 shows a differentiator circuit. Recall that complex waveforms are made of the fundamental frequency plus a large number of harmonics. When a complex waveform is applied to a differentiator, each frequency is affected differently. The ratio of the capacitive reactance (X_C) to R is different for each harmonic. This results in each harmonic being shifted in phase and reduced in amplitude by a different amount. The net result is distortion of the original waveform. Figure 40-12 shows what happens to a square wave applied to a differentiator. Figure 40-13 shows the effects of different RC time constants.

■ **FIGURE 40-11**

Differentiator circuit.

© 2014 Cengage Learning

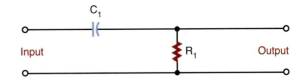

Input C_1 R_1 Output

■ **FIGURE 40-12**

Result of applying a square wave to a differentiator circuit.

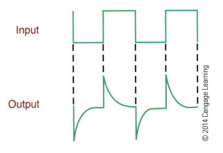

Input

Output

© 2014 Cengage Learning

■ **FIGURE 40-13**

Effects of different time constants on a differentiator circuit.

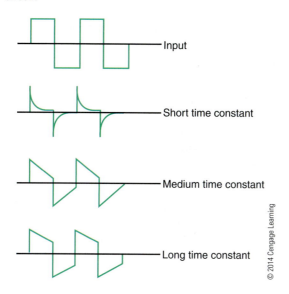

Input

Short time constant

Medium time constant

Long time constant

© 2014 Cengage Learning

An integrator circuit is similar to a differentiator except that the output is taken across the capacitor (Figure 40-14). Figure 40-15 shows the result of applying a square wave to an integrator. The integrator changes the waveform in a different way than the differentiator. Figure 40-16 shows the effects of different RC time constants.

Another type of circuit that can change the shape of a waveform is a **clipping circuit**, or **limiter circuit** (Figure 40-17). A clipping circuit can be used to

■ **FIGURE 40-14**

Integrator circuit.

© 2014 Cengage Learning

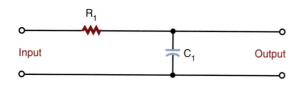

Input R_1 Output

C_1

■ **FIGURE 40-15**

Result of applying a square wave to an integrator circuit.

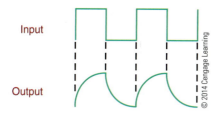

Input

Output

© 2014 Cengage Learning

■ **FIGURE 40-16**

Effects of different time constants on an integrator circuit.

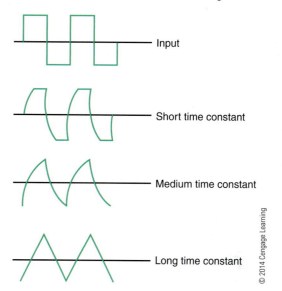

Input

Short time constant

Medium time constant

Long time constant

© 2014 Cengage Learning

■ **FIGURE 40-17**

Basic series diode clipping circuit.

NI Multisim™

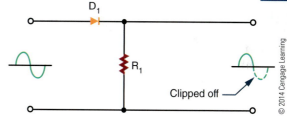

D_1

R_1

Clipped off

© 2014 Cengage Learning

square off the peaks of an applied signal, obtain a rectangular waveform from a sine-wave signal, eliminate positive or negative portions of a waveform, or keep an input amplitude at a constant level. The diode is forward biased and conducts during the positive portion of the input signal. During the negative portion of the input signal, the diode is reverse biased and does not conduct. Figure 40-18 shows the effect of reversing the diode: The positive portion of the input signal is clipped off. The circuit is essentially a half-wave rectifier.

By using a bias voltage, the amount of signal that is clipped off can be regulated. Figure 40-19 shows a biased series clipping circuit. The diode cannot conduct until the input signal exceeds the bias source. Figure 40-20 shows the result of reversing the diode and the bias source.

■ **FIGURE 40-18**

Result of reversing the diode in a clipping circuit.

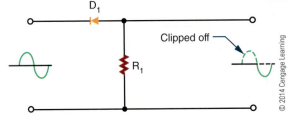

■ **FIGURE 40-19**

Biased series diode clipping circuit.

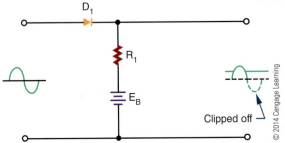

■ **FIGURE 40-20**

Result of reversing diode and bias source in a biased series clipping circuit.

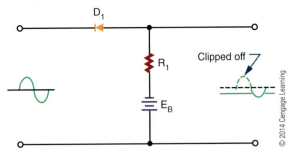

■ **FIGURE 40-21**

Shunt diode clipping circuit.

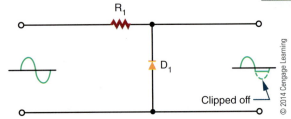

■ **FIGURE 40-22**

Result of reversing the diode in a shunt diode clipping circuit.

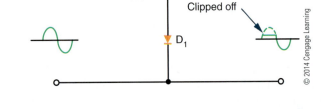

■ **FIGURE 40-23**

Biased shunt diode clipping circuit.

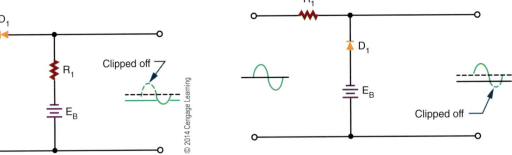

■ **FIGURE 40-24**

Effect of reversing the diode and bias source in a shunt diode clipping circuit.

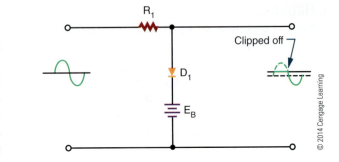

A shunt clipping circuit performs the same function as the series clipper (Figure 40-21). The difference is that the output is taken across the diode. This circuit clips off the negative portion of the input signal. Figure 40-22 shows the effect of reversing the diode. A shunt clipper can also be biased to change the clipping level, as shown in Figure 40-23 and Figure 40-24.

If both the positive and the negative peaks must be limited, two biased diodes are used (Figure 40-25). This prevents the output signal from exceeding predetermined values for both peaks. With the elimination of both peaks, the remaining signal is generally square

FIGURE 40-25

Clipping circuit used to limit both positive and negative peaks.

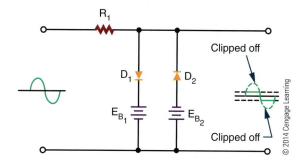

FIGURE 40-26

Alternate circuit to limit both positive and negative peaks.

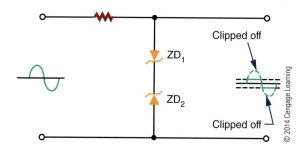

FIGURE 40-27

Diode clamper circuit.

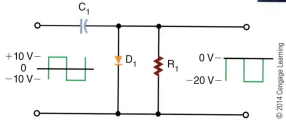

40-2 **QUESTIONS**

1. Draw schematic diagrams of the following RC networks:
   a. Differentiator
   b. Integrator
2. What are the functions of integrator and differentiator circuits?
3. Draw schematic diagrams of the following circuits:
   a. Clipper
   b. Clamper
4. What are the functions of clipper and clamper circuits?
5. What are applications of the following circuits?
   a. Differentiator
   b. Integrator
   c. Clipper
   d. Clamper

shaped. Therefore, this circuit is often referred to as a square-wave generator.

Figure 40-26 shows another clipping circuit that limits both positive and negative peaks. Therefore, the output is clamped to the breakdown voltage of the zeners. Between the two extremes, neither zener diode will conduct, and the input signal is passed to the output.

Sometimes it is desirable to change the DC reference level of a waveform. The **DC reference level** is the starting point from which measurements are made. A **clamping circuit** can be used to clamp the top or bottom of the waveform to a given DC voltage. Unlike a clipper, or limiter, circuit, a clamping circuit does not change the shape of the waveform. A diode clamper (Figure 40-27) is also called a DC restorer. This circuit is commonly used in radar, television, telecommunications, and computers. In the circuit shown, a square wave is applied to an input signal. The purpose of the circuit is to clamp the top of the square wave to 0 volts, without changing the shape of the waveform.

40-3 SPECIAL-PURPOSE CIRCUITS

The prefix *mono-* means "one." A **monostable multivibrator** has only one stable state. It is also called a **one-shot multivibrator** because it produces one output pulse for each input pulse. The output pulse is generally longer than the input pulse. Therefore, this circuit is also called a pulse stretcher. Typically, the circuit is used as a gate in computers, electronic control circuits, and communication equipment.

Figure 40-28 shows a schematic diagram of a monostable multivibrator. The circuit is normally in its stable state. When it receives an input trigger pulse,

■ **FIGURE 40-28**

Monostable multivibrator.

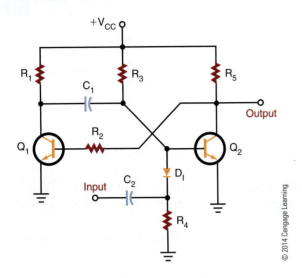

© 2014 Cengage Learning

■ **FIGURE 40-29**

Basic flip-flop circuit.

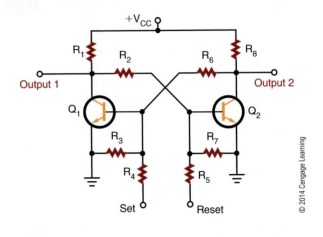

© 2014 Cengage Learning

■ **FIGURE 40-30**

Basic Schmitt trigger circuit.

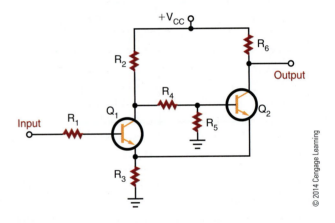

© 2014 Cengage Learning

it switches to its unstable state. The length of time the circuit is in the unstable state is determined by the RC time constant of resistor R_3 and capacitor C_1. Capacitor C_2 and resistor R_4 form a differentiator circuit, which is used to convert the input pulse to a positive spike and a negative spike. Diode D_1 allows only the negative spike to pass through to turn on the circuit.

A **bistable multivibrator** is a multivibrator having two stable states (*bi-*, meaning "two"). This circuit requires two inputs to complete one cycle. A pulse at one input sets the circuit to one of its stable states. A pulse at the other input resets it to its other stable state. This circuit is often called a **flip-flop** because of its mode of operation (Figure 40-29).

A basic flip-flop circuit produces a square or rectangular waveform for use in gating or timing signals or for on–off switching operations in binary counter circuits. A binary counter circuit is essentially two transistor amplifiers, with the output of each transistor coupled to the input of the other transistor. When an input signal is applied to the set input, transistor Q_1 turns on, which turns transistor Q_2 off. When transistor Q_2 turns off, it places a positive potential on the base of transistor Q_1, holding it on. If a pulse is applied to the reset input, it causes transistor Q_2 to conduct, turning off transistor Q_1. Turning off transistor Q_1 holds transistor Q_2 on.

Discrete versions of the flip-flop find little application today. However, integrated circuit versions of the flip-flop find wide application. It is perhaps the most important circuit in digital electronics, used for frequency division, data storage, counting, and data manipulation.

Another bistable circuit is the **Schmitt trigger** (Figure 40-30). One application of the Schmitt trigger is to convert a sine wave, sawtooth, or other irregularly shaped waveform to a square or rectangular wave. The circuit differs from a conventional bistable multivibrator in that a common-emitter resistor (R_3) replaces one of the coupling networks. This provides additional regeneration for quicker action and straighter leading and trailing edges on the output waveform.

40-3 QUESTIONS

1. What is a monostable multivibrator?
2. Draw a schematic diagram of a one-shot multivibrator.
3. What is a bistable multivibrator?
4. Draw a schematic diagram of a flip-flop.
5. How does a Schmitt trigger differ from a conventional bistable multivibrator?

SUMMARY

- Waveforms can be changed from one shape to another using various electronic circuits.
- The frequency domain concept holds that all periodic waveforms are made of sine waves.
- Periodic waveforms have the same waveshape in all cycles.
- Sine waves are the only waveform that cannot be distorted by RC, RL, or LC circuits.
- According to the frequency domain concept, waveforms consist of the fundamental frequency plus combinations of even or odd harmonics, or both.
- A square wave consists of the fundamental frequency plus an infinite number of odd harmonics.
- A sawtooth wave consists of the fundamental frequency plus even and odd harmonics crossing the zero reference line 180º out of phase with the fundamental.
- Periodic waveforms are measured from any point on a cycle to the same point on the next cycle.
- The pulse width is the length of the pulse.
- The duty cycle is the ratio of the pulse width to the period.
- The rise time of a pulse is the time it takes to fall from 10% to 90% of its maximum amplitude.
- The fall time of a pulse is the time it takes to fall from 90% to 10% of its maximum amplitude.
- Overshoot, undershoot, and ringing are undesirable in a circuit and exist because of imperfect circuits.
- An RC circuit can be used to change the shape of a complex waveform.
- If the output is taken across the resistor in an RC circuit, the circuit is called a differentiator.
- If the output is taken across the capacitor in an RC circuit, the circuit is called an integrator.
- Clipping circuits are used to square off the peaks of an applied signal or to keep an amplitude constant.
- Clamping circuits are used to clamp the top or bottom of a waveform to a DC voltage.
- A monostable multivibrator (one-shot multivibrator) produces one output pulse for each input pulse.
- Bistable multivibrators have two stable states and are called flip-flops.
- A Schmitt trigger is a special-purpose bistable multivibrator.

CHAPTER 40 SELF-TEST

1. Describe the frequency domain concept of waveforms.
2. What is the relationship between the fundamental frequency and the first harmonic?
3. Calculate the duty cycle of a waveform that has a period of 0.1 MHz and a pulse width of 50 kHz.
4. Why do such problems as overshoot, undershoot, and ringing occur in waveshaping circuits?
5. Describe where integrator and differentiator waveshaping circuits are used.
6. What waveshaping circuit can be used to change a sine wave into a rectangular waveform?
7. How can the DC reference level of a signal be changed?
8. Explain the difference between monostable and bistable circuit functions.
9. Of what significance is the flip-flop?
10. What function does a Schmitt trigger serve?

Section 5

1. Build an electronic circuit such as the one shown in the back of the lab manual. Break the circuit down into basic circuits. Write a description of how the device works, using your own words. Use an electronic simulation program to help explain how the circuit works.

2. Design and build an electronic circuit that will blink lights and/or emit sound. Verify the circuit's operation with an electronic simulation program before building the circuit.

3. Design and build an audio amplifier that has a minimum gain of 100. Verify the circuit's operation using an electronic simulation program.

4. To play a portable CD player requires three 1.5-volt batteries. When it is turned on, it is discovered that the batteries are discharged. The batteries are always running down because of the amount of playtime. The task is to build a power supply that can be used with 120V AC. Design a regulated power supply that will convert 120V AC to the required DC voltage. The current draw for the unit must be determined through measuring with an ammeter. Model the solution in an electronic simulation program.

SECTION 6

DIGITAL ELECTRONIC CIRCUITS

DIGITAL ELECTRONIC CIRCUITS

Francis Bacon (1561–1626)

Bacon invented a "bilateral alphabet code," a binary system that used the symbols A and B rather than 0 and 1 to develop ciphers and codes.

George Boole (1815–1864)

Boole was a founder of the computer science field; he invented Boolean logic, which is the basis of modern digital computer logic.

Gary Boone and Michael J. Cochran

In 1971, Boone and Cochran created the first microcontroller, also referred to as a microcomputer.

William Henry Eccles (1875–1966) and Frank Wilfred Jordan (1882–?)

In 1918, Eccles and Jordan invented the first electronic flip-flop, initially called the Eccles–Jordan trigger circuit.

Frank Gray (1887–1969)

In 1947, Gray introduced the term *reflected binary code*, which was later named Gray-to-Gray code by others who used it.

IBM

IBM coined the term *hexadecimal* consisting of the Greek prefix *hex* and the Latin suffix *decimal*. The reason for not using an all-Latin name is that the name would have been sexadecimal.

Maurice Karnaugh (1924–)

In 1953, Karnaugh refined the Veitch diagram into the Karnaugh map method.

Gottfried Leibniz (1646–1716)

In 1679, Leibniz invented the binary system, of critical importance in the development of modern computers.

Emanuel Swedenborg (1688–1772)

In 1718, Swedenborg wrote a manuscript on a new arithmetic that changes at the number 8 instead of at the usual number 10.

Edward W. Veitch (1924–)

In 1952, Veitch developed a graphical procedure for simplifying the optimization of logic circuits, referred to as the Veitch diagram.

Binary Number System

OBJECTIVES

After completing this chapter, the student will be able to:

- Describe the binary number system.
- Identify the place value for each bit in a binary number.
- Convert binary numbers to decimal, octal, and hexadecimal numbers.
- Convert decimal, octal, and hexadecimal numbers to binary numbers.
- Convert decimal numbers to 8421 BCD code.
- Convert 8421 BCD code numbers to decimal numbers.

KEY TERMS

number system is nothing more than a code. For each distinct quantity, there is an assigned symbol. When the code is learned, counting can be accomplished. This leads to arithmetic and higher forms of mathematics.

The simplest number system is the binary number system. The binary system contains only two digits, 0 and 1. These digits have the same value as in the decimal number system.

The binary number system is used in digital and microprocessor circuits because of its simplicity. Binary data are represented by binary digits, called bits. The term **bit** is derived from *b*inary dig*it*.

41-1 BINARY NUMBERS

The decimal system is called a **base-10** system because it contains 10 digits, 0 through 9. The **binary number system** is a **base-2** system because it contains two digits, 0 and 1. The position of the 0 or 1 in a binary number indicates its value within the number. This is referred to as its place value or weight. The place value of the digits in a binary number increases by powers of two.

Place Value

	32	16	8	4	2	1
Powers of 2:	2^5	2^4	2^3	2^2	2^1	2^0

Counting in binary starts with the numbers 0 and 1. When each digit has been used in the 1's place, another digit is added in the 2's place and the count continues with 10 and 11. This uses up all combinations of two digits, so a third digit is added in the 4's place and the count continues with 100, 101, 110, and 111. Now a fourth digit is needed in the 8's place to continue, and so on. Figure 41-1 shows the binary counting sequence.

To determine the largest value that can be represented by a given number of places in base 2, use the following formula:

$$\text{Highest number} = 2^n - 1$$

where n represents the number of bits (or number of place values used).

FIGURE 41-1

Decimal number and equivalent binary table.

Decimal Number	Binary Number				
	2^4	2^3	2^2	2^1	2^0
	16	8	4	2	1
0	0	0	0	0	0
1	0	0	0	0	1
2	0	0	0	1	0
3	0	0	0	1	1
4	0	0	1	0	0
5	0	0	1	0	1
6	0	0	1	1	0
7	0	0	1	1	1
8	0	1	0	0	0
9	0	1	0	0	1
10	0	1	0	1	0
11	0	1	0	1	1
12	0	1	1	0	0
13	0	1	1	0	1
14	0	1	1	1	0
15	0	1	1	1	1
16	1	0	0	0	0
17	1	0	0	0	1
18	1	0	0	1	0
19	1	0	0	1	1
20	1	0	1	0	0
21	1	0	1	0	1
22	1	0	1	1	0
23	1	0	1	1	1
24	1	1	0	0	0
25	1	1	0	0	1
26	1	1	0	1	0
27	1	1	0	1	1
28	1	1	1	0	0
29	1	1	1	0	1
30	1	1	1	1	0
31	1	1	1	1	1

EXAMPLE: Two bits (two place values) can be used to count from 0 to 3 because:

$$2^n - 1 = 2^2 - 1 = 4 - 1 = 3$$

Four bits (four place values) are needed to count from 0 to 15 because:

$$2^n - 1 = 24 - 1 = 16 - 1 = 15$$

1. What is the advantage of the binary number system over the decimal number system for digital circuits?
2. Where are binary numbers used?
3. How is the largest value of a binary number determined for a given number of place values?
4. What is the maximum value of a binary number with each of the following?
   a. 4 bits
   b. 8 bits
   c. 12 bits
   d. 16 bits
5. Write the following numbers in binary:
   a. 3_{10}
   b. 7_{10}
   c. 16_{10}
   d. 31_{10}

41-2 **BINARY AND DECIMAL CONVERSION**

As stated, a binary number is a weighted number with a place value. The value of a binary number can be determined by adding the product of each digit and its place value. The method for evaluating a binary number is shown by the following example:

EXAMPLE:

	Place Value					
	32	16	8	4	2	1
Binary:	1	0	1	1	0	1

Value:
$$1 \times 32 = 32$$
$$0 \times 16 = 0$$
$$1 \times 8 = 8$$
$$1 \times 4 = 4$$
$$0 \times 2 = 0$$
$$+ 1 \times 1 = 1$$
$$101101_2 = 45_{10}$$

The number 45 is the decimal equivalent of the binary number 101101.

Fractional numbers can also be represented in binary form by placing digits to the right of the binary zero point, just as decimal numbers are placed to the right of the decimal zero point. All digits to the right of the zero point have weights that are negative powers of two, or fractional place values.

Powers of 2	Place Value
$2^5 =$	32
$2^4 =$	16
$2^3 =$	8
$2^2 =$	4
$2^1 =$	2
$2^0 =$	1

decimal point

$$2^{-1} = \frac{1}{2^1} = \frac{1}{2} = 0.5$$

$$2^{-2} = \frac{1}{2^2} = \frac{1}{4} = 0.25$$

$$2^{-3} = \frac{1}{2^3} = \frac{1}{8} = 0.125$$

$$2^{-4} = \frac{1}{2^4} = \frac{1}{16} = 0.0625$$

EXAMPLE: Determine the decimal value of the binary number 111011.011.

Binary Number	Place Value	Value
1	$\times$ 32	= 32
1	$\times$ 16	= 16
1	$\times$ 8	= 8
0	$\times$ 4	= 0
1	$\times$ 2	= 2
1	$\times$ 1	= 1
0	$\times$ 0.5	= 0
1	$\times$ 0.25	= 0.25
+ 1	$\times$ 0.125	= 0.125
111011.011_2		$= 59.375_{10}$

In working with digital equipment, it is often necessary to convert from binary to decimal form and vice versa. The most popular way to convert decimal numbers to binary numbers is to progressively divide the decimal number by 2, writing down the remainder after each division. The remainders, taken in reverse order, form the binary number.

EXAMPLE: To convert 11_{10} to a binary number, progressively divide by 2 (LSB = least significant bit).

$11 \div 2 = 5$ with a remainder of 1   LSB
$5 \div 2 = 2$ with a remainder of 1
$2 \div 2 = 1$ with a remainder of 0
$1 \div 2 = 0$ with a remainder of 1

($1 \div 2 = 0$ means that 2 will no longer divide into 1, so 1 is the remainder.) Decimal 11 is equal to 1011 in binary.

The process can be simplified by writing the numbers in an orderly fashion as shown here for converting 25_{10} to a binary number.

EXAMPLE:

```
2   25              LSB
   2  12      1
      2  6     0
         2  3    0
            2  1   1
               0  1
```

Decimal number 25 is equal to binary number 11001.

Fractional numbers are done a little differently: The number is multiplied by 2 and the carry is recorded as the binary fraction.

EXAMPLE: To convert decimal 0.85 to a binary fraction, progressively multiply by 2.

$0.85 \times 2 = 1.70 = 0.70$ with a carry of 1   LSB
$0.70 \times 2 = 1.40 = 0.40$ with a carry of 1
$0.40 \times 2 = 0.80 = 0.80$ with a carry of 0
$0.80 \times 2 = 1.60 = 0.60$ with a carry of 1
$0.60 \times 2 = 1.20 = 0.20$ with a carry of 1
$0.20 \times 2 = 0.40 = 0.40$ with a carry of 0

Continue to multiply by 2 until the needed accuracy is reached. Decimal 0.85 is equal to 0.110110 in binary form.

EXAMPLE: Convert decimal 20.65 to a binary number. Split 20.65 into an integer of 20 and a fraction of 0.65 and apply the methods previously shown.

```
2   20              LSB
   2  10      0
      2  5     0
         2  2    1
            2  1   0
               0  1
```

Decimal 20 = Binary 10100

and

$0.65 \times 2 = 1.30 = 0.30$ with a carry of 1   LSB
$0.30 \times 2 = 0.60 = 0.60$ with a carry of 0
$0.60 \times 2 = 1.20 = 0.20$ with a carry of 1
$0.20 \times 2 = 0.40 = 0.40$ with a carry of 0
$0.40 \times 2 = 0.80 = 0.80$ with a carry of 0
$0.80 \times 2 = 1.60 = 0.60$ with a carry of 1
$0.60 \times 2 = 1.20 = 0.20$ with a carry of 1

Decimal 0.65 = Binary 0.1010011

Combining the two numbers results in $20.65_{10} = 10100.1010011_2$. This 12-bit number is an approximation, because the conversion of the fraction was terminated after 7 bits.

41–2   QUESTIONS

1. What is the value of each position in an 8-bit binary number?
2. What is the value of each position to the right of the decimal point for eight places?
3. Convert the following binary numbers to decimal numbers:
   a. 1001_2
   b. 11101111_2
   c. 11000010_2
   d. 10101010.1101_2
   e. 10110111.0001_2
4. What is the process for converting decimal numbers to binary digits?
5. Convert the following decimal numbers to binary form:
   a. 27_{10}
   b. 34.6_{10}
   c. 346_{10}
   d. 321.456_{10}
   e. 7465_{10}

41–3   OCTAL NUMBERS

Years ago, computers used **octal numbers** for representing binary numbers. Using octal numbers allows the reading of large binary numbers by breaking the

■ **FIGURE 41-2**

Decimal and binary equivalent of octal numbers.

Decimal Number	Binary Number	Octal Number
0	00 000	0
1	00 001	1
2	00 010	2
3	00 011	3
4	00 100	4
5	00 101	5
6	00 110	6
7	00 111	7
8	01 000	10
9	01 001	11
10	01 010	12
11	01 011	13
12	01 100	14
13	01 101	15
14	01 110	16
15	01 111	17
16	10 000	20
17	10 001	21
18	10 010	22
19	10 011	23
20	10 100	24
21	10 101	25
22	10 110	26
23	10 111	27
24	11 000	30
25	11 001	31
26	11 010	32
27	11 011	33
28	11 100	34
29	11 101	35
30	11 110	36
31	11 111	37

© 2014 Cengage Learning

binary number into groups of three. This speeded up the entering and reading of binary numbers from a computer, thus resulting in reduced error.

Figure 41-2 shows octal numbers with their decimal and binary equivalents. The advantage is that octal numbers are converted directly from 3 bits of binary numbers. The **octal number system** is referred to as **base 8**. There are only eight digits (0–7) and then it recycles like the decimal system (0–9).

To convert binary to an octal number requires dividing the binary number into groups of three, starting from the right, as shown by the following example:

EXAMPLE:

Binary number 100101001110000111010_2
Separate into groups of three:
$100\ 101\ 001\ 110\ 000\ 111\ 010_2$

Convert to octal: $100\ 101\ 001\ 110\ 000\ 111\ 010_2$
$$4\quad 5\quad 1\quad 6\quad 0\quad 7\quad 2_8$$
Octal equivalent is: 4516072_8

To convert an octal number to a binary equivalent requires reversing the process and converting the octal number to binary groups of three, as shown by the following example:

EXAMPLE:

Octal number: 1672054_8
Separate the numbers: $1\quad 6\quad 7\quad 2\quad 0\quad 5\quad 4_8$
Convert to binary: $001\ 110\ 111\ 010\ 000\ 101\ 100_2$
Binary equivalent: 00111011101000101100_2
Lead zeros can be dropped resulting in:

$$1110111010000101100_2$$

Like binary numbers, each octal number is a weighted number with a place value. The value of an octal number can be determined by adding the product of each digit and its place value. Figure 41-3 shows octal place values. The method for converting an octal number to a decimal is shown by the following example:

EXAMPLE

Octal number: 6072_8
Value: $\quad 6 \times 512 = 3072$
$\qquad\ \ 0 \times 64\ \ = 0$
$\qquad\ \ 7 \times 8\ \ \ = 56$
$\qquad\underline{\ \ 2 \times 1\ \ \ = 2}$
$\qquad 6072_8\ \ \ = 3130_{10}$

To convert a decimal number to an octal equivalent requires dividing the number by eight, as shown by converting 2382_{10} to an octal number in the following example:

■ **FIGURE 41-3**

Place values of octal numbers.

Powers of 8	Place Value
8^0	1
8^1	8
8^2	64
8^3	512
8^4	4096
8^5	32768
8^6	262144
8^7	2097152

© 2014 Cengage Learning

EXAMPLE

Decimal number: 2382_{10}

$2832 \div 8 = 354$	with remainder of	0 (LSB)
$354 \div 8 = 44$	with remainder of	2
$44 \div 8 = 5$	with remainder of	4
$5 \div 8 =$ will not divide, therefore yields		5 (most significant bit [MSB])

Resulting in $2382_{10} = 5420_8$

41-3 QUESTIONS

1. Why are octal numbers used with computers?
2. Convert the following binary numbers to octal numbers:
   a. 1010111111110010001000_2
   b. 1111011010000110101_2
   c. 1001110011100010101001_2
3. Convert the following octal numbers to binary numbers:
   a. 75634201_8
   b. 36425107_8
   c. 17536420_8
4. Convert the following octal numbers to decimal numbers:
   a. 653_8
   b. 4721_8
   c. 75364_8
5. Convert the following decimal numbers to octal numbers:
   a. 453_{10}
   b. 1028_{10}
   c. 32047_{10}

41-4 HEXADECIMAL NUMBERS

The **hexadecimal number system** is used primarily for entering and reading data of microprocessor-based 4-, 8-, 16-, 32-, and 64-bit systems. Using hexadecimal numbers allows the breaking of a binary number into groups of four to reduce chance of error when entering data.

The hexadecimal number system is referred to as **base 16**. There are 16 digits: 0, 1, 2, 3, 4, 5, 6, 7, 8, 9, A, B, C, D, E, and F. Figure 41-4 shows

■ FIGURE 41-4

Decimal and binary equivalent of hexadecimal numbers.

Decimal Number	Binary Number	Hexadecimal Number
0	0 0000	0
1	0 0001	1
2	0 0010	2
3	0 0011	3
4	0 0100	4
5	0 0101	5
6	0 0110	6
7	0 0111	7
8	0 1000	8
9	0 1001	9
10	0 1010	A
11	0 1011	B
12	0 1100	C
13	0 1101	D
14	0 1110	E
15	0 1111	F
16	1 0000	10
17	1 0001	11
18	1 0010	12
19	1 0011	13
20	1 0100	14
21	1 0101	15
22	1 0110	16
23	1 0111	17
24	1 1000	18
25	1 1001	19
26	1 1010	1A
27	1 1011	1B
28	1 1100	1C
29	1 1101	1D
30	1 1110	1E
31	1 1111	1F

© 2014 Cengage Learning

hexadecimal numbers with their decimal and binary equivalents. The advantage of hexadecimal numbers is that they are converted directly from 4 bits of binary numbers.

To convert binary to a hexadecimal number requires dividing the binary number into groups of four, starting from the right, as shown by the following example:

EXAMPLE:

Binary number 1001010011100000111010_2
Separate into groups of four:
$1\ 0010\ 1001\ 1100\ 0011\ 1010_2$
Convert to hexadecimal:
$1\ 0010\ 1001\ 1100\ 0011\ 1010_2$
$1\quad 2\quad 9\quad\ C\quad\ 6\quad\ A_{16}$
Hexadecimal equivalent is $129C6A_{16}$

To convert a hexadecimal number to a binary equivalent requires reversing the process and converting the hexadecimal number to binary groups of four, as shown by the following example:

EXAMPLE:

Hexadecimal number: $6A7F4D2C_{16}$
Separate the numbers: 6   A   7   F   4   D   2 C_{16}
Convert to binary:
0011 1010 0111 1111 0100 1101 0010 1100_2
Binary equivalent:
$00111010011111110100110100101100_2$
Lead zeros can be dropped, resulting in:

$$11101001111111010011010 0101100_2$$

Like binary numbers, each hexadecimal number is a weighted number with a place value. The value of a hexadecimal number can be determined by adding the product of each digit and its place value. Figure 41-5 shows hexadecimal place values. The method for converting a hexadecimal number to a decimal is shown by the following example:

EXAMPLE:

Hexadecimal number: 4AC916

$$
\begin{array}{llll}
\text{Value:} & 4 \times 4096 & = 16384 \\
(10)\ & \text{A} \times 256 & = 2560 \\
(12)\ & \text{C} \times 16 & = 192 \\
& 9 \times 1 & = 9 \\
\hline
& 4AC9_{16} & = 19145_{10}
\end{array}
$$

To convert a decimal number to a hexadecimal equivalent requires dividing the number by 16 as

■ FIGURE 41-5

Place values of hexadecimal numbers.

Powers of 16	Place Value
16^0	1
16^1	16
16^2	256
16^3	4096
16^4	65536
16^5	1048576
16^6	16777216
16^7	268435456

© 2014 Cengage Learning

shown by converting 41929_{10} to an octal number in the following example:

EXAMPLE:

Decimal number: 41929_{10}

$41929 \div 16 = 2620$	with remainder of	9 (LSB)
$2620 \div 16 = 163$	with remainder of	12 (C)
$163 \div 16 = 10$	with remainder of	3

$10 \div 16 =$ will not divide, therefore yields
10 (A) (MSB)
Resulting in $41929_{10} = A3C9_{16}$

41-4   QUESTIONS

1. Why are hexadecimal numbers used with computers?
2. Convert the following binary numbers to hexadecimal numbers:
   a. $10101111111001000100 0_2$
   b. 1111011010000110101_2
   c. $1001110011100010110 1_2$
3. Convert the following hexadecimal numbers to binary numbers:
   a. $42C1_{16}$
   b. $5B07_{16}$
   c. $A75C642E_{16}$
4. Convert the following hexadecimal numbers to decimal numbers:
   a. $6E53_{16}$
   b. $A7C1_{16}$
   c. $7F3BE_{16}$
5. Convert the following decimal numbers to hexadecimal numbers:
   a. 45374_{10}
   b. 32047_{10}
   c. 67326_{10}

41-5   BCD CODE

An **8421 code** is a **binary-coded decimal (BCD)** code consisting of four binary digits. It is used to represent the digits 0 through 9. The 8421 designation refers to the binary weight of the 4 bits.

Powers of 2:	2^3	2^2	2^1	2^0
Binary weight:	8	4	2	1

The main advantage of this code is that it permits easy conversion between decimal and binary form. This is the predominant BCD code used and is the one referred to unless otherwise stated.

Each decimal digit (0 through 9) is represented by a binary combination as follows:

Decimal	8421 Code
0	0000
1	0001
2	0010
3	0011
4	0100
5	0101
6	0110
7	0111
8	1000
9	1001

Although 16 numbers (2^4) can be represented by four binary positions, the six code combinations above decimal 9 (1010, 1011, 1100, 1101, 1110, and 1111) are invalid in the 8421 code.

To express any decimal number in the 8421 code, replace each decimal digit by the appropriate 4-bit code.

EXAMPLE: Convert the following decimal numbers into a BCD code: 5, 13, 124, 576, and 8769.

$$5_{10} = 0101_{BCD}$$
$$13_{10} = 0001\ 0011_{BCD}$$
$$124_{10} = 0001\ 0010\ 0100_{BCD}$$
$$576_{10} = 0101\ 0111\ 0110_{BCD}$$
$$8769_{10} = 1000\ 0111\ 0110\ 1001_{BCD}$$

To determine a decimal number from an 8421 code number, break the code into groups of 4 bits. Then write the decimal digit represented by each 4-bit group.

EXAMPLE: Find the decimal equivalent for each of the following BCD codes: 10010101, 1001000, 1100111, 1001100101001, and 1001100001110110.

$$1001\ 0101_{BCD} = 95_{10}$$
$$0100\ 1000_{BCD} = 48_{10}$$
$$0110\ 0111_{BCD} = 67_{10}$$
$$0001\ 0011\ 0010\ 1001_{BCD} = 1329_{10}$$
$$1001\ 1000\ 0111\ 0110_{BCD} = 9876_{10}$$

1. What is the 8421 code and how is it used?
2. What is the advantage of BCD code?
3. What are the similarities between BCD code and the hexadecimal number system?
4. Convert the following decimal numbers into BCD code:
   a. 17_{10}
   b. 100_{10}
   c. 256_{10}
   d. 778_{10}
   e. 8573_{10}
5. Convert the following BCD codes into decimal numbers:
   a. $1000\ 0010_{BCD}$
   b. $0111\ 0000\ 0101_{BCD}$
   c. $1001\ 0001\ 0011\ 0100_{BCD}$
   d. $0001\ 0000\ 0000\ 0000_{BCD}$
   e. $0100\ 0110\ 1000\ 1001_{BCD}$

SUMMARY

- The binary number system is the simplest number system.
- The binary number system contains two digits, 0 and 1.
- The binary number system is used to represent data for digital and computer systems.
- Binary data are represented by binary digits called bits.
- The term *bit* is derived from **b**inary dig**it**.
- The place value of each higher digit's position in a binary number is increased by a power of 2.
- The largest value that can be represented by a given number of places in base 2 is $2^n - 1$, where n represents the number of bits.
- The value of a binary digit can be determined by adding the product of each digit and its place value.
- Fractional numbers are represented by negative powers of 2.
- To convert from a decimal number to a binary number, divide the decimal number by 2, writing down the remainder after each division. The remainders, taken in reverse order, form the binary number.
- Octal numbers allow reading large binary numbers by breaking the binary number into groups of three.

- The octal number system is referred to as base 8.
- Similar steps are used to convert octal numbers to decimal and decimal numbers to octal as with the binary number system.
- The hexadecimal number system breaks binary numbers into groups of four to reduce error when entering data.
- The hexadecimal number system is a base-16 system.

- The hexadecimal number system is used with microprocessor-based systems.
- As with octal and binary number systems, the hexadecimal number system uses similar steps for converting to and from decimal numbers.
- The 8421 code, a binary-coded decimal (BCD) code, is used to represent digits 0 through 9.
- The advantage of the BCD code is ease of converting between decimal and binary forms of a number.

CHAPTER 41 SELF-TEST

1. What represents the decimal equivalent of 0 through 27 in binary?
2. How many binary bits are required to represent the decimal number 100?
3. Describe the process for converting a decimal number to a binary number.
4. Convert the following binary numbers to their decimal equivalents:
   a. 100101.001011_2
   b. 111101110.11101110_2
   c. 1000001.00000101_2
5. Convert the following binary numbers to octal numbers:
   a. $110010010111001010100111000011101_2$
   b. $0011000110111110010011110011010_2$
   c. $1001101011000100110011110100010110_2$
6. Convert the following octal numbers to binary numbers:
   a. 653172_8
   b. 773012_8
   c. 033257_8
7. Convert the following octal numbers to decimal numbers:
   a. 317204_8
   b. 701253_8
   c. 035716_8
8. Convert the following decimal numbers to octal numbers:
   a. 687_{10}
   b. 9762_{10}
   c. 18673_{10}

9. Convert the following binary numbers to hexadecimal numbers:
   a. $110010010111001010100111000011101_2$
   b. $0011000110111110010011110011010_2$
   c. $1001101011000100110011110100010110_2$
10. Convert the following hexadecimal numbers to binary numbers:
    a. $7B23C67F_{16}$
    b. $D46F17C9_{16}$
    c. $78F3E69D_{16}$
11. Convert the following hexadecimal numbers to decimal numbers:
    a. $3C67F_{16}$
    b. $6F17C9_{16}$
    c. $78F3E69D_{16}$
12. Convert the following decimal numbers to hexadecimal numbers:
    a. 687_{10}
    b. 9762_{10}
    c. 18673_{10}
13. Describe the process for converting decimal numbers to BCD.
14. Convert the following BCD numbers to their decimal equivalents:
    a. $0100\ 0001\ 0000\ 0110_{BCD}$
    b. $1001\ 0010\ 0100\ 0011_{BCD}$
    c. $0101\ 0110\ 0111\ 1000_{BCD}$

Basic Logic Gates

OBJECTIVES

After completing this chapter, the student will be able to:

- Identify and explain the function of the basic logic gates.
- Draw the symbols for the basic logic gates.
- Develop truth tables for the basic logic gates.

KEY TERMS

All digital equipment, whether simple or complex, is constructed of only a few basic circuits. These circuits, referred to as logic elements, perform some logic function on binary data.

There are two basic types of logic circuits: decision-making and memory. Decision-making logic circuits monitor binary inputs and produce an output based on the status of the inputs and the characteristics of the logic circuit. Memory circuits are used to store binary data.

42-1 AND GATE

The **AND gate** is a logic circuit that has two or more inputs and a single output. The AND gate produces an output of 1, only when all its inputs are 1s. If any of the inputs are 0s, the output is 0.

Figure 42-1 shows the standard symbol used for AND gates. An AND gate can have any number of inputs greater than one. Shown in the figure are symbols representing the more commonly used gates of two, three, four, and eight inputs.

The operation of the AND gate is summarized by the table in Figure 42-2. Such a table, called a **truth table**, shows the output for each possible input. The inputs are designated A and B. The output is designated Y. The total number of possible combinations in the truth table is determined by the following formula:

$$N = 2^n$$

where:

N = the total number of possible combinations
n = the total number of input variables

EXAMPLE:

For two input variables, $N = 2^2 = 4$
For three input variables, $N = 2^3 = 8$
For four input variables, $N = 2^4 = 16$
For eight input variables, $N = 2^8 = 256$

The AND gate uses the rules of multiplication. The **AND function** is known for multiplication-like results. The output of an AND gate is represented by the equation $Y = A \cdot B$ or $Y = AB$. The AND function is typically represented by the two variables A and B as AB.

42-1 QUESTIONS

1. Under what conditions does an AND gate produce a 1 output?
2. Draw the symbol used to represent a two-input AND gate.
3. Develop a truth table for a three-input AND gate.
4. What logical operation is performed by an AND gate?
5. What is the algebraic output of an AND gate?

42-2 OR GATE

An **OR gate** produces a 1 output if any of its inputs are 1s. The output is a 0 if all the inputs are 0s. The output of a two-input OR gate is shown in the truth table in Figure 42-3. The total number of possible combinations is expressed by $N = 2^2 = 4$. The truth table shows all four combinations.

An OR gate performs the basic operation of addition. The algebraic expression for the output of an OR gate is $Y = A + B$. The plus sign designates the **OR function**.

Figure 42-4 shows the logic symbol for an OR gate. The inputs are labeled A and B, and the output is

FIGURE 42-1

Logic symbol for an AND gate.

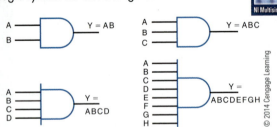

FIGURE 42-2

Truth table for a two-input AND gate.

Inputs		Output
B	A	Y
0	0	0
0	1	0
1	0	0
1	1	1

© 2014 Cengage Learning

© 2014 Cengage Learning

FIGURE 42-3

Truth table for a two-input OR gate.

Inputs		Output
B	A	Y
0	0	0
0	1	1
1	0	1
1	1	1

FIGURE 42-4

Logic symbol for an OR gate.

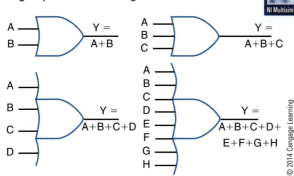

labeled Y. An OR gate can have any number of inputs greater than one. Shown in the figure are OR gates with two, three, four, and eight inputs.

42-2 QUESTIONS

1. What conditions produce a 1 output for an OR gate?
2. Draw the symbol used to represent a two-input OR gate.
3. Develop a truth table for a three-input OR gate.
4. What operation is performed by an OR gate?
5. What algebraic expression represents the output of an OR gate?

42-3 NOT GATE

The simplest logic circuit is the **NOT gate**. It performs the function called inversion, or complementation, and is commonly referred to as an inverter.

The purpose of the inverter is to make the output state the opposite of the input state. The two states associated with logic circuits are 1 and 0. A 1 state can also be referred to as a **high**, to indicate that the voltage is higher than in the 0 state. A 0 state can also be referred to as a **low**, to indicate that the voltage is lower than in the 1 state. If a 1, or high, is applied to the input of an inverter, a 0, or low appears on its output. If a 0, or low, is applied to the input, a 1, or high, appears on its output.

The operation of an inverter is summarized in Figure 42-5. The input to an inverter is labeled A and the Y output is labeled $\overline{A}$ (read "A NOT" or "NOT A"). The bar over the letter A indicates the complement of A. Because the inverter has only one input, only two input combinations are possible.

The symbol used to represent an inverter or NOT function is shown in Figure 42-6. The triangle portion of the symbol represents the circuit, and the circle, or "bubble," represents the circuit inversion, or complementary characteristic. The choice of symbol depends on where the inverter is used. If the inverter uses a 1 as the qualifying input, the symbol in Figure 42-6A is used. If the inverter uses a 0 as the qualifying input, the symbol in Figure 42-6B is used.

FIGURE 42-5

Truth table for an inverter.

Inputs	Output
A	Y
0	1
1	0

FIGURE 42-6

Logic symbol for an inverter.

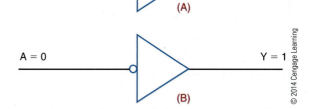

42–4 NAND GATE

A **NAND gate** is a combination of an inverter and an AND gate. It is called a NAND gate from the NOT-AND function it performs. The NAND gate is the most commonly used logic function. This is because it can be used to construct an AND gate, OR gate, inverter, or any combination of these functions.

The logic symbol for a NAND gate is shown in Figure 42-7. Also shown is its equivalency to an AND gate and an inverter. The bubble on the output end of the symbol means to invert the AND function.

Figure 42-8 shows the truth table for a two-input NAND gate. Notice that the output of the NAND gate is the complement of the output of an AND gate. Any 0 in the input yields a 1 output.

The algebraic formula for NAND-gate output is $Y = \overline{AB}$, where Y is the output and A and B are the inputs. NAND gates are available with two, three, four, eight, and thirteen inputs.

NAND gates are the most widely available gates on the market. The availability and flexibility of the NAND gate allows it to be used for other types of gates. Figure 42-9 shows how a two-input NAND gate can be used to generate other logic functions.

42–5 NOR GATE

A **NOR gate** is a combination of an inverter and an OR gate. Its name derives from its NOT-OR function. Like the NAND gate, the NOR gate can also be used to construct an AND gate, an OR gate, and an inverter.

The logic symbol for the NOR gate is shown in Figure 42-10. Also shown is its equivalency to an OR gate and an inverter. The bubble on the output of the symbol means to invert the OR function.

Figure 42-11 shows the truth table for a two-input NOR gate. Notice that the output is the complement of the OR-function output. A 1 occurs only when 0 is applied to both inputs. A 1 input produces a 0 output.

The algebraic expression for NOR-gate output is $Y = \overline{A + B}$, where Y is the output, and A and B are the inputs. NOR gates are available with two, three, four, and eight inputs.

■ FIGURE 42-7

Logic symbol for a NAND gate.

■ FIGURE 42-8

Truth table for a two-input NAND gate.

Inputs		Output
B	A	Y
0	0	1
0	1	1
1	0	1
1	1	0

◼ FIGURE 42-9

Using the NAND gate to generate other logic functions.

Logic	Logic Symbol	Logic Functions Using Only NAND Gates
Inverter	A ──▷∘── $\overline{A}$	A ──•──⊐∘── $\overline{A}$
AND	A, B ──⊐── AB	A, B ──⊐∘──•──⊐∘── AB
OR	A, B ──⊐── A+B	A ──•──⊐∘──, B ──•──⊐∘── ──⊐∘── A+B
NOR	A, B ──⊐∘── $\overline{A+B}$	A ──•──⊐∘──, B ──•──⊐∘── ──⊐∘──•──⊐∘── $\overline{A+B}$
XOR	A, B ──⊐D── A⊕B	A ──•──, B ──•── ──⊐∘──•──⊐∘──, ──⊐∘── ──⊐∘── A⊕B
XNOR	A, B ──⊐D∘── $\overline{A⊕B}$	A ──•──, B ──•── ──⊐∘──•──⊐∘──, ──⊐∘── ──⊐∘──•──⊐∘── $\overline{A⊕B}$

© 2014 Cengage Learning

◼ FIGURE 42-10

Logic symbol for a NOR gate.

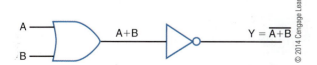

$$Y = \overline{A+B}$$

$$A+B \quad Y = \overline{A+B}$$

© 2014 Cengage Learning

◼ FIGURE 42-11

Truth table for a two-input NOR gate.

Inputs		Output
B	A	Y
0	0	1
0	1	0
1	0	0
1	1	0

© 2014 Cengage Learning

42-5 QUESTIONS

1. What is a NOR gate?
2. Why is the NOR gate useful in designing digital circuits?
3. Draw the symbol used to represent a NOR gate.
4. What is the algebraic expression for a NOR gate?
5. Develop a truth table for a three-input NOR gate.

42-6 EXCLUSIVE OR AND NOR GATES

A less common but still important gate is called an **exclusive OR gate**, abbreviated as **XOR**. An XOR gate has only two inputs, unlike the OR gate, which may have several inputs. However, the XOR is similar to the OR gate in that it generates a 1 output if either input is a 1. The exclusive OR is different when both inputs are 1s or 0s. In that case, the output is a 0.

The symbol for an XOR gate is shown in Figure 42-12. Also shown is the equivalent logic circuit. Figure 42-13 shows the truth table for an XOR gate. The algebraic output is written as $Y = A \oplus B$. It is read as "Y equals A exclusive or B."

■ FIGURE 42-12

Logic symbol for an exclusive OR gate.

© 2014 Cengage Learning

$$Y = A \oplus B$$

■ FIGURE 42-13

Truth table for an exclusive OR gate.

Inputs		Output
B	A	Y
0	0	0
0	1	1
1	0	1
1	1	0

© 2014 Cengage Learning

■ FIGURE 42-14

Logic symbol for an exclusive NOR gate.

© 2014 Cengage Learning

$$Y = \overline{A \oplus B}$$

■ FIGURE 42-15

Truth table for an exclusive NOR gate.

Inputs		Output
B	A	Y
0	0	1
0	1	0
1	0	0
1	1	1

© 2014 Cengage Learning

The complement of the XOR gate is the **XNOR (exclusive NOR) gate**. Its symbol is shown in Figure 42-14. The bubble on the output implies inversion or complement.

Also shown is the equivalent logic circuit. Figure 42-15 shows the truth table for an XNOR gate. The algebraic output is written as $Y = \overline{A \oplus B}$, read as "Y equals A exclusive nor B."

42-6 QUESTIONS

1. What is the difference between an OR gate and an XOR gate?
2. Draw the symbol used to represent an XOR gate.
3. Develop a truth table for an XOR gate.
4. Draw the symbol used to represent an XNOR gate.
5. Write the algebraic expressions for XOR and XNOR gates.

42-7 BUFFER

A **buffer** is a special logic gate that isolates conventional gates from other circuitry and provides a high driving current for heavy circuit loads or fan-out. Fan-out refers to the number of input gates connected to a single logic gate output. Buffers provide noninverting input and output. A 1 in provides a 1 out and a 0 in provides a 0 out. Figure 42-16 shows the schematic symbol for a basic buffer with its truth table.

Another type of buffer is the **three-state buffer** shown with its truth table in Figure 42-17. The three-state buffer has the usual 1 and 0 output states but it also has a third state, which is referred to as a high-impedance state. This state provides an open circuit between the input circuitry and the output and is controlled by the EN line. The EN line represents an enable/disable control input. Figure 42-18 shows how the three-state buffer operates.

■ **FIGURE 42-16**

Logic symbol for a buffer and truth table.

Input	Output
A	Y
0	0
1	1

© 2014 Cengage Learning

■ **FIGURE 42-17**

Logic symbol for a three-state buffer and truth table.

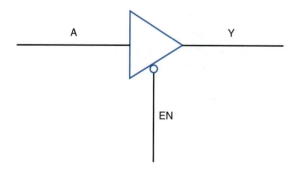

Input		Output
A	**EN**	**Y**
0 or 1	0	0 or 1
0 or 1	1	No output

© 2014 Cengage Learning

■ **FIGURE 42-18**

How a three-state buffer functions.

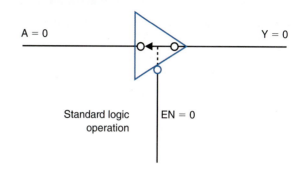

A = 0   Y = 0

Standard logic operation   EN = 0

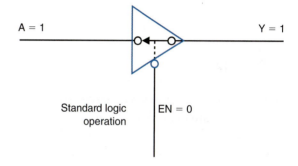

A = 1   Y = 1

Standard logic operation   EN = 0

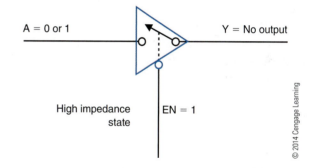

A = 0 or 1   Y = No output

High impedance state   EN = 1

© 2014 Cengage Learning

42–7 QUESTIONS

1. How does a buffer differ from an inverter?
2. What function does a buffer serve?
3. Draw the symbols for a buffer and a three-state buffer.
4. Develop a truth table for a three-state buffer.
5. What conditions must exist for no output from a three-state buffer?

SUMMARY

- An AND gate produces a 1 output when all of its inputs are 1s.
- An AND gate performs the basic operation of multiplication.
- An OR gate produces a 1 output if any of its inputs are 1s.
- An OR gate performs the basic operation of addition.
- A NOT gate performs the function called inversion or complementation.
- A NOT gate converts the input state to an opposite output state.
- A NAND gate is a combination of an AND gate and an inverter.

- A NAND gate produces 1 output when any of the inputs are 0s.
- A NOR gate is a combination of an OR gate and an inverter.
- A NOR gate produces a 1 output only when both inputs are 0s.
- An exclusive OR (XOR) gate produces a 1 output only if both inputs are different.
- An exclusive NOR (XNOR) gate produces a 1 output only when both inputs are the same.
- A buffer isolates conventional gates from other circuitry.
- A buffer provides a high current for heavy loads or fan-outs.
- A three-state buffer has a high-impedance third state.

CHAPTER 42 SELF-TEST

1. Draw the schematic symbol for a six-input AND gate.
2. Develop the truth table for a four-input AND gate.
3. Draw the schematic symbol for a six-input OR gate.
4. Develop the truth table for a four-input OR gate.
5. What is the purpose for the NOT circuit?
6. How does an inverter for an input signal differ from the inverter for an output signal?
7. Draw the schematic symbol for an eight-input NAND gate.
8. Develop the truth table for a four-input NAND gate.
9. Draw the schematic symbol for an eight-input NOR gate.
10. Develop the truth table for a four-input NOR gate.
11. What is the significance of the XOR gate?
12. The XNOR gate has what maximum number of inputs?
13. Draw the schematic symbol for a buffer.
14. What purpose does the buffer serve?
15. Draw the schematic symbol for a three-state buffer.

Simplifying Logic Circuits

OBJECTIVES

After completing this chapter, the student will be able to:

- Explain the function of Veitch diagrams.
- Describe how to use a Veitch diagram to simplify Boolean expressions.
- Explain the function of a Karnaugh map.
- Describe how to simplify a Boolean expression using a Karnaugh map.

KEY TERMS

43 Boolean algebra

43 Boolean expression

43-1 Veitch diagram

43-2 Karnaugh map

igital circuits are being used more and more in electronics, not only in computers but also in applications such as measurement, automatic control, robotics, and in situations requiring decisions. All of these applications require complex switching circuits that are formed from the five basic logic gates: the AND, OR, NAND, and NOR gates and the inverter.

The significant point about these logic gates is that they only have two operating conditions. They are either ON (1) or OFF (0). When logic gates are interconnected to form more complex circuits, it is necessary to obtain the simplest circuit possible.

Boolean algebra offers a means of expressing complex switching functions in equation form. A **Boolean expression** is an equation that expresses the output of a logic circuit in terms of its input. Veitch diagrams and Karnaugh maps provide a fast and easy way to reduce a logic equation to its simplest form.

43–1 VEITCH DIAGRAMS

Veitch diagrams provide a fast and easy method for reducing a complicated expression to its simplest form. They can be constructed for one, two, three, or four variables. Figure 43-1 shows several Veitch diagrams.

To use a Veitch diagram, follow these steps, as illustrated in the example:

1. Draw the diagram based on the number of variables.
2. Plot the logic functions by placing an X in each square representing a term.
3. Obtain the simplified logic function by looping adjacent groups of X's in groups of eight, four, two, or one. Continue to loop until all X's are included in a loop.
4. "OR" the loops with one term per loop. (Each expression is pulled off the Veitch diagram and "OR"ed using the + symbol, for example, ABC + BCD.)
5. Write the simplified expression.

EXAMPLE: Reduce $AB + \overline{A}B + A\overline{B} = Y$ to its simplest form.

Step 1. Draw the Veitch diagram. There are two variables, A and B, so use the two-variable chart.

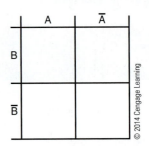

Step 2. Plot the logic function by placing an X in each square representing a term.

$$AB \quad + \quad \overline{A}B \quad + \quad A\overline{B}$$

1st	2nd	3rd
term	term	term

Plot 1st term AB.

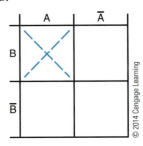

Plot 2nd term $\overline{A}B$.

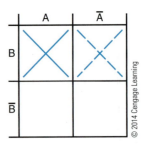

■ **FIGURE 43-1**

Two-, three-, and four-variable Veitch diagrams.

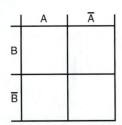
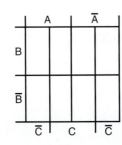
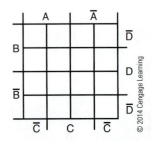

Plot 3rd term $A\overline{B}$.

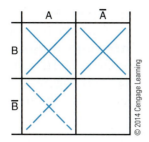

Step 3. Loop adjacent groups of X's in the largest group possible. Start by analyzing the chart for the largest groups possible. The largest group possible here is two.

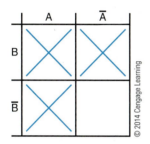

One possible group is the one indicated by the dotted line.

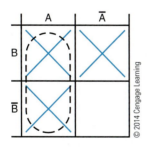

Another group is the one indicated by this dotted line.

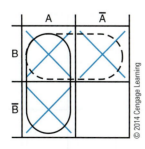

Step 4. "OR" the groups: Either A or B = A + B.
Step 5. The simplified expression for $AB + \overline{A}B + A\overline{B} = Y$ is $A + B = Y$ obtained from the Veitch diagram.

EXAMPLE: Find the simplified expression for

$$ABC + AB\overline{C} + A\overline{B}\,\overline{C} + \overline{A}\,\overline{B}\,\overline{C} = \overline{Y}$$

Step 1. Draw a three-variable Veitch diagram.

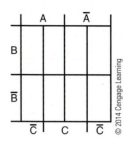

Step 2. Place an X for each term on the Veitch diagram.

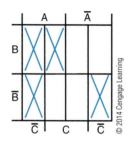

Step 3. Loop the groups.

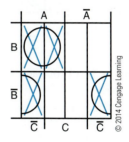

Step 4. Write the term for each loop, one term per expression:

$$AB, \overline{B}\,\overline{C}.$$

Step 5. The simplified expression is: $AB + \overline{B}\,\overline{C} = Y$. Notice the unusual looping on the two bottom squares. The four corners of the Veitch diagram are considered connected as if the diagram were formed into a ball.

EXAMPLE: Find the simplified expression for

$$\overline{A}BCD + \overline{A}B\overline{C}D + \overline{A}\overline{B}\overline{C}D + A\overline{B}CD +$$
$$A\overline{B}\overline{C}D + \overline{A}\overline{B}\overline{C}\overline{D} = Y$$

Step 1. Draw a four-variable Veitch diagram.

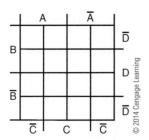

Step 2. Place an X for each term on the Veitch diagram.

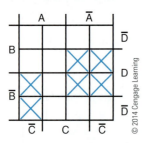

Step 3. Loop the groups.

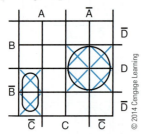

Step 4. Write the term for each loop, one term per expression: $\overline{A}D$, $A\overline{B}\overline{C}$.

Step 5. "OR" the terms to form the simplified expression: $\overline{A}D + A\overline{B}\overline{C} = Y$.

43-1 QUESTIONS

1. What is the function of a Veitch diagram?
2. How many variables can be represented on a Veitch diagram?
3. List the steps for using a Veitch diagram.
4. Simplify the following three-variable expressions using Veitch diagrams:
   a. $A\overline{B}C + \overline{A}\overline{B}C + AB\overline{C} + A\overline{B}\overline{C} + \overline{A}B\overline{C} = Y$
   b. $\overline{A}BC + A\overline{B}\overline{C} + AB\overline{C} + ABC + AB + \overline{A}B\overline{C} = Y$
5. Simplify the following four-variable expressions using Veitch diagrams:
   a. $ABCD + A\overline{B}CD + \overline{A}BC\overline{D} + \overline{A}\overline{B}CD + AB\overline{C}\overline{D} + \overline{A}\overline{B}C\overline{D} + AB\overline{C}D = Y$
   b. $A\overline{B} + \overline{A}BD + \overline{B}CD + \overline{B}C + \overline{A}BCD = Y$

43-2 KARNAUGH MAPS

Another technique for reducing complex Boolean expressions is **Karnaugh maps**, which Maurice Karnaugh developed in 1953. His technique is similar to using Veitch diagrams. Figure 43-2 shows several Karnaugh maps. It can be cumbersome to solve problems with five or six variables with either Veitch diagrams or Karnaugh maps. Those types of problems are best solved by other techniques or computer software designed specifically for them.

■ **FIGURE 43-2**

Two-, three-, and four-variable Karnaugh maps.

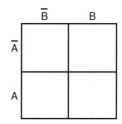

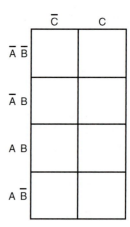

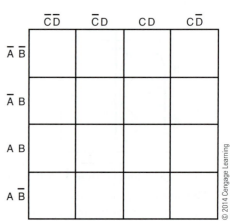

To use a Karnaugh map, follow these steps:

1. Draw the diagram based on the number of variables. As Figure 43-2 shows, a two-variable map requires $2^2 = 4$ squares, a three-variable map requires $2^3 = 8$ squares, and a four-variable map requires $2^4 = 16$ squares—the same requirements as a Veitch diagram.
2. Plot the logic functions by placing a "1" in each square representing a term.
3. Obtain the simplified logic function by looping adjacent groups of 1s in groups of eight, four, or two. Continue to loop until all 1s are included.
4. "OR" the loops with one term per loop. (Each expression is pulled off the Karnaugh map and "OR"ed using the "+" symbol, for example, ABC + BCD.)
5. Write the simplified expression.

EXAMPLE: Reduce $AB + \overline{A}B + A\overline{B} = Y$ to its simplest form.

Step 1. Draw the Karnaugh map. There are two variables, A and B, so use the two-variable map.

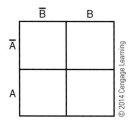

Step 2. Plot the logic function by placing a "1" in each square representing a term.

$\quad AB$ – first term
$\quad \overline{A}B$ – second term
$\quad A\overline{B}$ – third term

Plot 1st term AB.

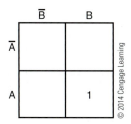

Plot 2nd term $\overline{A}B$.

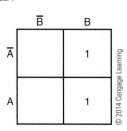

Plot 3rd term $A\overline{B}$.

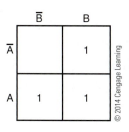

Step 3. Loop adjacent groups of 1s in the largest group possible. Start by analyzing the map for the largest groups possible. The largest group here is two.

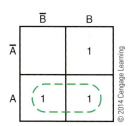

One possible group is the one indicated by the dotted line.

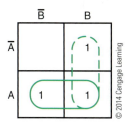

Another group is the one indicated by this dotted line.

Step 4. "OR" the groups: either A or B = A + B.

Step 5. The simplified expression for $AB + \overline{A}B + A\overline{B} = Y$ is $A + B = Y$ obtained from the Karnaugh map.

EXAMPLE: Find the simplified expression for

$$ABC + AB\overline{C} + A\overline{B}\,\overline{C} + \overline{A}\,\overline{B}\,\overline{C} = Y$$

Step 1. Draw the Karnaugh map. There are three variables, A, B, and C, so use the three-variable map.

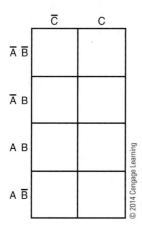

Step 2. Place a "1" in each square representing a term on the map.

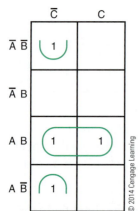

Step 3. Loop adjacent groups of 1s in the largest group possible.

Step 4. Write the term for each loop, one term per expression:

$$AB, \overline{B}\overline{C}.$$

Step 5. The simplified expression is: $AB + \overline{B}\overline{C} = Y$.

EXAMPLE: Find the simplified expression for $\overline{A}BC\overline{D} + \overline{A}B\overline{C}D + \overline{A}BCD + A\overline{B}\overline{C}\overline{D} + A\overline{B}\overline{C}D + \overline{A}\overline{B}CD = Y$

Step 1. Draw the Karnaugh map. There are four variables, A, B, C, and D, so use the four-variable map.

Step 2. Place a "1" in each square representing a term on the map.

Step 3. Loop adjacent groups of 1s in the largest group possible.

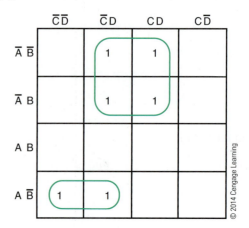

Step 4. Write the term for each loop, one term per expression: $\overline{A}D$, $A\overline{B}\,\overline{C}$.

Step 5. The simplified expression is: $\overline{A}D + A\overline{B}\,\overline{C} = Y$.

43-2 QUESTIONS

1. What is the function of a Karnaugh map?
2. What is the maximum number of variables that can be represented with a Karnaugh map?
3. List the steps for using a Karnaugh map.
4. Simplify the following three-variable expressions using Karnaugh maps:
   a. $A\overline{B}\,\overline{C} + \overline{A}\,\overline{B}C + A\overline{B}C + A\overline{B}\,\overline{C} + \overline{A}\overline{B}\,\overline{C} = Y$
   b. $\overline{A}\overline{B}C + A\overline{B}\,\overline{C} + A\overline{B}\,\overline{C} + ABC + AB + \overline{A}\overline{B}\,\overline{C} = Y$
5. Simplify the following four-variable expressions using Karnaugh maps:
   a. $ABCD + \overline{A}\overline{B}CD + \overline{A}BC\overline{D} + \overline{A}\overline{B}\overline{C}D + A\overline{B}\,\overline{C}\overline{D} + \overline{A}\overline{B}CD + AB\overline{C}\overline{D} = Y$
   b. $A\overline{B} + \overline{A}BD + \overline{B}\,\overline{C}D + \overline{B}\,\overline{C} + \overline{A}BCD = Y$

SUMMARY

- Veitch diagrams provide a fast and easy way to reduce complicated expressions to their simplest form.
- Veitch diagrams can be constructed from two, three, or four variables.
- The simplest logic expression is obtained from a Veitch diagram by looping groups of one, two, four, or eight X's and "OR"ing the looped terms.
- Like Veitch diagrams, Karnaugh maps provide a fast and easy method to reduce complex Boolean expressions to their simplest form.
- Karnaugh maps can be constructed for two, three, or four variables.
- The simplest logic expression is obtained from a Karnaugh map by looping groups of one, two, four, or eight 1s and "OR"ing the looped terms.

CHAPTER 43 SELF-TEST

1. Describe the procedure for using the Veitch diagram to simplify logic circuits.
2. Simplify the following Boolean expressions by using a Veitch diagram:
   a. $\overline{A}\overline{B}\,\overline{C} + A\overline{B}\,\overline{C} + AB\overline{C} + A\overline{B}C + \overline{A}\overline{B}C = Y$
   b. $\overline{A}\overline{B}\,\overline{C} + A\overline{B}C + A\overline{B}\,\overline{C} + A\overline{B}C + ABC + \overline{A}\overline{B}C = Y$
   c. $\overline{A}BC\overline{D} + \overline{A}\overline{B}CD + \overline{A}CD + A\overline{B}C + \overline{A}B + \overline{A}\overline{B}C\overline{D} = Y$
   d. $\overline{A}\overline{B}\overline{C}D + A\overline{B}\overline{C}\overline{D} + \overline{A}B\overline{C}\overline{D} + A\overline{B}\overline{C}D + \overline{A}\overline{B}CD + \overline{A}\overline{B}CD + ABCD = Y$
   e. $ABC\overline{D} + \overline{A}\overline{B}\overline{C}D + \overline{A}\overline{B}\overline{C}D + \overline{A}\overline{B}CD + \overline{A}\overline{B}CD + \overline{A}\overline{B}CD + AB\overline{C}\overline{D} + AB\overline{C}\overline{D} + \overline{A}\overline{B}C\overline{D} = Y$
3. Describe the procedure for using a Karnaugh map to simplify logic circuits.
4. Simplify the following Boolean expressions by using a Karnaugh map:
   a. $\overline{A}\overline{B}\,\overline{C} + A\overline{B}\,\overline{C} + AB\overline{C} + A\overline{B}C + \overline{A}\overline{B}C = Y$
   b. $\overline{A}\overline{B}\,\overline{C} + A\overline{B}C + A\overline{B}\,\overline{C} + A\overline{B}C + ABC + \overline{A}\overline{B}C = Y$
   c. $\overline{A}BC\overline{D} + \overline{A}\overline{B}CD + \overline{A}CD + A\overline{B}C + \overline{A}B + \overline{A}\overline{B}C\overline{D} = Y$
   d. $\overline{A}\overline{B}\overline{C}D + A\overline{B}\overline{C}\overline{D} + \overline{A}B\overline{C}\overline{D} + A\overline{B}\overline{C}D + \overline{A}\overline{B}CD + \overline{A}\overline{B}CD + ABCD = Y$
   e. $AB\overline{C}\overline{D} + \overline{A}\overline{B}\overline{C}D + \overline{A}\overline{B}\overline{C}D + \overline{A}\overline{B}\overline{C}D + \overline{A}\overline{B}CD + \overline{A}\overline{B}CD + AB\overline{C}D + AB\overline{C}D + \overline{A}\overline{B}C\overline{D} = Y$

CHAPTER 44

Sequential Logic Circuits

OBJECTIVES

After completing this chapter, the student will be able to:

- Describe the function of a flip-flop.
- Identify the basic types of flip-flops.
- Draw the symbols used to represent flip-flops.
- Describe how flip-flops are used in digital circuits.
- Describe how a counter and shift register operates.
- Identify the different types of counters and shift registers.
- Draw the symbols used to represent counters and shift registers.
- Identify applications of counters and shift registers.

KEY TERMS

Sequential logic circuits consist of circuits requiring timing and memory devices. The basic building block for sequential logic circuits is the flip-flop. Flip-flops can be wired together to form counters, shift registers, and memory devices.

The flip-flop belongs to a category of digital circuits called multivibrators. A multivibrator is a regenerative circuit with two active devices. It is designed so that one device conducts while the other device is cut off. Multivibrators can store binary numbers, count pulses, synchronize arithmetic operations, and perform other essential functions in digital systems.

There are three types of multivibrators: bistable, monostable, and astable. The bistable multivibrator is called a flip-flop.

44-1 FLIP-FLOPS

A flip-flop is a bistable multivibrator whose output is either a high or low voltage, a 1 or a 0. The output stays high or low until an input called a trigger is applied.

The basic flip-flop is the **RS flip-flop**. It is formed by two cross-coupled NOR or NAND gates (Figure 44-1). The RS flip-flop has two outputs, Q and $\overline{Q}$, and two controlling inputs, R (Reset) and S (Set). The outputs are always opposite or complementary: If Q = 1, then $\overline{Q}$ = 0, and vice-versa.

To understand the operation of the circuit, assume that the Q output, R input, and S input are all low. The low on the Q output is connected to one of the inputs of gate 2. The S input is low. The output of gate 2 is high. This high is coupled to the input of gate 1,

holding its output to a low. When the Q output is low, the flip-flop is said to be in the RESET state. It remains in this state indefinitely, until a high is applied to the S input of gate 2. When a high is applied to the S input of gate 2, the output of gate 2 becomes a low and is coupled to the input of gate 1. Because the R input of gate 1 is a low, the output changes to a high. The high is coupled back to the input of gate 2, ensuring that the Q output remains a low. When the Q output is high, the flip-flop is said to be in the SET state. It remains in the SET state until a high is applied to the R input, causing the flip-flop to RESET.

An "illegal" or "unallowed" condition occurs when a high is applied to both the R and S inputs simultaneously. In this case, the Q and $\overline{Q}$ outputs both try to go low, but Q and $\overline{Q}$ cannot be in the same state at the same time without violating the definition of flip-flop operation. When the highs on the R and S inputs are removed simultaneously, both of the outputs attempt to go high. Because there is always some difference in the gates, one gate dominates and becomes high. This forces the other gate to remain low. An unpredictable mode of operation exists and therefore the output state of the flip-flop cannot be determined.

Figure 44-2 shows the truth table for operation of an RS flip-flop. Figure 44-3 is a simplified symbol used to represent an RS flip-flop.

Another type of flip-flop is called a **clocked flip-flop**. It is different from the RS flip-flop in that an additional input is required for operation. The third input is called the clock or trigger. Figure 44-4 shows a logic diagram for a clocked flip-flop. A high at either input of the flip-flop portion activates the flip-flop, causing it to change states. The portion labeled "steering gate" steers or directs the clock pulses to either input gate.

■ **FIGURE 44-1**

Basic flip-flop circuit.

NI Multisim™

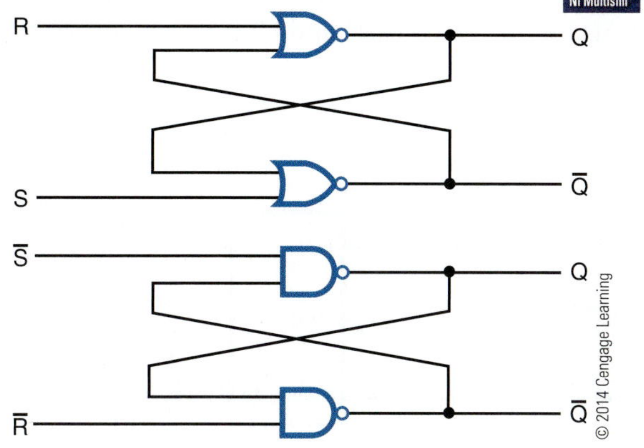

© 2014 Cengage Learning

■ **FIGURE 44-2**

Truth table for an RS flip-flop.

Inputs		Outputs	
S	R	Q	$\overline{Q}$
0	0	NC	NC
0	1	0	1
1	0	1	0
1	1	0	0

NC = No change

© 2014 Cengage Learning

■ **FIGURE 44-3**

Logic symbol for an RS flip-flop.

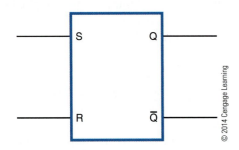

■ **FIGURE 44-5**

Logic symbol for a positive-edge-triggered RS flip-flop.

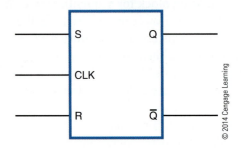

■ **FIGURE 44-4**

Logic circuit for a clocked RS flip-flop.

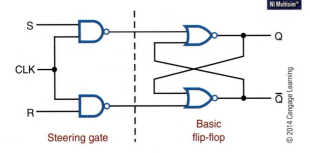

The clocked flip-flop is controlled by the logic state of the S and R inputs when a clock pulse is present. A change in the state of the flip-flop occurs only when the leading edge of the clock pulse is applied. The leading edge of the clock pulse is a positive-going transition (low to high). This means that the pulse goes from a zero voltage level to a positive voltage level. This is referred to as positive-edge-triggered (the edge of the pulse is what triggers the circuit).

As long as the clock input is low, the S and R inputs can be changed without affecting the state of the flip-flop. The only time that the effects of the S and R inputs are felt is when a clock pulse occurs. This is referred to as synchronous operation. The flip-flop operates in step with the clock. Synchronous operation

is important in computers and calculators when each step must be performed in an exact order. Figure 44-5 shows the logic symbol used to represent a positive-edge-triggered RS flip-flop.

A **D flip-flop** is useful where only 1 data bit (1 or 0) is to be stored. Figure 44-6 shows the logic diagram for a D flip-flop. It has a single data input and a clock input. The D flip-flop is also referred to as a delay flip-flop. The D input is delayed one clock pulse from getting to the output (Q). Sometimes the D flip-flop has a PS (preset) input and CLR (clear) input. The preset input sets output Q to a 1 when a low or 0 is applied to it. The clear input clears the Q output to a 0 when it is enabled by a low or 0. D flip-flops are wired together to form shift registers and storage registers. These registers are widely used in digital systems.

The **JK flip-flop** is the most widely used flip-flop. It has all the features of the other types of flip-flops. It is also edge triggered, accepting data only at the J and K inputs that are present at the active clock edge (high to low or low to high). This allows the accepting of input data on J and K at a precise instant. The logic diagram and symbol for the JK flip-flop is shown in Figure 44-7. J and K are the inputs. The significant feature of the JK flip-flop is that when both the J and K inputs are high, repeated clock pulses cause the output to toggle or change state. The two asynchronous inputs, PS (preset) and CLR (clear), override the

■ **FIGURE 44-6**

Logic circuit and symbol for the D flip-flop.

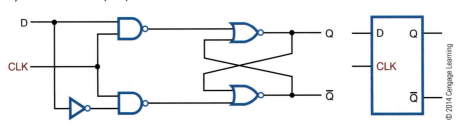

■ FIGURE 44-7

Logic circuit and symbol for the JK flip-flop.

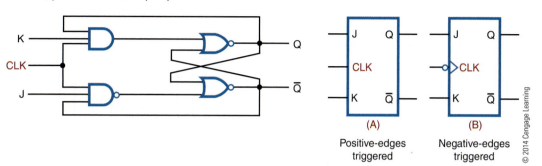

(A)
Positive-edges triggered

(B)
Negative-edges triggered

© 2014 Cengage Learning

synchronous inputs, the J and K data inputs, and the clock input. JK flip-flops are widely used in many digital circuits, especially counter circuits. Counters are found in almost every digital system.

A **latch** is a device that serves as a temporary buffer memory. It is used to hold data after the input signal is removed. The D flip-flop is a good example of a latch device. Other types of flip-flops can also be used.

A latch is used when inputting to a seven-segment display. Without a latch, the information being displayed is removed when the input signal is removed. With the latch, the information is displayed until it is updated.

Figure 44-8 shows a 4-bit latch. The unit has four D flip-flops enclosed in a single IC package. The E (enable) inputs are similar to the clock input of the D flip-flop. The data are latched when the enable line drops to a low, or 0. When the enable is high, or 1, the output follows the input. This means that the output will change to whatever state the input is in; for example, if the input is high, the output will become high; if the input is low, the output will become low. This condition is referred to as a **transparent latch**.

44-1 QUESTIONS

1. What is a flip-flop?
2. What are the different types of flip-flops?
3. What is meant by a clocked flip-flop?
4. What is the difference between an asynchronous input and a synchronous input?
5. What is a latch?

44-2 COUNTERS

A **counter** is a logic circuit that can count a sequence of numbers or states when activated by a clock input. The output of a counter indicates the binary number stored in the counter at any given time. The number of counts or states through which a counter progresses before returning to its original state (recycling) is called the **modulus** of the counter.

A flip-flop can act as a simple counter when connected as shown in Figure 44-9. Assuming that initially

■ FIGURE 44-8

Four-bit latch.

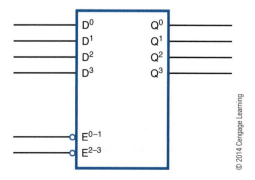

© 2014 Cengage Learning

■ FIGURE 44-9

JK flip-flop set up for counting.

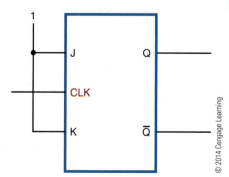

© 2014 Cengage Learning

■ FIGURE 44-10

Input and output waveforms of JK flip-flop set up as a counter.

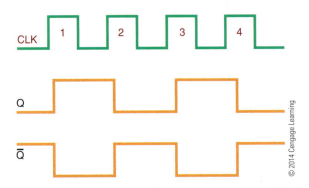

© 2014 Cengage Learning

■ FIGURE 44-11

Two-stage counter.

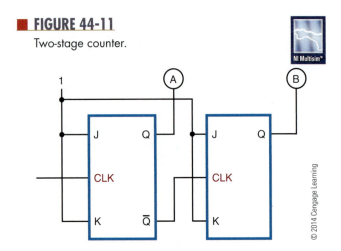

© 2014 Cengage Learning

the flip-flop is reset, the first clock pulse causes it to set (Q = 1). The second clock pulse causes it to reset (Q = 0). Because the flip-flop has set and reset, two clock pulses have occurred.

Figure 44-10 shows the output waveform of the flip-flop. Notice that the Q output is high (1) after every odd clock pulse and low (0) after every even clock pulse. Therefore, when the output is high, an odd number of clock pulses have occurred. When the output is low, either no clock pulses or an even number of clock pulses has occurred. In this case, it is not known which has occurred.

A single flip-flop produces a limited counting sequence, 0 or 1. To increase the counting capacity, additional flip-flops are needed. The maximum number of binary states in which a counter can exist depends on the number of flip-flops in the counter. This can be expressed as

$$N = 2^n$$

where: N = maximum number of counter states
n = number of flip-flops in counter

Binary counters fall into two categories based on how the clock pulses are used to sequence the counter. The two categories are **asynchronous** and **synchronous**.

Asynchronous means "not occurring at the same time." With respect to counter operations, asynchronous means that the flip-flops do not change states at the same time. This is because the clock pulse is not connected to the clock input of each stage. Figure 44-11 shows a two-stage counter connected for asynchronous operation. Each flip-flop in a counter is referred to as a **stage**.

Notice that the $\overline{Q}$ output of the first stage is coupled to the clock input of the second stage. The second stage only changes state when triggered by the transition of the output of the first stage. Because of the delay through the flip-flop, the second flip-flop does not change state at the same time the clock pulse is applied. Therefore, the two flip-flops are never simultaneously triggered, which results in asynchronous operation.

Asynchronous counters are commonly referred to as **ripple counters**. The input clock pulse is first felt by the first flip-flop. The effect is not felt by the second flip-flop immediately because of the delay through the first flip-flop. In a multistage counter, the delay is felt through each flip-flop, so the effect of the input clock pulse "ripples" through the counter. Figure 44-12 shows a three-stage binary counter and timing chart for each of the stages. A truth table shows the counting sequence.

Synchronous means "occurring at the same time." A **synchronous counter** is a counter in which each stage is clocked at the same time. This is accomplished by connecting the clock input to each stage of the counter (Figure 44-13). A synchronous counter is also called a **parallel counter** because the clock input is connected in parallel to each flip-flop.

A synchronous counter operates as follows. Initially the counter is reset with both flip-flops in the 0 state. When the first clock pulse is applied, the first flip-flop toggles and the output goes high. The second flip-flop does not toggle because of the delay from the input to the actual changing of the output state. Therefore, there is no change in the second flip-flop output state. When the second clock pulse is applied, the first flip-flop toggles and the output goes low. Because there is a high from the output of the first stage, the second

■ FIGURE 44-12

Three-stage binary counter.

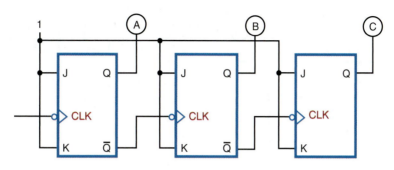

Number of Clock Pulses	Binary Count Sequence			Decimal Count
	C	**B**	**A**	
0	0	0	0	0
1	0	0	1	1
2	0	1	0	2
3	0	1	1	3
4	1	0	0	4
5	1	0	1	5
6	1	1	0	6
7	1	1	1	7
8	0	0	0	0

Count sequence

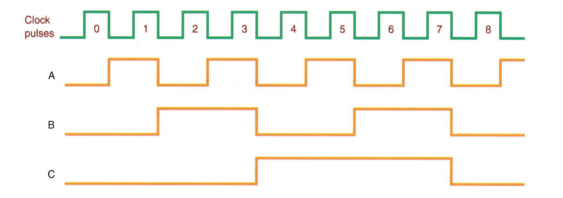

stage toggles and its output goes high. After four clock pulses, the counter recycles to its original state. Figure 44-14 shows a timing chart for this sequence of events with a two-stage synchronous counter.

Figure 44-15 shows a three-stage binary counter and timing chart. Figure 44-16 shows a four-stage synchronous counter and logic symbol.

One application of a counter is frequency division. A single flip-flop produces an output pulse for every two input pulses. Therefore, it is essentially a divide-by-two device with an output one-half the frequency of the input. A two-stage binary counter is a divide-by-four device with an output equal to one-fourth the input clock frequency. A four-stage binary

FIGURE 44-13

Two-stage synchronous counter.

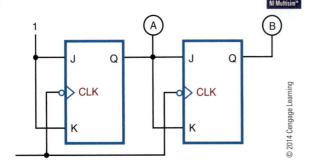

FIGURE 44-14

Input and output waveforms for the two-stage synchronous counter.

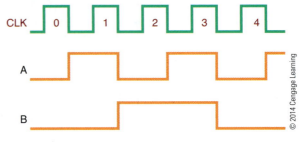

FIGURE 44-15

Three-stage binary counter and timing chart.

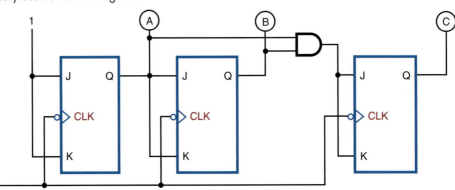

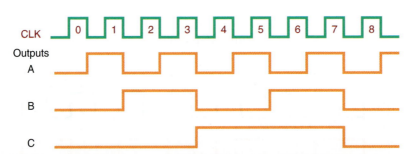

Number of Clock Pulses	Binary Count Sequence			Decimal Count
	C	B	A	
0	0	0	0	0
1	0	0	1	1
2	0	1	0	2
3	0	1	1	3
4	1	0	0	4
5	1	0	1	5
6	1	1	0	6
7	1	1	1	7
8	0	0	0	0

■ FIGURE 44-16

Logic symbol for a four-stage synchronous counter.

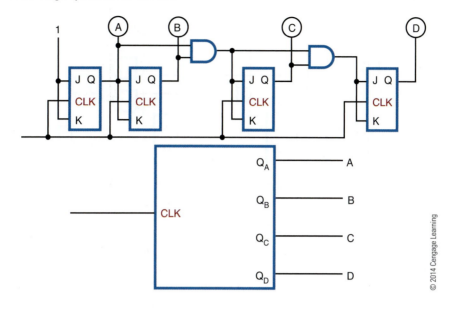

© 2014 Cengage Learning

■ FIGURE 44-17

A counter as a frequency divider.

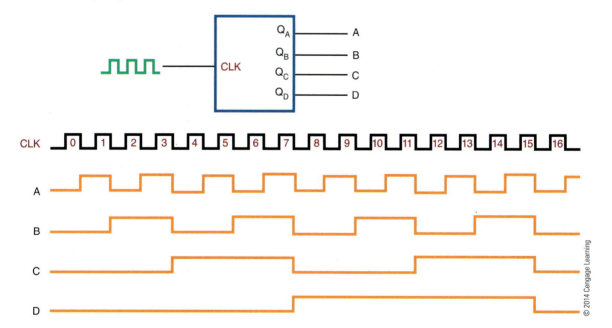

© 2014 Cengage Learning

counter is a divide-by-16 device with the output equal to one-sixteenth the input clock frequency (Figure 44-17).

A binary counter with n stages divides the clock frequency by a factor of 2n. A three-stage counter divides the frequency by 8 (2^3), a four-stage counter by 16 (2^4), a five-stage counter by 32 (2^5), and so on. Notice that the modulus of the counter is the same as the division factor.

Decade counters have a modulus of 10, or 10 states in their counting sequence. A common decade counter is the BCD (8421) counter, which produces a

■ **FIGURE 44-18**

Synchronous BCD decade counter.

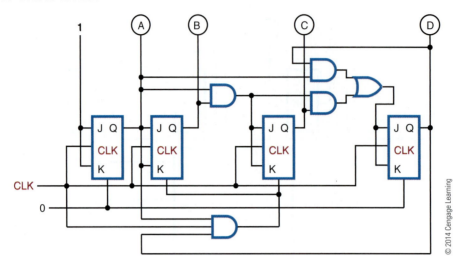

■ **FIGURE 44-19**

Logic symbol for a decade counter.

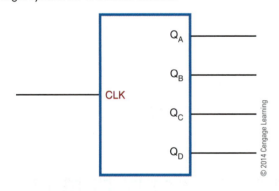

■ **FIGURE 44-20**

Logic symbol for an up–down counter.

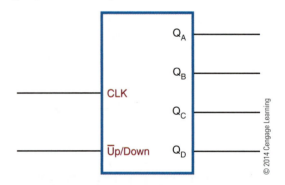

binary-coded-decimal sequence (Figure 44-18). The AND and OR gates detect the occurrence of the ninth state and cause the counter to recycle on the next clock pulse. The symbol for a decade counter is shown in Figure 44-19.

An **up–down counter** can count in either direction through a certain sequence. It is also referred to as a bidirectional counter. The counter can be reversed at any point in the counting sequence. Its symbol is shown in Figure 44-20. An up–down counter can consist of any number of stages. Figure 44-21 shows the logic diagram for a BCD up–down counter. The inputs to the JK flip-flops are enabled by the up–down input qualifying the up or down set of the AND gates.

Counters can be stopped after any sequence of counting by using a logic gate or combination of logic gates. The output of the gate is fed back to the input of

the first flip-flop in a ripple counter. If a 0 is fed back to the JK input of the first flip-flop (Figure 44-22), it prevents the first flip-flop from toggling, thereby stopping the count.

44–2 QUESTIONS

1. What function does a counter serve?
2. How many counting sequences are available with an eight-stage counter?
3. How does an asynchronous counter operate?
4. How does a synchronous counter differ from an asynchronous counter?
5. How can a counter be made to stop at any count desired?

FIGURE 44-21

Logic diagram for a BCD up–down counter.

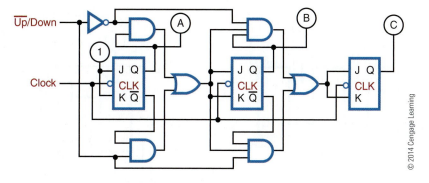

FIGURE 44-22

A low applied to the JK input of the first flip-flop prevents it from toggling, stopping the count.

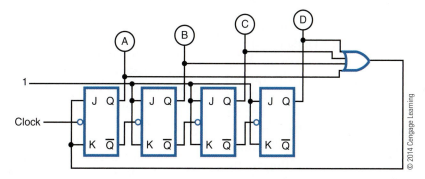

44-3 SHIFT REGISTERS

A **shift register** is a sequential logic circuit widely used to store data temporarily. Data can be loaded into and removed from a shift register in either a parallel or serial format. Figure 44-23 shows the four different methods of loading and reading data in a shift register. Because of its ability to move data 1 bit at a time from one storage medium to another, the shift

register is valuable in performing a variety of logic operations.

Shift registers are constructed of flip-flops wired together. Flip-flops have all the functions necessary for a register: They can be reset, preset, toggled, or steered to a 1 or 0 level. Figure 44-24 shows a basic shift register constructed from four flip-flops. It is called a 4-bit shift register because it consists of four binary storage elements. A significant feature of the shift register is that it can move data to the right or left a number of bit positions. This is equivalent to multiplying or dividing a number by a specific factor. The data are shifted

FIGURE 44-23

Methods of loading and reading data in a shift register.

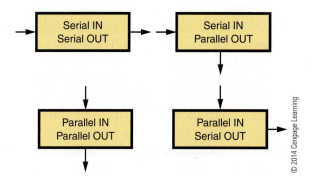

FIGURE 44-24

Shift register constructed from four flip-flops.

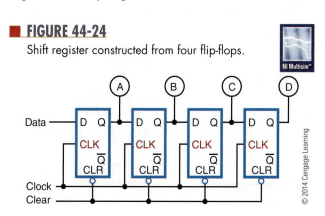

■ **FIGURE 44-25**

A typical shift register constructed of JK flip-flops.

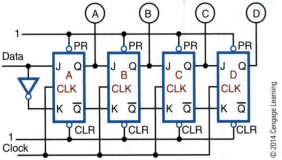

© 2014 Cengage Learning

1 bit position at a time for each input clock pulse. The clock pulses have full control over the shift register operations.

Figure 44-25 shows a typical 4-bit shift register constructed of JK flip-flops. The serial data and their complements are applied to the JK inputs of the A flip-flop. The other flip-flops are cascaded, with the outputs of one connected to the inputs of the next. The toggles of all the flip-flops are connected together, and the clock pulse is applied to this line. Because all the flip-flops are toggled together, this is a synchronous circuit. Also, the clear inputs of each flip-flop are tied together to form a reset line. Data applied to the input are shifted through the flip-flops 1 bit position for each clock pulse. For example, if the binary number 1011 is applied to the input of the shift register and a shift pulse is applied, the number stored in the shift register is shifted out and lost while the external number is shifted in. Figure 44-26 shows the sequence of events for storing a number in the shift register.

One of the most common applications of a shift register is serial-to-parallel or parallel-to-serial data

■ **FIGURE 44-26**

Storing a number in a shift register.

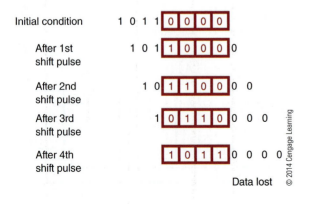

© 2014 Cengage Learning

■ **FIGURE 44-27**

Loading a shift register using parallel input.

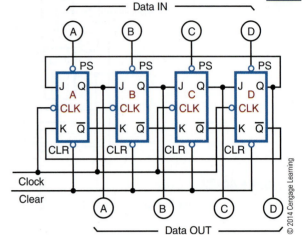

© 2014 Cengage Learning

conversion. Figure 44-27 shows how a shift register can be loaded by a parallel input. For parallel operations, the input data are preset into the shift register. Once the data are in the shift register, they can be shifted out serially, as discussed earlier.

For serial-to-parallel data conversion, the data are initially shifted into the shift register with clock pulses. Once the data are in the shift register, the outputs of the individual flip-flops are monitored simultaneously, and the data are routed to their destination.

Shift registers can perform arithmetic operations such as multiplication or division. Shifting a binary number stored in the shift register to the right has the same effect as dividing the number by some power of 2. Shifting the binary number stored in the shift register to the left has the same effect as multiplying the number by some power of 2. Shift registers are a simple and inexpensive means of performing multiplication and division of numbers.

Shift registers are often used for temporary storage. They are capable of storing one or more binary words. There are three requirements for this application of a shift register: First, it must be able to accept and store data. Second, it must be able to retrieve or read out the data on command. Third, when the data are read, they must not be lost. Figure 44-28 shows the external circuitry required to enable a shift register to read and maintain the data stored in it. The read/write line, when high, allows new data to be stored in the shift register. Once the data are stored, the read/write line goes low, enabling gate 2, which allows the data to recirculate while reading out the data.

■ FIGURE 44-28

Shift register circuitry for maintaining and reading data.

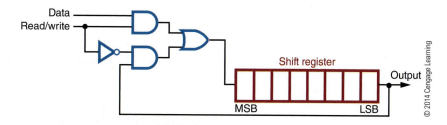

44-3 QUESTIONS

1. What is the function of a shift register?
2. What is a significant feature of a shift register?
3. From what are shift registers constructed?
4. What is a common application of a shift register?
5. What arithmetic operations can a shift register perform, and how does it perform them?

44-4 MEMORY

Memory serves to store digital data on a temporary or permanent basis. Because of the different ways that the stored data are used in digital systems, many different types of memories have evolved, each designed for a particular application.

Memory is built from storage registers. Storage registers store the digital data, such as programs or data, on a short term in digital systems. The flip-flop is the basic building block for storage registers.

A 4-bit register stores 4 bits, referred to as a **nibble**; an 8-bit register stores 8 bits, called a **byte**; and a larger register stores a **word**. Typical word sizes are 16, 32, 64, and 128 bits. Registers with 4, 8, and 16 bits are shown in Figure 44-29. By looking at the Q output, the status of the data stored can be checked in any of these registers. The complementary data are also available by looking at the $\overline{Q}$ output.

Figure 44-30 shows how data are stored in a larger memory configuration. A matrix of storage elements capable of storing a 1 or a 0 is organized into words that a decoder can locate by a specific address.

■ FIGURE 44-29

Four-, 8-, and 16-bit registers.

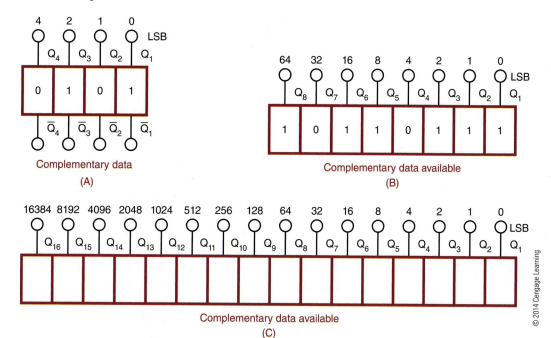

(A)

(B)

(C)

FIGURE 44-30

Large memory array (8 bit).

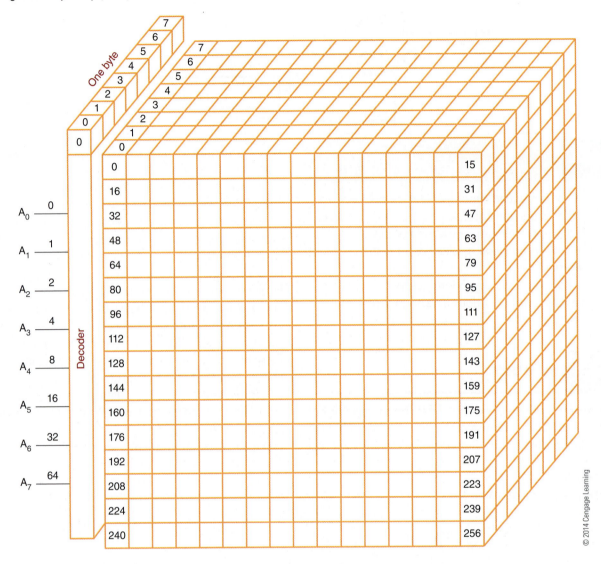

The technology used in making the memory integrated circuits (chips) is based on either bipolar (also referred to as transistor-to-transistor logic [TTL]) or metal-oxide semiconductor (MOS) transistors. Bipolar memory is faster than MOS memory, but MOS memory can be packed with a higher density, providing more memory locations in the same amount of area required by bipolar memory.

Data are stored digitally in two types of memory: random-access memory and read-only memory. **Random-access memory (RAM)** is used for the temporary storage of programs, data, control information, and so on. It provides random access to the stored data, with all storage locations being reached in the same amount of time. RAM has the capability to both read and write data. It is also volatile. This means that the stored data will be lost if the power is shut off.

Two types of RAM are available: static and dynamic. **Static RAM (SRAM)** uses a bistable circuit as the storage element. It loses data when the power is removed. **Dynamic RAM (DRAM)** is different because it uses capacitors to store the data. The drawback is that the capacitors discharge over time and must be recharged or refreshed periodically to maintain the stored data. Even with the additional support circuit requirement, SRAMs are still cheaper and have a higher density, reducing the circuit board area required.

Read-only memory (ROM) allows data to be permanently stored and allows the data to be read

■ FIGURE 44-31

Rom family chart.

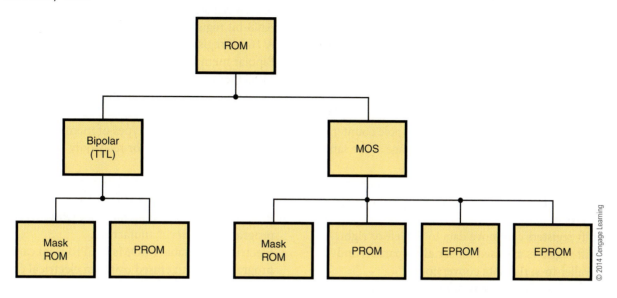

from memory at any time, with no provision for data to be written to it. ROMs are considered nonvolatile memory. ROMs use the same decoder circuitry as RAM. The stored data are retained even when the power is removed. ROMs may be programmed during manufacture to be either permanent or replaceable by the user, but they are not easily altered and require the use of a special device called a *burner*.

Figure 44-31 shows the ROM family chart. The mask-programming process takes place during the manufacturing process. The transistor-emitter connection is either made or not made through a mask during manufacture. The mask does the programming.

Programmable read-only memory (PROM) can be programmed after manufacture. The program is entered into the PROM by blowing a fusible link. Initially, all emitter links are connected as 1s in all storage locations. The emitter links where 0s occur are blown open with a high-current surge. PROMs cannot be reprogrammed.

Electrically erasable PROMs (EEPROMs) can be programmed and erased by using an electrical charge. By applying a high voltage (around 21 volts), a single byte or the entire chip can be erased in about 10 milliseconds. Erasing can be done with the chip in the circuit. EEPROMs are used where data needs to be maintained for later use, such as the custom guide in a television satellite system.

44–4 QUESTIONS

1. What is memory?
2. What are the two types of technology used to make memory chips?
3. What are the two types of memory devices?
4. What is meant by the term *volatile*?
5. Describe a situation in which it would be advantageous to convert from EPROM memory design to the mask ROM.

SUMMARY

- A flip-flop is a bistable multivibrator whose output is either a high or a low.
- Types of flip-flops include these listed here:
  - RS flip-flop
  - Clocked RS flip-flop
  - D flip-flop
  - JK flip-flop
- Flip-flops are used in digital circuits such as counters.
- A latch is a temporary buffer memory.

- A counter is a logic circuit that can count a sequence of numbers or states.
- A single flip-flop produces a count sequence of 0 or 1.
- The maximum number of binary states a counter can have depends on the number of flip-flops contained in the counter.
- Counters can be either asynchronous or synchronous.
- Asynchronous counters are called ripple counters.
- Synchronous counters clock each stage at the same time.
- Shift registers are used to store data temporarily.
- Shift registers are constructed of flip-flops wired together.
- Shift registers can move data to the left or right.
- Shift registers are used for serial-to-parallel and parallel-to-serial data conversions.
- Shift registers can perform multiplication and division.

- Memory stores data temporarily or permanently.
- Memory is built from shift registers.
- A 4-bit register stores a nibble.
- An 8-bit register stores a byte.
- A 16-bit register stores a word.
- Bipolar memory is faster than MOS memory.
- MOS memory can pack a higher density, providing more memory locations in the same area for bipolar memory.
- Data is stored into two types of memory: random-access memory (RAM) or read-only memory (ROM).
- Two types of RAM are available: static RAM (SRAM) or dynamic RAM (DRAM).
- Programmable read-only memory (PROM) can be programmed after manufacture.
- Electrically erasable PROMs (EEPROMs) can be programmed and erased using an electrical charge.

CHAPTER 44 SELF-TEST

1. Describe how an RS flip-flop changes states from a high on the Q output to a high on the $\overline{Q}$ output.
2. What is the major difference between the D flip-flop and the clocked RS flip-flop?
3. What components make up a counter, and how is it constructed?
4. Draw a schematic for a counter that will count to 10 and then repeat.
5. How does a shift register differ from a counter?
6. For what functions or applications can the shift register be used?
7. Define bit, nibble, byte, and word.
8. What is meant by complementary data?
9. What is the advantage of MOS memory or bipolar memory?
10. Describe the difference between static and dynamic RAM.
11. What is the difference between RAM and ROM?
12. Which type of ROM can be reprogrammed?

Combinational Logic Circuits

Combinational logic circuits are circuits that combine the basic components of AND gates, OR gates, and inverters to produce more sophisticated circuits. The output of a combinational logic circuit is a function of the states of the inputs, the types of gates used, and the interconnection of the gates. The most common of the combinational circuits are decoders, encoders, multiplexers, and arithmetic circuits.

■ FIGURE 45-1

Decimal-to-binary encoder.

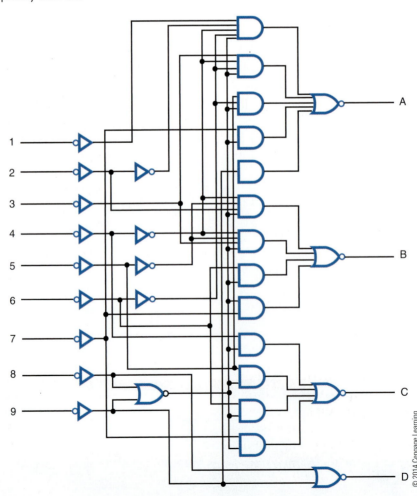

■ FIGURE 45-2

Decimal-to-binary priority encoder.

45-1   ENCODERS

An **encoder** is a combinational logic circuit that accepts one or more inputs and generates a multibit binary output. Encoding is the process of converting any keyboard character or number as input to a coded output such as a binary or BCD form.

Figure 45-1 shows a **decimal-to-binary encoder**. Its function is to take a single digit (0 to 9) as input and to output a 4-bit code representation of the digit. This is referred to as a 10-line-to-4-line encoder. That is, if the digit 4 on the keyboard is entered, this produces a low output, or a 0, on line 4, which produces the 4-bit code 0100 as an output.

Figure 45-2 shows a **decimal-to-binary priority encoder**. The priority function means that if two keys are pressed simultaneously, the encoder produces a BCD output corresponding to the highest-order decimal digit appearing on the input. For example, if both

■ FIGURE 45-3

Logic symbol for a decimal-to-binary priority encoder.

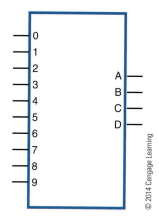

■ FIGURE 45-4

Binary-to-decimal decoder.

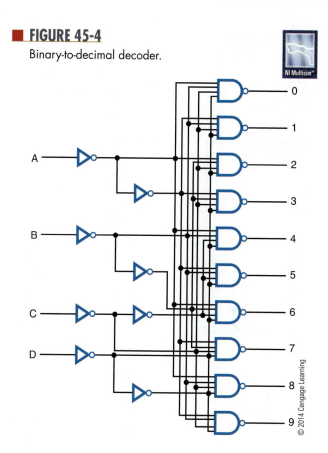

a 5 and a 2 are applied to the encoder, the BCD output is 1010, or the inverse of decimal 5. This type of encoder is built into a single integrated circuit and consists of approximately 30 logic gates. Figure 45-3 shows the symbol for a priority encoder.

This type of encoder is used to translate the decimal input from a keyboard to an 8421 BCD code. The decimal-to-binary encoder and the decimal-to-binary priority encoder are found wherever there is keyboard input, including in calculators, cell phones, computer keyboard inputs, electronic typewriters, telephones, and teletypewriters (TTY).

45-1   QUESTIONS

1. What is encoding?
2. What does an encoder accomplish?
3. What is the difference between a normal encoder and a priority encoder?
4. Draw the logic symbol for a decimal-to-binary priority encoder.

45-2   DECODERS

A **decoder** is one of the most frequently used combinational logic circuits. It processes a complex binary code into a recognizable digit or character. For example, it might decode a BCD number into one of the 10 possible decimal digits. The output of such a decoder is used to operate a decimal number readout or display. This type of decoder is called a 1-of-10 decoder or a 4-line-to-10-line decoder.

Figure 45-4 shows the 10 NAND gates required for decoding a 4-bit BCD number to its approximate output (one decimal digit). When all the inputs to a NAND gate are 1, its output is 0. All other outputs from the NAND gates in the decoder are 1s. Rather than draw all the logic gates each time the circuit is used, the symbol shown in Figure 45-5 is used.

■ FIGURE 45-5

Logic symbol for a binary-to-decimal decoder.

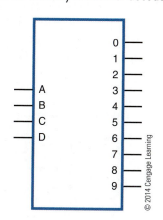

■ **FIGURE 45-6**

Logic symbols for a 1-of-8 decoder (A) and a 1-of-16 decoder (B).

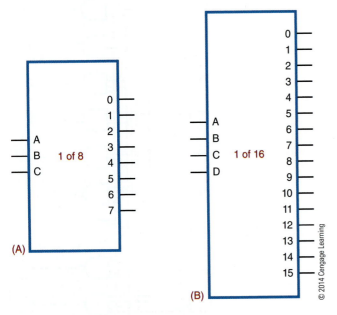

(A)

(B)

© 2014 Cengage Learning

■ **FIGURE 45-7**

Seven-segment display configuration.

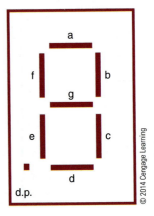

© 2014 Cengage Learning

Two other types of decoder circuits are the 1-of-8 (octal, or base-8) decoder and the 1-of-16 (hexadecimal, or base-16) decoder (Figure 45-6). The 1-of-8 decoder accepts a 3-bit input word and decodes it to one of eight possible outputs. The 1-of-16 decoder activates 1 of 16 output lines by a 4-bit code word. It is also called a 4-line-to-16-line decoder.

A special type of decoder is the standard 8421 BCD-to-seven-segment decoder. It accepts a BCD input code and generates a special 7-bit output code to energize a seven-segment decimal readout display (Figure 45-7). The display consists of seven LED segments that are lit in different combinations to produce each of the 10 decimal digits, 0 through 9 (Figure 45-8). Besides seven-segment LED displays, there are incandescent and liquid crystal (LCD) displays.

Each of these displays operates on the same principle. A segment is activated by either a high or low voltage level. Figure 45-9 shows two types of LED

displays: a common anode and a common cathode. In each case, the LED segment has to be forward biased for light to be emitted. For a common cathode, a high (1) lights up the segment; a low (0) does not.

Figure 45-10 shows the decoding logic circuit required to produce the output for a seven-segment display with a BCD input. Referring to Figure 45-7, notice that segment a is activated for digits 0, 2, 3, 5, 7, 8, and 9; segment b is activated for digits 0, 1, 2, 3, 4, 7, 8, and 9; and so on. Boolean expressions can be formed to determine the logic circuitry needed to drive each segment of the display. The logic symbol for a BCD-to-seven-segment decoder is shown in Figure 45-11. This represents the circuit contained in an integrated circuit.

45-2 QUESTIONS

1. What is a decoder?
2. What are the uses of decoders?
3. Draw the logic symbol for a 1-of-10 decoder.
4. What is the purpose of a seven-segment decoder?
5. What codes can be used in decoders?

■ **FIGURE 45-8**

Using the seven-segment display to form the 10 decimal digits.

© 2014 Cengage Learning

■ FIGURE 45-9

Differences between the two types of LED displays: a common cathode (A), and a common anode (B).

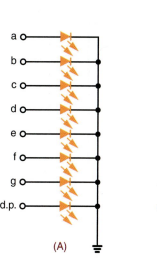
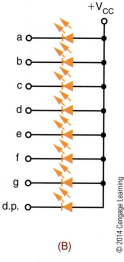

(A)

(B)

© 2014 Cengage Learning

■ FIGURE 45-11

Logic symbol for a BCD-to-seven-segment decoder.

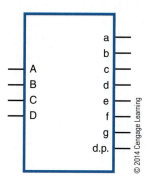

© 2014 Cengage Learning

■ FIGURE 45-12

A single-pole, multiposition switch can be used as a multiplexer.

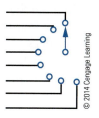

© 2014 Cengage Learning

■ FIGURE 45-10

Binary-to-seven-segment display decoder.

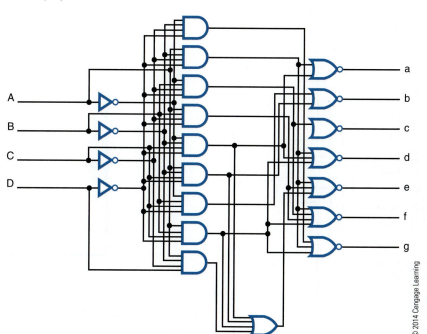

© 2014 Cengage Learning

45–3 MULTIPLEXERS

A **multiplexer** is a circuit used to select and route any one of several input signals to a single output. An example of a nonelectronic circuit multiplexer is a single-pole, multiposition switch (Figure 45-12).

Multiposition switches are widely used in many electronic circuits. However, circuits that operate at high speed require the multiplexer to switch at high speed and to be automatically selected. A mechanical switch cannot perform this task satisfactorily. Therefore, multiplexers used to perform high-speed switching are constructed of electronic components.

Multiplexers handle two basic types of data: analog and digital. For analog applications, multiplexers are built of relays and transistor switches. For digital applications, multiplexers are built from standard logic gates.

Digital multiplexers allow digital data from several individual sources to be routed through a common line for transmission to a common destination. A basic multiplexer has several input lines with a single output line. The input lines are activated by data selection input that identifies the line the data are to be received on. Figure 45-13 shows the logic circuit for an eight-input multiplexer. Notice that there are three input-control lines, labeled A, B, and C. Any of the eight input lines can be selected by the proper expression of the input-control line. The symbol used to represent a digital multiplexer is shown in Figure 45-14.

Figure 45-15 shows the symbol for a 1-of-16 multiplexer. Notice that there are four input-control lines to activate the 16 data input lines.

In addition to data line selection, a common application of a multiplexer is parallel-to-serial data

■ FIGURE 45-14

Logic symbol for an eight-input multiplexer.

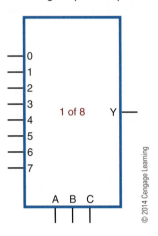

■ FIGURE 45-15

Logic symbol for a 16-input multiplexer.

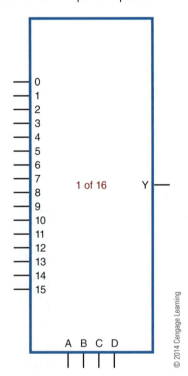

■ FIGURE 45-13

Logic circuit for an eight-input multiplexer.

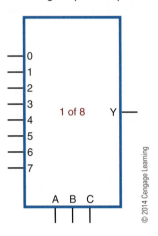

conversion. A parallel binary word is applied to the input of the multiplexer. Then, by sequencing through the enabling codes, the output becomes a serial representation of the parallel input word.

Figure 45-16 shows a multiplexer set up for parallel-to-serial conversion. A 3-bit binary input word from a counter is used to select the desired input. The parallel input word is connected to each of the input

■ **FIGURE 45-16**

Using a multiplexer for parallel-to-serial conversion.

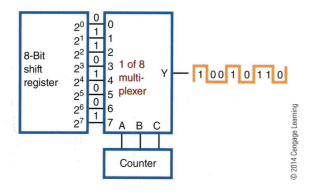

© 2014 Cengage Learning

lines of the multiplexer. As the counter is incremented, the input select code is sequenced through each of its states. The output of the multiplexer is equal to the parallel signal applied.

45-3 QUESTIONS

1. What is a multiplexer?
2. How are multiplexers used?
3. Draw a logic diagram for a multiplexer.
4. What types of data can a multiplexer handle?
5. How can a multiplexer be set up for parallel-to-serial conversion?

45-4 ARITHMETIC CIRCUITS ADDER

An **adder** is the primary computation unit in a digital computer. Few routines are performed by a computer in which the adder is not used. Adders are designed to work in either serial or parallel circuits. Because the parallel adder is faster and used more often, it is covered in more detail here.

To understand how an adder works, it is necessary to review the rules for adding:

$$
\begin{array}{cccc}
0 & 0 & 1 & 1 \\
+\,0 & +\,1 & +\,0 & +\,1 \\
\hline
0 & 1 & 1 & \text{Carry 1}\ \ 0
\end{array}
$$

Figure 45-17 shows a truth table based on these rules. Note that the Greek letter sigma (Σ) is used to represent the sum column. The carry column is

■ **FIGURE 45-17**

Truth table constructed using addition rules.

Inputs		Outputs	
B	A	Σ	C_0
0	0	0	0
0	1	1	0
1	0	1	0
1	1	0	1

© 2014 Cengage Learning

represented by C_0. These terms are used by industry when referring to an adder.

The sum column of the truth table is the same as the output column in the truth table of an exclusive OR gate (Figure 45-18). The carry column is the same as the output in a truth table for an AND gate (Figure 45-19).

Figure 45-20 shows an AND gate and an exclusive OR gate connected in parallel to provide the necessary logic function for single-bit addition. The carry output (C_0) is produced with the AND gate, and the sum

■ **FIGURE 45-18**

Truth table for an exclusive OR gate.

B	A	Y
0	0	0
0	1	1
1	0	1
1	1	0

© 2014 Cengage Learning

■ **FIGURE 45-19**

Truth table for an AND gate.

B	A	Y
0	0	0
0	1	0
1	0	0
1	1	1

© 2014 Cengage Learning

■ **FIGURE 45-20**
Half-adder circuit.

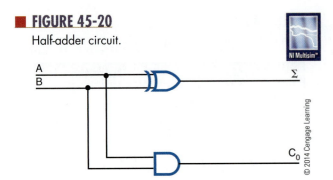

output (Σ) is produced with the XOR gate. The inputs A and B are connected to both the AND gate and the XOR gate. The truth table for the circuit is the same as the truth table developed using the binary addition rules (Figure 45-17). Because the circuit does not take into account any carries, it is referred to as a **half adder**. It can be used as the LSB adder for a binary addition problem.

An adder that takes into account the carry is called a **full adder**. A full adder accepts three inputs and generates a sum and carry output. Figure 45-21 shows the truth table for a full adder. The C_1 input represents the carry input. The C_0 output represents the carry output.

Figure 45-22 shows a full adder constructed of two half adders. The results of the first half adder are "OR"ed with the second half adder to form the carry output. The carry output is a 1 if both inputs to the first XOR gate are 1s or if both inputs to the second XOR gate are 1s. Figure 45-23 shows the symbols used to represent a half adder and a full adder.

■ **FIGURE 45-21**
Truth table for a full adder.

Inputs			Outputs	
C_{IN}	B	A	Σ	C_0
0	0	0	0	0
0	0	1	1	0
0	1	0	1	0
0	1	1	0	1
1	0	0	1	0
1	0	1	0	1
1	1	0	0	1
1	1	1	1	1

© 2014 Cengage Learning

■ **FIGURE 45-22**
Logic circuit for a full adder using two half adders.

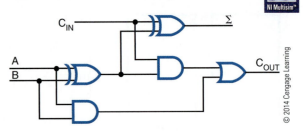

■ **FIGURE 45-23**
Logic symbols for half adder (A) and full adder (B).

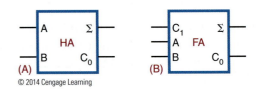

© 2014 Cengage Learning

A single full adder is capable of adding two single-bit numbers and an input carry. To add binary numbers with more than 1 bit, additional full adders must be used. Remember, when one binary number is added to another, each column that is added generates a sum and a carry of 1 or 0 to the next higher-order column. To add two binary numbers, a full adder is required for each column. For example, to add a 2-bit number to another 2-bit number, two adders are required. Two 3-bit numbers require three adders, two 4-bit numbers require four adders, and so on. The carry generated by each adder is applied to the input of the next higher-order adder. Because no carry input is required for the least significant position, a half adder is used.

Figure 45-24 shows a 4-bit parallel adder. The least significant input bits are represented by A_0 and B_0. The next higher-order bits are represented by A_1 and B_1

■ **FIGURE 45-24**
Four-bit parallel adder.

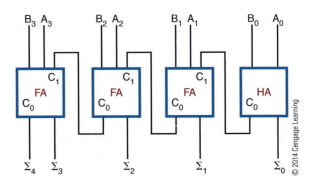

■ FIGURE 45-25

Truth table constructed using subtraction rules.

Inputs		Outputs	
B	**A**	**D**	B_0
0	0	0	0
0	1	1	0
1	0	1	1
1	1	0	0

© 2014 Cengage Learning

■ FIGURE 45-26

Logic circuit for a half subtractor.

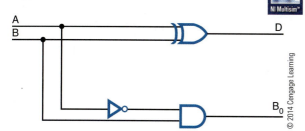

© 2014 Cengage Learning

and so on. The output sum bits are identified as Σ_0, Σ_1, Σ_2, and so on. Note that the carry output of each adder is connected to the carry input of the next higher-order adder. The carry output of the final adder is the most significant bit of the answer.

Subtractor

A **subtractor** allows subtraction of two binary numbers. To understand how a subtractor works, it is necessary to review the rules for subtraction.

$$\begin{array}{ccccc} 0 & \text{Borrow 1} & 0 & 1 & 1 \\ -\underline{0} & & -\underline{1} & -\underline{0} & -\underline{1} \\ 0 & & 0 & 0 & 0 \end{array}$$

Figure 45-25 shows a truth table based on these rules. The letter D represents the difference column. The borrow column is represented by B_0. Notice that the difference output (D) is 1 only when the input variables are not equal. Therefore, the difference can be expressed as the exclusive OR of the input variables. The output borrow is generated only when A is 0 and B is 1. Therefore, the borrow output is the complement of A "AND"ed with B.

Figure 45-26 shows a logic diagram of a **half subtractor**. It has two inputs and generates a difference and a borrow output. The difference is generated by an XOR gate and the borrow output is generated by an AND gate with A and B inputs. The A is achieved by using an inverter on the variable A input.

However, a half subtractor is identified as such because it does not have a borrow input. A **full subtractor** does. It has three inputs and generates a difference and a borrow output. A logic diagram and truth table for a full subtractor are shown in Figure 45-27.

■ FIGURE 45-27

Logic circuit (A) and truth table (B) for a full subtractor.

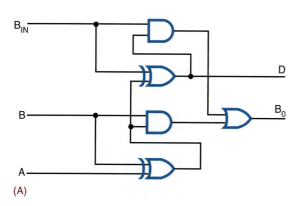

(A)

Input			Output	
B_{IN}	**B**	**A**	**D**	B_0
0	0	0	0	0
0	0	1	1	0
0	1	0	1	1
0	1	1	0	0
1	0	0	1	1
1	0	1	0	0
1	1	0	0	1
1	1	1	1	1

(B)

© 2014 Cengage Learning

■ **FIGURE 45-28**

Logic symbols for half subtractor (A) and full subtractor (B).

© 2014 Cengage Learning

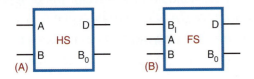

■ **FIGURE 45-29**

Four-bit subtractor.

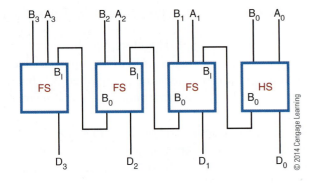

© 2014 Cengage Learning

■ **FIGURE 45-30**

Truth table for a comparator.

Input		Output
B	**A**	**Y**
0	0	1
0	1	0
1	0	0
1	1	1

© 2014 Cengage Learning

■ **FIGURE 45-31**

Comparing two 2-bit numbers.

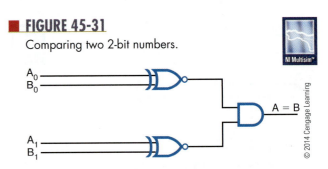

© 2014 Cengage Learning

Figure 45-28 shows the symbols used to represent a half-subtractor and full subtractor.

A half subtractor can handle only two 1-bit numbers. To subtract binary numbers with more than 1 bit, additional full subtractors must be used. Keep in mind that if a 1 is subtracted from a 0, a borrow must be made from the next higher-order column. The output borrow of the lower-order subtractor becomes the input borrow of the next higher-order subtractor.

Figure 45-29 shows a block diagram of a 4-bit subtractor. A half subtractor is used in the least significant bit position because there is no input borrow.

Comparator

A **comparator** is used to compare the magnitudes of two binary numbers. It is a circuit used simply to determine whether two numbers are equal. The output not only compares two binary numbers but also indicates whether one is larger or smaller than the other.

Figure 45-30 shows a truth table for a comparator. The only time an output is desired is when both bits being compared are the same. The output column represents an exclusive OR with an inverter, also known as an exclusive NOR (XNOR) gate. An XNOR gate is essentially a comparator, because its output is a 1 only if the two inputs are the same. To compare numbers containing 2 bits or more, additional XNOR gates are necessary. Figure 45-31 shows a logic diagram of

a comparator for checking two 2-bit numbers. If the numbers are equal, a 1 is generated from the XNOR gate. The 1 is applied to the AND gate as a qualifying level. If both XNOR gates produce a 1 for the inputs to the AND gate, that identifies the numbers as equal and generates a 1 for the output of the AND gate. However, if the inputs to the XNOR gate are different, the XNOR gate generates a 0, which disqualifies the AND gate. The output of the AND gate is then a 0. Figure 45-32 shows a logic diagram of a comparator for checking two 4-bit numbers. Figure 45-33 is the diagram used to represent a 4-bit comparator.

■ **FIGURE 45-32**

Comparing two 4-bit numbers.

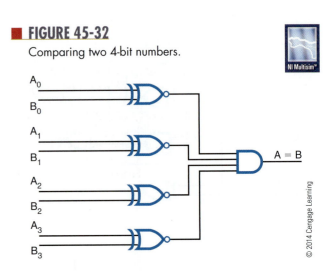

© 2014 Cengage Learning

■ FIGURE 45-33

Diagram of a 4-bit comparator.

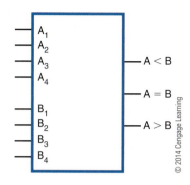

© 2014 Cengage Learning

45-4 QUESTIONS

1. What are the rules for binary addition?
2. What is the difference between a half adder and a full adder?
3. When is a half adder used?
4. What are the rules for binary subtraction?
5. Draw a block diagram of a 4-bit subtractor.
6. What is the function of a comparator?
7. Draw a logic diagram of a comparator.

45-5 PROGRAMMABLE LOGIC DEVICES (PLDS)

In Chapter 43, complex switching functions were reduced to the simplest circuit possible using Veitch diagrams or Karnaugh maps. For example:

$$\overline{A}BCD + \overline{A}\overline{B}CD + \overline{A}B\overline{C}D + A\overline{B}\overline{C}D + A\overline{B}\overline{C}D +$$

$$A\overline{B}C\overline{D} = Y$$

reduced to its simplest form became:

$$\overline{A}D + A\overline{B}\overline{C} = Y$$

To implement this expression using conventional logic gates would require several ICs. As the complexity of the circuit increases, the number of ICs becomes excessive.

A solution that is becoming popular requires the use of **programmable logic devices (PLDs)** to implement the logic function. PLDs can be selected from any of the IC families depending on the speed, power, and logic functions required.

There are three basic forms of PLDs available. They are programmable read-only memory (PROM), **programmable array logic (PAL®)**, and **programmable logic array (PLA)**. PROMs are used primarily as a storage device and are not well adapted to implementing complex logic equations.

PALs have several multi-input AND gates connected to the input of an OR gate and inverter (Figure 45-34). The inputs are set up as a fusible-link

■ FIGURE 45-34

Programmable array logic (PAL®) architecture.

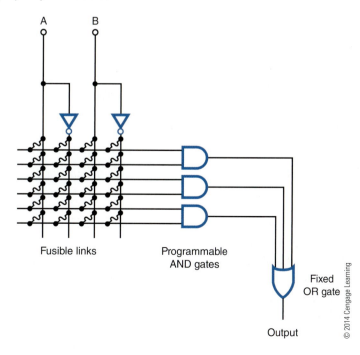

Fusible links   Programmable
AND gates

Fixed
OR gate

Output

© 2014 Cengage Learning

Programmable logic array (PLA) architecture.

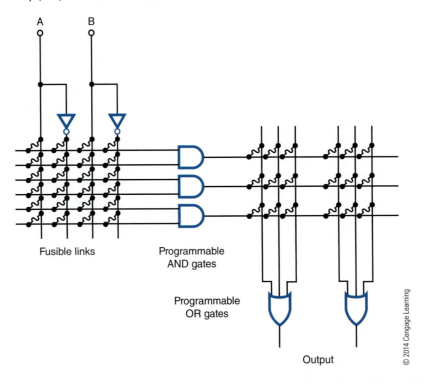

Fusible links

Programmable
AND gates

Programmable
OR gates

Output

© 2014 Cengage Learning

array. By blowing specific fuses in an array, the PAL can be programmed to solve a variety of complex logic equations.

The PLA is similar to the PAL but has the additional flexibility of having its AND gates connected to several programmable OR gates. This results in a device that is harder to program than the PAL and the additional level of logic gates increases the overall propagation delay time. Figure 45-35 shows the detailed interconnections. The interconnections start with all fusible links in place. The programmer decides which fuses will be left intact and which will be blown. The links are blown to represent a logic equation.

The main difference between PALs and PLAs is that PALs have fixed (hard-wired) OR gates and PLAs have programmable OR gates.

45-5 QUESTIONS

1. What is the purpose of a programmable logic device?
2. What are the three forms of programmable logic devices?
3. Explain how to program a PAL.
4. Draw the PAL schematic circuit that will produce the following expression:

$$Y = \overline{A}B + A\overline{B} + \overline{A}\overline{B}$$

5. Describe how a PAL differs from a PLA.

SUMMARY

- An encoder accepts one or more inputs and generates a multibit binary output.
- A decimal-to-binary encoder takes a single digit (0 through 9) and produces a 4-bit output code that represents the digit.
- A priority encoder accepts the higher-order key when two keys are pressed simultaneously.
- Decimal-to-binary encoders are used for keyboard encoding.
- A decoder processes a complex binary code into a digit or character that is easy to recognize.
- A BCD-to-seven-segment decoder is a special-purpose decoder to drive seven-segment displays.
- A multiplexer allows digital data from several sources to be routed through a common line for transmission to a common destination.
- Multiplexers can handle both analog and digital data.
- Multiplexers can be hooked up for parallel-to-serial conversion of data.

- The truth table for the adding rules of binary numbers is equivalent to the truth table for an AND gate and an XOR gate.
- A half adder does not take into account the carry.
- A full adder takes the carry into account.
- To add two 4-bit numbers requires three full adders and one half adder.
- The truth table for the subtracting rules of binary numbers is equivalent to the truth table for an AND gate with an inverter on one of the inputs and an XOR gate.
- A half subtractor does not have a borrow input.
- A full subtractor has a borrow input.
- A comparator is used to compare the magnitudes of two binary numbers.
- A comparator generates an output only when the 2 bits being compared are the same.
- A comparator can also determine whether one number is larger or smaller than the other.
- Programmable logic devices (PLDs) are used to implement complex logic functions.
- Three types of PLDs are PROMs, PALs, and PLAs.
- PLAs are similar to PALs except that PALs have fixed OR gates and PLAs have programmable OR gates.

CHAPTER 45 SELF-TEST

1. Why are encoders necessary in logic circuits?
2. What type of encoder is required for keyboard input?
3. Why are decoders important in logic circuits?
4. What are the applications for the different types of decoders?
5. Briefly describe how a digital multiplexer works.
6. For what applications can a digital multiplexer be used?
7. Draw a schematic using logic symbols for a half adder and full adder tied together for 2-bit addition.
8. Explain the operation of the adder drawn in question 7.
9. Design a PLA that would satisfy the following expression:

   $A B \overline{C} + A \overline{B} C + \overline{A} B \overline{C} + A \overline{B} \overline{C} + A B C + \overline{A} B \overline{C} = Y$

10. Which is easier to program, a PAL or a PLA, and why?

Microcomputer Basics

OBJECTIVES

After completing this chapter, the student will be able to:

- Identify the basic blocks of a digital computer.
- Explain the function of each block of a digital computer.
- Describe what a program is and its relationship to both digital computers and microprocessors.
- Identify the basic registers in a microprocessor.
- Explain how a microprocessor operates.
- Identify the instruction groups associated with microprocessors.
- Identify the purpose of microcontrollers.
- Describe the function of microcontrollers in everyday life.

KEY TERMS

46	digital computer
46-1	arithmetic logic unit (ALU)
46-1	central processing unit (CPU)
46-1	control unit
46-1	input/output (I/O)
46-1	instruction register
46-1	interrupt
46-1	microprocessing unit (MPU)
46-1	program
46-2	accumulator
46-2	arithmetic instructions
46-2	compare instructions
46-2	complement instruction
46-2	condition-code register
46-2	data movement instructions
46-2	input/output (I/O) instructions

46-2	logic instructions
46-2	masking
46-2	microprocessor
46-2	miscellaneous instructions
46-2	program-control instruction
46-2	program counter
46-2	rotate and shift instructions
46-2	stack
46-2	stack instructions
46-2	stack pointer
46-3	microcontroller
46-3	PDIP
46-3	PLCC
46-3	programmable interface controller (PIC)
46-3	QFN
46-3	reduced instruction set computer (RISC)
46-3	SOIC
46-3	SSOP

The greatest application of digital circuits is in digital computers. A **digital computer** is a device that automatically processes data using digital techniques. Data are pieces of information. Processing refers to the variety of ways that data can be manipulated.

Digital computers are classified by size and computing power. The largest computers are called *mainframes*. These computers are expensive, having extensive memory and high-speed calculating capabilities. Smaller-scale computers—the *minicomputer* and the *microcomputer*—are more widely used. Even though they represent a small percentage of the total computer dollars invested, small-scale computers represent the largest number of computers in use. The microcomputer is the smallest and least expensive of the digital computers that still retains all the features and characteristics of a computer.

Computers are also classified by function, the most common function being data processing. Industry, business, and government use computers to maintain records, perform accounting tasks, keep inventory, and provide a wide variety of other data processing functions. Computers can be general purpose or special purpose. General-purpose computers are flexible and can be programmed for any task. Special-purpose, or dedicated, computers are designed to perform a single task.

46-1 COMPUTER BASICS

All digital computers consist of five basic blocks or sections: control, arithmetic logic unit (ALU), memory, input, and output (Figure 46-1). In some cases the input and output blocks are a single block identified as **input/output (I/O)**. Because the control unit and the ALU are closely related and difficult to separate, they may be collectively referred to as the **central processing unit (CPU)** or **microprocessing unit (MPU)**.

The **control unit** decodes each instruction that enters the computer. It then generates the necessary pulses to carry out the functions specified. If, for example, an instruction requires two numbers to be added together, the control unit sends pulses to the ALU to perform the addition. If the instruction requires a word to be stored in memory, control sends the necessary pulses to memory to store the data.

FIGURE 46-1

Basic blocks of a digital computer.

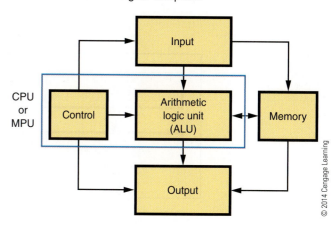

© 2014 Cengage Learning

Modern computers utilize a means of incorporating several instructional commands into a single input instruction. This is accomplished by a program stored in memory. When the control unit decodes the instruction, it causes a sequence of instructions to be executed. The control unit varies from computer to computer. Basically, the control unit consists of an address register, an instruction register, an instruction decoder, a program counter, a clock, and circuitry for generating the control pulses (Figure 46-2).

The **instruction register** stores the instruction word to be decoded. The word is decoded by the instruction decoder, which sends the appropriate logic signal to the control pulse generator. The control pulse generator produces a pulse when the appropriate clock signal is given. The output of the control pulse generator enables other circuitry in the computer to carry out the specific instruction.

The program counter keeps track of the sequence of instructions to be executed. The instructions are stored in a program in memory. To begin the program, the starting address (specific memory location) of the program is placed in the program counter. The first instruction is read from memory, decoded, and performed. The program counter then automatically moves to the next instruction location. Each time an instruction is fetched and executed, the program counter advances one step until the program is completed.

Some instructions specify a jump or branch to another location in the program. The instruction register contains the address of the next instruction location and it is loaded into the address register.

■ FIGURE 46-2

Control unit of a computer.

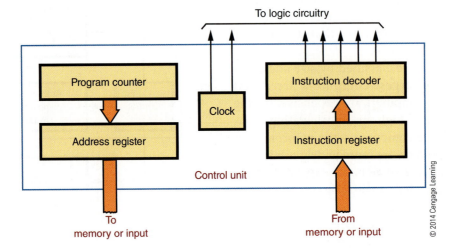

■ FIGURE 46-3

Arithmetic logic unit (ALU).

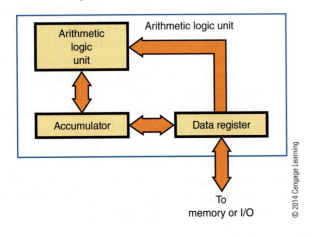

The **arithmetic logic unit (ALU)** performs math logic and decision-making operations. Most ALU can do addition and subtraction. Multiplication and division are programmed in the control unit. The ALU can perform logic operations such as inversion, AND, OR, and exclusive OR. It can also make decisions by comparing numbers or test for specific quantities such as 0s, 1s, or negative numbers.

Figure 46-3 shows an arithmetic logic unit. It consists of arithmetic logic circuitry and an accumulator register. All data to the accumulator and the ALU are sent via the data register. The accumulator register can be incremented (increased by one), decremented (decreased by one), shifted right (one position), or shifted left (one position). The accumulator is the same size as the memory word; the memory word is 32 bits wide, and the accumulator is also 32 bits wide in a 32-bit microprocessor.

The arithmetic logic circuitry is basically a binary adder. Both addition and subtraction can be done with the binary adder as well as logic operations. To add two binary numbers, one number is stored in the accumulator register and the other is stored in the data register. The sum of the two numbers is then placed in the accumulator register, replacing the original binary number.

Memory is the area where programs are stored. **Programs** contain the instructions that tell the computer what to do. A program is a sequential set of instructions to solve a particular problem.

A computer memory is simply a number of storage registers. Data can be loaded into the registers and then taken out, or "read out" to perform some operation, without losing the register content. Each register or memory location is assigned a number called an address. The address is used to locate data in memory.

Figure 46-4 shows a typical memory layout. The memory registers retain the binary data. This memory, based on its ability to store (write) or retrieve (read) data, is usually referred to as random-access read or writes memory (RAM). Based on the ability of being able to read only data or instruction from the memory, it is referred to as read-only memory (ROM).

■ FIGURE 46-4

Memory layout for a computer.

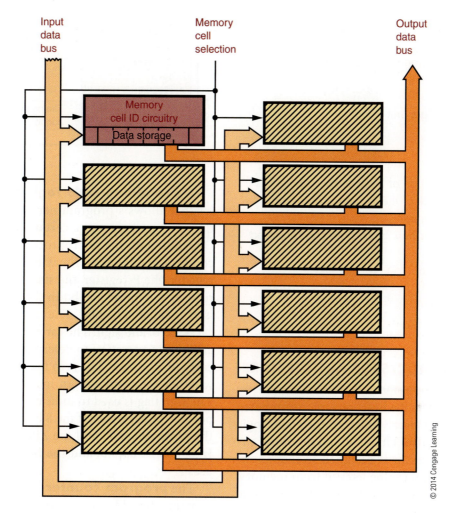

The memory address register allows access to specific memory locations by the memory address decoder. The size of the memory address register determines the maximum memory size for a computer. For example, a memory address register of 16 bits allows a maximum number of 2^{16} or 65,536 memory locations.

A word to be stored in memory is located in the data register and then placed in the desired memory location. To read data from memory, the memory location is determined, and data at the memory location are loaded into the shift register.

The input and output units of a computer allow it to receive and transmit information from and to the world outside the computer. An operator or peripheral equipment enters data into a computer through the input unit. Data from the computer are passed to external peripheral equipment through the output unit.

The input and output units are under the control of the CPU. Special I/O instructions are used to transfer data in and out of the computer.

Most digital computers can perform I/O operations at the request of an interrupt. An **interrupt** is a signal from an external device requesting service in the form of receiving or transmitting data. The interrupt results in the computer leaving the current program and jumping to another program. When the interrupt request is accomplished, the computer returns to the original program.

46–1 QUESTIONS

1. Draw and label a block diagram of a digital computer.
2. What is the function of the following blocks in a digital computer?
   a. Control
   b. Arithmetic logic unit
   c. Memory
   d. Input
   e. Output
3. What is the function of ROM in a computer?
4. What keeps track of the sequence of instructions to be executed?
5. What determines how much data can be stored in a computer?
6. Define a program.

46–2 MICROPROCESSOR ARCHITECTURE

A **microprocessor** is the heart of a microcomputer. It contains four basic parts: registers, arithmetic logic unit, timing and control circuitry, and decoding circuitry. A microprocessor is designed so that an instruction, or program, can be fetched from memory, placed in the instruction register, and decoded. The program affects all the timing, control, and decoding circuitry. The program allows the operator to route data in or out of various registers or into the arithmetic logic unit. The registers and the arithmetic logic unit are used by the microprocessor for data or information manipulation.

Each microprocessor is different in its architecture and its instruction set. Figure 46-5 shows the basic parts of many of the 8-bit microprocessors. Because the names and number of registers vary from one microprocessor to the next, they are shown and identified separately.

The **accumulator** is the register most often used in the microprocessor. It is used to receive or store data from memory or an I/O device. It also works closely with the arithmetic logic unit. The number of bits in the accumulator determines the microprocessor's word size. In an 8-bit microprocessor, the word size is 8 bits.

■ **FIGURE 46-5**

Parts of an 8-bit microprocessor.

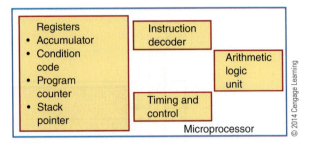

© 2014 Cengage Learning

The **condition-code register** is an 8-bit register that allows a programmer to check the status of the microprocessor at a certain point in a program. Depending on the microprocessor, the name of the condition-code register may be the processor-status register, the P-register, the status register, or the flag register. An individual bit in the condition-code register is called a flag bit. The most common flags are carry, zero, and sign. The carry flag is used during an arithmetic operation to determine whether there is a carry or a borrow. The zero flag is used to determine whether the results of an instruction are all zeros. The sign flag is used to indicate whether a number is positive or negative. Of the 8 bits in the code register, the Motorola 6800 and the Zilog Z80 use 6 bits; the Intel 8080A uses 5; and the MOS Technology 6502 uses 7.

The **program counter** is a 16-bit register that contains the address of the instruction fetched from memory. While the instruction is being carried out, the program counter is incremented by one for the next instruction address. The program counter can only be incremented. However, the sequence of the instructions can be changed by the use of branch or jump instructions.

The **stack pointer** is a 16-bit register that holds the memory location of data stored in the stack. The stack is discussed further later in the chapter.

Most microprocessors have the same basic set of instructions with different machine codes and a few unique instructions. The basic instructions fall into nine categories:

1. Data movement
2. Arithmetic
3. Logic
4. Compare and test
5. Rotate and shift

6. Program control
7. Stack
8. I/O
9. Miscellaneous

Data movement instructions move data from one location to another within the microprocessor and memory (Figure 46-6). Data are moved 8 bits at a time, in a parallel fashion (simultaneously), from the source to the destination specified. Microprocessor instructions use a symbolic notation that refers to how the data are moving. In the 6800 and 6502 microprocessors, the arrow moves from left to right. In the 8080A and the Z80, the arrow moves from right to left. In either case, the message is the same. The data move from the source to the destination.

Arithmetic instructions affect the arithmetic logic unit. The most powerful instructions are add, subtract, increment, and decrement. These instructions allow the microprocessor to compute and manipulate data. They differentiate a computer from a random logic circuit. The result of these instructions is placed in the accumulator.

Logic instructions are those instructions that contain one or more of the following Boolean operators: AND, OR, and exclusive OR. They are performed 8 bits at a time in the ALU, and the results are placed in the accumulator.

Another logic operation is the **complement instruction**. This includes both 1s and 2s complement. Because complementing is done with additional circuitry, it is not included in all microprocessors. The 6502 has neither complement instruction. The 8080A has a 1s complement instruction. The 6800 and the Z80 have both 1s and 2s complement. Complementing provides a method of representing signed numbers. Complementing numbers allow the ALU to perform subtract operations using an adder circuit. Therefore, the MPU can use the same circuits for addition and subtraction.

Compare instructions compare data in the accumulator with data from a memory location or another register. The result of the comparison is not stored in the accumulator, but a flag bit might change as a result of it. Masking or bit testing may be used to perform comparison. **Masking** is a process of subtracting two numbers and allowing only certain bits through. The mask is a predetermined set of bits that is used to determine whether certain conditions exist within the MPU. There is a disadvantage with the masking procedure because it uses an AND instruction and therefore destroys the content of the accumulator. The bit-testing procedure, although it also uses an AND instruction, does not destroy the content of the accumulator. Not all microprocessors have a bit-testing instruction.

Rotate and shift instructions change the data in a register or memory by moving the data to the right or left one bit. Both instructions involve use of the carry bit. The difference between the instructions is that the rotate instruction saves the data and the shift instruction destroys the data.

Program-control instructions change the content of the program counter. These instructions allow the microprocessor to skip over memory locations to

■ **FIGURE 46-6**

Data movement instructions.

Description	Mnemonic	Notation	Source	Destination
Load Accumulator	LDA	M → A	Memory	Accumulator
Load X-register	LDX	M → X	Memory	X-register
Store Accumulator	STA	A → M	Accumulator	Memory
Store X-register	STX	X → M	X-register	Memory
Transfer accumulator to X-register	TAX	A → X	Accumulator	X-register
Transfer X-register to accumulator	TXA	X → A	X-register	Accumulator

execute a different program or to repeat a portion of the same program. The instructions can be unconditional, where the content of the program counter changes, or conditional, where the state of a flag bit is first checked to determine whether the content of the program counter should change. If the condition of the flag bit is not met, the next instruction is executed.

Stack instructions allow the storage and retrieval of different microprocessor registers in the **stack**. The stack is a temporary memory location that is used for storing the contents of the program counter during a jump to a subroutine program. The difference between a stack and other forms of memory is the method in which the data is accessed or addressed. A push instruction stores the register content, and a pull instruction retrieves the register content. There is an advantage with the stack because data can be stored into it or read from it with single-byte instructions. All data transfers are between the top of the stack and the accumulator. That is, the accumulator communicates only with the top location of the stack.

In the 6800 and 6502 microprocessors, the content of the register is stored in the stack, and then the stack pointer is decremented by 1. This allows the stack pointer to point to the next memory location where data can be saved. The *stack pointer* is a 16-bit register that is used to define the memory location that acts as the top of the stack. When the pull instruction is used, the stack pointer is incremented by 1, and the data are retrieved from the stack and placed in the appropriate register. In the 8080A, the top of the stack contains the pointer to the last memory location. The push instruction first decrements the stack pointer by 1 and then stores the register content in the stack.

Input/output (I/O) instructions deal specifically with controlling I/O devices. The 8080A, 8085, and Z80 have I/O instructions. The 6800 and 6502 do not have specific I/O instructions. If a microprocessor uses an I/O instruction to deal with external devices, the technique is called an *isolated I/O*.

Some instructions do not fall in any of the categories mentioned. These instructions are grouped together and are called **miscellaneous instructions**. Among these instructions are those used to enable or disable interrupt lines, clear or set flag bits, or allow the microprocessor to perform BCD arithmetic. Also included is the instruction to halt or break the program sequence.

46-2 QUESTIONS

1. What are the basic parts of a microprocessor?
2. What registers are located in the microprocessor?
3. What are the major categories of microprocessor instructions?
4. What determines the difference between a computer and a random logic circuit?
5. What are the uses of miscellaneous instructions?

46-3 MICROCONTROLLERS

Micro identifies the device as small. *Controller* identifies the device as being used to control objects, processes, or events.

Any device that measures, stores, controls, calculates, or displays information has a microcontroller inside. In today's society, **microcontrollers** control many appliances (e.g., microwaves, toasters, and stoves), operate high-tech toys, run the engines in automobiles, and play music in greeting cards. Many of the devices they control are taken for granted today.

A microcontroller is a single-chip computer. It contains limited memory, I/O interfacing, and a central processing unit (CPU) on the chip (Figure 46-7). This makes it ideal for monitoring and controlling functions. Because it is on a chip, the microcontroller and its support circuits are often built into the device they control.

Microcontrollers are designed for machine-control applications and do not require human interaction to operate. For example, toasters and microwave ovens have one or two fixed, repetitive programs. A microcontroller does not require any human interface device such as a keyboard, monitor, or mouse.

One example of a microcontroller-integrated circuit is the 8051, an 8-bit processor with ROM and RAM located on the chip as well as the I/O circuitry. The 8051 is a popular chip that is being produced and developed by several companies. Intel's version of the 8051 chip is the MCS 251, which is 15 times faster than the MCS 51. A block diagram of the MCS 51 is shown in Figure 46-8.

■ **FIGURE 46-7**

Block diagram of a microcontroller.

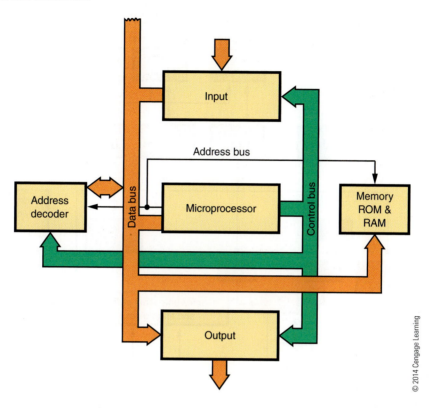

© 2014 Cengage Learning

The MCS 51 features 4 kbytes of EPROM/ ROM; 128 bytes of RAM; 32 input/out lines; two 16-bit timer/counters; five-source, two-level interrupt structures; a full-duplex serial port; and an on-chip oscillator and clock circuitry.

The 68HC11 is a powerful 8-bit, 16-bit address microcontroller developed by Motorola, now produced by Freescale Semiconductor. It has an instruction set that is similar to the older 68xx (6801, 6805, and 6809) family. Depending on the application, the 68HC11 comes with a variety of features including built-in windowed EPROM/ EEPROM/ OTPROM (one-time programmable), static RAM, digital I/O, timers, A/D converter, PWM generator, and synchronous and asynchronous communication channels. Freescale Semiconductor offers a low-cost evaluation board to explore the capabilities of the 68HC11.

The Microelectronics Division of General Instrument developed the first version of a **programmable interface controller (PIC)**. PICs are now a family of microcontrollers made by Microchip Technology that use **reduced instruction set computer (RISC)**. RISC is based on the strategy that fewer basic instructions allow for higher performance. Basing microcontrollers on this architecture is a strategy that was derived from the original PIC1640 microcontroller.

PIC microcontrollers are widely available, inexpensive, and do not require a lot of prerequisites before running in an application. The PIC development environment is freely available and up to date. There is a broad range of different PIC microcontrollers, to fit any application. The hardware programming is easy and does not require a lot of additional equipment or programmers. The various PIC packages available include **PDIP** (plastic dual in-line package), **SSOP** (shrink small outline package), **SOIC** (small outline package), **PLCC** (plastic leaded chip carrier), and **QFN** (quad flat no-lead package) (Figure 46-9). PIC packages allow for convenient prototyping without a lot of effort.

■ **FIGURE 46-8**

Microcontroller block diagram of the MCS 51.

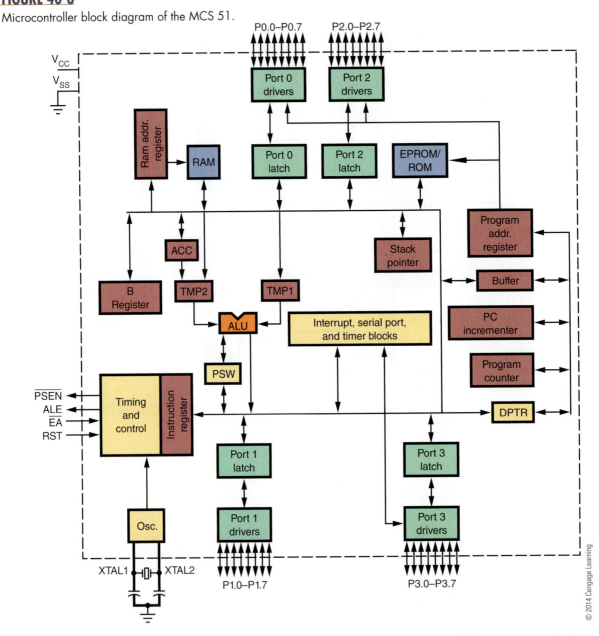

■ **FIGURE 46-9**

Various PIC package outlines.

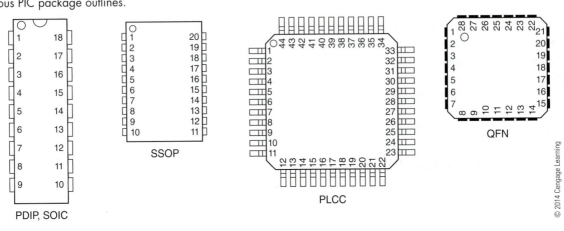

46-3 QUESTIONS

1. What does a microcontroller do?
2. Define microcontroller.
3. What were microcontrollers designed for?
4. What is a PIC microcontroller?
5. What is the difference between an MCS51, a 68HC11, and a PIC microcontroller?

SUMMARY

- Digital computers consist of a control section, an arithmetic logic unit, memory, and an I/O section.
- The control section decodes the instructions and generates pulses to operate the computer.
- The arithmetic logic unit performs math, logic, and decision-making operations.
- Memory is an area where programs and data are stored while waiting for execution.
- The I/O units allow data to be transmitted in and out of the computer.
- The control section and arithmetic logic unit may be included in a single package called a microprocessor.
- A program is a set of instructions arranged in a sequential pattern to solve a particular problem.
- A microprocessor contains registers, an arithmetic logic unit, timing and control circuitry, and decoding circuitry.
- Instructions for a microprocessor fall into nine categories:
  - Data movement
  - Arithmetic
  - Logic
  - Compare and test
  - Rotate and shift
  - Program control
  - Stack
  - I/O
  - Miscellaneous
- Microcontrollers are single-chip devices that are used to control objects, processes, and events.
- Any device that measures, stores, controls, calculates, or displays information has a microcontroller in it.
- A microcontroller is a single-chip computer.
- Microcontrollers are often built into the device they control.
- Microcontrollers are designed for machine control applications.
- Microcontrollers, once programmed, do not require human interface.
- Examples of microcontrollers include the MCS 251, 68HC11, and PICs.

CHAPTER 46 SELF-TEST

1. Describe how a computer operates.
2. By what means does a computer that is interfaced to the real world manipulate data transmitted from an external device?
3. What is the difference between a microcomputer and a microprocessor?
4. What is the function of the microprocessor?
5. What is the difference between a microcontroller and a microprocessor?
6. Define the following microprocessor instructions: data movement, arithmetic, logic, and I/O.
7. Describe how microcontrollers are used.
8. Identify five home devices or appliances that would use microcontrollers.
9. Where would microcontrollers be used in a current automobile?
10. What is meant by a microcontroller using RISC architecture?

PRACTICAL APPLICATIONS

1. Design and build a circuit that will output your birthday or any other important date on a seven-segment display. Verify the circuit works properly using Multisim.

2. Design and build a circuit that can be programmed to count up or down to 199 using a seven-segment display. Verify the circuit works properly using Multisim.

3. Design and build an electronic circuit that can be programmed to output to a seven-segment display. Verify the circuit's operation using Multisim.

4. Design and build a circuit that can be programmed to control a small DC motor to run forward or in reverse. Use Multisim to verify that the circuit works.

5. Design and build a circuit that will allow a computer to sample data from its serial, parallel, or USB port.

PRACTICAL APPLICATIONS

Robin Adler and Jan Hofland
In 1969, Adler and Hofland published an article in the HP Journal titled, "Logic Pulser and Probe: A New Digital Troubleshooting Team."

Army Air Corps (1926–1947)
After World War II, technicians would strip the copper braid from coaxial cable, dip it in solder flux and then used it for desoldering, creating the desoldering braid.

Paul Eisler (1907–1992)
In 1936, Eisler invented the printed circuit board to use in a radio. The printed circuit board was first used on a large scale in 1948.

IBM (1911–)
In 1911, the Computing Tabulating Recording Company was formed through a merger of three companies; it adopted the name International Business Machines (IBM) in 1924. In 1960s, IBM developed the surface mount technology that became widely used in late 1980s.

Radio (1907–)
In the early days, amateurs would hammer small nails to a bread cutting board and solder electronic components to them, creating the first electronics breadboard.

Ernst Sachs (1890–1977)
In 1921, Sachs invented the first electric soldering iron.

Unknown
In 1962, the logic probe was introduced to look into failed integrated logic circuits, helping to identify the basic reason for a chip to fail.

Carl E. Weller (1911–1994)
In 1941, Weller invented a transformer-type soldering gun that heated up rapidly.

Robert L. Zahn
In 1983, Zahn invented a variable radius lead former for axial lead electronic component.

Project Design

OBJECTIVES

After completing this chapter, the student will be able to:

- Explain the ideation design process.
- Discuss brainstorming and how Pareto voting is associated with it.
- Identify the purpose for a printed circuit board prototype.
- Discuss the advantages of using a PCB prototype.
- Explain why a printed circuit board design could be copyrighted.
- Identify ways to protect a printed circuit board design.
- Describe what function a project timeline serves.
- Identify the steps required to create a timeline.
- Describe how budgets are used when constructing a project.
- Discuss the steps for creating a budget.
- Explain what purpose a production schedule serves for manufacturing a product.
- Identify two charts used for developing a production schedule.
- Discuss why industry wants technicians to use reference manuals.
- Describe the purpose of a cross-reference guide.

KEY TERMS

47-1 printed circuit board (PCB) prototype

47-1 schematic diagram

47-2 copyright

47-3 project timeline

47-4 budget

47-5 Gantt chart

47-5 Project Evaluation and Review Technique (PERT) chart

47-5 production schedule

47-6 cross-reference guide

47-6 manufacturer reference manual

This chapter serves as an introduction to the procedures in fabricating a printed circuit board from a schematic diagram. It takes a lot of planning on the front end to make it look effortless on the output.

Issues arise on such things as copyright, building the first prototype, determining whether the project is cost effective, and so on. In the design phase, many references are available to help an electronics technician. Keep in mind that this book is aimed at developing a background in electronics, not in designing and building a circuit from scratch.

47-1 DEVELOPING A PROTOTYPE

In electronics, a prototype refers to build a working example of an electronic circuit. Before building a prototype, there are several steps that must take place first. To develop a new electronic circuit means it must serve a need or a want. Ideas that could lead to innovation include the following:

- Combining different circuits to create a new circuit
- Creating a new approach to a circuit design
- Developing a do-nothing circuit
- Improving on a circuit that already exists
- Solving a problem that exists
- Targeting a specific problem

Ideas for a new electronic circuit can come about through an ideation and brainstorming.

Ideation is the process of developing and expanding ideas. It is a technique used to generate ideas to meet a need or solve a problem. Ideation includes all steps from identifying an idea, to developing the idea, to building a working prototype of the circuit (Figure 47-1).

Brainstorming is a team effort for reaching a solution for a specific problem by generating a list of possible solutions by the team members (Figure 47-2). The procedure for brainstorming includes these steps:

- Identify a recording secretary
- Begin brainstorming, establishing a time when to stop
- Identify brief ideas for possible solutions to the problem

- Do not explain or criticize possible solutions
- Record all ideas produced during the session

As with everything, brainstorming has rules, which include these:

- Do not hold back when a thought strikes
- Do not judge or criticize
- Keep responses short
- Listen and offer variations to other ideas
- Participate rather than just sitting

When working with a team, ideas can go in two distinct directions. Ideas can expand outward with a broad list of ideas, and this is referred to as divergent thinking. Ideas can also be focused narrowly inward to evaluate and eliminate in order to select an idea, and this is referred to as convergent thinking.

Divergent thinking allows a group to generate many new ideas in a short timeframe. It encourages and suggests a number of possible solutions by using a concept that no idea is outrageous and that there are no bad ideas. Divergent thinking allows the largest number of ideas that are not tied to any constraints, limitations, or judgment.

Convergent thinking will then evaluate each idea carefully to determine whether it has merit, looks at possible limitations, evaluates the facts, determines the costs, and selects the best solution. This thinking technique can help narrow many ideas to a single idea.

Pareto voting can quickly reduce a number of ideas to a manageable list. It is based on the principle that 20% of the listed ideas of a group will satisfy 80% of the group. To use Pareto voting:

1. List all ideas clearly on a chart.
2. Take the total number of ideas and multiply by 0.20. If a fraction exists, round to the next whole number. This indicates the number of votes for each team member.
3. Identify the criteria for voting on the ideas.
4. Each team member places a check by those ideas that represent his or her personal choice. Team members can use their votes on the same idea or split their votes among all the ideas.
5. The ideas with the greatest number of votes have the greatest support by the team and will satisfy 80% of the team.

■ FIGURE 47-1

Ideation process—The prototype is checked against the original problem and the idea selected.

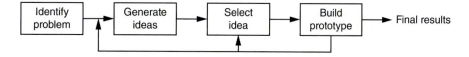

■ **FIGURE 47-2**

Brainstorming—There is no feedback on the ideas generated.

© 2014 Cengage Learning

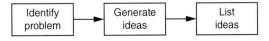

The list of ideas is then shrunk to a manageable list, and convergent thinking helps to shrink the list down to one idea to develop into a feasible prototype.

To summarize the ideation design process (Figure 47-3) requires several steps:

- Identify a problem through needs and wants
- Generate ideas to a possible solution
- Select the best solution
- Prototype the solution
- Test and evaluate the solution
- Redesign the solution if the solution does not satisfy or solve the problem

Prototyping and modeling are often used interchangeably and can lead to confusion. There is no general agreement on what constitutes a prototype.

A **printed circuit board (PCB) prototype** is an early example of the PCB built to test a circuit and to learn from it. One of the first steps in the design of a PCB prototype is to determine that the electronic

■ **FIGURE 47-3**

The ideation design process.

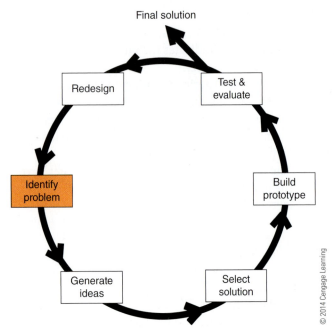

■ **FIGURE 47-4**

One of the first steps in building an electronic circuit is to breadboard the circuit as shown here with actual components or using a virtual breadboard.

© 2014 Cengage Learning

circuit to be built works through breadboarding (Figure 47-4). This can be done on a physical breadboard or virtually on an electronic circuit simulation program (Chapter 6: Software for Electronics).

So what should the PCB prototype look like? First, it depends on the ability to read a **schematic diagram**. Second, it depends on the overall goal, to make the smallest PCB, to fit a certain shape, to fit a certain location, and so on. If possible, it's a good idea to start with paper dolls (Chapter 48) and progress from there.

Taking an electronic circuit and creating a PCB prototype can be satisfying and provides an opportunity to tap into an individual's creativity. A challenge faced when engaging in electronic circuit PCB prototyping is underestimating the time it will take to complete a layout. It should be noted that the phase of redesign and correction often take as long as the initial design process.

National Instrument—Ultiboard and New Waves Concept—Circuit Wizard will both make PCB layouts from a schematic diagram and also can test the layout before printing out a copy.

Finally, when its time to make the PCB prototype, a hand-drawn board can provide a good idea that the layout will work.

A PCB prototype provides other advantages, as well:

1. It enables testing and refining the PCB prototype design.
2. It makes it possible to test the design with various components in order to improve performance.

Making a prototype by hand is a great way to start bringing the electronic circuit to life. Remember, there are no rules, so the prototyping stage is a great time to use untapped creative ability and to explore all the possibilities. Don't be limited by any preconceived notions with regards to electronic circuit fabrication.

Once the prototype is built, it must be evaluated to determine whether it meets the needs of the problem identified. It has to be tested under a variety of conditions that the final PCB and circuit will be exposed to. If it does not pass this stage, the solution goes back into the loop for redesigning by changing the circuit component values or the circuit. Once the solution passes the testing and evaluating step and the documentation is complete, the prototype design process is done.

To benefit from the prototype process:

- Document all steps leading up to the final electronic circuit prototype.
- Take photos for a final presentation.
- Make a detailed schematic drawing for the final electronic circuit solution, using electronic circuit simulation software. Include the date and the names of all team members in a box in the lower right-hand corner of the drawing.
- Draw the top and bottom views of the electronic circuit printed circuit board. In the top view, include component placement. Electronic circuit simulation software that includes PCB fabrication can be used, or use a good hand-drawn PCB layout.
- Test and evaluate electric circuit thoroughly to ensure that it satisfies the solution requirements in all environments.
- Finalize the electronic circuit prototype by installing it in an enclosure if required.
- Develop a parts list that includes all costs in fabricating the printed circuit board.
- Develop a timeline on what it would take to mass-produce the electronic circuit.
- Write a summary of the purpose of the electronic circuit.

A presentation of the final solution must be presented to explain what the circuit is used for, why the circuit was prototyped, and how the circuit works. Software for completing the steps in different types of presentations is covered in Chapter 6.

Options for making a presentation include the following:

- Slide show of the steps involved in building the prototype
- Video showing the circuit working

- A handout explaining the evolution of the electronic circuit solution
- A PowerPoint that includes all the steps above and that is posted to a website
- An oral presentation covering all the steps listed above

47-1 QUESTIONS

1. What is the difference between ideation and brainstorming?
2. What is the difference between divergence and convergence thinking?
3. What does Pareto voting accomplish?
4. What is the first step in making a PCB prototype?
5. What is one challenge when designing a PCB prototype?

47-2 COPYRIGHT ISSUES

The U.S. copyright laws apply to first invented, not first filed. For intellectual property, they provide an extensive range for creative and industrial property that includes artwork and software.

An original copy showing a printed circuit board (PCB) layout may be copyrighted (Figure 47-5). Therefore, a PCB design (not including schematic diagrams; Figure 47-6) is protected by **copyright**. However, the copyright is weak because the printed circuit board is identified as a useful article under Section 101, and the industrial design copyright does not protect it in the United States.

Another way of looking at it is that PCBs are declared utilitarian works. It is the functionality of the PCB to be protected, not the layout of the PCB. Should the layout be awarded copyright, the layout is protected only in the aesthetic aspects under the law. It is then only protected to the extent that the aesthetic aspects are separable from the functionality of the PCB.

Because PCB layouts are dictated solely by function, and not by aesthetics, the copyright protection is weak or nonexistent. There is no coverage for PCB layouts or for schematic diagrams other than where a schematic diagram contains entirely proprietary silicon and the PCB layout is necessary to the added function of the silicon.

In lieu of enforceable copyright, the best way to protect PCBs is through continually upgrading

■ **FIGURE 47-5**

An original copy showing a PCB layout can be copyrighted, but the copyright is weak unless the circuit is made in silicon as an IC.

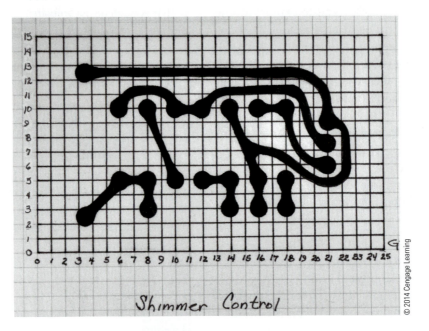

© 2014 Cengage Learning

■ **FIGURE 47-6**

The schematic diagram is not protected by the copyright law.

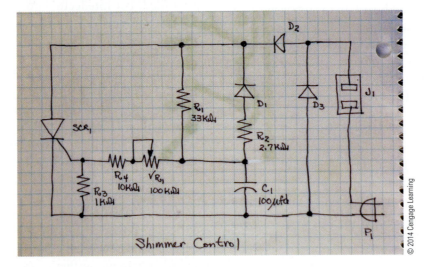

© 2014 Cengage Learning

performance advantage over the competition. Another technique is to provide physical securement by potting or encapsulating the PCB with epoxy that cannot be seen through or removed without destroying the boards. One other choice is to cast them in silicon and seek protection under the Circuit Layout Act.

The Circuit Layout Act of 1989 provides protection for eligible layouts of integrated circuits and computer chips. A layout is the plan showing the three-dimensional location of the electronic components of an integrated circuit or computer chip.

The Circuit Layout Act gives the originator of an original circuit layout certain exclusive rights in relation to that layout, including the right to copy the layout and the right to make an integrated circuit in accordance with the layout.

47-3   DEVELOPING A TIMELINE

Developing a **project timeline** is key for any project building strategy. However, developing and sticking to a timeline is not an easy task. Problems and unexpected complications can occur at any time, quickly throwing a project off track.

Keeping a *project timeline* is an important resource for planning and a useful tool for staying on task. A project timeline allows the group to:

- Identify any problems before they can delay the project.
- Act on earlier notice of potential delays or changes.
- Generate status reports on which tasks are complete or running behind.
- Track actual time spent on the project, which is useful with developing future timelines.

At first, developing accurate timelines may appear to be tough even if it feels like guessing on how the project will go. Each timeline represents a history of the project completion data that can be used later to estimate the actual time needed for similar projects in the future; Figure 47-7.

The first step in building a timeline is to identify the tasks to be accomplished. The tasks will fall into five categories.

1. Group assignments
2. Developing the timeline
3. Planning the project
4. Building the project
5. Presentation of the project

The next step to create the timeline requires:

1. Discussing with the group which tasks needs to be completed for the project.

2. Aligning the task into the building blocks of the project timeline.
3. Filling in between the tasks with sequential steps that must be completed to move from one task to the next.

When estimating the time required for accomplishing each task, consider which team member are involved. Let the responsible team members define the deadlines. Each team member will have additional duties and responsibilities. Taking into account other activities, missing days, etc. will require adjustments in the suggested timeline. Remember that the timeline is an estimate and will need to adjust because some tasks may be completed earlier than suggested, while others may take longer over the course of the project. Make the changes right on the timeline and keep it as accurate as possible.

A project can take several weeks to complete. Typically, they are often at the front end of other activities, such as lab work, testing, or other outside influences. The following can help to establish a clear project timeline:

- Outline the due dates of key markers and the deadline to have the project completed.
- Include enough time for the team to complete necessary research if needed.
- Identify resources and when they will be needed along the project timeline. If more information is needed then identify it and note when.
- Include deadlines in the timeline for tasks to be completed to prevent gridlock on equipment needed for certain tasks.

A computer software program such as Microsoft Excel may include templates for time-management schedules as shown in Figure 47-8 and they could be useful to help layout a timeline.

■ **FIGURE 47-7**

A timeline can represent the history of a project that helps to estimate actual time on a project.

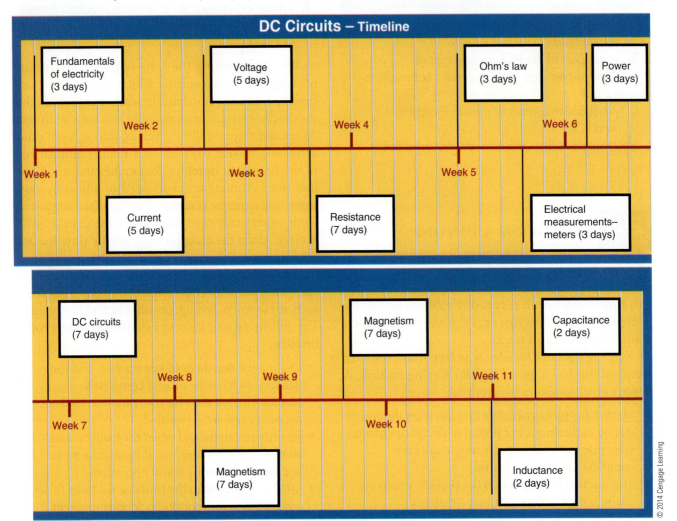

© 2014 Cengage Learning

47-4 DEVELOPING A BUDGET

A **budget** is part of a financial plan created during the initial planning stage when building a project. A budget is similar to any kind of budget used for personal finances.

Budgets can be an effective tool to help determine whether an idea is practical and cost-effective. Maintaining a budget on a regular basis can help to track expenses to apply to future budget planning.

Steps for budget planning are listed here:

1. Identify goals.
2. Review what is available.
3. Define the costs.
4. Create the budget.

Creating a budget for a project requires keeping a list of all components and estimated costs for PCB fabrication (Figure 47-9). The schematic diagram is helpful in developing the components list, and an Internet search can produce the cost for a PCB fabrication when it is not being fabricated in-house.

Compile the information collected and develop a spreadsheet. Remember, in developing a budget, some adjustments may have to be made. A budget should be a tool to be used, not a document to be created and then forgotten. By using a working budget, the budget will become more accurate over time, helping to make informed decisions.

■ FIGURE 47-8

An example of a spreadsheet template that can be used for a time-management schedule.

Source: Microsoft™ Excel™ © 2014 Cengage Learning

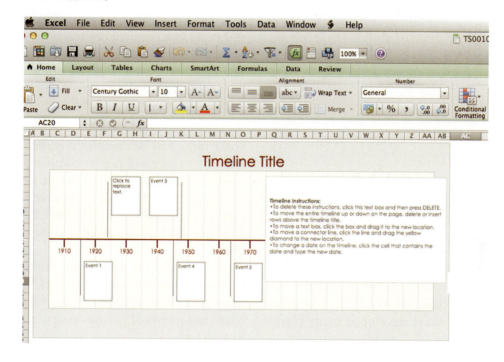

■ FIGURE 47-9

A Bill of Materials will help in creating a budget for estimating the cost of PCB fabrication.

Description	QTY	Cost	Total
470 Ω Resistor (1/4W)	2	$0.03	$0.06
1 KΩ Resistor (1/4W)	1	$0.03	$0.03
11 KΩ Potentiometer	1	$1.19	$1.19
100 μF Electrolytic Capacitor	1	$0.08	$0.08
NE555 Timer	1	$0.25	$0.25
Amber LED	1	$0.15	$0.15
Red LED	1	$0.15	$0.15
Battery Clip	1	$0.79	$0.79
1.9" × 2.7" PCB Board	1	$0.20	$0.20
		Total	$2.90

© 2014 Cengage Learning

1. What is the function of building a budget?
2. How can a budget be used in developing a project?
3. What are the steps for budget planning?
4. Where can the information for component parts be generated?
5. What purpose does the budget serve once it is created?

47–5 DEVELOPING A PRODUCTION SCHEDULE

Manufacturing facilities have a need to develop and maintain **production schedules** to monitor the progress on the production line. Over the years, management and workers have developed many clever and practical methods for controlling production schedules. Manufacturing organizations can generate and update production schedules, which are specific plans that identify when certain controllable activities should take place.

The production schedule coordinates activities to increase productivity and minimize operating expenses. A production schedule can also help with the following tasks:

- Identify resource conflicts.
- Control the release of jobs to the shop.
- Ensure that required raw materials are ordered in time.
- Determine whether delivery promises can be met.
- Identify time periods available for preventive maintenance.

The production schedule is not written in concrete. Some tasks may be completed earlier than expected, and other tasks will not be completed on time. The schedule helps keep track no matter what happens.

Two different kinds of designs are used to generate production schedules: the **Project Evaluation and Review Technique (PERT) chart** and the **Gantt chart**.

A PERT chart is a flowchart of the tasks required in the production phase of a project. The relationships among activities are clearly shown, and completion times and names of persons assigned can be attached to each task.

With the exception of the beginning and end of the chart, each task is preceded and followed by another task. The tasks can also branch out and travel their own paths and then rejoin the main path at some later point.

Any tasks such as review or completion can also be indicated. Figure 47-10 provides an idea of what

■ **FIGURE 47-10**

An example of a general-purpose PERT chart that can help plan a project.

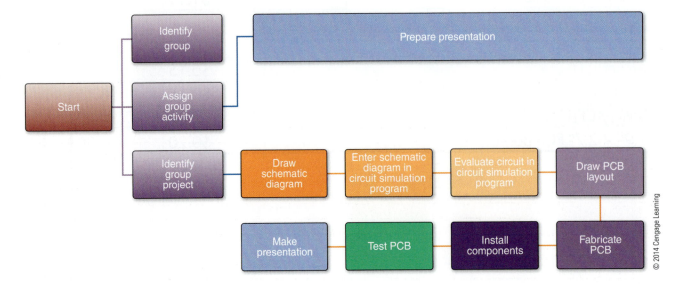

© 2014 Cengage Learning

a PERT chart looks like. Each chart will be different because it will start at the production phase of your project and contain a lot more detail than this one.

Use the following steps to create a PERT chart:

1. Set up a recording board (e.g., a flip-chart, chalk or white board, etc.).
2. Make a list of tasks involved in the pre-production, production, and post-production phases of the project.
3. Align the tasks in sequential order.
4. Draw the PERT chart on the recording board. Be prepared to make adjustments to the tasks once the sequence is laid out.
5. Next, estimate the time to complete each task and enter it next to the task on the list.
6. Readjust the sequence of tasks as necessary.
7. Determine who is responsible for each task on the list, and add his or her name next to that task on your PERT chart.
8. Continue to readjust sequence until all team members agree to the PERT chart accuracy.
9. Once the PERT chart is complete, redraw it and make copies for all team members.

The Gantt chart is a timeline chart that clearly shows when each task is to begin, the amount of time necessary to complete each task, and those tasks that will be going on at the same time. There can be more than one level of Gantt charts. For example, one chart could display the overall production phase from beginning to end; another chart could focus on activities for a short period of time; and another chart might show a task for a specific week. Figure 47-11 shows a two-week chart.

Use the following steps to create a Gantt chart:

1. Identify all tasks that need to be completed during a production cycle, including any pre- and post-production activities.
2. Put these tasks in chronological order.
3. Create estimates for how long each task will take, based on research and staff input, and add the estimates to the list.
4. Create a chart with time across the top and tasks chronologically down the right side. Draw a horizontal line across the chart for each task to indicate how long each task will take.
5. Add information to your chart across each task line, such as the people responsible for each task or the materials or cost associated with each task.
6. Post copies of the production schedule, and confirm that all staff members will work when they are scheduled.

If the PERT chart is created first, it can be used to create a Gantt chart, which can speed up the process.

47-5 QUESTIONS

1. What is the purpose of a production schedule?
2. How can a production schedule coordinate activities to increase productivity?
3. What are two types of production schedule charts?
4. What is a PERT chart?
5. What is a Gantt chart?

FIGURE 47-11

A Gantt chart covering a two-week period in the construction of a printed circuit board.

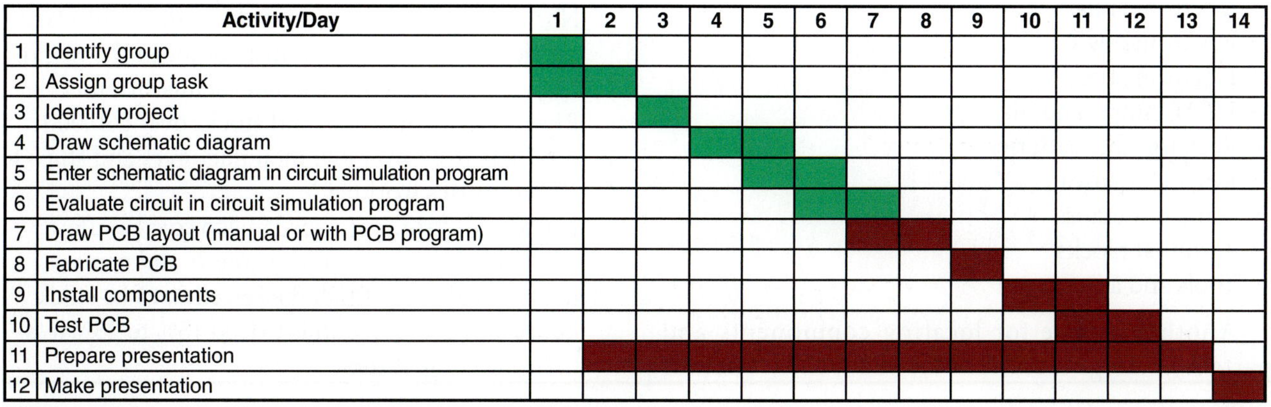

	Activity/Day	1	2	3	4	5	6	7	8	9	10	11	12	13	14
1	Identify group	■													
2	Assign group task		■												
3	Identify project			■											
4	Draw schematic diagram				■										
5	Enter schematic diagram in circuit simulation program					■									
6	Evaluate circuit in circuit simulation program						■								
7	Draw PCB layout (manual or with PCB program)							■							
8	Fabricate PCB								■						
9	Install components										■	■			
10	Test PCB												■		
11	Prepare presentation		■	■	■	■	■	■							
12	Make presentation														■

© 2014 Cengage Learning

47-6   REFERENCE MATERIALS AND MANUALS

The electronic industry wants an electronic technician to know where to find information rather than to memorize it. They want this person to have the ability to do the job rather than to spend time trying to memorize a lot of information that is in the reference manuals. However, they also want the electronic technician to know how and where to find the necessary information.

Many reference books provide this electronics knowledge. Manufacturer reference books are set up more for designers of electronic circuits. The electronics technician should be familiar with these **manufacturer reference manuals**. Manufacturer's reference manuals are updated annually. Before throwing out an older reference manual, check that the newer manual covers the same material. Many times a product line is discontinued, and the older reference manuals are the only documentation available.

The purpose of **cross-reference guides** (Figure 47-12) is to assist searching for a replacement part by manufacturer, substituting components with devices that are similar in function. It is strongly recommended to review the data sheets when locating a substitute device.

A partial listing of electronic components that can be substituted with different manufacturers is presented here:

- Battery holders
- Batteries
- Cable tie
- Electronic alarms
- Electronic buzzers
- Electronic capacitor
- Electronic plugs and connectors
- Electronic potentiometer
- Electronic relay
- Electronic switch
- Electronic transducers
- LEDs, lamps, and displays
- Rechargeable batteries
- Resettable fuses
- Semiconductors
- Terminal blocks
- Tools and accessories

Another source for locating components and parts is catalogs from vendors that sell electronics

■ FIGURE 47-12

An example of a cross-reference published by NTE. These guides are no longer published, as the information is now on the Internet.

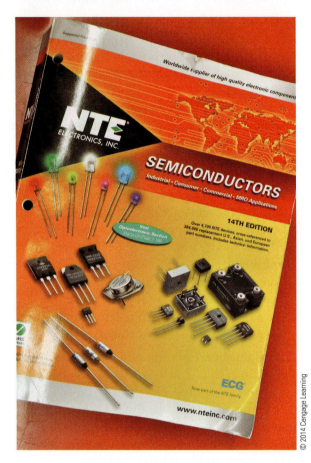

components, parts, and equipment. Listed here are three vendors (Figure 47-13) that offer catalogs and sell on the Internet:

- Digi-Key Corporation (www.digikey.com)—the fourth largest electronic component distributor in North America, Digi-Key ranks as the eighth largest electronic component distributor in the world. Digi-Key is located in a single facility in Thief River Falls, Minnesota, and stocks over 500,000 products from over 470 manufacturers and ships to 170 countries worldwide.
- Jameco Electronics (www.Jameco.com)—offers more than 50,000 popular name-brand components but also offers a large selection of in-stock house and generic brands priced 15% to 25% less than the competition.

■ FIGURE 47-13

Three vendors that offer catalogs and sell on the Internet.

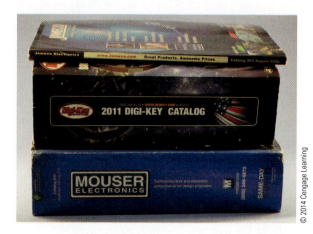

© 2014 Cengage Learning

- Mouser Electronics (http://www.mouser.com)— was founded by Jerry Mouser, a physics instructor, who needed components for a newly formed electronics program at his school. Mouser prides itself on a vast selection of quality products; competitive pricing; fast, accurate delivery; and excellent service.

There are many suppliers of electronics components and parts, and these three vendors are just a sampling of suppliers available.

47-6 QUESTIONS

1. What is the electronics industry looking for when hiring new electronics technicians?
2. What is one source of electronics reference manuals?
3. List five electronic devices and/or components that can be substituted.
4. What should be reviewed when substituting components?
5. What is another source for locating electronics parts and components?

SUMMARY

- Ideation is the process of developing and expanding ideas.
- Brainstorming is a team effort for reaching a solution to a specific problem.
- Pareto voting can reduce many ideas to a problem's solution to a few.
- A printed circuit board (PCB) prototype is a sample of the PCB built to test a circuit and to learn from it.
- A PCB prototype provides the following advantages: It enables testing and refining the PCB prototype design and makes it possible to test the design with various components in order to improve performance.
- A PCB design, not the schematic diagrams, can be protected by a copyright.
- Ways to protect PCBs are through continually upgrading performance, potting or encapsulating the PCB with epoxy that cannot be seen through or removed, or casting them in silicon and seeking protection under the Circuit Layout Act.
- A project timeline is an important resource for planning and a useful tool for staying on task.
- Steps to develop a timeline include identifying which tasks need to be completed, aligning tasks into the building blocks, and filling in between the tasks with sequential steps to move from one task to the next.
- A budget is a financial plan created during initial planning to help determine whether an idea is practical and cost-effective.
- Steps used to create a budget include identifying goals, looking at resources, defining costs, and creating the budget.
- A production schedule coordinates activities to increase productivity and minimize operating expenses.
- Production schedules are created using PERT charts and Gantt charts.
- The electronics industry wants electronic technicians to be able to perform; they also want them to know where to find pertinent information.
- Cross-reference guides assist in searching for replacement parts by manufacturer.

CHAPTER 47 SELF-TEST

1. Using Multisim samples, select the Wien-Bridge Oscillator circuit (Figure 47-14) and develop a PCB prototype. *Note: The first step is to redraw the circuit into a proper schematic diagram and document all steps required to complete the prototype.*

2. Using the schematic diagram in problem 1, justify why copyright issues are not violated.

3. Is a PCB prototype layout capable of being copyrighted?

4. Develop a timeline for problem 1.
   Note: Include steps from gathering of the parts to completely assembling the PCB.

5. In problem 4, how much time is required to complete the task?

6. Using problem 1, assemble a budget for what it would cost to build the PCB.

7. Using the budget created in problem 6, determine whether the project is cost-effective.

8. Develop a production schedule using a PERT chart for problem 1.

9. From the PERT chart in problem 8, develop a Gantt chart.

10. Using the Internet, find the least expensive components to construct the circuit in problem 1.

■ **FIGURE 47-14**

Wien-Bridge Oscillator circuit.

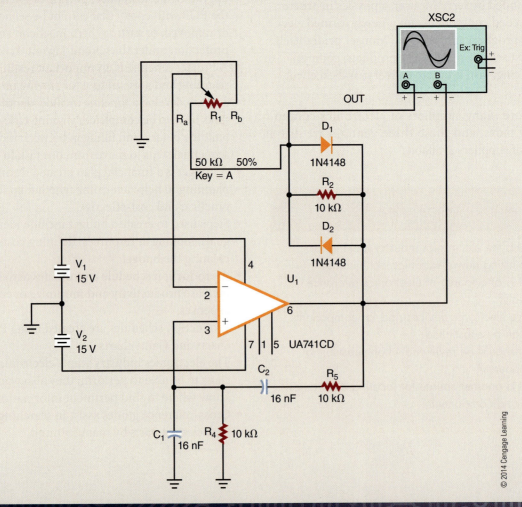

Printed Circuit Board Fabrication

OBJECTIVES

After completing this chapter, the student will be able to:

- Describe the fundamental process for making a printed circuit board.
- Design and lay out a printed circuit board from a schematic diagram.
- Discuss how to transfer a design to a copper-clad board using the hand transfer, direct transfer, or screenprinting technique.
- Identify techniques to remove the excess copper from a copper-clad board on which a design is formed.
- Explain how to drill the appropriate holes in an etched printed circuit board.

KEY TERMS

n the past, all **printed circuit boards (PCBs)** had **traces** on only one side. Today, printed circuit boards have traces on both sides and are referred to as double-sided PCB. Complex boards may have four or more trace layers, with each layer insulated from the other by board material. Special **multilayer PCB** may have eight or more trace layers. This chapter focuses on how simple single and double-sided printed circuit boards are fabricated.

48-1   FUNDAMENTALS

A printed circuit board (PCB) consists of an insulating base material that supports the copper traces, as shown in Figure 48-1. The insulating base material is typically epoxy fiberglass, but ceramic and Teflon® are also used as base material in special cases.

Phenolic base material was used in the past for applications in which performance was not critical because it was inexpensive. The standard base thickness was typically 1/16 inch, or 0.062 inch (Figure 48-2).

A majority of printed circuit boards made in the United States start with a solid copper foil that is bonded to the insulating base material, which is

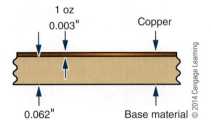

■ FIGURE 48-2
Standard board thickness.

■ FIGURE 48-3
Example of epoxy fiberglass printed circuit board material.

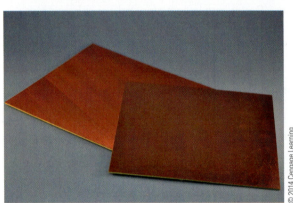

■ FIGURE 48-1
Copper clad board which is made into a printed circuit board.

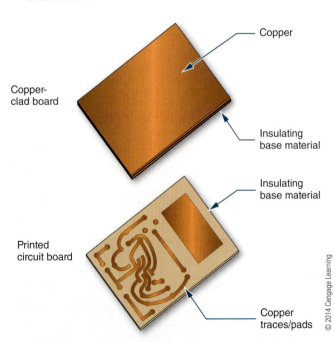

typically epoxy fiberglass (Figure 48-3). The thickness of the copper on a printed circuit board is designated in ounces. One ounce refers to the weight of copper spread over 1 square foot. One-ounce copper plating is approximately 0.003 inch thick. Two-ounce copper is 0.006 inch thick, and so on.

The desired traces are printed onto the copper with a material called resist. The board is then placed in a chemical tray with a chemical for etching, referred to as an etchant, to remove the copper not covered by the resist, as shown in Figure 48-4. The etchants often used for etching small quantities of printed circuit boards are either ferric chloride or ammonium persulfate. A major concern with etchants is disposal according to current environmental standards.

Commercial printed circuit board fabricators generally use different chemicals that are less expensive, such as ammonium chloride. They heat the bath and add agents to improve performance.

After washing and cleaning the etched board, the result is a copper circuit pattern on an insulating base—a printed circuit board (Figure 48-5).

■ **FIGURE 48-4**

Example of an etched printed circuit board.

© 2014 Cengage Learning

■ **FIGURE 48-6**

X-ray view of a printed circuit board layout from the component side.

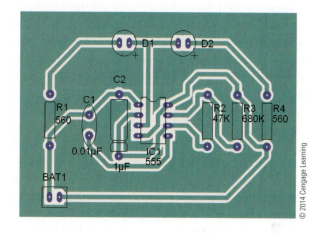

© 2014 Cengage Learning

■ **FIGURE 48-5**

Printed circuit board that has been etched and resist removed.

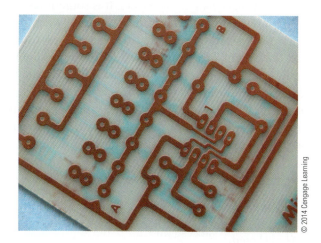

© 2014 Cengage Learning

■ **FIGURE 48-7**

Examples of the component layer on printed circuit boards.

© 2014 Cengage Learning

Before the arrival of **surface mount technology (SMT)**, there were two designated layers for through-hole printed circuit boards: the **component side** and the **solder side**. With the implementation of SMT, components are now placed on both sides of the printed circuit board, which are referred to as top and bottom. Both conventional through-hole and SMT layouts are viewed from the component side, or top. This makes the bottom layer similar to an X-ray view through the insulating base material, which can be very confusing (Figure 48-6).

There are also two additional layers of a printed circuit board after it is etched. These additional layers are nonconductive and form the component layout layer and the solder mask layer. The component layout layer

acts as a map when assembling and troubleshooting the printed circuit board. This layout layer consists of text and graphic symbols that identify each component with its proper orientation. A good component layout layer can save a tremendous amount of time (Figure 48-7).

The **solder mask** controls or indicates where the **solder** will be placed on the printed circuit board pads during a process called wave soldering. This solder only connects the components to the solder pads. When making only one or two boards with a simple design, a solder mask may not be needed. Additionally, the solder mask adds corrosion protection to the printed circuit board traces.

In theory, with only through-hole passive components and discrete transistors, a single layer is all that is necessary (Figure 48-8). Most radio-frequency (RF) printed circuit boards have traces on only one side, although a second layer is used as a **ground plane** to improve performance, as shown in Figure 48-9. Standard integrated circuit pin outs used with analog ICs—such as op-amps, timers, and so on—can get by with a single layer. This is because there is plenty of room for traces to pass between the leads of the passive components.

When digital logic or other designs that use high-pin-count chips are used, two trace layers are required. Digital IC chips with a large number of pins are impossible to connect to other chips with a different layout without crossing and shorting the traces. Therefore, a second layer is required. In theory, only two layers are ever necessary with high-pin-count components.

In practice, more than two layers are used for dense, highly populated boards or for performance considerations. Two layers are inadequate with densely populated printed circuit boards because the parts are so close together and the traces and holes take up too much space. There simply isn't room for all the traces, holes, and vias on two sides. Vias are connections between one layer and another (Figure 48-10). If the density could be reduced, the two-layer theory would work.

For high-speed or high-precision designs, more than two layers are used to improve performance. With four trace layers, the two middle ones are usually power and ground. This evens out the flow of power and reduces voltage fluctuations at different points on the printed circuit board. The ground plane layer provides controlled impedance for high-speed signals. The close proximity of the ground and power planes also acts as a capacitor to reduce noise. One problem with multilayer printed circuit boards is the reliability of connections between the layers. It is sometimes difficult to detect which layer is defective.

Two sides can be etched at the same time; therefore, the cost of a **double-sided PCB** is only slightly higher than that of a **single-sided PCB**. Multilayer boards are built up by laminating layers together. A four-layer board costs about twice as much as a double-sided printed circuit board. The printed circuit board is often the most expensive part of a design.

■ **FIGURE 48-8**

Single-sided printed circuit board.

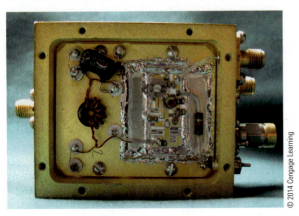

© 2014 Cengage Learning

■ **FIGURE 48-9**

Ground plane on a printed circuit board.

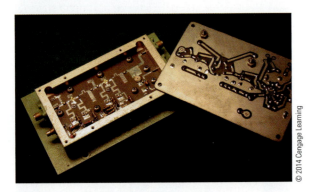

© 2014 Cengage Learning

■ **FIGURE 48-10**

VIAs connect the top to the bottom trace of a printed circuit board.

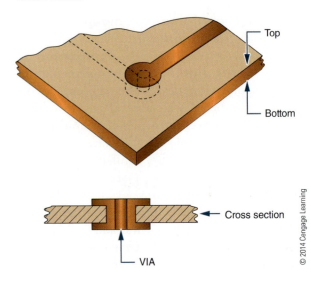

Top

Bottom

Cross section

VIA

© 2014 Cengage Learning

1. What are the materials that make up a printed circuit board?
2. How is the copper on a printed circuit board designated?
3. What is the function of resist?
4. Describe the process for making a printed circuit board.
5. What is the significance of surface mount technology (SMT)?

48-2 LAYING OUT PRINTED CIRCUIT BOARDS

Once a good schematic diagram is confirmed through breadboarding, either physical or virtual, the printed circuit board can be laid out. The necessary electronic components should be collected for sizing the design layout. Leave the components from the breadboard together and use them only as a last resort. A working circuit on the breadboard can be used for troubleshooting in the event there is a problem with the printed circuit board design.

Once the decision to design a printed circuit board has been made, a number of questions must be addressed before the printed circuit board is laid out:

1. What is the power source? Will it be part of the board design or external to it?
2. What oddly shaped or sized devices will be mounted? Devices such as relays, switches, and large capacitors must be identified.
3. How are the discrete components attached? They have to be identified along with sockets, pins, connectors, and mounting devices.
4. What are the testing and troubleshooting considerations? They must be taken into account to make the board accessible for test measurements of all signals.
5. Will the board be mounted in an enclosure? How much space it will use and how it will be attached must be taken into consideration.

Once these questions have been answered, identify which components will be mounted on the printed circuit board and which ones will be off the board.

One approach to a printed circuit board design layout is to use **paper dolls** of each component that is to be mounted on the printed circuit board (Figure 48-11A). Each paper doll is the exact size of the base of the component being used and permits correct component placement.

Using the paper dolls, lay out the printed circuit board in such a way that the paper dolls duplicate the schematic diagram. Exact component and lead placement are not critical at this stage. Next, consider the trace placements (Figure 48-11B) for proper electrical connections. During the layout process, the printed circuit board will be viewed from the component side or top. Three issues address the best component placement:

1. Wasted board space—Keep components in a horizontal or vertical plane, parallel to the board edge, to achieve a tighter layout. Avoid diagonal placements.
2. Insertion of polarized components—Try to group polarized components facing the same direction to aid assembly.
3. Ease of troubleshooting—Components should be easy to find; try to keep input on the left side of the board, output on the right, and power from top to bottom of the board.

■ FIGURE 48-11

Using paper dolls for laying out a printed circuit board design.

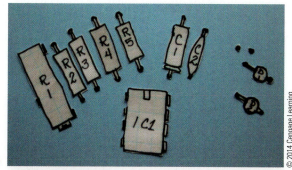

(A)

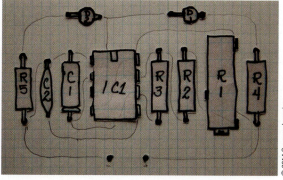

(B)

© 2014 Cengage Learning

With the exception of discrete logic gates and multiple op-amp packages, arrange the IC pins in physical order, as viewed from the top. Start with the IC that has the most pins, and then place the other parts that follow the simple schematic. The goal is to lay out a printed circuit board, with no crossed traces, that resembles the physical relationship of the schematic diagram, which becomes useful in troubleshooting.

Transfer the paper doll components to a 0.100-inch grid graph paper (10 lines per inch). Component orientation should be horizontal or vertical to the edge of the board. Identify the top and bottom layers with text component side and solder side. The solder side is viewed from the top like an X-ray view, as shown in Figure 48-12. Traces are then laid out left to right or top to bottom, with exceptions permitted. Traces may be routed between component leads, but not between the leads of semiconductor devices. At this stage, it may be necessary to reposition parts on the layout to avoid crossing traces. Once the paper doll components are placed and the traces are laid out, the decision to make a single-sided or double-sided printed circuit board can be identified. If no traces cross, a single-sided printed circuit board is optimal; otherwise, a double-sided printed circuit board should be used.

Another choice for crossed traces is to make a single-sided board with **jumpers**, as shown in Figure 48-13, but keep the jumpers to a minimum. Rearranging artwork can often eliminate the need for a jumper wire. Some guidelines for do's and don'ts for trace layouts are shown in Figure 48-14. Rules are made to be broken, even when designing printed circuit boards. When a printed circuit board is designed that has a connection

■ FIGURE 48-13

Single-sided printed circuit board using jumpers.

Jumper

© 2014 Cengage Learning

■ FIGURE 48-14

Guidelines for trace layouts.

Do   Don't

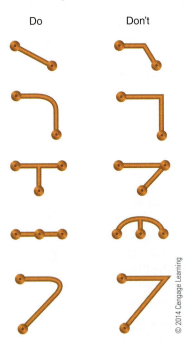

© 2014 Cengage Learning

on one edge, the input and output connections are on the same edge of the board. When this occurs, separate them on opposite ends of the connection. If this is not possible, separate them by putting other traces in the space between them.

Pads are placed on the printed circuit board when a connection is to be made. A pad is used to connect a lead from a component or connect a wire from an off-board component. Generally, each lead has its own

■ FIGURE 48-12

Solder side as viewed from the top.

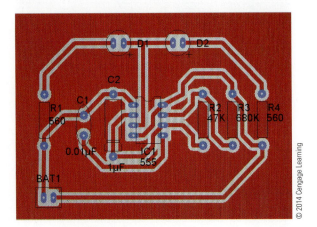

© 2014 Cengage Learning

■ FIGURE 48-15

Two leads connecting to the same point will have a pad for each lead.

© 2014 Cengage Learning

■ FIGURE 48-16

0.007-inch trace versus 0.020-inch trace.

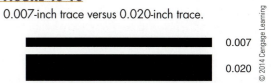

© 2014 Cengage Learning

mounting pad. Two leads connected to the same point each have their own mounting pad. Multiple pads are permitted (Figure 48-15). Each component should have enough mounting room. It is difficult to modify the board or component for errors after a printed circuit board is made.

Double-sided printed circuit boards should always have some identification on the top and bottom layers to aid in proper orientation. In the past, optical targets or registration marks were used to aid in differentiation between the top and bottom traces. Presently, this step is omitted by identifying pin 1 of an IC with a square pad.

Printed circuit boards are easier to manufacture if the traces are wide and far apart—this also makes the board more reliable and provides better performance. Most printed circuit board manufacturers will work with 0.007-inch traces and spaces (Figure 48-16).

Traces that are too thin are fragile and have higher resistance than thick traces, and they handle less current. Estimating how much current a trace can handle is important. Figure 48-17 shows a table comparing trace width and current capacity for a 1-ounce-per-square-foot copper foil. These values are reasonably linear for different trace cross-section areas.

Traces that are placed close together may short out and can electrically short to each other. The traces can also act as antennas and capacitors. High-speed digital lines next to high-impedance analog lines will cause problems. Long traces have the same problems as long wires. They pick up noise and have noticeable impedance. Even though a digital design only runs at 1 MHz, the changing state will cause transients in the GHz range.

Guidelines for laying out traces are shown in Figure 48-18. Try to limit using angles of 90° in traces. They leave a point that causes problems by acting like a one-turn inductor. Additionally, the point will not be well supported and will easily lift off the base material of a printed circuit board.

■ FIGURE 48-17

Printed circuit board trace information for 1-ounce copper clad board (based on a trace length of 10 inches).

Trace Width in Inches	Trace Width in Millimeters	Resistance /MM in $\mu\Omega$	1 OZ Copper 0.00135" Thick		Current in Amperes at 20°C
			cm^2	AWG	
0.02	0.508	989.7	0.000174	35	3.00
0.04	1.016	494.9	0.000348	32	5.00
0.06	1.524	329.9	0.000523	30	6.50
0.08	2.032	247.4	0.000697	29	7.00
0.10	2.540	197.9	0.000871	28	8.00
0.12	3.048	165.0	0.001045	27	9.25
0.14	3.556	141.4	0.001219	26	10.00
0.16	4.064	123.7	0.001394	26	11.00
0.18	4.572	110.0	0.001568	25	11.75
0.20	5.080	99.0	0.001742	25	12.50

© 2014 Cengage Learning

■ FIGURE 48-18

Trace layout Do's and Don'ts.

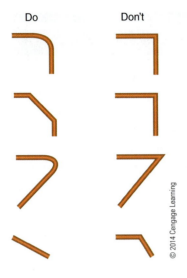

© 2014 Cengage Learning

Using two 45° angles instead of one 90° angle, or using a curve, will save space. This is especially true when there are a lot of traces running parallel to each other. The more space saved, the smaller and less expensive the printed circuit board will be. Printed circuit boards are purchased by the square inch.

Proper grounding is critical. A ground and power plane are ideal, but often this is simply too expensive, as it doubles the cost of a printed circuit board.

In laying out a printed circuit board, it is best to start with a wide ring of 0.100- to 0.248-inch around the printed circuit board in order to form a ground. This provides large ground areas around the printed circuit board mounting holes, which are typically located near the edge of the printed circuit board. This setup allows a star washer and a metal screw to make a good chassis–ground connection. It also means that ground traces only have to go to any board edge, simplifying the layout. If there is any concern regarding ground loops with more than one ground current flowing in one conductor, it can be solved by cutting the loop.

In this manner, the current flows in the trace can be controlled. If a design is fairly simple, spend the extra time to make it a single-sided printed circuit board, and use the second side as a ground plane. There are always more connections to ground than to any other circuit node.

When the layout is finished, increase the width of the power and ground traces as much as possible. In large areas without traces, fill them with solid copper connected to power or ground. More copper often helps and very rarely hurts printed circuit board performance and reduces etching time.

Proper placement of traces is something that is learned through practice. The only way to do it is to actually hand-route printed circuit boards. A common and effective technique is to run horizontal traces on one layer and vertical traces on the other layer. Obviously, this doesn't work for single-layer boards. This tends to create boards with a lot of **vias**. A via is a plated through-hole used to pass a conductor from one side of the printed circuit board to the other.

Basic rules for laying out a printed circuit board include those listed here:

- Avoid running parallel traces closer than necessary.
- Segregate analog and digital areas.
- Put ground traces next to sensitive analog lines to act as shields. Do the same for high-speed clock lines to reduce **electromagnetic interference (EMI)** and crosstalk.
- Pay close attention to parts placement. It's amazing how this can simplify routing.

It is useful to use a color-coded system to simplify the presentation of a double-sided printed circuit board during the design process. This process is standard with computer-generated designs. It can also be done with colored pencils. A typical color-code system is as follows:

- Black—board outline, holes, pads, and hardware
- Green—component electrical symbols and outlines
- Red—connecting traces on the component side
- Blue—connecting traces on the trace side
- Yellow—through holes

Surface mount technology (SMT) has added more facets to printed circuit board layout. The parts are very small, and the lead pitch or spacing between the centers of adjacent leads can be very small. It is not possible to run any traces between the leads. This makes routing more difficult and increases the number of vias.

It's virtually impossible to use a single-side layout with high-pin-count SMT ICs. The pad sizes and shapes are different for different parts. You will need to refer to the manufacturer's specifications. With SMT designs, testing and repair need more consideration because it's hard to probe tiny, closely spaced pins. Boards should be laid out to make servicing easy.

Autorouter programs are designed to use the computer to lay out a printed circuit board. Autorouters should not be relied on unless they are "smart." Typical low-cost routers only connect points together. They don't consider the length of the trace, which traces are analog and which are digital, which lines are sensitive, which are ground loops, and so on. Subtle points can make a significant difference in how a printed circuit board performs. It's easier to lay out a printed circuit board manually than to root out elusive errors made by a program. Autorouter programs include National Instrument's Ultiboard, New Wave Concepts' Circuit Wizard, and Cadence's OrCAD PCB Designer.

Getting a printed circuit board made commercially requires a design to be sent via the Internet to a service. It requires the following files in **Gerber format** for a standard double-sided printed circuit board:

1. Top trace layout
2. Bottom trace layout
3. Silkscreen (if used)
4. Solder mask (if used)
5. Aperture file
6. Drill file

The Gerber format is a standard printed circuit board format supported by virtually all layout software. The Gerber files define the trace placement, but they don't specify the physical sizes. The **aperture file** specifies the physical sizes with a short list of "flash codes." Additionally, the Gerber files don't specify the actual size of the holes to be drilled in the printed circuit board. The **drill file** informs the manufacturer of where the holes are and what size they are supposed to be. Drilling and etching are two separate procedures. When these files are sent, they include a short "read me" file that relates the file names to the physical parts of the printed circuit board to save time and reduce confusion.

Proper printed circuit board layout requires common sense, patience, and attention to detail.

48-2   QUESTIONS

1. What is necessary to collect before laying out a printed circuit board, and why?
2. What decision must be made before laying out a printed circuit board?
3. What approach can be used in laying out a printed circuit board?
4. What issues must be addressed in placing components on a printed circuit board layout?
5. How can cross traces be addressed on a single-sided board?
6. On double-sided board layouts, how are the sides identified?
7. Why are thin traces undesirable on a printed circuit board layout?
8. What guidelines are essential in laying out a printed circuit board?
9. What purpose do ground and power planes serve?
10. How are traces run on a double-sided board?
11. What consideration should be regarded when laying out a printed circuit board?

48-3   TRANSFERRING DESIGNS

There are several techniques used to transfer the artwork to the copper-clad board. The simplest and first technique is to **hand-draw** the design on the board with a resist pen or any other permanent marker with a very fine tip.

Take the original artwork and transfer it to the back of the graph paper. Place the component side of the artwork to the surface of a light table or hold the paper to a window with light on the opposite side. Start by placing Xs where each pad is located, and then connect the pads with the trace lines. On the side just drawn, write "BOTTOM."

Wrap the artwork around a piece of copper-clad board that has been properly cut to size. Clean the board with steel wool or with powder cleanser and water before wrapping the artwork around the board. The copper side of the board should be toward the artwork, with the side identified as bottom facing outward.

Using a scribe, locate the center of each pad where an X was placed on the bottom side. Before removing the artwork, run a finger over each pad to ensure it was marked with the scribe.

Remove the artwork, being careful not to get fingerprints on the copper.

With a resist pen or other fine-tip permanent marker, such as a black Sharpie, draw the pads at each of the scribe marks. Using the bottom-view artwork, connect the pads with traces. Go over each of the traces and pads several times to build up a thick layer of resist.

A second technique for creating a resist involves using the computer to make the artwork with a CAD or drawing program. Draw the design from the component side and then reverse it to form the trace or bottom side. Print out a positive image 1:1 on transparency or special film that can then be transferred to a clean, properly sized copper-clad board with an iron and pressure.

A third technique involves printing out two positive copies of the artwork 1:1 on transparency film on a copying machine. Align the two pieces of film and tape them together. They form a positive film transparency. If a higher density is needed from the film, align and tape additional pieces of film. Sensitize a clean copper-clad board with positive resist according to the manufacturer's instructions. Using a contact print frame, sandwich the film and sensitized board together with the bottom side outward. Expose the sandwiched board and film to an ultraviolet light source or sunlight. Develop the board in the appropriate developer; rinse and dry the board. This technique forms a very high-quality resist.

A fourth technique involves using the positive film transparency developed in the previous discussion to create a screen for screenprinting the printed circuit board. A good graphic arts book will cover the making of a screen and the screenprinting process in detail. Use resist inks or oil base inks when screenprinting the layout on the boards. This technique provides the capability of making many PCBs with the same design.

48-3   QUESTIONS

1. List the techniques for transferring a design to a printed circuit board.
2. Why is it essential to label the top and bottom of a printed circuit board?
3. What can serve as a resist with hand-drawn printed circuit boards?
4. How can the density be increased when using transparency film?
5. What purpose does screenprinting serve with printed circuit board fabrication?

48-4   ETCHING PRINTED CIRCUIT BOARDS

There are two major techniques to remove copper from a copper-clad board to form a printed circuit board. The first technique involves using a mild acid such as **ferric chloride** or **ammonium persulfate**. The other technique uses a CAD-type program and a **computer numerical control (CNC) machine** to remove the copper and is referred to as a nonchemical process. Environmental and safety issues are favoring the use of nonchemical copper removal techniques.

Using either ferric chloride or ammonium persulfate acid, the copper can be removed by the tray method or with a spray etcher.

CAUTION!

When etching, always wear safety glasses, rubber gloves, and an apron.

To use the tray method, the copper-clad board with applied resist is etched in a glass tray by placing the board onto the surface of the acid. The copper side of the board should face toward the acid. Float it on the acid for approximately 15 to 20 minutes. The acid laden with copper sinks to the bottom, and fresh acid replaces it. In this fashion, no agitation is required.

A different technique uses a 1-inch square of sponge to dab the acid on the printed circuit board. The acid is absorbed in the sponge by pouring a small amount on it. The sponge is dabbed continuously on the copper on the board until the copper is removed. Boards with thinner copper etch faster.

CAUTION!

Always wear safety glasses, rubber gloves, and an apron when using this technique.

Another method to remove the excess copper from the board is to use a commercial or shop-built spray etcher (Figure 48-19). These etchers can etch the board in approximately two minutes. A spray etcher is the best way to etch double-sided boards.

Computer programs such as New Wave Concepts' PCB Wizard provide the appropriate codes to drive a CNC machine to remove the copper to form a printed

■ FIGURE 48-19

Examples of commercial spray etchers.

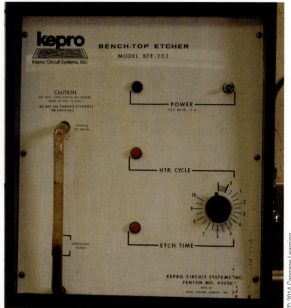

(A)

(B)

© 2014 Cengage Learning

■ FIGURE 48-20

Computer numerical control (CNC) machine used for printed circuit board fabrication.

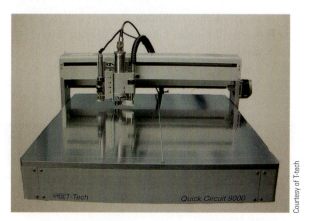

Courtesy of T-tech

circuit board. Figure 48-20 shows a CNC machine used for making printed circuit boards.

48-4 QUESTIONS

1. What are two ways to remove the copper to form a printed circuit board?
2. What acids can be used safely for removing copper to make a printed circuit board?
3. What safety precautions should be used when working with acid?
4. What are three techniques for removing copper using acid?
5. How do computer programs drive a CNC machine?

48-5 PREPARING THE ETCHED PRINTED CIRCUIT BOARD

After the printed circuit board has been etched, the resist must be removed. Remove the printed circuit board from the acid, rinse it thoroughly in running water, and dry it. Soaking the board in a solvent, scrubbing it with powder cleanser, or using steel wool to rub it off can remove the resist. After the resist is removed from the printed circuit board, be sure to avoid getting fingerprints on the board.

The next step is optional in fabricating a printed circuit board. Using a tin-plating solution, the printed circuit board can be tin-plated to stop oxidation of the copper, as shown in Figure 48-21. The printed circuit

■ FIGURE 48-21

Example of tin-plating.

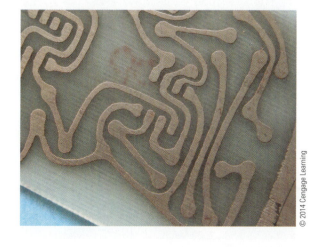

© 2014 Cengage Learning

■ FIGURE 48-22

Example of traces shorted together.

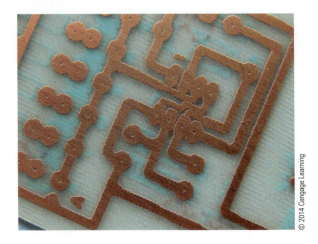

© 2014 Cengage Learning

board is immersed in the solution for a set amount of time identified by the manufacturer.

Any defects resulting from the etching process should be taken care of at this time. Types of defects include traces that are **bridged** or **shorted** together or traces that are **open** (Figure 48-22). Traces that are shorted together can be separated using a knife. Traces that are open need to be bridged with wire and solder.

■ FIGURE 48-23

Hi-speed drill for drilling holes in a printed circuit board.

© 2014 Cengage Learning

■ FIGURE 48-24

Removing burrs from drilled holes.

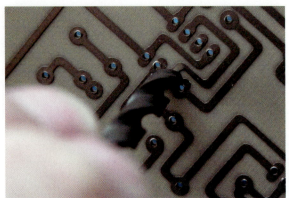

© 2014 Cengage Learning

Use a high-speed drill for drilling all holes (Figure 48-23). Use a backer board under the printed circuit board to avoid breakout of the backside. A number 59 drill bit is used for all holes. Holes that need to be drilled larger are then drilled. **Burrs** that form around the holes can be removed by hand using a larger drill bit and a twisting motion (Figure 48-24).

If the printed circuit boards are going to be soldered using a wave soldering station, a solder mask is applied to the bottom of the board. The solder mask allows the solder to flow on only the points that require solder. The solder mask can be screenprinted on the printed circuit board at this time.

48-4   QUESTIONS

1. How can the resist be removed from a printed circuit board?
2. What has to be avoided once the resist is removed from the printed circuit board?
3. What is the purpose of tin plating?
4. What types of defects can occur when fabricating a printed circuit board?
5. How are the holes made in a printed circuit board?
6. What is the function of the solder mask?

SUMMARY

- A printed circuit board (PCB) consists of an insulating base material, which supports the copper traces.

- The insulating base material is typically phenolic or epoxy fiberglass.
- The thickness of the copper is designated in ounces and refers to the weight of 1 ounce of copper spread over 1 square foot.
- The desired traces are printed onto the copper with a material called resist.
- The etchants most often used for etching small quantities of printed circuit board are ferric chloride and ammonium persulfate.
- Commercial printed circuit board fabricators use different chemicals that are less expensive, such as ammonium chloride that is heated with additional agents to improve performance.
- SMT allows components to be placed on both sides of the printed circuit board.
- Two additional layers, the silkscreen layer and the solder mask layer, are applied to a printed circuit board after it is etched.
- In theory, a single layer is all that is necessary with through-hole passive components and discrete transistors.
- Most RF boards have traces on only one side, with a second layer used as a ground plane to improve performance.
- More than two layers on a printed circuit board are used for dense, highly populated boards or for performance considerations.
- A number of decisions must be made before a printed circuit board can be laid out.
- A printed circuit board design can be laid out using paper dolls of each component.
- During the layout process, the printed circuit board is viewed from the component side, or top.
- The goal in laying out a printed circuit board is to have no crossed traces and to have the layout resemble a physical relationship of the schematic diagram, which is useful in troubleshooting.
- Jumpers can be used for crossed traces on a single-sided board.
- A pad is used to connect a lead from a component or wire from an off-board component.

- Printed circuit boards are easier to manufacture if the traces are wide and far apart.
- Traces that are too thin are fragile, have higher resistance than thick traces, and handle less current.
- Traces that are placed close together may short out and can be electrically coupled to each other.
- Avoid using angles of less than 90° in laying out traces.
- When the layout is finished, increase the width of the power and ground traces as much as possible.
- Use a color-coded system to simplify the presentation of a double-sided printed circuit board during the design process.
- Surface mount technology (SMT) adds more facets to printed circuit board layout.
- Autorouters are designed to use the computer to lay out a printed circuit board.
- Getting a printed circuit board made commercially requires a design to be sent to a service.
- The Gerber format is a standard printed circuit board format that defines the trace placement.
- The aperture file defines the physical size with a short list of "flash codes."
- The drill file informs the manufacturer of where the holes are located and what size they are.
- Transferring the artwork to the copper-clad board can be done by hand, with transparency film, and through screenprinting.
- Copper can be removed from a copper-clad board to form a printed circuit board by using a mild acid such as ferric chloride or ammonium persulfate or by using a CAD-type program with a computer numerical control (CNC) machine.
- After the printed circuit board has been etched, the resist must be removed from it.
- The printed circuit board can be treated with a tin-plated solution to stop oxidation of the copper.
- Types of defects in fabricating a printed circuit board include traces that are bridged or shorted together and traces that are open.
- Use a high-speed drill for drilling all holes in a printed circuit board.

CHAPTER 48 SELF-TEST

1. What is the function of a printed circuit board?
2. Redraw the following sketch as a proper schematic diagram.

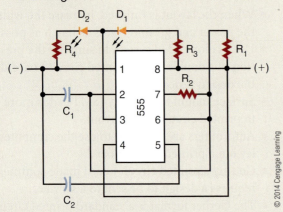

© 2014 Cengage Learning

3. Using the schematic shown in problem 2, manually lay out a single-sided printed circuit board that has no jumpers.
4. Redraw the following sketch as a block diagram.

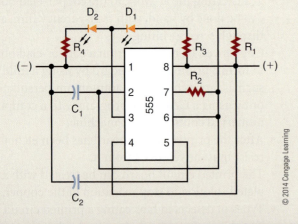

© 2014 Cengage Learning

5. List the basic rules for laying out a printed circuit board.
6. What is a Gerber file and what does it contain?
7. Describe the hand-draw process for transferring a design to a blank printed circuit board.
8. When etching a printed circuit using the tray method, what is the best way to place the board in the tray?
9. How are shorts and opens repaired on a printed circuit board?
10. Locate an MSDS for ammonium persulfate acid, and identify the proper storage procedure and whether it is okay to heat it.

Printed Circuit Board Assembly and Repair

OBJECTIVES

After completing this chapter, the student will be able to:

- Describe the items in an electronics technician's toolbox, and explain their function.
- Identify the different types of test equipment used for troubleshooting.
- Identify the purpose of different types of soldering irons and tips.
- Describe proper soldering iron tip care.
- Describe how to solder components on an etched printed circuit board.
- Identify the appearance of a properly soldered connection.
- Understand and observe all safety precautions associated with the fabrication, assembly, and soldering of a printed circuit board.

KEY TERMS

An **electronics technician** uses a variety of tools to build and repair electronic circuits on printed circuit boards. These tools, coupled with electronic test equipment, allow the technician to analyze any electronic circuit for proper operation or to identify faulty circuit components and repair them. The right tools make the job easier for an electronics technician.

49-1 ELECTRONICS TECHNICIAN TOOLBOX

The installation and repair of components in a printed circuit board may be accomplished through the use of tools designed specifically for the task. Figure 49-1 shows some of the tools in an electronics technician's toolbox.

Figure 49-2 shows two types of printed circuit board holders for holding a printed circuit board while it is being assembled or repaired. The fixtures should provide good support and have the ability to be rotated.

An illuminated magnifier, shown in Figure 49-3, provides a means of detecting improper connections or broken traces on the printed circuit board. It is also helpful when you are soldering SMT components to a printed circuit board.

Figure 49-4 shows a special tool used to assist with the forming of leads of resistors for installing in a printed circuit board.

■ FIGURE 49-1

Tools used by an electronics technician for printed circuit board fabrication and repair.

Item	Picture	Description
Spudgers	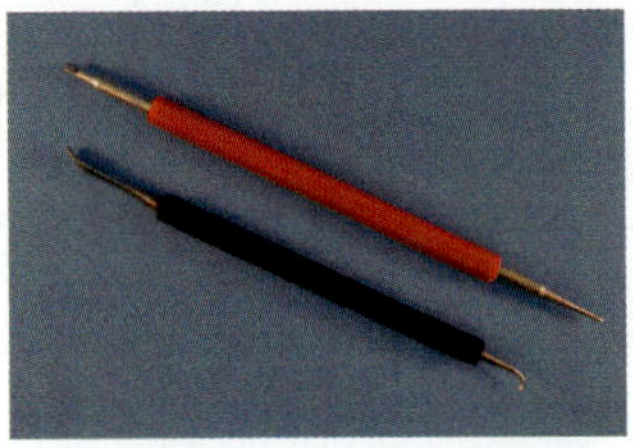	Used for lifting wires and pushing/aligning leads.
Acid brush	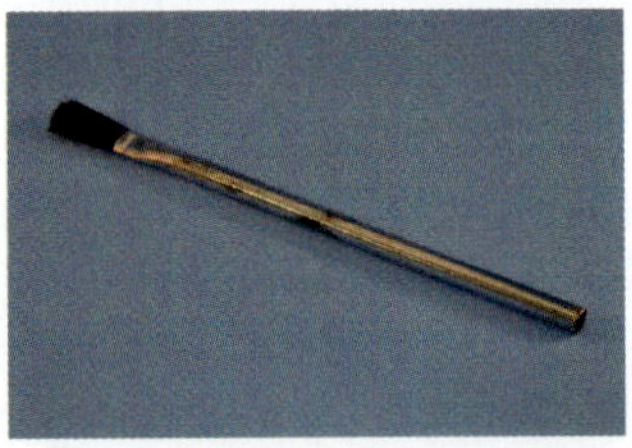	Used for applying solvent to PCB for cleaning.
Wire stripper	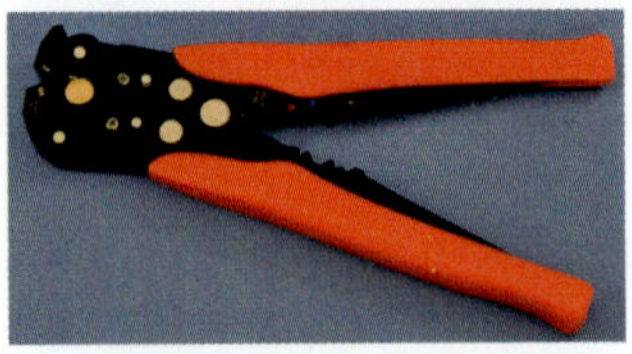	Used to strip ends of insulated leads.
Surgical scissors	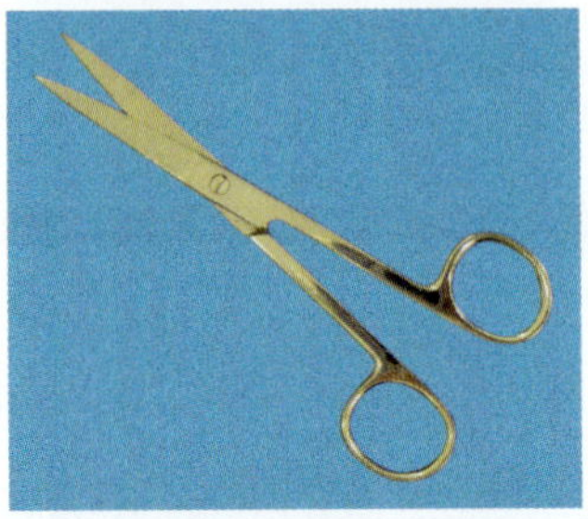	Used for general cutting

■ **FIGURE 49-1** *(Continued)*

Tools used by an electronics technician for printed circuit board fabrication and repair.

Item	Picture	Description
X-Acto knife	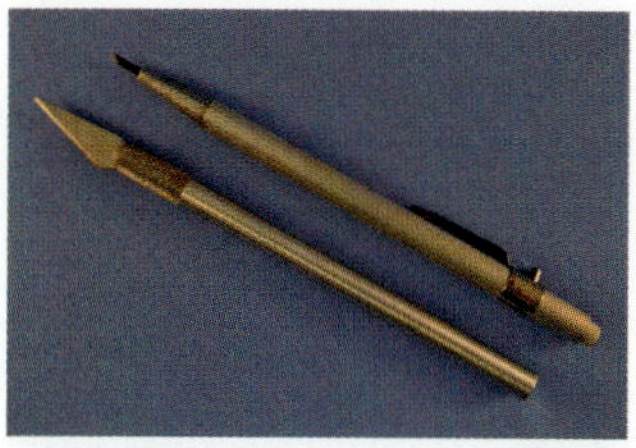	Used for general cutting, removal of bridges on PCB
Tweezers	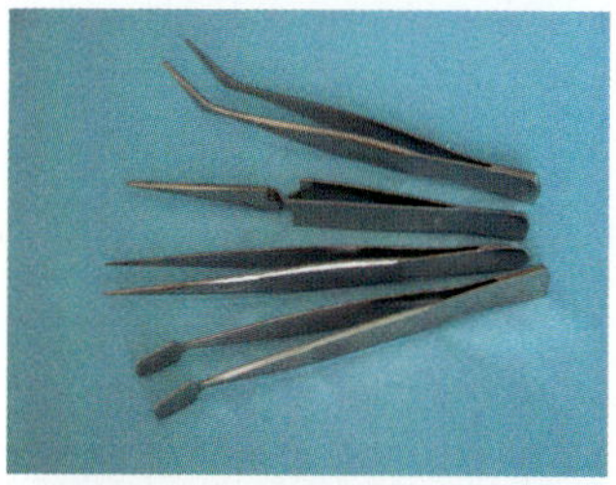	Used as an aid to lift small components.
Flat nose pliers (Duck bill)	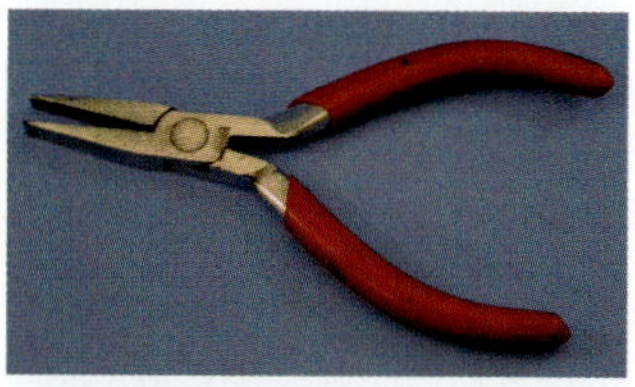	Used to straighten component leads.
Needle Nose Pliers	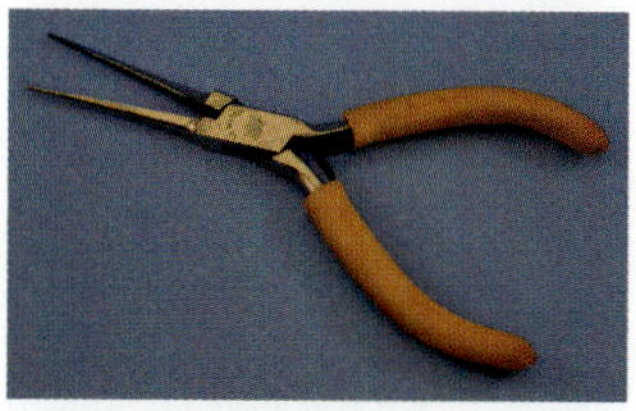	Used to form leads
Round Nose Pliers	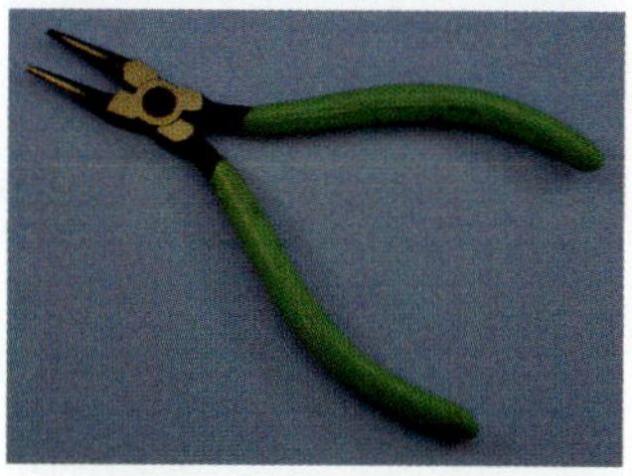	Used to form leads

■ **FIGURE 49-1** *(Continued)*
Tools used by an electronics technician for printed circuit board fabrication and repair.

Item	Picture	Description
Diagonal Cutters (Flush)		Used to cut leads flush to PCB and general lead trimming
Screwdrivers (Flat blasé, Phillips)	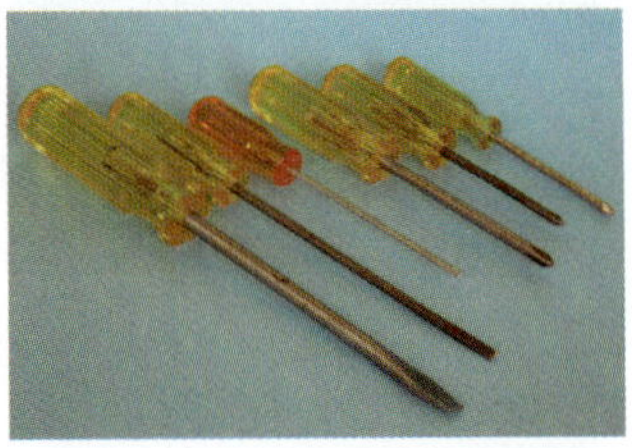	Used to install/remove screws
Nut Drivers	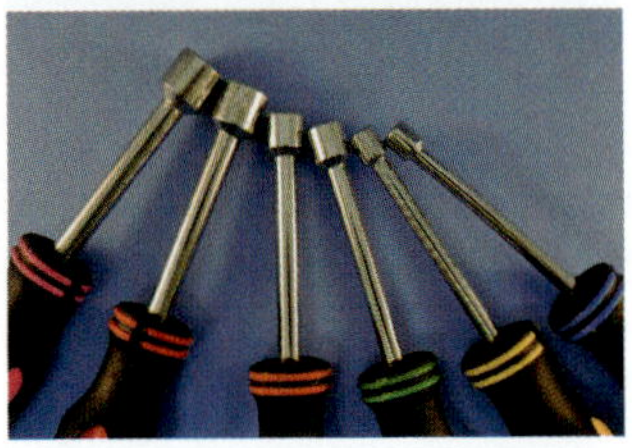	Used to install/remove bolts and nuts
1/16" And 1/8" Chisel Tip For Soldering Iron	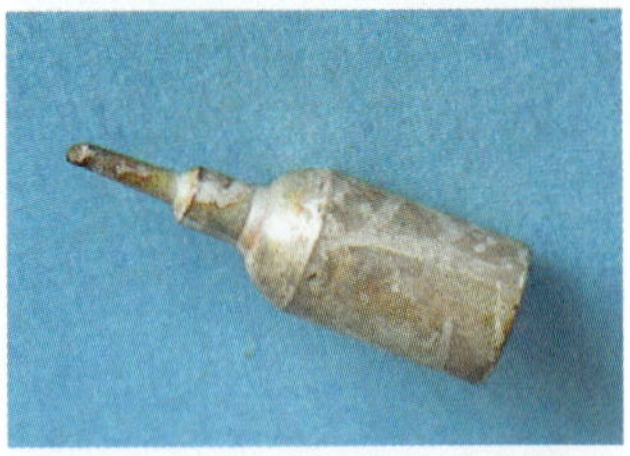	Options for components to be soldered
Heat Sinks	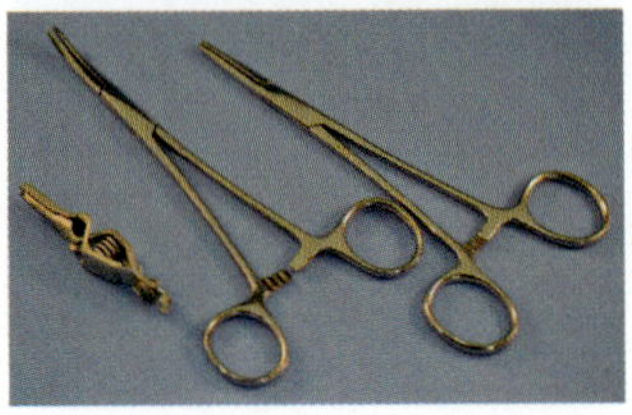	Used to remove excess heat with semiconductors

■ **FIGURE 49-1** *(Continued)*
Tools used by an electronics technician for printed circuit board fabrication and repair.

Item	Picture	Description
Alcohol And Flux Bottle	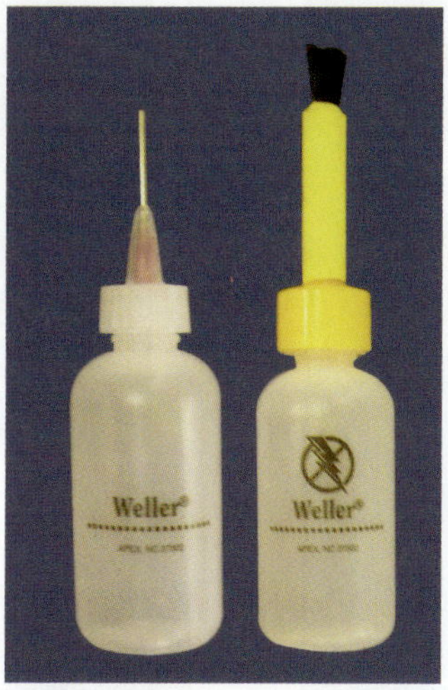	Used to hold alcohol and flux
Soldering Iron	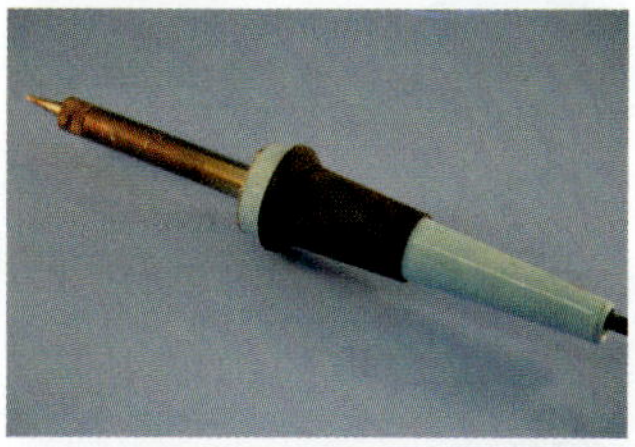	Used to solder
Desoldering Braid		Used for solder removal
Manual Desoldering Tool	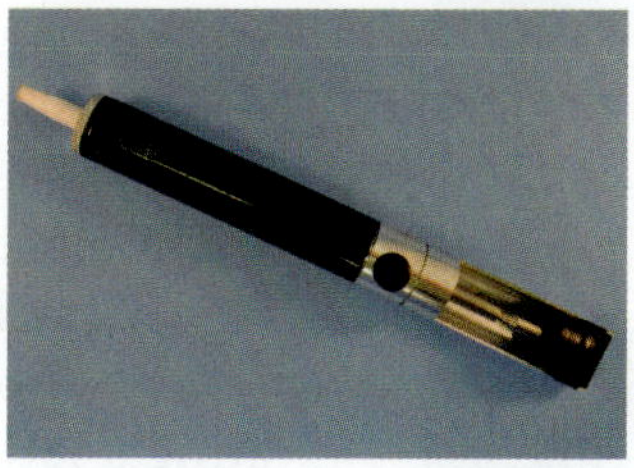	Used for solder removal

■ **FIGURE 49-2**

Examples of printed circuit board holders.

■ **FIGURE 49-3**

Illuminated magnifier.

■ **FIGURE 49-4**

Component lead former tool.

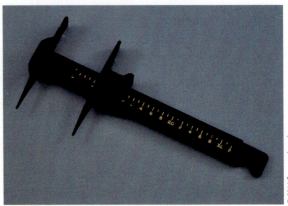

49-1 QUESTIONS

1. List the tools that would be included in an electronics technician's toolbox.
2. Where might duckbill pliers be used?
3. What type of device is used to hold a printed circuit board?
4. Where would an illuminated magnifier be used?
5. How can the leads of a resistor be accurately formed for insertion into the printed circuit board?

49-2 ELECTRONIC TEST EQUIPMENT

Electronics technicians use a variety of test equipment for determining proper operation of a circuit, aligning is components, and troubleshooting it. Figure 49-5 shows a sample of test equipment available to electronics technicians and explains their function.

An **isolated power supply** is important when powering a circuit, to prevent it from generating a lethal shock potential. Power supplies are available that generate DC voltage and/or AC voltage. Both fixed and variable voltage power supplies are available. The power supply is used to supply a circuit with the correct voltage for testing and troubleshooting.

■ FIGURE 49-5

Test equipment used by an electronics technician for determining proper operation of or for troubleshooting a circuit.

Item	Picture	Description
Isolated Power Supply	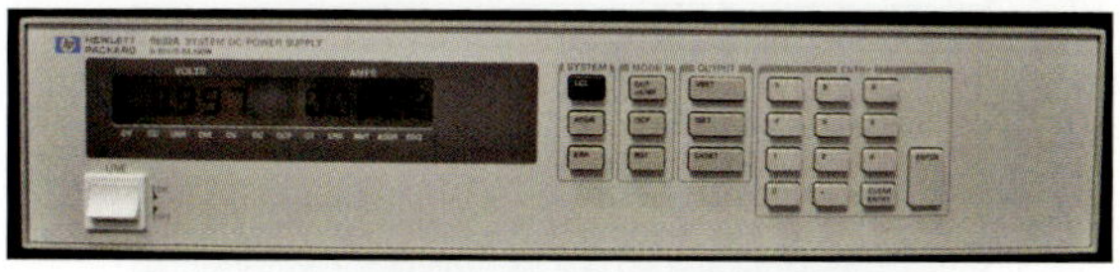	Used to power a PCB under test.
Analog Multimeter (VOM)		Used to measure and/or align voltage, resistance, and current of a PCB under test.
Digital Multimeter (DMM)		Used to measure and/or align voltage, resistance, and current of a PCB under test.
Oscilloscope	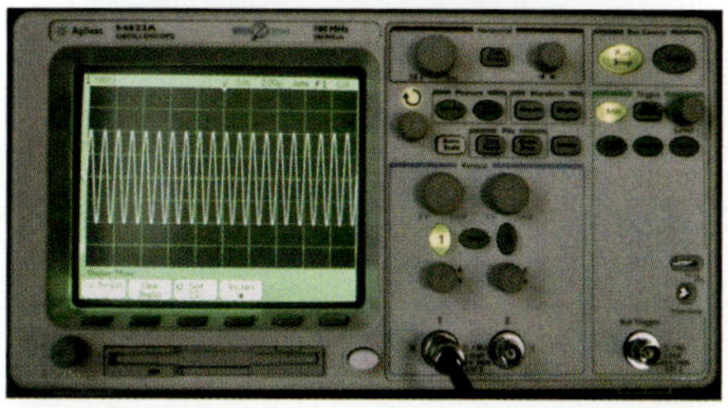	Used to measure and/or align amplitude and frequency and to compare waveforms.
Frequency Counter	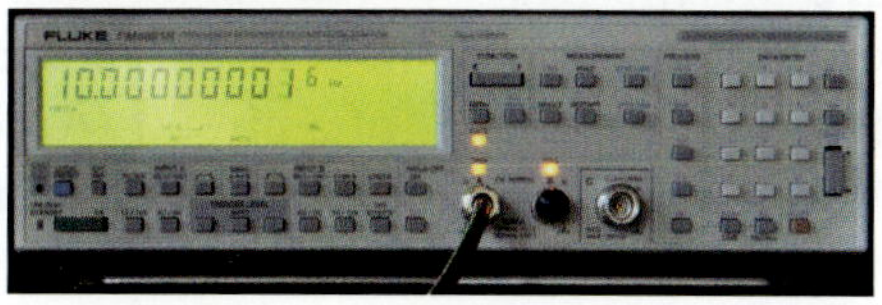	Used to measure and/or align frequency and period of recurring signals.

■ **FIGURE 49-5 (Continued)**

Test equipment used by an electronics technician for determining proper operation of or for troubleshooting a circuit.

Item	Picture	Description
Signal Generator	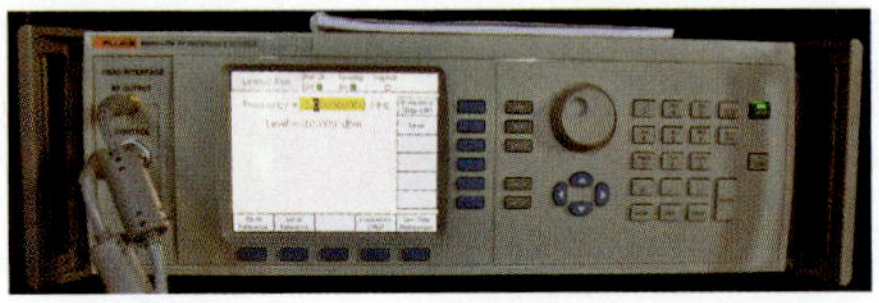	Used to generate a sinusoidal waveform for troubleshooting and aligning an electronic circuit.
Function Generator	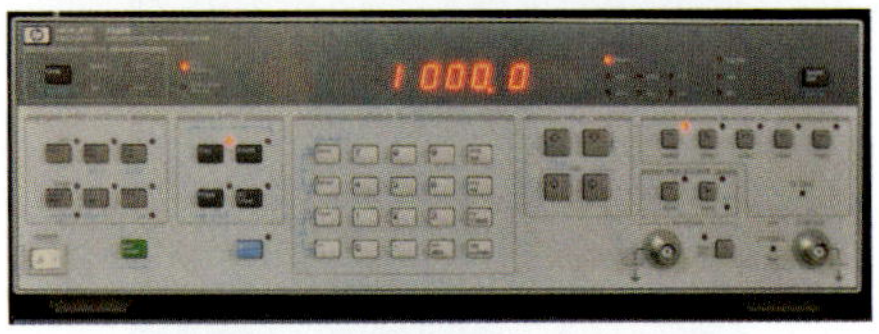	Used to generate a sinusoidal, triangular, or square waveform for troubleshooting and aligning an electronic circuit.
Logic Probes	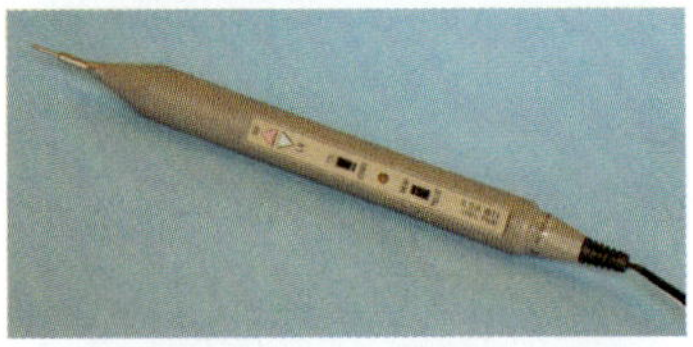	Used to detect a high or a low clock signal in digital electronic circuits.
Logic Pulser	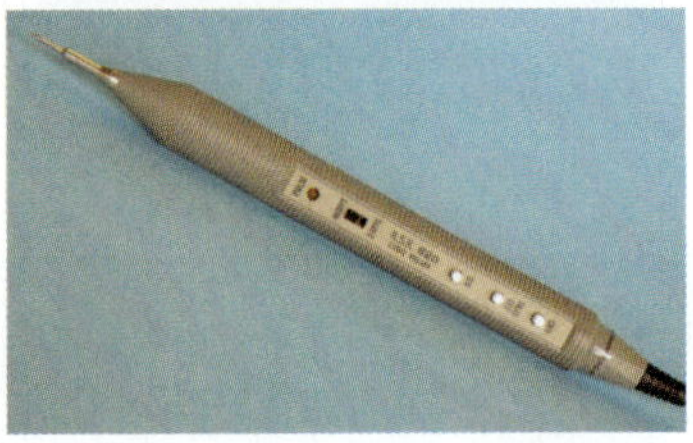	Used to inject a digital signal into a digital circuit for troubleshooting.

© 2014 Cengage Learning

Analog and digital **multimeters** are the first pieces of equipment used to analyze a circuit. To prevent possible circuit damage, the meter should have a sensitivity of more than 20,000 ohms per volt on the voltage scales. The meter should also not pass more than 1 milliampere of current on the resistance scales. Static DC and resistance checks usually locate power failures or defects that exhibit large deviations from the normal on the printed circuit board. However, these methods are time-consuming and inadequate when the defect is intermittent in the circuit.

An oscilloscope is one of the most useful pieces of test equipment available to the electronics technician. It can identify defects in a circuit while it is powered up; this includes changing parameters of a circuit. An oscilloscope displays the instantaneous voltage on the vertical axis against time on the horizontal axis and can detect both DC and AC signals. Some oscilloscopes allow the comparison of two waveforms of the same frequency to determine the phase shift. The oscilloscope probe on most oscilloscopes allows switching between settings of 1X and 10X (Figure 49-6). When the signal exceeds the capability of the oscilloscope on the 1X setting, the 10X setting multiplies the voltage setting of the oscilloscope by 10 and makes viewing the signal possible.

FIGURE 49-6

Oscilloscope probe.

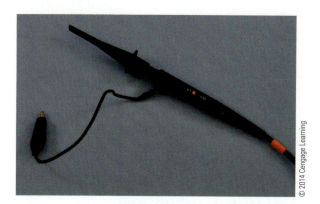

Frequency counters are useful for measuring the frequency and period of a recurring signal because they provide a digital readout of the frequency. Frequency measurements are an important part of preventive and corrective maintenance of electronic equipment.

Signal generators are used in the testing and aligning of radio transmitters, receivers, and amplifiers. They are also used for troubleshooting various electronic devices that require an input signal. Signal generators produce an alternating voltage of a desired frequency and amplitude for testing or measuring a circuit. They are classified by their frequency into the following categories: audio generator, video signal generator, radio-frequency (RF) generator, frequency-modulated (FM) RF generators, and special types of signal generators that combine several or all of the frequency ranges.

A **function generator** is a type of signal generator and is a common piece of electronic test equipment. The function generator is capable of generating fundamental waveforms such as sine waves, triangular waveforms, and square waveforms. The voltage, symmetry, and time characteristics of a waveform can be adjusted to meet a specific application.

A **logic probe** contains LEDs that produce a visual indication of what is occurring at a specific point in a digital circuit. A logic probe can indicate a high, low, clock, or lack of a signal. It can be used with a **logic pulser**, which looks similar to the logic probe. The pulser injects a logic level into a circuit, which can be picked up by the logic probe. The logic probe is useful for checking signals of various input and output pins of a circuit or an IC.

49–2 QUESTIONS

1. Why does an electronics technician use test equipment with a printed circuit board?
2. Why use an isolated power supply when working on or testing a circuit?
3. Discuss the difference on an oscilloscope settings for a 1X probe and a 10X probe.
4. How does a function generator differ from a signal generator?
5. What type of circuit would a logic probe be used for?

49–3 SOLDER AND SOLDERING IRONS

Soldering is the joining of two pieces of metal with a solder alloy having a melting point below 800° Fahrenheit (427°C). Solder used to include a combination of tin and lead in ranges from pure tin to pure lead and including all proportions in between. In electrical soldering, the alloy mix is usually 60% tin and 40% lead (60/40) or 63% tin and 37% lead (63/37) (Figure 49-7). For soldering **surface mounted devices**, a solder is available with 62% tin, 28% lead, and 2% silver (62/36/2). To prevent lead poisoning, solder manufacturers now produce solder with a very low lead content or without lead.

Characteristics of alloys of tin and lead are plotted against temperature in Figure 49-8. This graph allows one to see that only an alloy of 63/37 has a

FIGURE 49-7

Spool of solder.

■ **FIGURE 49-8**

Solder alloy versus temperature.

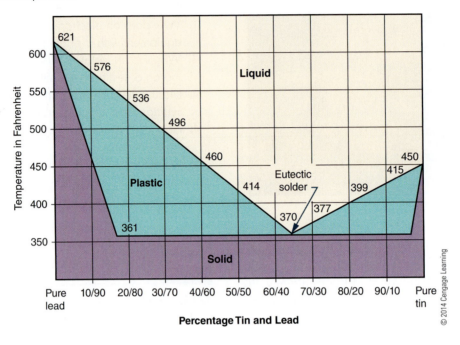

© 2014 Cengage Learning

eutectic point (a single melting point). All other combinations start melting at one temperature, pass through a plastic stage, and then become a liquid at a higher temperature. Any physical movement of the components being soldered while the solder is in the plastic stage results in a fractured solder connection. Such a connection appears dull and grainy, and is mechanically weaker and less reliable. Therefore, 63/37 or 60/40 solder is commonly used in electronics because it does not remain in the plastic stage very long.

Flux helps the solder alloy flow around the connections. Flux cleans the component leads of oxide and film, allowing the solder to adhere. Soldering flux is usually included in the solder in the central core. **Rosin flux** is always used in electrical soldering because it is noncorrosive.

The primary tool used for installing, removing, and/or replacing components on a printed circuit board is the **soldering iron**. An improper choice in selecting a soldering iron or **soldering iron tip** may result in damage to the component and/or the printed circuit board. Select only a low-wattage **pencil soldering iron** (Figure 49-9) with an appropriate tip for the soldering being done. **Solder guns** (Figure 49-10) are inappropriate for printed circuit

board fabrication. They are designed to supply heat to large electrical connections and will ruin the small detail of a printed circuit board.

The soldering iron is usually specified by wattage. In most situations, a 25- to 40-watt soldering iron is adequate. Wattage represents the amount of heat capacity available at the tip. Soldering irons of all wattages usually run at the same temperature.

■ **FIGURE 49-9**

Example of low-wattage pencil soldering iron.

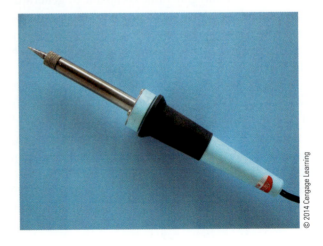

© 2014 Cengage Learning

■ **FIGURE 49-10**

Solder gun.

■ **FIGURE 49-11**

Several types of pencil soldering irons.

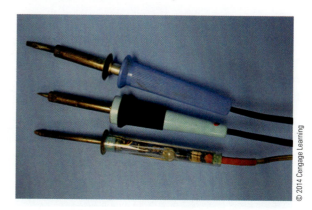

A low-wattage soldering iron tends to cool faster during the soldering process than a high-wattage soldering iron. The handle of the soldering iron should be comfortable to hold and should be well insulated to prevent heat buildup. Soldering irons can produce static voltage spikes that will destroy many integrated circuit components. A grounded soldering iron tip is a good safety measure. Figure 49-11 shows several types of soldering irons suitable for printed circuit boards. The least expensive is the pencil soldering iron, and with proper care it will last many years. The **temperature-controlled soldering station** provides precise tip temperature through a closed loop circuit and maintains a constant tip temperature throughout the soldering process. The cordless soldering iron is similar to the pencil soldering iron except it uses a battery for its power source.

Ironclad tips are the best selection for soldering printed circuit boards. An unplated tip provides better heat transfer but wears rapidly and requires frequent cleaning and tinning. The ironclad tip requires only an occasional cleaning and tinning.

Soldering iron tips come in a wide variety of shapes and sizes to meet a specific application. Many soldering irons have replaceable tips. The tip selected should provide maximum contact area between the surfaces to be joined and also allow accessibility to the connection. Figure 49-12 shows the most popular shapes of soldering iron tips. Typically, a small bevel tip is preferred for general printed circuit board work.

■ **FIGURE 49-12**

Soldering iron tips.

Soldering Iron Tip			Description
Type	**End View**	**Side View**	
Conical		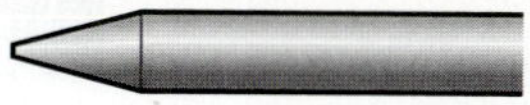	Used for general assembly and repair work.
Pyramid		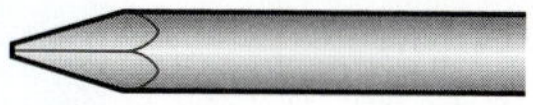	Used for general assembly and repair work.
Bevel		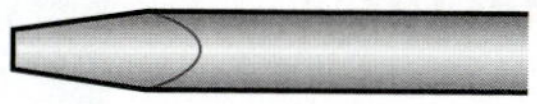	Allows rapid transfer of heat. Used for printed circuit boards.
Chisel		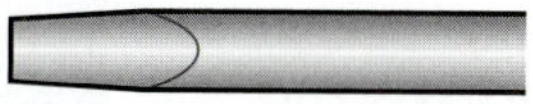	Allows large areas to be heated rapidly. Used for tinning wires, soldering terminals.

© 2014 Cengage Learning

■ FIGURE 49-13

Idling soldering iron.

Maintaining the soldering iron tip is essential for proper soldering. Use an anti-seize compound when installing the tip to prevent it from becoming frozen. Make sure the tip is fully installed in the heating element. Clean the tip with steel wool before heating it, then heat the tip and apply solder as it warms up. Applying solder to the tip is called **tinning**. Rest the soldering iron on a stand to remove excess heat from non-temperature-controlled soldering irons, as shown in Figure 49-13; this is referred to as **idling**. Wipe the tip of the soldering iron across a damp sponge or tip cleaner to clean it of any oxidation before soldering.

49-3 QUESTIONS

1. What is the primary tool used for installing a component on a printed circuit board?
2. A soldering iron used for printed circuit board work should be of what wattage?
3. Why should the soldering iron tip be grounded?
4. What is the best tip for soldering on a printed circuit board?
5. How is the tip of a soldering iron tinned?

49-4 SOLDERING A PRINTED CIRCUIT BOARD

Use a systematic method for installing components on the printed circuit board. One technique is to install all the resistors first, then the capacitors, ICs, and so on.

With axial lead components, the leads have to be bent at a right angle. Grasp the ends of the component between the thumb and index finger, and pull the body outward while maintaining a grip on the ends of the leads (Figure 49-14). The component should now fit between the pad mounting holes. Position components so the values can be easily read. Arrange resistors so the value can be read top to bottom or left to right. Observe polarity of all polarized components, and orientate them properly on the printed circuit

■ FIGURE 49-14

Manually bending component leads.

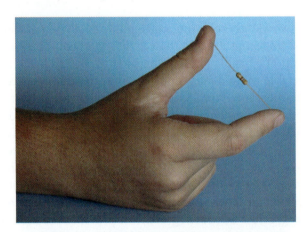

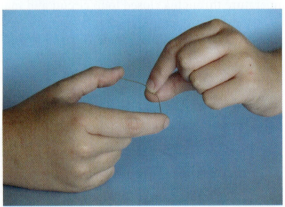

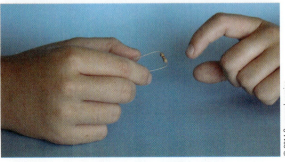

board. Pay close attention to installing ICs on the printed circuit board to ensure that pin 1 is lined up with its appropriate location on the board. Use ¼-inch **leg warmers** on LEDs and small transistors to avoid overheating (Figure 49-15). After the component leads have been inserted, they should be bent to a 45° angle to keep them from falling out before soldering.

Once the printed circuit board is assembled, the next area to consider is the off-board components (potentiometers, lamps/LEDs, switches, transformers, etc.). A **wiring diagram** will make the installation clear and

easy to understand. The two types of drawings are the pictorial diagram (Figure 49-16) and the functional diagram (Figure 49-17). There is no standard color code for wiring a printed circuit board. It is best to use different color wires for each **discrete component** and note the colors on the wiring diagram. Discrete components are components that are not mounted on the PCB. There typically aren't enough colors for each signal to have its own color, but different colors can be used to denote different classes of signals.

When constructing a diagram, one set of signals that should be color-coded is power and ground. A different color for the positive voltage, negative voltage, and ground should be used, and these colors should not be any of the colors that have been used for other signals. It would be nice to use "standard" colors for these, but unfortunately there are several competing standards. Red is frequently used to indicate the positive supply voltage. Most automotive and electronic wiring diagrams use black to denote ground. Electronics wiring that uses black for ground often uses blue for negative supply voltage. Green is also frequently used to denote ground.

■ **FIGURE 49-15**

Example of leg warmers.

© 2014 Cengage Learning

■ **FIGURE 49-16**

Example of a pictorial diagram.

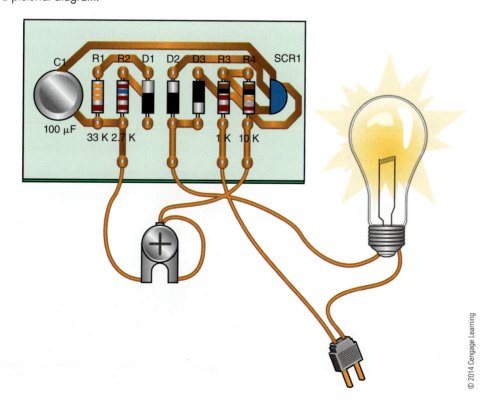

© 2014 Cengage Learning

■ FIGURE 49-17

Example of a printed circuit board functional diagram.

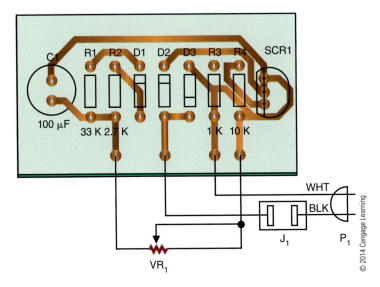

In electronics, the basic goal of soldering is to electrically and mechanically join two circuit components. For soldering to be successful and reliable, the solder must adhere to the mating surface. The application of solder to a base metal requires the surface to be clean and free of contamination. If the surface to be soldered is contaminated, the solder tends to ball up and not adhere. Sources of contamination include greases, oils, and dirt. Aging also causes surface contamination by the formation of an oxide film.

The best soldering techniques can be outlined simply. First, wipe the tip of the soldering iron on a damp sponge. Then place a small amount of solder on the tip and apply the tip to the connection to be soldered. Solder should not be applied until the connection becomes hot enough to melt the solder. How long this takes is quickly learned by a few trials. When the solder has melted and flowed into a contoured fillet, remove the solder. Keep the tip of the soldering iron on the connection for a few seconds longer, and then remove it. Do not disturb the newly made connection until it has had time to solidify. A good solder connection will be bright and shiny (Figure 49-18).

Handle components and the printed circuit board as little as possible, because fingers may be dirty and oily. If there is a question about the cleanliness of a part, clean it using alcohol, steel wool, or a scouring powder. When using scouring powder, ensure that the parts are rinsed thoroughly. If steel wool is used, use a lint-free cloth to remove all pieces of the steel wool from the part or printed circuit board.

Soldering on a printed circuit board requires careful attention because of the heat sensitivity on the pads and traces. Too much heat with too much pressure can cause a pad to lift from the board substrate. Always apply the tip with a light touch—practice will determine approximately how long it needs to be held. The soldering iron should rest on the junction of the lead and pad rather than being pressed to it (Figure 49-19).

■ FIGURE 49-18

Example of a good solder connection.

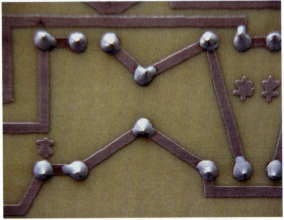

■ **FIGURE 49-19**

Place the soldering iron tip at the junction of lead and the pad with no pressure.

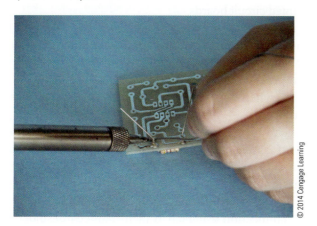

To summarize the proper soldering technique:

1. Wipe tip on a damp sponge or tip-cleaning apparatus.
2. Tin the tip by applying solder to it.
3. Apply the tip to both the component lead and pad.
4. Apply solder to the back of the connection; the heat of the tip will draw the solder to it.
5. Remove the solder.
6. Remove the tip from the connection.
7. Let the connection cool.
8. Trim the excess leads flush to the printed circuit board.

During the soldering process, a flux residue forms on the printed circuit board. The residue is noncorrosive in most cases. However, it should be removed, as it is sticky and foreign particles can attach to it. Also, removing the flux makes the inspection of the board easier and gives a more professional appearance to the board. There are several **defluxers** on the market that make removing easy. Unfortunately, they are not environmentally friendly because they contain trichlorotrifluoroethane as their main ingredient.

To use the **flux cleaner**, spray the surface to be cleaned in a well-ventilated area. Allow the chemical to saturate the board for 10 to 15 seconds, and then clean it with a medium-bristle brush such as an acid brush. Allow the components to dry completely before powering up the circuit.

Once the skill of soldering is mastered, the skill of **desoldering** is next. Desoldering requires special implements: desoldering braid, desoldering bulb, desoldering pump, combination bulb and soldering iron, and desoldering station.

In desoldering, the first step is to solder. Applying more solder on the connection ensures the connection is heated up before removing the solder. Do not heat a connection with very little solder, as it will result in the pad lifting from the substrate.

A **desoldering braid** consists of a braided copper material with flux imbedded in it that uses capillary action to pull up the hot solder (Figure 49-20). Using the widest braid practical, place it on the pad to be desoldered. Place a hot soldering iron on top of the braid. The solder will melt and be wicked into the braid. When the braid gets saturated, remove it, clip off the saturation end with diagonal cutters, and reapply until the solder is removed. Use a hot soldering iron—the braid acts like a heat sink, cooling the soldering iron tip.

The **desoldering bulb** uses a bulb and a Teflon tip. Squeeze the bulb and apply the tip to the melted solder; releasing the bulb draws up the molten solder (Figure 49-21).

■ **FIGURE 49-20**

Desoldering braid.

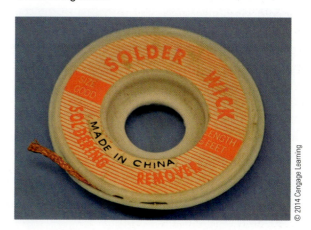

■ **FIGURE 49-21**

Desoldering bulb.

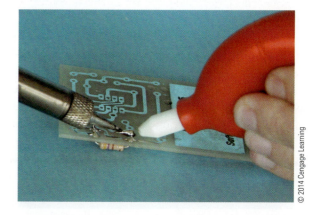

A variation of the desoldering bulb is a spring-loaded **desoldering pump** (Figure 49-22) that works by depressing a plunger until it locks. To suck up solder, press the release button. The spring forces the plunger up rapidly, creating a vacuum. The combination rubber bulb and soldering iron uses a hollow tip to suck up the solder (Figure 49-23), thus eliminating the rush to get the solder sucker to the pad before the solder rehardens.

A professional **desoldering station** uses a foot-operated vacuum pump and a hollow-tip soldering iron to draw up molten solder. It is the most expensive tool used for desoldering a component from a printed circuit board pad.

Any of the desoldering tools discussed will do the job of removing solder from a printed circuit board connection. Each requires different techniques and fits different budgets.

■ **FIGURE 49-22**
Solder pump.

■ **FIGURE 49-23**
Combination desoldering bulb and soldering iron with a hollow tip.

The mounting or removal and replacement of components for surface mount technology (SMT) printed circuit boards are different from in conventional printed circuit boards.

With SMT printed circuit boards, in addition to the board being smaller, the components are much smaller and must be handled with tweezers. Mount the printed circuit board in a vise. This allows stability when using a magnifying glass to work on the circuit.

The proper method for removing SMT components is with a rework station. To desolder SMT components manually, proceed as described in the next paragraphs.

For an IC, remove as much solder as possible from the connection using 0.030-inch-diameter desoldering braid. The only solder remaining is holding the component leads to the printed circuit board. Wipe the soldering iron clean and heat the lead and pad while using an X-Acto knife to gently pry up the lead. When the lead releases, remove the soldering iron and allow the lead to cool. Repeat on each lead until the IC is removed from the printed circuit board.

CAUTION!

Too much prying pressure can lift the pad from the printed circuit board.

For SMT resistors and capacitors, add solder to the pad to increase the thermal mass to prevent it from cooling too fast. Alternately heat each end and use tweezers to remove the component while the solder is molten. Use desoldering braid to remove the excess solder.

The solder used should be 63/37 or a silver bearing solder of 62/36/2 with a diameter of 0.020 inch to 0.015 inch. A noncorrosive liquid flux with a drop dispenser and a light-duty spray defluxer is required for applying and cleaning flux on the SMT leads. Clear tape is useful for taping down components before soldering.

After the printed circuit board is completely assembled and soldered, apply power. If the circuit works properly, enjoy. Otherwise, troubleshoot by first removing power from the printed circuit board. Inspect all traces for bridging or breaks and all solder connections for shorts. Check for improper solder connections. Correct all problems and retest.

1. What is an effective method for installing the components on a printed circuit board?
2. What should be observed when placing nonpolarized components on a printed circuit board?
3. What function do leg warmers serve on LEDs and small transistors?
4. What is the guideline for colored wires when installing discrete components?
5. What is the best solder mixture for effective printed circuit board soldering?
6. What is the best tip to use for general printed circuit board soldering?
7. List the steps for effective soldering on a printed circuit board.
8. What tools are required for desoldering a component on a printed circuit board?
9. How can the flux residue be removed from the soldered printed circuit board?
10. What is different when soldering SMD components as compared to full-size components?

- Removing the solder too soon results in not enough solder.
- Too much solder is caused by not removing the solder soon enough.
- **Bridging** (inadvertently connecting two pads) is caused by not removing the solder soon enough.
- Damage to a component is caused by too much heat—the iron is too large or held on too long.

49–5 **QUESTIONS**

1. Describe what a properly soldered connection should look like.
2. Draw a picture of a properly soldered component on a printed circuit board.
3. What are the two most common types of soldering defects?
4. What causes the most common defects on a printed circuit board?
5. List other types of defects that may occur when soldering.

49–5 **ANALYZING SOLDERED CONNECTIONS**

The surface of a properly soldered connection should be smooth, well feathered at the edges, and bright and shiny with a slight concave configuration (Figure 49-24).

Two of the most common soldering defects include **cold solder connections** and **disturbed** or **fractured connections**. Cold solder connections are the result of withdrawing the soldering iron too soon. The solder does not become liquid and leaves a connection that is dull gray in appearance. A fractured or disturbed connection is the result of movement of the component leads or wire before the solder sets. The fractured joint is evident by a granulated and frosty appearance with many cracks showing.

Other types of defects include these:

- **Rosin connection** is caused by the flux and oxidation particles staying between the solder pad and lead. It is the result of not enough heat.
- **Wicking** is caused by the capillary action of stranded wire. It is the result of too much heat.
- **Voids** are caused by impurities in the surface when soldering. They are the result of too much heat.

■ **FIGURE 49-24**

Characteristics of a solder connection.

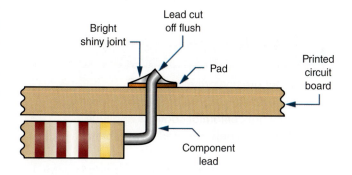

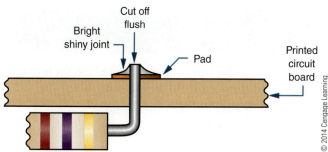

49–6   PROTECTIVE COATINGS

Conformal coatings are nonconductive materials applied in a thin layer on printed circuit boards. They provide environmental and mechanical protection to significantly extend the life of the components and circuitry. Dipping, spraying, or flow coating are traditional methods for applying conformal coatings.

Conformal coatings protect electronic printed circuit boards from abrasion, corrosion, dust, fungus, moisture, and environmental stresses. Common conformal coatings include acrylic, epoxy, parylene, silicone, and urethane. Most conformal coatings consist of a synthetic resin developed in a volatile solvent. When they are properly applied to a clean surface, the solvents evaporate, leaving a thin film of solid resins. Conformal coatings have one or more of the following characteristics to protect the printed circuit board effectively:

1. Good thermal heat conductivity to carry heat away from the components
2. Inorganic composition to prevent fungus growth
3. Low moisture absorption
4. Low shrinkage factors during application and curing to prevent coating from applying strains or stresses to the components or leads
5. Resilience, hardness, and strength to support and protect the components
6. Qualities of electrical insulation

49–6   QUESTIONS

1. What are conformal coatings?
2. What is the purpose of conformal coatings?
3. How are conformal coatings applied?
4. What materials are used for conformal coatings?
5. What are the characteristics of a good conformal coating?

SUMMARY

- A toolbox for working on a printed circuit board should contain the following tools:
  - Plastic and orangewood spudgers
  - Bristle and acid brushes
  - Wire strippers
  - Surgical scissors
  - X-Acto® knife
  - Tweezers
  - Flat-nose pliers (duckbill)
  - Needlenose pliers
  - Round-nose pliers
  - Diagonal cutters (flush)
  - ¹⁄₁₆-inch and ¹⁄₈-inch chisel tip for soldering iron
  - Heat sinks
  - Alcohol and flux bottle
  - Soldering iron
  - Solder wick
  - Manual desoldering tool
- A printed circuit board holding device is useful for holding a printed circuit board.
- An illuminated magnifier helps to detect improper connections or defects on a printed circuit board.
- Leads of resistors and ICs can be shaped with lead-forming tools.
- An isolated power supply is essential to prevent a lethal shock potential.
- The multimeter is the first piece of test equipment used to analyze a circuit.
  - The voltmeter should have a sensitivity of 20,000 ohms per volt.
  - The ohmmeter should not pass more than 1 milliampere of current.
- An oscilloscope is useful for detecting defects in a circuit.
- Frequency counters are used for measuring the frequency and period of a recurring signal.
- Signal generators are used to test and align radio transmitters, receivers, and amplifiers.
- Function generators produce sine, triangular, and square waveforms.
- A logic probe provides a visual indication of what is occurring in a digital circuit.
- A logic pulser injects a logic level into a circuit for detection by the logic probe.
- The soldering iron is the primary tool for installing, removing, and replacing components on a printed circuit board.
- Soldering guns should not be used on printed circuit boards.
- Tinning is the process of applying solder to the soldering iron tip.
- There is no standard color code for wiring discrete components to a printed circuit board.

- The basic goal of soldering is to electrically and mechanically join circuit components to the printed circuit board.
- In electrical soldering, an alloy mix of 60% tin and 40% lead (60/40) or 63% tin and 37% lead (63/37) is used.
- Soldering surface mounted devices requires a solder with 62% tin, 36% lead, and 2% silver (62/36/2).
- Flux helps the solder alloy flow around the connections by cleaning the component leads of oxide and film, allowing the solder to adhere.

- A properly soldered connection will be bright and shiny.
- Desoldering requires special tools and materials to make the task easier.
- In desoldering, the first step is to apply more solder.
- SMT components are much smaller and must be handled with tweezers; use a magnifying glass and mount the board in a vise.
- Conformal coatings are nonconductive materials applied in a thin layer on printed circuit boards to provide environmental and mechanical protection.

CHAPTER 49 SELF-TEST

1. Refer to the tools identified in an electronics technician's toolbox; what tools would be used to remove an IC from a printed circuit board?
2. What is the purpose of a printed circuit board?
3. What pieces of test equipment would be used to test an amplifier circuit on a printed circuit board?
4. How would a logic pulser be used?
5. What is the function of flux when soldering?
6. Where would a solder gun be used?
7. Which would be better for soldering a double-pole, double-throw switch to a printed circuit board: a temperature-controlled soldering station or a cordless soldering iron?
8. What type of tip would be used for the soldering iron in question 7?
9. Explain the procedure for idling a soldering iron.
10. What is another characteristic of conformal coating that is not covered in this chapter?

Basic Troubleshooting

OBJECTIVES

After completing this chapter, the student will be able to:

- Develop common troubleshooting techniques.
- Identify the effects of a short in a circuit.
- Identify the effects of an open in a circuit.
- Describe the effects of aging on component tolerances in a circuit.
- Discuss why documents should be kept for a circuit.

KEY TERMS

50-2 component placement error

50-2 troubleshooting

50-2 wiring error

50-3 data recording log

50-4 power of observation

50-5 documentation

Troubleshooting is the logical process of isolating a defective component in an electrical circuit. This chapter discusses troubleshooting techniques to effectively isolate a defective component in a circuit. The techniques apply to analog series, parallel, and series-parallel circuits as well as complex digital circuits.

50–1 TOOLS FOR TROUBLESHOOTING

An electronics technician needs to know the circuit details for effective troubleshooting. When troubleshooting a circuit, only make a measurement when the normal value is known. A schematic diagram will help to minimize the number of measurements made.

The ohmmeter function of a multimeter (Figure 50-1) is useful in troubleshooting for determining an open circuit with an infinite resistance or a short circuit with 0 resistance. When using the ohmmeter for measuring individual component resistance in a circuit, one lead from the component must be removed from the circuit for an accurate reading.

The voltmeter function of a multimeter (Figure 50-2) is useful for measuring the voltage drop across a component without removing it from the circuit. Due to the loading effect of a voltmeter because it is inserted in parallel across the component, the voltage reading may be in error. The loading resistance of a voltmeter should be 20 MΩ/volt to minimize the loading effect.

The ammeter function of a multimeter (Figure 50-3) measures the current flow for the entire circuit or the flow through individual components. The ammeter must be inserted in the circuit in series and requires the removal of one lead of a component from the circuit. The ammeter is not the first piece of equipment to use in troubleshooting unless measuring the total current flow of the circuit.

■ **FIGURE 50-1**

Ohmmeter.

■ **FIGURE 50-2**

Voltmeter.

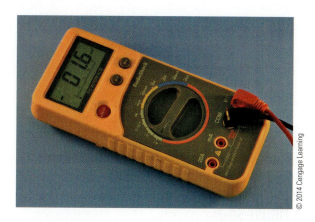

■ **FIGURE 50-3**

Ammeter.

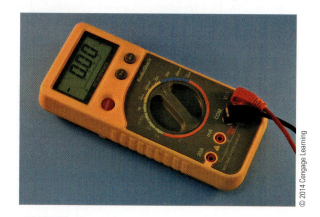

■ **FIGURE 50-4**

Oscilloscope.

Though many problems can be located using just a multimeter, an oscilloscope (Figure 50-4) is essential for advanced troubleshooting. The basic oscilloscope requirement is a dual trace, 10–20 MHz minimum vertical; a delayed sweep is desirable but not essential. A good set of proper 10X/1X probes is also necessary. A storage scope or digital scope might be desirable for certain tasks but is not essential for basic troubleshooting.

■ **FIGURE 50-5**

Isolation transformer.

© 2014 Cengage Learning

■ **FIGURE 50-6**

Tools for successful repairing of a printed circuit board.

© 2014 Cengage Learning

■ **FIGURE 50-7**

Types of wiring diagrams.

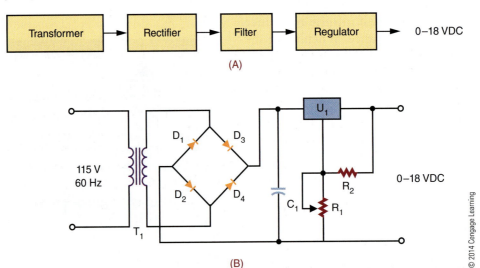

(A)

(B)

© 2014 Cengage Learning

An isolation transformer (Figure 50-5) isolates the circuit that does not use a transformer for its power supply. If the circuit is not isolated from the AC line, it can become a shock hazard.

A soldering iron, solder, and desoldering tools (Figure 50-6) are necessary for removing and replacing components in a printed circuit board.

Before starting to troubleshoot an electronic circuit, a good wiring diagram of the circuit is required. There are two types of wiring diagrams: a block diagram (Figure 50-7a) and a schematic diagram (Figure 50-7b). A block diagram lays out the components of the system and how they relate to each other. It helps to gain an understanding of the basic system and how it works. A schematic diagram provides more detail than the block diagram by using schematic symbols to show the individual components of a circuit and how they are interconnected.

50–1 QUESTIONS

1. What pieces of equipment are used for taking measurements when troubleshooting?
2. What must be done when measuring resistance and current in a circuit?
3. What precaution must be taken when using a voltmeter for troubleshooting?
4. What is an oscilloscope used for in troubleshooting?
5. Why is an isolation transformer essential to troubleshooting?
6. What function does a schematic diagram serve in troubleshooting?

50–2 ISOLATION TECHNIQUES FOR EFFECTIVE TROUBLESHOOTING

Effective **troubleshooting** is a step-by-step process that requires an electronics technician to work through four steps to verify, isolate, and eliminate problems in a circuit.

Step 1 in locating a problem for a malfunctioning circuit or a recently assembled printed circuit board is to visually check the printed circuit board over thoroughly before power is applied. Check the circuit for construction errors. The two types of construction errors are improper component placement and wiring errors for discrete components. Discrete components are components not mounted on the printed circuit board.

Component placement errors include the following: wrong component or component not in the correct location, components mounted in backwards, or multi-lead components with leads inserted improperly. **Wiring errors** include the following: a wire installed in the wrong location, a wire missing, a wire that is broken (open), and a wire that is installed with a faulty solder connection.

Step 2 requires applying power to the circuit to determine whether an electrical problem exists. Depending on the type of circuit, indicators will light, sound will emit, and so on. If the circuit appears to perform properly, let the circuit warm up and settle in its normal operating mode. The purpose is to determine whether anything is defective with the circuit. If the circuit appears to be operating properly, then it can be calibrated or adjusted if required, and then move on to step 4.

Step 3 is to troubleshoot the circuit to find out what is wrong with the circuit, causing it not to work properly, and then determine what has to be done. Electronics technicians use their eyesight to *look* for trouble. If a component receives excess current, it may split open or char. Technicians also *smell* for burnt insulations from components that receive excess current, such as transformers, coils, or components with insulated leads. They *listen* for sounds such as a snap, crackle, or pop that indicate loose wires, bridged connections, or frying components. They *touch* components to determine whether they are running hot or cold. An electronics technician's senses of sight, smell, hearing, and touch are only a beginning. To solve more difficult problems, meters and oscilloscopes are used once the problem is located in a specific part of the circuit. Use this data to isolate and repair the defect(s).

Step 4 is to determine whether the circuit will work when it is exposed to the environment it is designed to operate in.

50–2 QUESTIONS

1. What are the four steps for effective troubleshooting?
2. What should be looked for when visually inspecting an assembled board?
3. What is the function of step 2 in the troubleshooting process?
4. How do electronics technicians use their senses for troubleshooting?
5. What is the function of step 4 in the troubleshooting process?

50–3 COMMON TYPES OF DEFECTS

Figure 50-8 shows a chart indicating common types of defects during printed circuit board assembly. The chart indicates that the most common types of defects are open and short circuits. An open, or a break, in the circuit has an infinite resistance. An open in a circuit may not have infinite ohms, but the resistance will be so high relative to the other circuit resistances that it can be assumed to be infinite. When measuring voltage with a voltmeter, connect the negative side of the voltmeter to ground. Probe various test points with the positive lead. If the open circuit is between a monitored point and ground, the voltmeter will read the full supply voltage; otherwise the voltmeter will read 0 (Figure 50-9).

A short circuit has 0 ohms resistance. In reality, the short may not be 0 ohms but is lower than what should be measured and is assumed to be zero resistance. A *short circuit* is the term used to identify an unintended connection or a defective component with 0 ohms resistance. When measuring voltage across a shorted component, the voltage will read 0 volts. When measuring the resistance of a shorted component, the resistance will be 0 ohms. Shorts in power equipment are dangerous because they can result in electrical fires due to a high current flow, especially if the circuit is not fused (Figure 50-10).

When taking measurements in a circuit, check the measured values against known values. Make a **data**

■ FIGURE 50-8

Common defects of printed circuit board assembly.

PCB Assembly Defects	Percentage
Open	34
Short, solder bridges	15
Missing parts	15
Misoriented parts	9
Marginal solder connection	9
Solder balls, voids	7
Bad components	6
Wrong components	5

© 2014 Cengage Learning

■ FIGURE 50-9

Open on a printed circuit board.

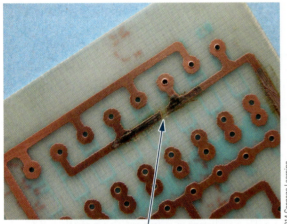

Open

© 2014 Cengage Learning

■ FIGURE 50-10

Shorts on a printed circuit board.

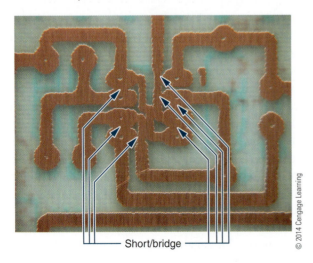

Short/bridge

© 2014 Cengage Learning

recording log and record any measurements taken, as shown in Figure 50-11.

Resistors may increase or decrease in value through age or circuit defects. If a resistor increases in resistance, the results will be similar to that of an open circuit. If a resistor decreases in resistance, the results will be similar to that of a short circuit.

Traces on a printed circuit board can also develop defects. If a trace becomes shorted to another trace through soldering, the circuit may not work or components may be damaged—or worse, the board may catch fire. If a trace opens, it acts like a resistor that has opened and the circuit will not work or will only partially work.

50-3   QUESTIONS

1. In a series circuit, what is the voltage measured across a shorted resistor?
2. Why are shorts dangerous in a circuit?
3. How is an open in a circuit determined with a voltmeter?
4. What happens to resistors in a circuit to cause them to change value?
5. What can happen to traces on a printed circuit board to cause a defect?

50-4   TROUBLESHOOTING TIPS

Never start troubleshooting a circuit with test equipment. Start with the **power of observation**. A technician's senses and thinking can help to isolate a problem quickly.

■ **FIGURE 50-11**
Parts/equipment data recording chart for future referencing.

**Project/
equipment:** _____ **Date:** _____

visual (Check all components for improper placement and orientation, wiring errors, and faulty soldering connections.)
Record all faults located and corrected.

Defective Component	Action Taken
1.	
2.	
3.	
4.	
5.	

Power-Up (Check circuit/device for proper operation.) Record all faults; otherwise perform any calibrations and adjustments.

Faults	Calibrations/Adjustments Completed	Measured Values
1.		
2.		
3.		
4.		
5.		

Troubleshoot (After determining a problem exists, use the appropriate test equipment to identify the fault.) Record all
faults; otherwise perform any calibrations and adjustments. Waveforms may be attached with proper
registration/identification and record any comments.

Problem	Voltage/Current Measured	Waveforms	Comments
1.			
2.			
3.			
4.			
5.			

Operation (Run the circuit for an extended period of time to insure no problems occur.) Record any problems;
otherwise enjoy.

Problem	Comments
1.	
2.	
3.	
4.	
5.	

Operation (Record any comments, thoughts, ideas for future reference.)

Comments

Many problems have simple solutions. When troubleshooting, do not immediately assume a problem is complex. It could be a bad connection or a blown fuse. Remember that problems with the most impact usually have the simplest solutions. Difficult problems to solve are those that are intermittent or hard to reproduce.

If a problem is difficult to solve, step away from it for a while. Sometimes, just thinking about the problem leads to a different approach or solution. Never work on a problem when tired because it is both dangerous and nonproductive.

Whenever working on equipment, make notes and diagrams or take photographs for use when it is time to reassemble the unit. Most connectors are keyed against incorrect insertion or interchange of cables, but not all. Apparently identical screws may be of differing lengths or may have slightly different thread types. Little parts may fit in more than one place or orientation. Pill bottles, film canisters, egg cartons, and plastic ice cube trays provide a handy place for sorting and storing screws and other small parts during disassembly.

The workspace should be open, well lit, electrostatic discharge (ESD) protected, and situated so that dropped parts can be easily found. The best location should be dust free and allow suspending troubleshooting without having to place everything in a box for storage.

Take the precautions of touching a ground point first to prevent electrostatic discharge into a component. Components such as MOSFETs and ICs are vulnerable to ESD.

WARNING

When using an isolation transformer, a live chassis should not be considered a safe ground point.

50-4 QUESTIONS

1. What is the first thing to do when troubleshooting a circuit?
2. What types of problems are the most difficult to solve?
3. What should be done when disassembling a piece of equipment?
4. How can parts be stored when disassembling a piece of equipment?
5. Describe the ideal work location for troubleshooting.

50-5 DOCUMENTATION

The most important document when troubleshooting a circuit is the schematic diagram. It allows the electronics technician to read the circuit like a book. It shows not only how the circuit is connected but also voltage drops in a circuit. A good schematic also shows waveforms at various test points in the circuit.

When a circuit is assembled, it's a good idea to take a copy of the schematic and record voltages measured for future reference. Also, record any waveforms that will help in future troubleshooting.

The secret of successful troubleshooting is good **documentation**. Keep all the documents of a circuit, including a data-recording log, together for future reference.

50-5 QUESTIONS

1. What is the most important document for troubleshooting?
2. What function does a schematic diagram serve?
3. What should be done after a circuit is assembled?
4. What is the secret of successful troubleshooting?
5. What should be done after a circuit is assembled and running?

SUMMARY

- Only make measurements when the normal value is known.
- An ohmmeter is useful for detecting opens and shorts.
- When checking the resistance of a component, one lead must be removed from the printed circuit board.
- A voltmeter can measure the voltage drop across a component.
- Voltmeters should have a loading resistance of 20 MΩ/volt.
- An ammeter can be used to measure individual components or total current flow in a circuit.
- An oscilloscope is useful for detecting the presence or absence of a signal.

- An isolation transformer isolates a circuit from the AC line to prevent electrical shock.
- Schematic diagrams are necessary for troubleshooting a circuit.
- The four steps for effective troubleshooting are as follows:
  - Check the circuit for construction errors before applying power.
  - Apply power and check for normal operation.
  - Troubleshoot using sight, smell, hearing, and touch to isolate the problem to a specific part of the circuit.
  - Expose the circuit to the environment it is designed to operate in.
- A short circuit has 0 ohms resistance.
- An open circuit has an infinite resistance.
- When making measurements in a circuit, check the measured values against known values.
- Resistors may increase or decrease in value through age or circuit defects.
- Traces on a printed circuit board can become shorted to another trace through soldering and may damage the printed circuit board or components.
- Traces may become open through manufacturing and thus not work properly.
- A schematic diagram is the most important document when troubleshooting a circuit.
- Good schematic diagrams will show waveforms at various test points in the circuit.
- When a circuit is assembled, it is a good idea to record voltages measured for future reference.
- Never start troubleshooting with test equipment; always start with analytical thinking.
- Problems with the most impact usually have the simplest solutions.
- Difficult problems to solve include intermittent and hard-to-reproduce problems.
- If a problem is hard to solve, stepping away to think about it may lead to a different approach or solution.
- The secret of successful troubleshooting is good documentation for future reference.

CHAPTER 50 SELF-TEST

1. What can be used to prevent electrical shock when troubleshooting a circuit?
2. Describe how to troubleshoot a circuit that does not work.
3. How can test equipment effectively be used for troubleshooting?
4. How can electronic technicians extend their senses when troubleshooting?
5. What is the difference between a short circuit and an open circuit?
6. What is the best way to identify missing parts on a printed circuit board?
7. What does an open trace on a printed circuit board act like?
8. Why would an electronics technician want to keep a log of repairing a printed circuit board or electronic device?
9. What does an electronics technician need to do when faced with a problem that is difficult to solve?
10. How can a schematic diagram effectively be used for troubleshooting?

Section 7 Procedures

1. Form a team of three team members and select a problem listed below.

2. Determine each team member's role.

3. Determine team members' skills.

4. Brainstorm possible solutions.

5. Use the six-step design loop as a basis to focus on the task.

6. Maintain a design portfolio or lab notebook that includes all steps relevant to the final design. Each team member should keep his or her own individual journal.

7. Model the solution in Multisim or Circuit Wizard 2.

8. Build a prototype using Ultiboard or Circuit Wizard 2 to develop a PCB board design. (Use Section 1, Chapter 6; and Section 7, Chapters 48 through 50 as a guide).

9. Evaluate the prototype, using Appendix 14: Activity Assessment Rubric.

Problem

1. Using the schematic in Figure 17-18 of a real-world voltage divider, lay out a printed circuit board design and proof it with an electronic simulation program.

2. Design a printed circuit board for the voltage tripler shown in Figure 36-36. Check on the Internet for the Multisim circuit files for component values. Layout a printed circuit board by hand.

3. Fabricate a printed circuit board from one of the printed circuit board designs developed in activity 1 or 2.

4. Using a printed circuit board from an electronic device that is inoperative, desolder all the components and draw a schematic of the circuit located on the printed circuit board.

5. Take apart an electronic device to determine how it was constructed. Record each step in the disassembly of the device. (Follow proper safety precautions.)

6. Using an assembled circuit board, draw a schematic by following the traces on the circuit board. It is best to start with a single-sided printed circuit board.

7. Take the schematic that was drawn for activity 6 and draw it in an electronic simulation program to determine whether the schematic works properly.

8. On an electronic device that does not work, use the steps for troubleshooting to determine why it is inoperative.

9. To play a portable CD player requires three 1.5-volt batteries. When it is turned on, it is discovered the batteries are discharged. The batteries are always running down because of the amount of playtime. Build a power supply that can be used with 120V AC. Design a regulated power supply that will convert 120V AC to the required DC voltage. The current draw for the unit must be determined through measuring with an ammeter.

GLOSSARY

A

accumulator The most used register in a microprocessor.

active filter A filter without inductors, using an integrated circuit (IC).

adder A logic circuit that performs addition of two binary numbers.

alternating current (AC) Current that changes direction in response to change in the voltage polarity.

alternation The two halves of a cycle.

alternator AC generator.

ammeter A meter used to measure the amount of current flowing in a circuit.

ammonium persulphate Mild acid used to etch printed circuit boards. Highly toxic.

ampere (A) A measure of current flow.

ampere-hour A rating used to determine how well a battery delivers power.

amplification Providing an output signal that is larger than the input.

amplitude The maximum value of a sine wave or a harmonic in a complex waveform measured from zero.

analog meter A meter that uses a graduated scale with a pointer.

analog schematic diagram Electronics circuits are drawn with lines to each lead of a component.

AND function Performs the basic logic operation of multiplication.

AND gate One of the three primary gates and provides a high output only when all its inputs are high.

antistatic workstation Designed to provide a ground path for static charges.

aperture file Specifies the physical size of a printed circuit board in Gerber format.

apparent power Obtained by multiplying the source voltage and current and measured in units of volt-amps.

apprenticeship A program provided by an employer in which an apprentice learns a craft or trade through hands-on experience by working with a skilled worker.

arithmetic instructions Allow the microprocessor to compute and manipulate data.

arithmetic logic unit (ALU) Performs math logic and decision-making operations in a computer.

armature The moving portion of a magnetic circuit.

artificial magnet A magnet that is created by rubbing a piece of soft iron against another magnet.

associate CET exam A first step in getting certified and focuses on basic electronics. The exam is available to students in electronics or technician programs and good for four years.

astable multivibrator A free-running multivibrator.

asynchronous Not occurring at the same time.

atom The smallest basic unit of matter.

atomic number Represents the number of protons in the nucleus of an atom.

atomic weight Is based on the mass of an atom (the number of protons and neutrons).

audio amplifier Amplifies AC signals in the audio range of 20 to 20,000 hertz.

automation mechanic Assembles and maintains automated or computerized devices.

automotive mechanic Needs knowledge of electronics.

autoplace A subroutine for placing components on a PCB layout automatically.

autorouter A subroutine in a CAD program that automatically routes printed circuit board traces automatically.

autotransformer Used to step the applied voltage up or down without isolation.

average power The power consumed in an AC circuit; it is found by using the power formula with the rms values for current and voltage. The value will fall approximately midway between the peak value and 0.

B

barrier voltage A voltage created at the junction of P-type and N-type material when joined together.

base In a transistor, a semiconductor section of either N or P material and is represented by the letter B. Also, a term used to indicate the number of digits used in a number system.

base 2 See *binary number system*.

base 8 See *octal number system*.

base 10 See *decimal number system*.

base 16 See *hexadecimal number system*.

battery A combination of two or more cells.

BCD See *binary coded decimal*.

bench electronics technician A technician involved with repairing, testing, and/or assembling electronic circuits and/or systems at a workbench.

Berkley SPICE Simulation Program with Integrated Circuit Emphasis developed by the University of California at Berkley for simulating analog, digital, and mixed analog and digital electronic circuits.

bias voltage The external voltage applied across the P–N junction of a device.

binary coded decimal (BCD) A code of four binary digits to represent decimal numbers 0 through 9.

binary number system A number system with two digits, 0 and 1 and is referred to as base 2.

bipolar transistor A semiconductor device capable of amplifying voltage or power.

bistable action A multivibrator locking onto one of two stable states.

bistable multivibrator A multivibrator with two stable states.

bit A term derived from the words *binary digit*.

block diagram Represents how the subcircuit components are connected for an electronic circuit using simplified interconnected blocks.

blocking oscillator A relaxation oscillator that uses a capacitor discharge to block (cut off) the transistor. The RC time constant determines how long the transistor is cut off.

bode plotter A computer simulation tool that measures a signal's voltage gain or phase shift.

body resistance A measurement of body resistance, it represents the inverse of current through the body.

Boolean algebra Developed by George Boole and used to analyze binary logic circuits.

Boolean expression Algebraic expressions that obey the rules of Boolean algebra with 1s and 0s to help analyze digital logic circuits.

breadboard A platform for building a prototype of an electronic circuit for testing, modifying, and troubleshooting.

breadboarding The process of constructing a temporary electronic circuit for testing and/or modifying a circuit design.

breakdown voltage The point at which a breakdown occurs in a zener diode when a reverse bias voltage is applied.

bridge rectifier Operates both half cycles with full voltage out.

bridging Occurs when two traces in close proximity touch or when too much solder is used on a connection and solder flows to another pad or trace.

budget A financial plan created during the initial planning stages when building a project.

Butler oscillator A two-transistor circuit that uses a tank circuit and a crystal. The tank circuit is tuned to the crystal frequency with a small voltage placed across the crystal to relieve stress on it.

buffer A logic gate that isolates conventional gates from other circuitry.

bypass capacitor A capacitor placed across the emitter resistor in an amplifier, allowing the AC signal to bypass the resistor.

byte A group of 8 bits.

C

CAD See *computer-aided design*.

Canadian Standard Association (CSA®) Similar to Underwriters Laboratories, with a very strict safety code to ensure a product is safe to use as intended.

capacitance (C) The ability of a device to store electrical energy in an electrostatic field.

capacitive reactance Opposition a capacitor offers to applied AC voltage.

capacitor A device that possesses capacitance.

capacitor filter Used in power supplies after the rectifier to convert pulsating DC voltage to a smoother DC voltage.

cardiopulmonary resuscitation (CPR) A procedure used to revive breathing and pulse in an accident victim.

cathode-ray tube (CRT) A vacuum tube with an electron gun to accelerate electrons in a beam to strike a phosphor coated view screen.

cell A basic unit that can produce electrical energy by using two dissimilar metals, copper and zinc, immersed in a salt, acid, or alkaline solution. Cells can be connected in series and/or parallel combinations to increase voltage and/or current.

center-tapped secondary A transformer having a center tap in the electrical center of the secondary winding.

certified electronics technician (CET) program Measures the degree of knowledge and technical skills of working electronics technicians.

channel The U-shaped region of a JFET.

chemical cell A cell used to convert chemical energy into electrical energy.

chip An integrated circuit. It derives from the fact that an integrated circuit is manufactured from a single chip of silicon crystal.

circuit breaker A device that performs the same function as a fuse but does not have to be replaced.

Circuit Wizard A software package that combines schematic circuit design, PCB design, simulation, and CAD/CAM outputs by New Wave Concepts.

clamp-on ammeter An ammeter used to measure current by clamping around a wire using the magnetic field for sensing current flow.

clamping circuit A circuit used to clamp the top or bottom of a waveform to a DC level.

Clapp oscillator A variation of the Colpitts oscillator, with an additional capacitor in series with the inductor in the tank circuit to allow tuning the oscillator frequency.

class A amplifier An amplifier biased so that current flows throughout the entire cycle.

class AB amplifier An amplifier biased so that current flows for less than a full but more than a half cycle.

class B amplifier An amplifier biased so that current flows for only one-half of the input cycle.

class C amplifier An amplifier biased so that current flows for less than half of the input cycle.

clipping circuit A circuit used to square off the peaks of an applied signal.

clocked flip-flop A flip-flop that has a third input called the clock or trigger.

closed circuit A complete path for current flow.

closed-loop mode An op-amp circuit that uses feedback.

CNC See *computer numerical control machine*.

coefficient of coupling The amount of mutual induction between two coils based on a number of 0 to 1.

cold solder connection Occurs when the soldering iron is withdrawn too soon.

collector In a transistor, a semiconductor section of either N or P material; it is represented by the letter C.

color code A system developed to place large numerical values on small components. Used with resistors and capacitors.

Colpitts oscillator Similar to the shunt-fed Hartley oscillator except it uses two capacitors in place of the tapped inductor in the tank circuit to make it more stable.

combustible A material that is able to catch fire and burn easily.

commercial Used to refer to any business involved in producing, transporting, or merchandising a commodity.

common-base circuit The base in a transistor circuit is common to both the input and output circuits.

common-collector circuit The collector in a transistor circuit is common to both the input and output circuits.

common-emitter circuit The emitter in a transistor circuit is common to both the input and output circuits.

common mode rejection A measure of an op-amp's ability to reject signals common to both inputs or noise.

common signals Used on digital schematics diagrams to avoid confusion by labeling rather than drawing every connection.

communications A process by which information is exchanged between team members.

commutator The part of an armature in a DC generator that converts the AC voltage to DC voltage.

comparator A logic circuit used to compare the magnitude of two binary numbers.

compare instructions Compare data in the accumulator with data from a memory location or another register.

complementary push–pull amplifier An amplifier that uses two transistors, one to push the current and the other to pull the current, to drive loudspeakers.

component placement error Placing the wrong component in a location on the PCB, mounting a component in backwards, or inserting multi-lead components improperly.

component side The side of a printed circuit board where the components are placed.

compound Combination of two or more elements.

computer-aided design (CAD) A computer program used to assist in the design of a product or object.

computer electronics technician A technician who focuses on computer maintenance and information technology supporting field electronics technicians.

computer engineer A career field that is broken down into hardware and software engineers.

computer numerical control (CNC) machine A milling machine that is controlled by a computer and is useful in the removal of copper to form a printed circuit board.

computer technician A technician who focuses on computer and information technology.

condition-code register A register that keeps track of the status of the microprocessor.

conductance The ability of a material to pass electrons.

conductor A material that contains a large number of free electrons.

conformal coating A nonconductive material applied to an assembled printed circuit board to provide environmental and mechanical protection.

consumer A person or organization that uses a commodity or service.

continuity test A test for an open, short, or closed circuit using an ohmmeter.

control unit Decodes each instruction as it enters the computer.

cooperative education (co-op): A program in which students spend part of the day in class and the balance of the day working full-time in a job related to their chosen career field.

copyright Provides legal protection for the creator of an original work for a limited time frame.

coulomb Unit of electrical charge (represents 6.24×10^{18} electrons).

counter A logic circuit that can count a sequence of numbers.

counter electromotive force (cemf) An effect produced by an inductance that opposes the applied voltage.

coupling Joining of two amplifier circuits together.

coupling network A technique to prevent one amplifier's bias voltage from affecting the operation of the second amplifier when amplifiers are connected together.

covalent bonding The process of sharing valence electrons.

cross-reference guide A document that provides information relevant to making a substitution for various semiconductor devices.

crowbar An over-voltage protection circuit used in a power supply.

crystal oscillator An oscillator that uses a crystal to determine the frequency of the output signal.

current (I) The slow drift of electrons in a circuit.

cutoff frequency The frequency above or below the frequencies passed or attenuated in a filter network.

cycle Two complete alternations of current with no reference to time.

D

D flip-flop A flip-flop that has only a single data input and a clock input.

Darlington arrangement Two transistors connected together to function as one.

data movement instructions Move data from one location to another within the microprocessor and memory.

DC amplifier See *direct-coupled amplifier*.

decade counter A counter with a modulus of 10.

decimal number system A number system with 10 digits, 0–9.

decimal-to-binary encoder Converts a single decimal digit (0–9) to a 4-bit binary code.

decimal-to-binary priority encoder Produces a binary output based on the highest decimal on the input.

decoder Processes complex binary codes into recognizable characters.

decoupling network A circuits that allows a DC signal to pass while attenuating or eliminating an AC signal.

deflection plates Used in a CRT to bend the electron beam to display a pattern on its phosphor-viewing screen.

defluxer A chemical substance that is used to dissolve flux to aid in its removal from a printed circuit board.

degenerative feedback A method of feeding back a portion of a signal to decrease amplification.

degree of rotation The angle to which the armature has turned.

Department of Transportation (DOT) Establishes the overall transportation policy for the United States of hazardous materials.

depletion mode A mode in which electrons are conducting until they are depleted by the gate bias voltage in a MOSFET.

depletion MOSFET A semiconductor device that conducts when zero bias is applied to gate.

depletion region A region located near the junction of N-type and P-type material when joined together.

desktop publishing Using a computer page layout program to create a wide range of printed matter using layout and design elements.

desoldering The removal of solder from a component lead and printed circuit board pad to aid in the removal of a component that has been soldered to a printed circuit board.

desoldering bulb A rubber bulb used to create a vacuum to suck up molten solder; used with a soldering iron.

desoldering pump Uses a spring-loaded plunger to create a vacuum to suck up molten solder; used with a soldering iron.

desoldering braid Uses a copper braid with flux to create a capillary action to pull up molten solder; used with a soldering iron.

DIAC Diode AC, a bidirectional triggering diode.

Dictionary of Occupational Titles (DOT) A publication produced by the U.S. Department of Labor that provides job listings and job descriptions.

difference amplifier An amplifier that subtracts one signal from another signal.

difference of potential The force that moves electrons in a conductor.

differential amplifier An amplifier with two separate inputs and one or two outputs.

differentiator A circuit used to produce a pip or peak waveform from a square wave.

digital computer A device that automatically processes data using digital techniques.

digital meter A meter that provides readout in a numeral display.

digital schematic diagram In digital circuits, the common signals are bundled together in a single line and labeled.

diode A semiconductor device that allows current to flow only one way.

DIP See *dual in-line package.*

direct-coupled amplifier Provides amplification of low frequencies to DC voltage.

direct current (DC) A current that flows in a circuit in only one direction.

discrete component Components that must be interconnected with other components to form an operational circuit.

discrimination The unequal treatment of an individual because of an employer's prejudices.

disposal The process of throwing away or getting rid of something.

disturbed connection The result of movement of the component lead or wire before the solder sets.

DMM Digital multimeter.

documentation Collection and organization of all paperwork, for future reference, when troubleshooting a circuit.

domains Groups of atoms arranged in a common direction to magnetic north and south.

doping A process of adding impurities to semiconductor material to increase the number of free electrons or holes.

double-sided printed circuit board A printed circuit board with copper traces and pads on both sides.

drill file Informs the printed circuit board manufacturer where the holes are located and what size the holes are.

dual in-line package (DIP) An integrated circuit package that has a row of pins along each of its longer sides for through-hole mounting on a PCB.

duty cycle Ratio of the pulse width to the period.

dynamic RAM (DRAM) Read/write memory that must be refreshed (read or written into) periodically to maintain the storage of information.

E

effective value A value of AC current that will produce same amount of heat as DC current.

electric circuit The path that current follows.

electrical charge (Q) Created by the displacement of electrons moving past a point that leads to current flow.

electrical engineer Person responsible for designing new products, writing performance specifications, and developing maintenance requirements.

electrical shock The result of a reflex response to the passage of electric current through the body that can be felt.

electrical symbols Used to represent electrical fixtures and devices in an architectural drawing.

electrically erasable PROM (EEPROM) PROMs that can be programmed and erased using an electrical charge and are used where data needs to be kept for later use.

electrician An individual who assembles, installs, and maintains air conditioning, heating, lighting, power, and refrigeration components.

electrocution The passage of a high electric current through the body, resulting in death.

electromagnet Magnet created by passing a current through a coil of wire.

electromagnetic induction Inducing a voltage from one coil into another coil.

electromotive force (emf) Also identified as a difference of potential.

electron A negative charge particle in orbit around the nucleus of an atom.

electronics circuit simulation Software that can model electronic circuit performance.

electronics engineer An electrical engineer involved in the designing of electronics circuits, devices, and systems.

electronics engineering A discipline of electrical engineering.

Electronics System Associate (ESA) exam A four-part exam available to anyone with an interest in electronics. Upon passing all four parts, the technician is given a lifetime certification.

electronics technician An individual who builds, installs, tests, and repairs electronic devices.

Electronics Technician Association International (ETAI) A professional organization, founded by electronics technicians for electronics technicians, that promotes recognized professional credentials as proof of a technician's skills and knowledge.

electrostatic discharge (ESD) The discharge of static electricity from an object to an object with a lesser charge; may be fatal to sensitive electronic components.

element The basic building block of nature.

ELI The AC current lags the voltage by 90° in an inductor (acronym).

ELVIS An educational laboratory with virtual instrumentation using a LabVIEW-based design as a prototype environment for the education community by National Instrument.

emitter In a transistor, a semiconductor section of either N or P material; it is represented by the letter E.

emitter follower Another name for a common emitter amplifier because the emitter voltage follows the base voltage.

encoder A logic circuit of one or more inputs that generates a binary output.

enhancement mode A mode in which electron flow is normally cut off until it is aided or enhanced by the bias voltage on the gate of a MOSFET.

enhancement MOSFET A device that only conducts when a bias is applied to the gate.

ESD See *electrostatic discharge*.

ethics A set of established standards based on core values such as honesty, respect, and trust.

even harmonics Frequencies that are even multiples of a fundamental frequencies.

exclusive NOR gate (XNOR) A two-input logic gate that produces a high output only when its inputs are identical.

exclusive OR gate (XOR) A two-input logic gate that produces a high output only when its inputs are a high and a low.

exponents Symbols written above and to the right of a mathematical expression.

externship An educational program lasting from one day to several weeks, intended to introduce students to the work requirements in a particular career field.

F

faceplate A plate that is marked centimeters (cm) along both horizontal and vertical axis and place in front of the phosphor screen of the CRT of an oscilloscope.

fall time The time it takes for a pulse to fall from 90% to 10% of maximum amplitude.

farad The basic unit of capacitance (F).

Faraday's law The basic law of electromagnetism. The induced voltage in a conductor is directly proportional to the rate at which the conductor cuts the magnetic lines of force.

Federal Communications Commission (FCC) The electronics communication licensing and enforcing agency in the United States.

ferric chloride acid A mild acid used to remove unwanted copper during the printed circuit board fabrication process.

ferromagnetic materials A ferrous material that responds to a magnetic field.

field It produces an electromagnetic field in a generator or motor when operating.

filter Converts pulsating DC into a smoother DC. In a power supply, a filter removes AC ripple. An electronic circuit that either passes or rejects a specific range of frequencies.

first harmonic Another name for fundamental frequency.

fixed capacitor A capacitor having a definite value that cannot be changed.

flammable A material that is easily ignited and capable of burning rapidly.

flat response Indicates that the gain of an amplifier varies only slightly within a stated frequency range.

flip-flop A bistable multivibrator whose output is either a high or a low. See *bistable multivibrator*.

flux A material used to remove oxide films from metallic surfaces. May be a liquid, paste, or located in the center of solder.

flux cleaner A chemical cleaner used to remove flux residue from a soldered connection.

flux lines Invisible lines of force that surround a magnet.

forward bias When a semiconductor device is connected to cause the current to flow. A positive voltage is applied to the P-material and a negative voltage is connected to the N-material, causing a current to flow across the junction.

fractured connection See *disturbed connection*.

free running Does not require an external trigger to keep running.

frequency Number of cycles that occur in a specific period of time.

frequency counter Measures frequency by comparing against known frequency.

frequency domain All periodic waveforms are made of sine waves.

full adder An adder circuit that takes into account a carry.

full-scale value The maximum value indicated by a meter.

full subtractor A subtractor circuit that takes into account the borrow input.

full-wave rectifier Operates during both half cycles with half the output voltage.

function generator A signal generator that is capable of producing fundamental waveforms such as sine, triangular, and rectangular or square.

fundamental frequency Represents the repetition rate of the waveform and is also referred to as the first harmonic.

fuse A protection device that causes an open circuit when the current flow through it exceeds a determined value.

G

Gantt chart A timeline chart that shows when each task of a project begins and the amount of time to complete the task and shows which tasks are going on at the same time.

generator A device used to generate electricity by means of a magnetic field.

Gerber file A standard printed circuit board format that is supported by all layout software.

germanium A gray-white element recovered from the ash of certain types of coal that is used in the manufacturing of semiconductor devices.

good work habits Unique methods to consistently accomplish common tasks.

graphic calculator The most powerful and versatile of the calculators. It is often equipped with special features for solving engineering problems and is capable of displaying graphs and being programmed.

graphical images Images prepared in computer illustration (drawing) programs or image editor programs for desktop publishing and web design.

graticule Mounted in front of the CRT of an oscilloscope to assist in measuring a signal.

ground The common point in electrical/electronic circuits.

ground fault circuit interrupt (GFCI) A fast-acting circuit breaker that is sensitive to very low levels of current leakage to ground to prevent serious injuries.

ground plane An entire printed circuit board layer that is isolated from connections and connected to ground potential.

grounding The process of making a connection from a circuit board or product to ground that provides an alternate path for carrying current to ground in the event of a failure.

guard Protects operators from preventable injuries with power tools.

H

half adder An adder circuit that does not take into account any carries.

half subtractor A subtractor circuit that does not take into account any borrowed input.

half-wave rectifier Operates only during one-half of the input cycle.

hand-drawn The simplest technique to transfer a design to a printed circuit board; uses a resist pen or permanent marker.

hand tool A nonpowered tool for performing work on a material or a physical system.

harassment Continually annoying an individual through verbal or physical conduct that creates an unpleasant or hostile environment.

harmonics Multiples of the fundamental frequency.

hazard identification A hazard assessment conducted in order to make the workplace as safe as possible.

hazardous material Any substance or material that could adversely affect the safety of employees or handlers during use or transportation.

hazardous waste Chemical substance that poses a substantial or potential threat to public health or the environment.

henry (H) The unit of inductance.

hertz (Hz) The unit of frequency that represents cycles per second.

hexadecimal number system A number system with 16 digits, 0–9 and A–F.

high A term used to identify that in a 1 state the voltage is higher that the 0 state in logic circuits.

high-pass filter A filter circuit that passes high frequencies while attenuating the low frequencies.

hole Represents the absence of an electron.

HTML Hypertext Markup Language, the main markup language used for displaying web pages on a web browser.

I

IBIS model See *input/output buffer information specification model.*

IC See *integrated circuit.*

ICE The AC current leads the voltage by 90° in a capacitor (acronym).

idling Placing a soldering iron on a stand to remove excess heat.

IEEE Institute of Electrical and Electronics Engineers.

IF amplifier See *intermediate frequency amplifier.*

impedance Opposition to current flow in an AC circuit (Z).

impedance matching A process of using a transformer to adjust the impedance of the source to the load to make them equal.

inductance The property of a coil or electric circuit that resists a change in current flow.

induction The effect one body has on another body without physical contact.

inductive reactance Opposition offered by an inductor in an AC circuit to current flow.

inductor (L) A device designed to have a specific inductance.

in phase When two signals change polarity at the same time.

input/output (I/O) Allows the computer to accept and produce data.

input/output instructions Control I/O devices.

input/output buffer information specification (IBIS) model A model that is used in place of SPICE models for high-speed circuits.

instantaneous rate The rate at which work is done in a circuit when the voltage causes electrons to move and is referred to as the electric power rate and measured in watts.

Institute of Electrical and Electronics Engineers (IEEE) An influential organization for electronics engineers.

instruction register Stores the instruction word to be decoded.

insulated A system that prevent the passage of electricity to or from a power tool by covering it in nonconducting material.

insulating Prevents electrical shock by separating the body from electrical source.

insulator A material that has few free electrons.

integrated circuit (IC) A device in which all of the components of an electronic circuit are fabricated on a single piece of semiconductor material, often called a chip, which is placed in a plastic or metal package with terminals for external connections.

integrator A circuit used to reshape a waveform.

intermediate frequency (IF) amplifier A single frequency amplifier.

International Certification Accreditation Council (ICAC) Accredits certifications to assure compliance, professional management and service to the public. This group is responsible for setting the standards for certification and licensing programs internationally.

International Electrotechnical Commission (IEC) A group responsible for keeping electronic symbols up to date.

International Society of Certified Electronics Technicians (ISCET) A group that prepares and tests technicians in the electronics and appliance service industry.

Internet A global information system that links together, through a unique Internet Protocol (IP) address, an extensive range of information, resources, and services.

internship An educational program that allows students to gain experience in a chosen career field by participating in supervised work experience.

interrupt Used in a microprocessor- ot microcontroller-based system to cause the system to stop what it is doing and perform another task.

intrinsic material Semiconductor material that is identified as pure material.

inversely proportional When one variable changes, the opposite occurs to the other variable.

inverter A logic circuit that changes its input to the opposite state for an output.

inverting amplifier Also referred to as a unity gain amplifier, it is used to invert the polarity of the input signal.

inverting input The negative input of an op-amp.

ionization The process of gaining or losing electrons.

iron-vane meter movement A meter movement that uses an iron vane and a stationary coil to measure low-frequency AC current or voltage.

isolated power supply A power supply that is disconnected from ground to prevent a shock hazard when connected to ungrounded circuits.

isolation transformer Used to prevent electrical shocks by disconnecting the circuit being worked on from the main circuit.

J

JK flip-flop A flip-flop that incorporates all the features of other flip-flops.

job interview A follow-up for a successful introduction with a letter of application and resume. The prospective employer meets the prospective employee personally to question him or her in detail further about qualifications and to answer questions about employment.

job shadowing A work experience program to help students see firsthand what a chosen career field is like.

Journeyman level CET The level required for being a fully certified electronics technician, achieved upon passing a selected Journeyman Level CET exam and good for four years.

jumper A piece of wire used to make an electrical connection between two points on a single sided printed circuit board.

junction FET (JFET) Field effect transistor that can provide amplification.

junction transistor See *bipolar transistor.*

K

Karnaugh map A graphing technique using a rectangular arrangement for simplifying complex Boolean expressions. Maurice Karnaugh perfected this technique as an improvement on Veitch diagrams.

key habits Take action, overcome fear of failure, and never give up.

Kirchhoff's current law The algebraic sum of all the currents entering and leaving a junction is equal to 0.

Kirchhoff's voltage law The algebraic sum of all the voltages around a closed circuit equals 0.

L

LabVIEW A graphical programming environment used by engineers and scientists to develop sophisticated measurements, tests, and control systems by National Instrument.

latch A logic circuit that serves as a temporary buffer memory and is changed by an external source.

LC oscillator An oscillator that uses a tank circuit of capacitors and inductors, connected in series or parallel, to determine the frequency of the circuit.

lead A connecting wire, such as a test lead, battery lead, and so forth.

leading edge Identifies the front edge of a waveform.

leg warmer A small piece of insulation placed on the leads of heat-sensitive components to prevent overheating when soldering.

letter of application An introductory letter written specifically to a prospective employer and usually sent with a resume.

light Electromagnetic radiation that is visible to the human eye.

light-emitting diode (LED) Converts electrical energy directly into light energy.

limiter circuit See *clipping circuit*.

load A device or circuit on which work is performed.

loading down A load placed on the output that affects the amount of output current.

logic instruction An instruction that contains one or more of the Boolean operators (AND, OR, and exclusive OR).

logic probe A digital instrument that can detect a high or low level or a group of pulses at a point in a digital circuit.

logic pulser A device that can inject a digital signal into a digital circuit for troubleshooting.

loudspeaker A device for converting audio frequency current into sound waves.

low Indicates a 0 state and indicates that the voltage is lower that a 1 state.

low-pass filter A filter circuit that passes low frequencies and attenuates high frequencies.

LSB The least significant bit in a number; it is located at the rightmost and is the smallest weight bit in a whole digital number.

M

magnet A piece of iron or steel that attracts other pieces of iron or steel.

magnetic field The field that surrounds a magnet.

magnetic induction The effect a magnet has on an object without physical contact.

magnetism The property of a magnet.

manufacturer's reference manual A document that provides a manufacturer's specification data sheet on semiconductor products for engineers and technicians.

masking An instruction used with microprocessors to compare two numbers by subtracting one from the other and allowing only certain bits through.

material safety data sheet (MSDS) A document containing information on health risks with exposure to chemicals or other potentially dangerous substances, which is attached when a chemical is purchased or acquired that contains a hazardous substance.

mathematic skills The knowledge required to reason with numbers.

matter Anything that occupies space.

memory Stores the program and data for operation by the computer.

memory function A temporary storage location on calculators to save data entry time.

mentoring A partnership between trade professionals who share knowledge, information, perspectives, and skills in their chosen career with less experienced colleagues or trainees.

metal-oxide semiconductor field effect transistor (MOSFET) Insulated gate FET that uses a metal gate that is electrically isolated from the semiconductor channel by a thin layer of oxide.

metrology electronics technician Performs equipment calibration and repair to National Institute of Standards and Technology (NIST) or other standards.

mho ($\mho$) The older unit of conductance that is now referred to as Siemens.

microampere One-millionth of an ampere.

microcontroller A single-chip computer designed for machine-control applications.

microfarad 0.000001 farad, or one-millionth of a farad.

microphone An energy converter that changes sound energy into corresponding electrical energy.

microprocessor Contains ALU, timing, control, and decoding circuitry for the computer.

milliampere One-thousandth of an ampere.

miscellaneous instruction An instruction used to enable or disable interrupt lines, clear or set flag bits, or allow the microprocessor to perform BCD arithmetic.

mixed-mode simulator Capable of simulating both analog and event driven digital at the same time with accurate information.

mixture A physical combination of elements and compounds.

modulus The maximum number of states in a counting sequence.

molecule The smallest part of a compound that retains the properties of the compound.

monostable multivibrator A multivibrator with only one stable state.

MOSFET See *metal-oxide semiconductor field effect transistor*.

moving coil meter movement A meter movement that uses a coil pivoting between the poles of a permanent magnet.

MSB The most significant bit; it is located at the leftmost and is the largest weight bit in a whole digital number.

MSDS See *material safety data sheet*.

multilayer board A printed circuit board that is fabricated from two or more printed circuits with traces on one or both sides laminated together.

multimeter A voltmeter, ammeter, and ohmmeter combined into a single meter.

multiplexer A circuit that selects and routes one of several inputs to a single output.

multivibrator A relaxation oscillator that has two temporary stable conditions.

Multisim An electronic schematic capture and simulation program that is part of a suite of electronic design programs by National Instrument.

mutual inductance An effect resulting when a magnetic field expands in the same direction as the current in the primary, aiding it and causing it to increase.

N

NAND gate A combination of an inverter and an AND gate.

National Electric Code (NEC®) Registered trademark of the National Fire Protection Association Incorporated located in Quincy, Massachusetts; it provides a summary of electrical wiring codes.

NEC® guide book A publication on electrical wiring codes that is produced by the National Fire Protection Association.

negative feedback Used in an amplifier circuit, it requires the feeding back of a small amount of the output signal to the input signal 180° out of phase in order to improve stability and signal reproduction.

negative ion An atom that has gained one or more electrons.

negative temperature coefficient As temperature increases, resistance decreases.

neutron Electrically neutral particle in the nucleus of an atom.

nibble A binary group of four bits.

noninverting amplifier An op-amp connected such that the input and output signals are in phase and have a gain greater than 1.

noninverting input The positive (+) input of an op-amp.

nonsinusoidal oscillator An oscillator that does not produce a sine wave output.

nonsinusoidal waveform Any waveform that does not have the properties of a sine wave.

NOR gate A combination of an OR gate and an inverter that provides a high out only when all inputs are low.

NOT gate One of the three primary gates, it is a single input gate where the output is the inverse of the input. This gate is also referred to as an *inverter*.

N-type material Semiconductor material doped with a pentavalent material.

nucleus The center of the atom, which contains the mass of the atom.

O

Occupational Outlook Handbook A document produced by the U.S. Department of Labor.

Occupational Safety and Health Administration (OSHA) A government agency in the

Department of Labor charged with maintaining a safe and healthy work environment in the workplace.

octal number system A number system with eight digits, 0–7.

odd harmonics Frequencies that are odd multiples of the fundamental frequency.

ohm (Ω) Unit of resistance.

ohmmeter A meter used to measure resistance.

Ohm's law The relationship among current, voltage, and resistance.

one-shot multivibrator See *monostable multivibrator*.

op-amp See *operational amplifier*.

open circuit A circuit that has infinite resistance because no current flows through it.

open-loop mode An op-amp circuit that does not use feedback.

operational amplifier A very high-gain DC amplifier.

optical coupler Used to isolate loads from their source.

OR function Performs the logic operation of addition.

OR gate One of the three primary gates, it provides a high output when any of its inputs are high.

oscillator A circuit that generates a repetitive AC signal.

oscilloscope Provides a visual display of what occurs in a circuit.

out of phase When two signals are not in phase.

overshoot Occurs when the leading edge of the waveform exceeds the maximum value.

over-voltage protection circuit Protects a load from voltage increases above a predetermined level.

P

packing groups Indicates the degree of risks associated with hazardous material when transported.

pad Located on a printed circuit board where connections are made to connect the lead from a component to the trace.

PAL See *programmable array logic*.

paper dolls Two-dimensional representations of circuit components that are the exact size of the base of the components and that are used to assist in component placement when laying out a printed circuit board.

parallel cells Multiple ells that are connected with all positive and all negative terminals connected together.

parallel circuit A circuit that provides two or more paths for current flow.

parallel counter A synchronous counter where all the clock inputs are connected in parallel.

passive filter A filter that use resistors, inductors, and capacitors.

PCB prototype An early example of a printed circuit board that is built to test a circuit and to learn from it.

peak inverse voltage (PIV) Maximum safe reverse voltage rating.

peak-to-peak value Represents the vertical distance from one peak to the other peak of a waveform.

peak value Absolute value of point on a waveform at the greatest amplitude.

pencil soldering iron A soldering iron that is held like a pencil.

pentavalent An atom with five valence electrons in its outer shell, used with making N-type material from pure silicon.

perfectionism A personality trait driven by the rejection of any accomplishments falling short of excellence.

period Time required to complete one cycle of a sine wave.

periodic waveform A waveform that occurs at regular intervals.

permanent magnet A magnet that retains magnetic properties.

PERT chart See *project evaluation* and *review technique chart*.

phase angle The exact phase shift between the input and the output.

phase-shift network Shifts the phase of the output signal with respect to the input.

phase-shift oscillator An oscillator using a conventional amplifier with a phase-shifting RC feedback network, with the frequency determined by the resistor and capacitor values in the RC network.

phosphor screen The inside of a CRT that, when struck by an electron beam, emits light.

photocell See *photoconductive cell*.

photoconductive cell Has internal resistance that changes with light intensity.

photodiode Used to control current flow by means of light energy.

phototransistor Works like a photodiode but produces higher output current.

photovoltaic cell A device used to convert light energy into electrical energy.

pi Represents a mathematical constant of 3.14.

PIC See *programmable interface controller*.

picofarad (pF) Represents 0.000,000,000,001 farad, or 1/1,000,000,000,000 farad.

Pierce oscillator An oscillator that is similar to the Colpitts oscillator except that the tank inductor is replaced with a crystal.

piezoelectric effect A process where pressure is applied to a crystal to produce a voltage. When voltage is applied to a crystal, it causes the crystal to vibrate at a specific frequency that is determined by the crystal.

PIN photodiode A photodiode with an intrinsic layer between the P and N regions.

PLA See *programmable logic array*.

plastic quad flat pack (PQFP) A high-density, surface mount package with leads on all four sides that provides the most variation of any package type.

PLD See *programmable logic device*.

portable Can be carried or moved easily.

positive feedback Feeding back a part of the output signal that is in phase.

positive ion An atom that has lost one or more electrons.

potential The ability to do work.

potentiometer A variable resistor used to control voltage.

power (P) Electrical power is the rate at which energy is dissipated by the resistance of a circuit. In mathematics, it is shown as an exponent and indicates how many times to multiply a number with itself.

power amplifier An audio amplifier designed to drive a specific load.

power curve A curve that results from taking the product of the sine curve for current and the sine curve for voltage. For resistive circuits, the power curve is a positive area because the load absorbs the power as heat. For purely reactive circuits the power curve has both positive and negative areas because the load absorbs no net power.

power dissipated The power dissipated by a component is determined by finding the product of the voltage drop across a component and the current flowing through the component. The power dissipated in an AC resistive circuit is the same as in DC circuits.

power factor The ratio of true power in watts to apparent power in volt-amps.

power of observation The most important step in troubleshooting by an electronic technician for verifying, isolating, and eliminating problems in an electronic circuit.

power tool A tool that is actuated by an electric motor, internal combustion engine, or compressed air.

PQFP See *plastic quad flat pack*.

presentation program A computer program used to display information in the form of a slide presentation.

primary cell A chemical cell that cannot be recharged.

primary winding In a transformer, the winding that receives the AC voltage.

printed circuit board (PCB) Consists of an insulating base material on which conductive traces of copper act as interconnecting wires. Traces may be located on one or both sides.

printed circuit board prototype An early example of a PCB built to test a circuit and to learn from it.

procrastination The process of continuing to put doing something off because of habitual laziness.

production schedule A manufacturing guide to monitor progress on a production line to increase productivity and minimize operating expenses.

program A list of computer instructions arranged sequentially to solve a problem.

program-control instructions Instructions that change the content of the program counter.

program counter Contains the instruction addresses for a program.

programmable array logic (PAL) A programmable logic device that has fusible-link multi-input AND gates connected to OR gates. By blowing specific fuses, a PAL can be programmed to solve a variety of complex logic equations.

programmable interface controller (PIC) Represents a family of microcontrollers that use reduced instruction set computer (RISC) for programming.

programmable logic array (PLA) Similar to a PAL, with programmable logic device that has fusible-link multi-input AND gates connected to additional fusible-link OR gates.

programmable logic device (PLD) Device similar to a PROM because it uses a fusible solution for the output.

programmable read-only memory (PROM) A device that can be programmed after it is manufactured by blowing a fusible link using a high current surge; it cannot be reprogrammed.

project evaluation and review technique (PERT) chart A flowchart of tasks required in the production phase of a project.

project timeline A manufacturing guide to monitor progress on a production line to increase productivity and minimize operating expenses.

PROM See *programmable read-only memory.*

proton Positively charged particle in the nucleus of an atom.

P-type material Semiconductor material doped with a trivalent material.

pulse width The duration the voltage is at its maximum or peak value until it drops to its minimum value.

push–pull amplifier Consists of two mirror image amplifiers that operate on different half cycles of the input signal.

Pythagorean theorem Relates the sides of a right triangle by the formula $a^2 = b^2 + c^2$ in electronics as $E_T^2 = E_R^2 + E_C^2$ or $E_T^2 = E_R^2 + E_L^2$.

Q

quasi-complementary amplifier An amplifier circuit that uses both NPN and PNP transistors in a symmetrical arrangement that permits push–pull operation but does not require high-power complementary output transistors like complementary amplifiers.

R

radio-frequency amplifiers Amplify signals of 10,000 hertz to 30,000 megahertz.

random access memory (RAM) Memory used to temporarily store programs, data, control, or information.

RC network A resistor-capacitor network used for filtering, decoupling, DC blocking, or coupling phase-shift circuit.

RC oscillator An oscillator that uses a resistor-capacitor network to determine the oscillator frequency.

reaction A chemical change that forms new substances.

read-only memory (ROM) Stores data permanently and allows the data to be read but has no provision for data to be written.

reciprocals multiplicative inverse found by putting a one over a whole number.

rectification The process of converting AC to DC.

rectifier circuit Converts an AC voltage to a DC voltage.

reference designator Identifies components in a schematic diagram and on a printed circuit board.

regenerative feedback A method of obtaining an increased output by feeding part of the output back to the input.

relaxation oscillator An oscillator that stores energy during part of the oscillation cycle.

relay An electromagnetic switch that opens and closes with an armature.

repelling A condition that occurs when like charges are place in proximity of other like charges or like magnetic field fields of two magnets are placed in close proximity to each other.

residual magnetism A magnetic field that remains after a magnetic source is removed.

resistance Opposition to electron flow in a circuit (R).

resistors Components manufactured to possess a specific value of resistance to the flow of current.

resonance When both the inductive reactance and the capacitive reactance in an AC circuit are equal at one particular frequency.

resonant circuit A circuit that contains both inductive and capacitive reactance balanced at a particular frequency.

résumé A summation of personal information including education, certifications, skills, and experiences, which is presented to prospective employers when seeking employment.

retentivity The ability of a material to retain a magnetic field after the magnetic force is removed.

reverse bias A semiconductor device connected so that current does not flow across the P–N junction.

reverse Polish notation (RPN) A notation for forming mathematical expressions where each operator is followed by its operands when entering data into an RPN calculator.

RF amplifiers See *radio-frequency amplifiers.*

rheostat A variable resistor used to control current.

ringing Dampened oscillations occurring as the transient response of a resonant circuit to a shocked excitation.

ripple counter An asynchronous counter where all the flip-flops except for the first flip-flop are triggered by the preceding flip-flop, resulting in a delay such that the input clock pulse ripples through the counter.

ripple frequency The frequency of pulses on the output of a rectifier circuit.

rise time The time it takes for a pulse to rise from 10% to 90% of maximum amplitude.

rms value Another term for effective value, root-mean-square.

root A term that refers to the square root or the nth root of a number.

ROM See *read-only memory.*

rosin connection A poor solder connection that is connected with rosin flux in place of solder.

rosin flux A noncorrosive flux used to clean impurities from surfaces prior to their being soldered.

rotate and shift instructions Change the data in a register or memory by moving the data to the left or right 1 bit.

rounding A technique used to reduce the number of significant digits by dropping the least significant digits until the desired number of digits remain.

RS flip-flop A flip-flop with a set and reset input.

S

safety Practicing awareness in the work environment to protect human life and the equipment used.

safety switch A switch requiring two different actions to activate a power tool.

sawtooth waveform A special case of a triangular waveform, with one long ramp and one with a shorter duration.

scaling A voltage divider that allows an instrument to read a voltage above its normal range.

schematic diagram A diagram that uses symbols to represent an electrical or electronic circuit.

schematic symbols Drawings used to represent electrical and electronic devices and components in a schematic drawing.

Schmitt trigger A modified bistable multivibrator that changes state when the input level reaches a specific triggering level.

scientific calculator A calculator that handles exponential, trigonometric, and other special functions and performs basic arithmetic operations in physics, mathematics, and engineering.

scientific notation Using powers of 10 to express large and small numbers.

secondary cell A chemical cell that can be recharged.

secondary winding The coil into which the voltage is induced in a transformer.

semiconductor A material that has four valence electrons.

senior electronics technician An in-house position for running the shop, requiring extensive experience within the organization employed.

series-aiding cells Connecting positive terminal to negative terminal of cells to increase the voltage level over a single cell.

series circuit A circuit that provides a single path for current flow.

series-fed Hartley oscillator An oscillator that uses a tapped inductor in the tank circuit, with DC current flowing through a portion of it.

series-opposing cells Connecting negative to negative terminal or positive to positive terminal of a cell.

series-parallel cells Used to increase current and voltage outputs above that of a single cell.

series-parallel circuit A combination of the series circuit and the parallel circuit.

series voltage regulator The voltage regulator is connected in series with the load.

sequential logic circuit Logic circuit that requires timing and memory devices such as flip-flops.

shell The orbit about the nucleus of an atom.

shift-register Constructed of flip-flops and used to store data temporarily.

short circuit A circuit made inactive by a low resistance path across the circuit. Has zero ohms resistance.

shorting stick Used to discharge high-voltage charges in equipment.

shunt-fed Hartley oscillator An oscillator that overcomes the problem of DC current in the tank circuit of the series-fed Hartley oscillator by using a coupling capacitor in the feedback line

shunt voltage regulator A voltage regulator connected in parallel with the load.

Siemens (S) The unit for measuring conductance, the reciprocal of resistance.

signal generator A piece of electrical test equipment that provides a calibrated frequency over a large range with a variable output.

silicon A mineral recovered from silicon dioxide, found extensively in the earth's crust and used in the manufacturing of semiconductor devices.

silicon control rectifier (SCR) A thyristor that controls AC or DC current flow in only one direction.

sine wave A sinusoidal waveform.

single-sided printed circuit board A printed circuit board that contains the copper traces and pads on only one side.

sinusoidal oscillator An oscillator that produces sine wave outputs.

sinusoidal waveform An alternating waveform that is most often referred as a sine wave. It is related to the trigonometric sine function.

slip rings Used to remove AC voltage from the armature in an alternator.

slow blow fuse A fuse that can withstand a brief period of overloading before it opens.

SMD See *surface mounted device*.

SMT See *surface mount technology*.

solar cell See *photovoltaic cell*.

solder A tin-lead alloy with a low melting temperature that is used to form a permanent bond and reliable electrical connections.

solder side Contains the copper traces for a single-sided printed circuit board.

solder sucker See *desoldering pump*.

soldering Forming an electrical connection using solder.

soldering gun A soldering iron that resembles a pistol and is capable of generating a high amount of heat that is used heavy-duty soldering.

soldering iron An electrical heating device that consists of a heating element that heats the tip to melt solder.

soldering iron tip Transfers the heat to the connection being soldered.

solderless breadboard A platform with an array of holes to accept component leads allowing electronic circuits to be assembled and modified quickly and easily without soldering.

solenoid A cylindrical coil with a movable plunger in the center.

spectrum analyzer Displays a waveform in a frequency domain.

SPICE An industry standard for electronic circuit simulation.

spreadsheet A computer program used for inputting, manipulating, organizing, and storing numbers and data.

square waveform A periodic waveform with flat tops and bottoms with a square shape.

stack An area of memory used for temporary storage.

stacked based Using stacks to permits the storage of results for later use.

stack instructions Allow the storage and retrieval of different microprocessor registers in the stack.

stage Each flip-flop in a digital counter.

stack pointer A register that holds memory locations of data stored in the stack.

static electricity Produced when two substances are rubbed together or separated, resulting in the transfer of electrons from one substance to another.

static RAM (SRAM) A temporary memory device that uses bistable circuits as the storage element but loses the data when the power is removed.

stationary Equipment fixed in and tending to stay in one place.

step-down transformer A transformer that produces a secondary voltage less than the primary one.

step-up transformer A transformer that produces a secondary voltage greater than the primary one.

strain gauge A gauge for transforming a strain applied into a proportional change in resistance.

substrate Forms the base for a semiconductor device.

subtractor A logic circuit that performs subtraction of two binary numbers.

summing amplifier Used when mixing audio signals together.

surface mount technology (SMT) Technology that uses surface mounted components on the top and/or bottom of a printed circuit board.

surface mounted device (SMD) A device for surface mounting as opposed to through-hole mounting.

synchronous A term that means occurring at the same time.

synchronous counter A digital counter that has all the flip-flops triggering at the same time by a clock pulse.

T

tank circuit Formed by connecting an inductor and a capacitor in parallel.

tap A connection point made in the body of the secondary of a transformer.

team A small group of people with similar skills committed to a common purpose.

teamwork The process of a group of people working together to achieve a common goal.

temperature-controlled solder station Provides a precise tip temperature through a closed loop circuit.

temporary magnet A magnet that retains only a small portion of its magnetic property.

thermal instability A condition that when temperatures increase, causing a transistor's internal resistance to vary, which results in changes in the bias circuit of an amplifier circuit, thus reducing the gain.

thermocouple A device used to convert heat into electrical energy.

thin small outline package A variation of the first surface mount package, it supports a wider body width.

three-state buffer A buffer that provides a noninverting of the input to the output (0 to 0 and 1 to 1) and a high impedance state controlled by an enable line.

thyristors A broad range of semiconductor devices used for electronically controlled switches.

time constant The time relationship in L/R and RC circuits.

time domain concept An analysis of waveforms with regards to changes made with time. An oscilloscope presents waveforms in the time domain, with the horizontal axis measuring time and the vertical axis measuring volts of the signal being measured.

tinning The process of applying solder to the soldering iron tip to keep it from oxidizing or to promote heat transfer. Also, the process of applying solder to stranded wire prior to making a connection.

TO (transistor outline) Most common transistor package identifier.

tolerance A measure of how much a resistor's value may vary and still be an acceptable value.

toolbox A case that varies with the craft of the owner, used to organize, carry, and store tools.

torque The tendency of twisting force to change the rotational motion of a power tool.

trace A substitute for wires in printed circuit board development.

trailing edge The back edge of a waveform.

transformer An AC electrical device for transferring and changing the value of AC electrical voltage using a changing magnetic field.

transient A temporary component of current existing in a circuit during adjustment to a load change, voltage source difference, or line impulse.

transistor See *bipolar transistor*.

transparent latch A D flip-flop with the enable line high (1) allowing the output to follow the input to hold the data for reading until updated.

TRIAC Triode AC semiconductor, a device that provides full control of an AC voltage.

triangular waveform A waveform consisting of a positive ramp and a negative ramp with the same slope.

trigonometric function A function used to relate the angles of a triangle.

trivalent An atom with three valence electrons in its outer shell, used with making P-type material from pure silicon.

truth table A table that lists all possible combinations of digital input and output levels for a gate.

TSOP Thin small outline package.

turns ratio Determined by dividing number of turns in secondary by number of turns in primary.

U

ultiboard Included as part of National Instrument's Circuit Design Suite and combined with Multisim for a complete circuit design suite that can simulate and validate an electronic circuit and automatically lay out a printed circuit board.

undershoot Occurs when the trailing edge exceeds its normal minimum value.

Underwriters Laboratories (UL®) is an independent product safety certification organization that certifies a product is safe for its intended use.

unity gain amplifier An amplifier with a gain of one that is used to invert the polarity of the input signal.

untuned amplifier An RF amplifier that provides amplification over a large RF range.

up–down counter A counter that can count in either direction.

V

vacuum desoldering pump A vacuum pump to suck molten solder through a hollow-tip soldering iron for desoldering.

valence Indication of an atom's ability to gain or lose electrons.

valence shell The outer electron shell of an atom.

Van de Graaf generator A device used to generate electricity by means of friction.

variable capacitor A capacitor whose value can be changed by either varying the space between its plates (trimmer capacitor) or by varying the amount of meshing between two sets of plates (tuning capacitor).

variable resistor See *potentiometer*.

varistor Protects circuits against excessive transient voltages from switching and induced lightning surges.

V_{CC} In schematic drawings, it represents the transistor DC voltage supply.

vector A graphic representation by an arrow of a quantity having magnitude and direction.

Veitch diagram A chart used to simplify complex Boolean expressions.

verilog A hardware description language (HDL) used to model electronic systems.

VHDL Very high-speed integrated circuit hardware descriptive language.

video amplifier Wideband amplifiers used to amplify video signals to 6 megahertz.

virtual breadboarding A surrogate environment where circuits are created by computers to simulate the real environment to test a circuit design.

virtual instruments Computer simulated virtual instruments that behave like real-world instruments.

virtual test equipment A surrogate environment where test equipment is created by computers to simulate real test equipment.

void A soldering defect caused by impurities in the surface when soldering and using too much heat.

volt (V) Unit for measuring voltage.

volt-ampere (VA) A rating used for transformers, similar to a power rating.

volt-ohm-milliammeter (VOM) An analog multimeter.

voltage (E) Another name for difference of potential.

voltage amplifier An audio amplifier used to produce a high gain.

voltage divider a circuit used to divide a higher voltage source to a lower voltage for a specific application.

voltage doubler Produces a DC output voltage approximately twice the peak value of the input.

voltage drop Occurs when current flows in a circuit.

voltage multiplier A circuit capable of producing higher DC voltages without a transformer.

voltage regulator Produces a constant output voltage regardless of change of load.

voltage rise The voltage applied to a circuit.

voltage source A device that supplies electrons from one end of a conductor (negative terminal) and removes them from the other end of the conductor (positive terminal).

voltage tripler Produces an output voltage approximately three times the peak value of the input.

voltmeter Device used to measure the voltage between two points in a circuit.

W

watt (W) Unit of power.

web browser A computer program used to locate, retrieve, and view HTML documents on the Internet.

web design Combines desktop publishing with HTML to create web pages for a website for displaying on a web browser.

Wheatstone bridge A balanced bridge circuit that can measure unknown resistance.

wicking A soldering defect caused by too much heat, resulting in capillary action pulling solder under the insulation of wires.

wiring diagram A diagram that shows how off-board components are connected to one another and to the printed circuit board.

wiring error Installing wires in the wrong location, missing a wire, or connecting a broken wire; also, a wire with a faulty soldering connection.

word A group of 16, 32, 64, or 128 bits.

word processing Using a word processor to enter text, format, make revisions, and print a document.

word processor A computer program used for entering text (capturing keystrokes) and data into a computer.

work To be employed; to have a job. The transfer of energy or power as the rate at which the work is done.

work ethic A belief in the moral value of work.

WYSIWYG "What you see is what you get." It is used in computer programs when designing page layout for desktop publishing or web design because it is easier than writing code.

Z

zener diode A semiconductor device designed with a predictable breakdown voltage when the device is connected in a circuit to operate in reverse bias.

zener region A region above breakdown, where a zener diodes operate.

APPENDICES

APPENDIX 1

Periodic Table of Elements

Group IA	IIA												IIIB	IVB	VB	VIB	VIIB	VIII
1 H Hydrogen																		**2 He** Helium
3 Li Lithium	**4 Be** Beryllium												**5 B** Boron	**6 C** Carbon	**7 N** Nitrogen	**8 O** Oxygen	**9 F** Fluorine	**10 Ne** Neon
11 Na Sodium	**12 Mg** Magnesium	IIIA	IVA	VA	VIA	VIIA		—VIIIA—		IB	IIB	**13 Al** Aluminum	**14 Si** Silicon	**15 P** Phosphorus	**16 S** Sulfur	**17 Cl** Chlorine	**18 Ar** Argon
19 K Potassium	**20 Ca** Calcium	**21 Sc** Scandium	**22 Ti** Titanium	**23 V** Vanadium	**24 Cr** Chromium	**25 Mn** Manganese	**26 Fe** Iron	**27 Co** Cobalt	**28 Ni** Nickel	**29 Cu** Copper	**30 Zn** Zinc	**31 Ga** Gallium	**32 Ge** Germanium	**33 As** Arsenic	**34 Se** Selenium	**35 Br** Bromine	**36 Kr** Krypton
37 Rb Rubidium	**38 Sr** Strontium	**39 Y** Yttrium	**40 Zr** Zirconium	**41 Nb** Niobium	**42 Mo** Molybdenum	**43 Tc** Technetium	**44 Ru** Ruthenium	**45 Rh** Rhodium	**46 Pd** Palladium	**47 Ag** Silver	**48 Cd** Cadmium	**49 In** Indium	**50 Sn** Tin	**51 Sb** Antimony	**52 Te** Tellurium	**53 I** Iodine	**54 Xe** Xenon
55 Cs Cesium	**56 Ba** Barium	**57 La** Lanthanum	**72 Hf** Hafnium	**73 Ta** Tantalum	**74 W** Tungsten	**75 Re** Rhenium	**76 Os** Osmium	**77 Ir** Iridium	**78 Pt** Platinum	**79 Au** Gold	**80 Hg** Mercury	**81 Tl** Thallium	**82 Pb** Lead	**83 Bi** Bismuth	**84 Po** Polonium	**85 At** Astatine	**86 Rn** Radon
87 Fr Francium	**88 Ra** Radium	**89 Ac** Actinium	**104 Unq** (Unniquadium)	**105 Unp** (Unnipentium)	**106 Unh** (Unnihexium)												

58 Ce Cerium	**59 Pr** Praseodymium	**60 Nd** Neodymium	**61 Pm** Promethium	**62 Sm** Samarium	**63 Eu** Europium	**64 Gd** Gadolinium	**65 Tb** Terbium	**66 Dy** Dysprosium	**67 Ho** Holmium	**68 Er** Erbium	**69 Tm** Thulium	**70 Yb** Ytterbium	**71 Lu** Lutetium
90 Th Thorium	**91 Pa** Protactinium	**92 U** Uranium	**93 Np** Neptunium	**94 Pu** Plutonium	**95 Am** Americium	**96 Cm** Curium	**97 Bk** Berkelium	**98 Cf** Californium	**99 Es** Einsteinium	**100 Fm** Fermium	**101 Md** Mendelevium	**102 No** Nobelium	**103 Lr** Lawrencium

■ APPENDIX 2
The Greek Alphabet

Greek Letter		Greek Name
A	α	alpha
B	β	beta
Γ	γ	gamma
Δ	δ	delta
E	ε	epsilon
Z	ζ	zeta
H	η	eta
Θ	θ	theta
I	ι	iota
K	κ	kappa
Λ	λ	lambda
M	μ	mu
N	ν	nu
Ξ	ξ	xi
O	o	omicron
Π	π	pi
P	ρ	rho
Σ	σ	sigma
T	τ	tau
Y	υ	upsilon
Φ	φ	phi
X	χ	chi
Ψ	ψ	psi
Ω	ω	omega

■ APPENDIX 3

Metric Prefixes Used in Electronics

Prefix	Symbol		Multiplication Factor
Yotta-	Y	10^{24}	1,000,000,000,000,000,000,000,000
Zetta-	Z	10^{21}	1,000,000,000,000,000,000,000
exa-	E	10^{18}	1,000,000,000,000,000,000
peta-	P	10^{15}	1,000,000,000,000,000
tera-	T	10^{12}	1,000,000,000,000
giga-	G	10^{9}	1,000,000,000
mega-	M	10^{6}	1,000,000
kilo-	k	10^{3}	1,000
hecto-	h	10^{2}	100
deka-	da	10^{0}	10
deci-	d	10^{-1}	0.1
centi-	c	10^{-2}	0.01
milli-	m	10^{-3}	0.001
micro-	μ	10^{-6}	0.000,001
nano-	n	10^{-9}	0.000,000,001
pico-	p	10^{-12}	0.000,000,000,001
femto-	f	10^{-15}	0.000,000,000,000,001
atto-	a	10^{-18}	0.000,000,000,000,000,001
zepto-	z	10^{-21}	0.000,000,000,000,000,000,001
yocto-	y	10^{-24}	0.000,000,000,000,000,000,000,001

APPENDIX 4

Electronics Abbreviations

Term	Abbreviation
Alpha	α
Alternating current	ac, AC
Ampere	A, amp
Beta	β
Capacitance	C
Capacitive reactance	X_c
Charge	Q
Conductance	G
Coulomb	C
Current	I, i
Cycles per second	Hz
Degrees celsius	°C
Degrees fahrenheit	°F
Direct current	dc, DC
Electromotive force	emf, E, e
Farad	F
Frequency	f
Giga	G
Henry	H
Hertz	Hz
Impedance	Z
Inductance	L
Inductive reactance	X_L

Term	Abbreviation
Kilo	k
Mega	M
Micro	μ
Milli	m
Nano	n
Ohm	Ω
Peak-to-peak	p-p
Period	T
Pico	p
Power	P
Reactance	X
Resistance	R, r
Resonant frequency	f_r
Root mean square	rms
Second	s
Siemens	S
Time	t
Var	var
Volt	V, v
Volt-ampere	VA
Wavelength	λ
Watt	W

■ APPENDIX 5

Common Reference Designators

Component	Designator
Battery	B
Bridge rectifier	BR
Capacitor	C
Cathode-Ray Tube	CRT
Diode	D
Fuse	F
Integrated Circuit	IC
Jack Connector	J
Inductor	L
Liquid Crystal Display	LCD
Light-Dependent Resistor	LDR
Light-Emitting Diode	LED
Speaker	SPK
Motor	M
Microphone	MIC
Neon Lamp	NE
Operational Amplifier	OP
Plug	P
Printed Circuit Board	PCB
Transistor	Q
Resistor	R
Relay	K
Silicon-Controlled Rectifier	SCR
Switch	S
Transformer	T
Test Point	TP
Integrated Circuit	U
Vacuum Tube	V
Transducer	X
Crystal	Y

© 2014 Cengage Learning

APPENDIX 6

DC and AC Circuit Formulas

	Series	**Parallel**
Resistance	$R_T = R_1 + R_2 + R_3 + \cdots R_n$	$\dfrac{1}{R_T} = \dfrac{1}{R_1} + \dfrac{1}{R_2} + \dfrac{1}{R_3} + \cdots \dfrac{1}{R_n}$
Capacitance	$\dfrac{1}{C_T} = \dfrac{1}{C_1} + \dfrac{1}{C_2} + \dfrac{1}{C_3} + \cdots \dfrac{1}{C_n}$	$C_T = C_1 + C_2 + C_3 + \cdots C_n$
Inductance	$L_T = L_1 + L_2 + L_3 + \cdots L_n$	$\dfrac{1}{L_T} = \dfrac{1}{L_1} + \dfrac{1}{L_2} + \dfrac{1}{L_3} + \cdots \dfrac{1}{L_n}$
Impedance	$Z_T = Z_1 + Z_2 + Z_3 + \cdots Z_n$	$\dfrac{1}{Z_T} = \dfrac{1}{Z_1} + \dfrac{1}{Z_2} + \dfrac{1}{Z_3} + \cdots \dfrac{1}{Z_n}$
Voltage	$E_T = E_1 + E_2 + E_3 + \cdots E_n$	$E_T = E_1 = E_2 = E_3 = \cdots E_n$
Current	$I_T = I_1 = I_2 = I_3 = \cdots I_n$	$I_T = I_1 + I_2 + I_3 + \cdots I_n$
Power	$P_T = P_1 + P_2 + P_3 + \cdots P_n$	$P_T = P_1 + P_2 + P_3 + \cdots P_n$

Formulas

Ohm's Law	$I = \dfrac{E}{R}$	Conductance	$G = \dfrac{1}{R}$
Impedance	$I = \dfrac{E}{Z}$	Reactance	$X_L = 2\pi fL$
Power	$P = IE$		$X_C = \dfrac{1}{2\pi fC}$
		Resonance	$F_R = \dfrac{1}{2\pi\sqrt{LC}}$

AC Values Chart

	Peak	RMS	AVG
Peak		$0.707 \times$ Peak	$0.637 \times$ Peak
RMS	$1.41 \times$ RMS		$0.9 \times$ RMS
AVG	$1.57 \times$ AVG	$1.11 \times$ AVG	

Miscellaneous

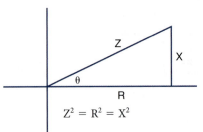

$$Z^2 = R^2 = X^2$$

$$\text{Sin } 0 = \frac{X}{Z}$$

$$\text{Cos } 0 = \frac{R}{Z}$$

$$\text{Tan } 0 = \frac{X}{R}$$

Conversion

Rectangular to Polar

$$R = \sqrt{X^2 + Y^2} \text{ and } 0 = \text{Arctan } \frac{Y}{X}$$

Polar to Rectangular

$$X = R \text{ Cos } 0 \text{ and } Y = R \text{ Sin } 0$$

Temperature

Fahrenheit to Celsius

$$C = (F - 32)(5/9)$$

Celsius to Fahrenheit

$$F = (C \times 9/5) + 32$$

Formula Shortcuts

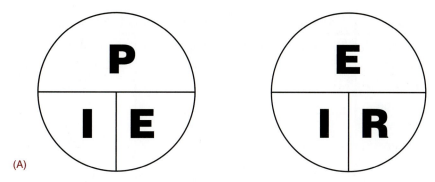

(A)

When two values are known, put your finger on the unknown values and solve.

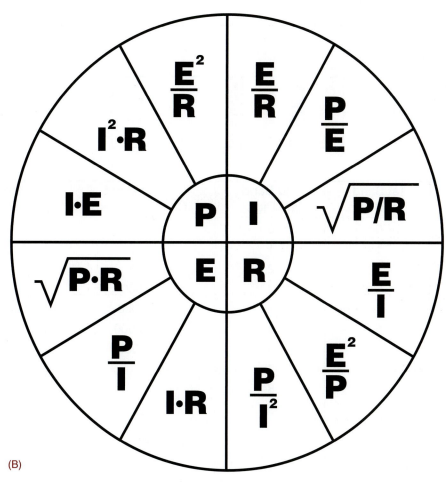

(B)

Solving for two resistors in parallel $\quad R_T = \dfrac{R_1 R_2}{R_1 + R_2}$

Solving multiple resistors in series with same value $\quad R_T = \dfrac{R_x}{N}$

$R_X =$ Value of one resistor
$N =$ Number of resistors

■ APPENDIX 8

Resistor Color Codes

Color	Value
Black	0
Brown	1
Red	2
Orange	3
Yellow	4
Green	5
Blue	6
Violet	7
Gray	8
White	9

Color	Tolerance
Gray	0.05%
Violet	0.10%
Blue	0.25%
Green	0.5%
Brown	1%
Red	2%
Gold	5%
Silver	10%
None	20%

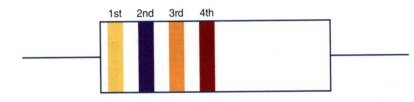

Color	1st Band 1st Digit	2nd Band 2nd Digit	3rd Band Number of Zeros	4th Band Tolerance
Black	0	0	—	—
Brown	1	1	0	1%
Red	2	2	00	2%
Orange	3	3	000	—
Yellow	4	4	0000	—
Green	5	5	00000	0.5%
Blue	6	6	000000	0.25%
Violet	7	7		0.10%
Gray	8	8		0.05%
White	9	9		—
Gold	—	—	x.1	5%
Silver	—	—	x.01	10%
None	—	—	—	20%

APPENDIX 9

Common Resistor Values

±2% and ±5% Tolerance	±10% Tolerance	±20% Tolerance
1.0	1.0	1.0
1.1		
1.2	1.2	
1.3		
1.5	1.5	1.5
1.6		
1.8	1.8	
2.0		
2.2	2.2	2.2
2.4		
2.7	2.7	
3.0		
3.3	3.3	3.3
3.6		
3.9	3.9	
4.3		
4.7	4.7	4.7
5.1		
5.6	5.6	
6.2		
6.8	6.8	6.8
7.5		
8.2	8.2	
9.1		

© 2014 Cengage Learning

APPENDIX 10

Capacitor Color Code

Color	Digit	Multiplier	Tolerance
Black	0	1	20%
Brown	1	10	1%
Red	2	100	2%
Orange	3	1000	3%
Yellow	4	10000	4%
Green	5	100000	5%
Blue	6	1000000	6%
Violet	7		7%
Gray	8		8%
White	9		9%
Gold		0.1	
Silver		0.01	10%
Body			20%

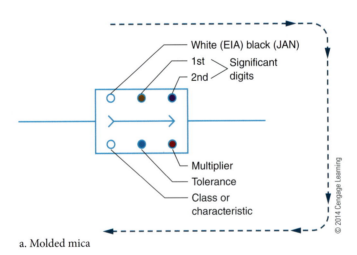

a. Molded mica

Note: EIA stands for Electronic Industries and Association, and
JAN stands for Joint Army-Navy, a military standard.

■ APPENDIX 10 (*continued*)
Capacitor Color Code

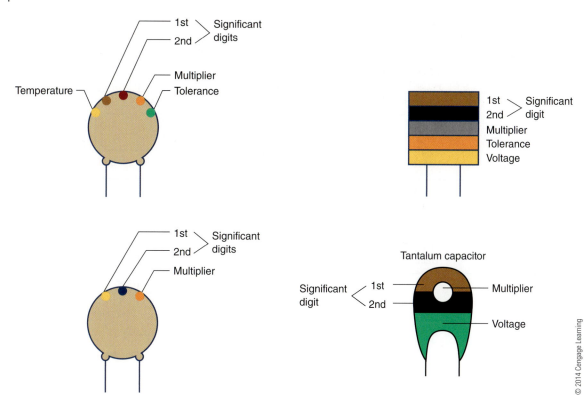

b. Disc ceramics

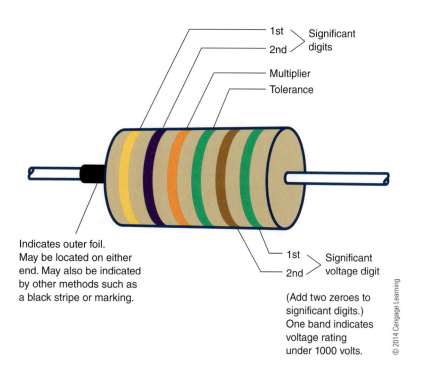

Indicates outer foil.
May be located on either
end. May also be indicated
by other methods such as
a black stripe or marking.

(Add two zeroes to
significant digits.)
One band indicates
voltage rating
under 1000 volts.

c. Axial Foil

APPENDIX 11
Electronics Symbols

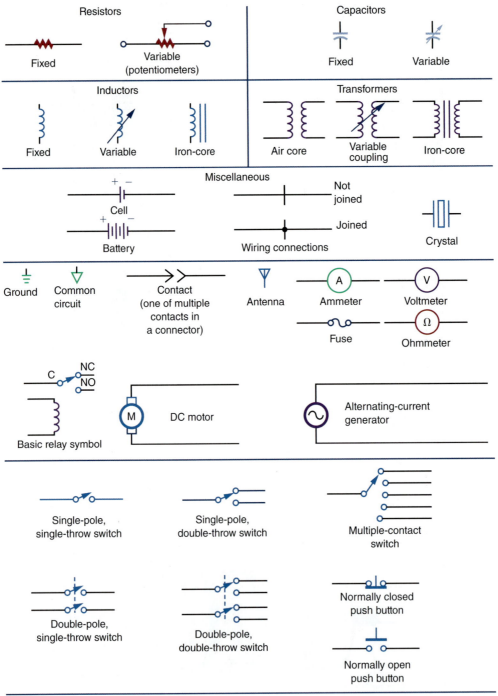

APPENDIX 12

Semiconductor Schematic Symbols

Name of Device	Circuit Symbol	Commonly used Junction Schematic	Major Applications
Diode			Rectification Blocking Detecting Steering
Zener diode			Voltage regulation
N-P-N transistor			Amplification Switching Oscillation
P-N-P transistor			Amplification Switching Oscillation
N-channel JFET			Amplification Switching Oscillation
P-channel JFET			Amplification Switching Oscillation
Enhancement N-channel MOSFET			Switching Digital applications
Enhancement N-channel MOSFET			Switching Digital applications
Depletion N-channel MOSFET			Amplification Switching
Depletion P-channel MOSFET			Amplification Switching
Silicon-controlled rectifier (SCR)			Power switching Phase control Inverters Choppers
TRIAC			AC switching Phase control Relay replacement
DIAC			Trigger
Photo transistor			Tape readers Card readers Position sensor Tachometers
Light-emitting diode (LED)			Indicator Light source Optical coupler Displays

APPENDIX 13

Digital Logic Symbols

Logic Function	American (Mil/Ansi) Symbol	Common German Symbol	International Electrotechnical Commission (IEC) Symbol	British (BS3939) Symbol
Buffer			1	1
Inverter (NOT gate)			1	1
2-input AND gate			&	&
2-input OR gate			≥1	≥1
2-input NAND gate			&	&
2-input NOR gate			≥1	≥1
2-input (exclusive OR) X-OR gate			= 1	= 1
2-input (exclusive NOR) X-NOR gate			= 1	= 1

© 2014 Cengage Learning

■ APPENDIX 14

Activity Assessment Rubric

Name: _____   **Task:** _____

Group No.: _____   **Responsibilities**

    Circuit Design: _____

    Computer Simulation: _____

    Circuit Fabrication: _____

Assessment Scale

    5 – *Exemplary* - Demonstrate excellence in the solution presented. _____

    4 – *Proficient* - Solution Fulfills all objectives. _____

    3 – *Partially Proficient*- Solution is acceptable but needs minor revisions. _____

    2 – *Not Yet Proficient* - Solution is unacceptable and need major revisions. _____

    1 – *Incomplete* - Solution was not turned in, is incomplete or not correct. _____

Circuit Design

1. The solution meets the criteria presented. _____
2. The solution is reduced to the best solution to meet the criteria. _____
3. A lag/journal was kept to reflex the different solutions. _____
4. A description was prepared to explain how the solution worked. _____
5. A parts list was generated that corresponds to the solution. _____

Computer Simulation

1. A Multisim solution was submitted with the final solution on a disk. _____
2. The solution represents the least number of parts. _____
3. A detail description of how the project work was included with the Multisim. _____
4. The Multisim solution includes original solution and the final solution. _____
5. A PCB was developed using the appropriate software. _____

Circuit Fabrication

1. A printed circuit board design is submitted that matches the solution. _____
2. Circuit is wired efficiently using a minimum of jumpers. _____
3. The PCB is logically designed with inputs and outputs separated. _____
4. The circuit operates to the specification of the final design. _____
5. A chart is prepared showing voltages and critical signals. _____

Class Presentation

1. An assembly manual was prepared and logically organized for presentation to the class. _____
2. The group provided a presentation that includes all relevant materials. _____
3. An electronic presentation (e.g. PowerPoint) was prepared and presented to the class. _____
4. A spreadsheet of all parts and budget was prepared for presentation. _____
5. All group members actively participated in presentation. _____
6. All questions were responded to in a professional manner. _____

Group work

1. Group members started work in a professional manner. _____
2. Group members worked together with consistent effort. _____
3. Group members actively contributed to the group effort. _____
4. All group members participated in a peer review process. _____
5. All group members were supportive of each other in the achievement of the final solution. _____

Total ____/26

Final Assessment _____

■ APPENDIX 15
Safety Test

Directions: Read each question and answer carefully, selecting the best answer for the question.

1. Why is it important to develop good work habits?
   a. To improve nonproductivity
   b. To increase chances of being successful
   c. To prevent learning new skills
   d. To reinforce bad habits

2. What is an employer responsible for providing?
   a. Nothing
   b. Safe work area, training, and protection equipment
   c. Safety awareness posters in the workshop
   d. Training and opportunity to purchase safety equipment

3. Why do young workers have a higher risk of being injured on a job?
   a. They were careless.
   b. They were aware of their rights.
   c. They were not properly trained or supervised.
   d. They were provided a safe work environment.

4. Who can legally perform electrical work?
   a. Anyone in school studying electricity
   b. Anyone with knowledge of electricity
   c. The oldest person on the jobsite
   d. Only licensed electricians

5. Under what conditions can an electrical shock occur?
   a. A worker becomes part of the circuit.
   b. A worker is wearing conductive clothing.
   c. A worker has on rubber-soled shoes.
   d. A worker is not sweating.

6. What is the most common cause of electrocution?
   a. Making contact with overhead wires
   b. Putting the ladder in storage
   c. Working on a wooden ladder
   d. Working too close to the ground

7. Electrical accidents occur when
   a. equipment has been properly maintained.
   b. following proper work practices occurs.
   c. there is no proper supervision.
   d. there is a proper training procedure.

8. How does body resistance affect current through the body?
   a. The higher the body resistance, the higher the current flows.
   b. The higher the body resistance, the lower the current flows.
   c. The lower the body resistance, the higher the current flows.
   d. The lower the body resistance, the lower the current flows.

9. Electric shock is the result of _____ through the body?
   a. a very high resistance
   b. current flow
   c. the applied power
   d. the applied voltage

10. What factors influence the effect of an electrical shock?
    a. Frequency, intensity, path, and time
    b. Frequency, resistance, time, and applied voltage
    c. Intensity, path, pressure, and resistance
    d. Path, pressure, resistance, and applied voltage

11. What determines the severity of an electrical shock?
    a. Amount of applied voltage received
    b. Amount of current received
    c. Application of CPR
    d. Applied body resistance

12. What is the first thing to do if someone has been severely shocked?
    a. Immediately begin CPR.
    b. Pull victim away from power source.
    c. Send for help.
    d. Switch off power source.

■ **APPENDIX 15** *(Continued)*
Safety Test

13. How can an accident be prevented when working with electricity?
  a. Ground the circuit to be worked on.
  b. Only wear conductive shoes in the work area.
  c. Try to only work in damp areas.
  d. Wear loose clothing.

14. Who is responsible for safety in the workshop?
  a. Everyone
  b. The safety officer
  c. The supervisor
  d. The technicians

15. What function does grounding serve?
  a. Prevents current from leaking through the insulation
  b. Provides an alternate path for current flow
  c. Provides a lower level for current to flow to with the circuit
  d. Shuts the circuit down when there is excess current flow

16. What function does the Underwriters Laboratory serve?
  a. Ensures that a product does what it is suppose to
  b. Ensures that the product is built to the highest quality
  c. Ensures that the product is safe to use
  d. Ensures that the product meets maximum safety standards

17. What does a ground-fault interrupter circuit accomplish?
  a. It is a delayed-action circuit breaker that is tripped by a line-to-load fault.
  b. It is a delayed-action circuit breaker that is tripped by a load-to-line fault.
  c. It is a fast-acting circuit breaker that is tripped by a large current flow.
  d. It is a fast-acting circuit breaker that is tripped by a small current flow.

18. What is the purpose of the *National Electrical Code*?
  a. It provides a standardized color code for wiring a circuit.
  b. It explains how to lay out a wiring diagram for a circuit.
  c. It provides residential wiring guidelines.
  d. It provides maximized safety standards for electrical wiring.

19. Electrical static discharge is the result of
_____.
  a. grounding a circuit
  b. static electricity
  c. transferring electricity to a flat surface
  d. rubbing two similar materials together

20. Antistatic workstations are designed to provide
_____.
  a. a ground path for any static charge
  b. a low resistance to current flow in a circuit
  c. a path for excess voltage to return to the source
  d. a wrist strap to provide current flow to a circuit

21. Every hazardous product in the workshop has an MSDS, which stands for _____.
  a. manufacturer's safety and data sheet
  b. manufacturer' security and document statement
  c. material safety and data sheet
  d. material security and document statement

22. How can an accident in the workshop be avoided?
  a. Develop safe work habits.
  b. Get training on each job by the person responsible for shop safety.
  c. Have a minimum of two technicians work every job.
  d. Have the person responsible for shop safety make a list of rules.

■ APPENDIX 15 (Continued)
Safety Test

23. What is the biggest hazard with hand tools?
   a. Carrying a tool incorrectly
   b. Holding a dull-edged tool incorrectly when using it
   c. Improper maintenance and using a tool improperly
   d. Not reading the safety requirements for a tool being used

24. When using a power tool, make sure _____.
   a. all adjustments are done with power applied for accuracy
   b. to remove the guards so you can see the work better, but replace them when you are done
   c. to only set up a special setup with the power applied
   d. the tool has a UL label

25. What is the first step in using a hand tool?
   a. Check to make sure the blade is dull on all cutting tools for safety.
   b. Get permission from the shop supervisor.
   c. Look at all the tools before selecting an alternate.
   d. Select the proper tool for the job.

26. When working with tools on high voltage (> 1000 V), the tools must have of _____.
   a. slip-on plastic handles installed
   b. the 1000 V rating symbol
   c. the handles of the tools plastic dipped
   d. the handles wrapped with vinyl electrical tape

27. Which organization has the responsibility of protecting workers on the jobsite?
   a. CSA
   b. DOT
   c. NEC
   d. OSHA

28. Who is responsible for the safe condition of tools and test equipment in the workshop?
   a. Each user
   b. The person responsible for safety
   c. The shop supervisor
   d. The employer

29. Which of the following is a safe practice to observe when working on electrical or electronic circuits?
   a. Always wear loose clothing.
   b. Never working on any circuit unless the power is off.
   c. Remove the ground connection from the circuit being worked on.
   d. Wear conductive shoes to provide a path for current flow.

30. The law requires that MSDSs be _____.
   a. destroyed after being read over by the shop supervisor
   b. kept on file for 30 days
   c. maintained and available on request
   d. used for training by the person responsible for shop safety

31. Who is responsible for reading the MSDS?
   a. The CEO of the company
   b. Everyone using the product
   c. The person responsible for shop safety
   d. The shop supervisor

32. The DOT categorizes hazardous materials for transporting by _____.
   a. chemical and physical properties
   b. the color of the product
   c. the degree that the product is affected by static electricity
   d. the size of container the material is stored in

Chapter 1 Careers in Electricity and Electronics

1. *Instructor-supervised activity*
2. *Instructor-supervised activity*
3. *Instructor-supervised activity*
4. *Instructor-supervised activity*
5. *Instructor-supervised activity*
6. *Instructor-supervised activity*
7. *Instructor-supervised activity*
8. *Instructor-supervised activity*
9. *Instructor-supervised activity*
10. *Instructor-supervised activity*
11.

Job Title	Salary	Benefits	Job Description

© 2014 Cengage Learning

12. *Instructor-supervised activity*
13. *Instructor-supervised activity*
14. *Instructor-supervised activity*

Chapter 2 Certification for the Electricity and Electronics Field

1. To certify electronic technicians
2. Certification makes the technician more employable.
3. Associate-Level CET exam or Electronics System Associate CET exam and Journeyman CET exam
4. Valid for only four-years
5. Passing an Associate-Level CET exam and a Journeyman-Level CET.
6. Bench ET, Computer ET, Metrology ET, Senior ET
7. *Instructor-supervised activity*
8. Read schematics and maintenance manuals, test and troubleshoot equipment, solder and unsolder components, research availability of parts and costs, and determine whether further assistance is needed
9. *Instructor-supervised activity*
10. *Instructor-supervised activity*

Chapter 3 Work Habits and Issues

1. Good work habits help to improve productivity and quality of work.
2. Develop goals, be on time with good attendance, stay on task, keep work area neat, work smart, and rest after task is completed.
3. Keep workspace clean, make use of technology, focus on one task at a time, streamline important tasks, create quiet time for planning, and keep paperwork under control.

4. Health and safety

5. Following safety instructions, using equipment carefully, and reporting hazards and injuries

6. Lack of supervision and training, inadequate work practices, poorly maintained equipment, hazardous workplace environment, and unauthorized electrical repairs

7. In the 1960s, diversity came about because of the need to comply with equal employment opportunities.

8. Title VII of the 1964 Civil Rights Act, the Age Discrimination in Employment Act of 1967 (ADEA), the Americans Disabilities Act of 1990 (ADA), and the Equal Pay Act of 1963

9. Any behavior that is unwelcomed or discriminatory

10. By providing good value for the pay; based on honesty, respect, and trust

11. To get team members to be comfortable working with each other

12. Team leader, assembler, data recorder, and media presenter

13. Communicate, accept responsibility, support the team, be a team member, and get involved

Chapter 4 Calculators for Electricity and Electronics

1. *Instructor-supervised activity*

2. 576

3. Enter numbers correctly, enter numbers in the correct order, and use the correct function keys

4. **a.** 85,040
   b. 1911
   c. 7,127,816
   d. 64.478
   e. 20.667

5. **a.** 546.52
   b. 546.5
   c. 547
   d. 500

Chapter 5 Electronic Circuit Diagrams

1.

Description	Symbol
Battery	—╢╟╢╟—
Resistor	—⟋⟍⟋⟍—
Capacitor	+ ⊣⊢ −
Wires not connected	+
Wires connected	+
Diode	—▶⊢—
NPN transistor	B ◯ C / E
LED	▶▶
Ground	⏚

© 2014 Cengage Learning

2.

Description	Reference Designator
Battery	V_{cc}
Resistor	R_2
Capacitor	C_3
Wires not connected	
Wires connected	
Diode	D_4
NPN transistor	Q_5
LED	D_6
Ground	

© 2014 Cengage Learning

3. Top of the diagram

4.

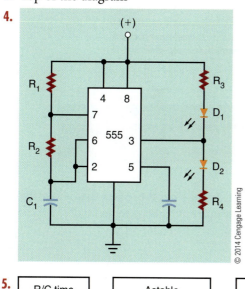

© 2014 Cengage Learning

5.

| R/C time constant | → | Astable multivibrator | → | Output indicator |

© 2014 Cengage Learning

6.

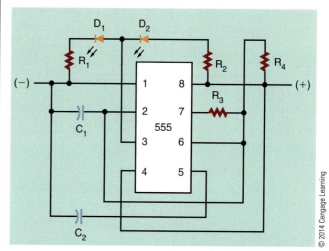

© 2014 Cengage Learning

7. It allows the schematic diagram to be quickly drawn and allows testing of the circuit for proper operation.

8. It allows a circuit to be tested or modified.

9. For developing a prototype of a circuit.

10. You can use actual components for building a circuit; the circuit can be quickly modified; and components can be swapped out easily.

Chapter 6 Software for Electronics

1. Mathematical algorithms

2. The user becomes part of the learning process by being engaged in creation, studying circuit operation, and evaluating performance.

3. Best-known simulation program.

4. Hardware description languages.

5. Multisim uses *what-if* experiments and simulation-driven instruments.

6. A prototyping workstation.

7. A graphical programming environment for creating virtual instrumentation.

8. Circuit Wizard combines schematic circuit design, simulation, PCB design, and CAD/CAM manufacturing.

9. The PCB can also be simulated to ensure it is properly laid out.

10.

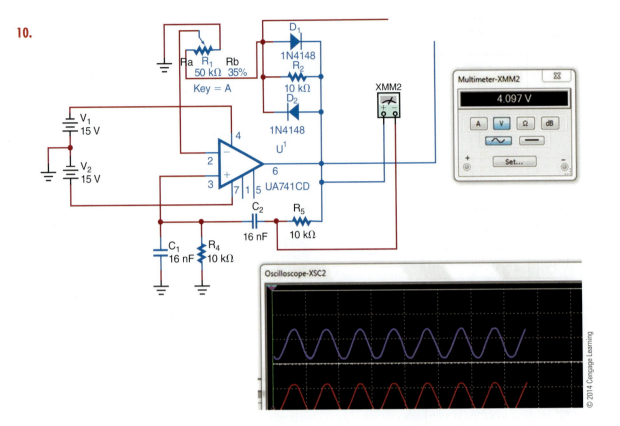

11. *Instructor-supervised activity*

12. *Instructor-supervised activity*

13. *Instructor-supervised activity*

14. *Instructor-supervised activity*

15. *Instructor-supervised activity*

16. *Instructor-supervised activity*

17. *Instructor-supervised activity*

Chapter 7 **Safety**

1. When the current flows through a person's chest region, the heart go into ventricular fibrillation that results in rapid and irregular muscle contractions leading to cardiac arrest and respiratory failure.

2. Insulation and grounding

3.

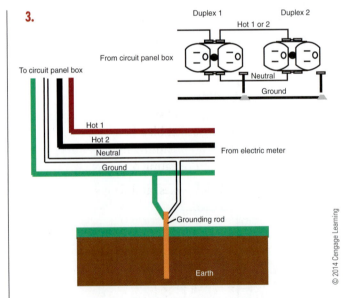

4. It is a fast-acting circuit breaker that is sensitive to a very low current leakage to ground.

5.

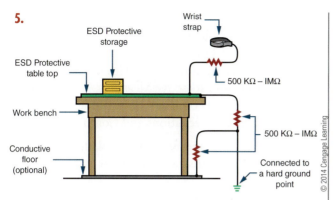

6. Use a wrist strap, check for ESD components, discharge package, minimize handling ESD components, minimize any movement when installing ESD components, avoid touching anything that creates a static charge, use a grounded soldering iron, and only use grounded test equipment.

7. Cardiopulmonary resuscitation is used for someone who has received a severe electric shock, to keep the person breathing and the heart pumping blood.

8. Keep one hand in your back pocket when working on live circuits; use isolation transformer; ensure all capacitors are discharged before troubleshooting; use grounded line cords; and keep hands off live circuits.

9. Wear safety glasses; never touch tip of soldering iron or change tip when soldering iron is hot; avoid contact with heating element; avoid too much pressure; avoid synthetic clothing; minimize handling assembled PCB to avoid cuts and punctures; avoid breathing fumes and vapors; and wash hands before eating or drinking.

10. With antistatic bags, boxes, and wraps

Chapter 8 **Tools and Equipment**

1. Use the proper tool for the job; keep cutting edge down when carrying; keep hands clean when using tools; use clamps when working with small pieces; avoid deformed tools; only use a file with a handle; never use plastic handled tools near a heat source; keep metal tools clear of electrical circuits; and cut one wire at a time when cutting wire.

2. Ensure all power tools have three prong plugs and/or insulated housing; make all adjustments with power off; double-check all special setups; keep all safety guards in position at all times; keep work area and power tools clean; avoid damaging power tools; keep power tools in good operating condition; and place power tools in a safe place when finished using them.

3. Always use the proper tool for the job; keep tools clean and proper condition; keep hands clean; only make adjustments with the power off; double-check all special setups; only work in a clean work area; avoid treating tools roughly; and keep tools in good condition with regular maintenance.

4. Make a storage rack that can sit on the workbench with the tools readily accessible.

5.

6. Safety glasses, dust mask, and rubber and/or leather gloves

7. How an electric drill generates torque

8. Operator's manual

9.

Tools	Safe Practices
Files	Only use with a proper handle installed
Screwdrivers	Use the proper blade width for the screw
Hammers	Check for loose handles
Chisels and Punches	Check for mushroom head
Electric Hand Tools	Unplug when changing bits and blades
Stationary Power Tools	Make all adjustments prior to applying power
Tables and Routers	Keep safety guard in place
Metal tools	Keep away from live electrical circuits

© 2014 Cengage Learning

10. Don't misuse the tools; don't forget maintenance of tools; don't use tool in an environment it was not intended for; don't forget eye protection; and don't forget hand protection.

11. Remove the power source from the device or equipment being worked on.

Chapter 9 Hazardous Materials

1. The Occupational Safety and Health Administration was formed December 29, 1970.

2. To prevent work-related illness and death by creating a better work environment for the workforce.

3. Manufacturer's Information; Sections: 1. Hazardous Materials, 2. Physical and Chemical Properties, 3. Fire and Explosion Hazard Data, 4. Reactivity Hazard Data, 5. Health Hazard Data, 6 Control and Protective Measures, 7. Spill or leak Procedures, 8. Hazardous Material Identification, and 9. Special Precautions and Comments.

4. Keep from freezing. Store in a cool, dry, well-ventilated area, away from incompatible substances.

5. Classes: 1. Explosives; 2. Gases; 3. Flammable and Combustible Liquids; 4. Flammable Solids, Spontaneous Combustibles, and Dangerous When Wet; 5. Oxidizers and Organic Peroxides; 6. Poison (Toxic) and Poison Inhalation Hazard; 7. Radioactive; 8. Corrosive; and 9. Miscellaneous.

6. Packing Group 3, Class 8, shipper must be trained and certified. Do not ship by air. By sea, shipper must be trained and certified.

7. Consumer commodity: a hazardous material packaged for distribution in quantities intended for retail sale to be consumed for personal use.

8. Everyone in the workspace.

9. In a flammable materials storage cabinet.

10. Dispose in accordance with all local, provincial, state, and federal regulations. Water runoff can cause environmental damage.

Chapter 10 Fundamentals of Electricity

1. An element is the basic building block of nature. An atom is the smallest particle of an element that retains the characteristics of the element. A molecule is the chemical combination of two or more atoms. A compound is the chemical combination of two or more elements and can be separated only by chemical means.

2. The number of free electrons that are available

3. How many electrons are in the valence shell; less than four—conductor; four—semiconductor; more than four—insulator

4. To understand how electricity flows or does not flow through various materials

5. 6.24×10^{18} electrons

6. The work in a circuit is the result of a difference of potential, also referred to as electromotive force (emf) or voltage.

7. Opposing the flow of electrons

8.

Term	Symbol	Unit
Current	I	Ampere (A)
Voltage	E	Volt (V)
Resistance	R	Ohm (Ω)

© 2014 Cengage Learning

9. The flow of electrons is current; the force that moves the electrons is voltage; and the opposition to the flow of electrons is resistance.

10. The resistance of a material is determined by its size, shape, and temperature.

Chapter 11 Current

1.
Given	Solution
I = ?	$I = \dfrac{Q}{t}$
Q = 7 coulombs	
t = 5 seconds	$I = \dfrac{7}{5}$
	I = 1.4 amperes

2. Electrons flow from the negative terminal of the potential through the conductor, moving from atom to atom, to the positive terminal of the potential.

3. **a.** $235 = 2.35 \times 10^2$

   b. $0.002376 = 2.376 \times 10^{-3}$

   c. $56323.786 = 5.6323786 \times 10^4$

4. **a.** *Milli-* means to divide by 1000 or to multiply by 0.001, expressed as 1×10^{-3}.

   b. *Micro* means to divide by 1,000,000 or to multiply by 0.000001, expressed as 1×10^{-6}.

5.

	Converts	To
a.	305 mA	0.305 A
b.	6 μA	0.006 mA
c.	17 V	17000 mV
d.	0.023 mV	0.000023 μV
e.	0.013 kΩ	13 Ω
f.	170 MΩ	170,000,000 Ω

© 2014 Cengage Learning

Chapter 12 Voltage

1. The actual work accomplished in a circuit (the movement of electrons) is the result of the difference of potential (voltage).

2. Electricity can be produced by chemicals, friction, heat, light, magnetism, and pressure.

553

3. Secondary cells are rated in ampere-hours.

4.

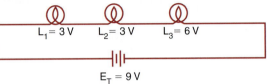

6 each 1.5 V @ 250 mA

9 V
+ 1 A −

© 2014 Cengage Learning

5. **Given**   **Solution**

$E_T = 9\,V$ Draw the circuit:

$L_1 = 3\,V$ rating

$L_2 = 3\,V$ rating

$L_3 = 6\,V$ rating

$L_1 = 3\,V$ $L_2 = 3\,V$ $L_3 = 6\,V$

$E_T = 9\,V$

© 2014 Cengage Learning

Half of the voltage would be dropped across L_1 and L_2, and the other half of the voltage would be dropped across L_3.

Therefore:

$L_1 + L_2 = 6\,V$ drop   L3 would drop 4.5 V

$L_3 = 6\,V$ drop   L2 would drop 2.25 V

$9 \times \dfrac{1}{2} = \dfrac{9}{2} = 4.5\,V$ L1 would drop 2.25 V

Total voltage = 9.00 V

6.

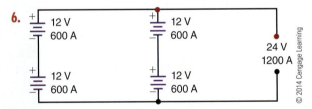

12 V
600 A

12 V
600 A

24 V
1200 A

12 V
600 A

12 V
600 A

© 2014 Cengage Learning

7. The energy is given up as heat.

8. The current flows from the negative terminal of the battery and returns to the positive terminal of the battery.

9. In the home, ground is used to protect the user from electrical shock. In an automobile, the ground serves as part of the complete circuit. In electronics, ground serves as the zero reference point against which all voltages are measured.

10.

	Formulas for	
	Series	**Parallel**
Current	$I_T = I_1 = I_2 = I_3 \ldots = I_n$	$I_T = I_1 + I_2 + I_3 \ldots + I_n$
Voltage	$E_T = E_1 + E_2 + E_3 \ldots + E_n$	$E_T = E_1 = E_2 = E_3 \ldots = E_n$

© 2014 Cengage Learning

Chapter 13 **Resistance**

1. The resistance of a material depends on the type of a material and its size, shape, and temperature. It is determined by measuring a 1-foot length of wire made of the material that is 1 mil in diameter and at a temperature of 20° Celsius.

2.

Given   **Solution**

Resistance = 2200 ohms   2200 × 0.10 = 220 ohms

Tolerance = ± 10%   2200 − 220 = 1980 ohms

2200 + 220 = 2420 ohms

Tolerance range is 1980 ohms

to 2420 ohms

3. a. Green, blue, red, gold

  b. Brown, green, green, silver

  c. Red, violet, gold, gold

  d. Brown, black, brown, none

  e. Yellow, violet, yellow, silver

4. RC0402D104T:

  RC = Chip resistor

  0402 = Size (0.04" × 0.02")

  D = Tolerance (± 0.5%)

  104 = Resistance (100,000 Ω)

  T = Packaging method

5. Potentiometers may have the actual value stamped on them or they may have an alphanumeric code.

© 2014 Cengage Learning

6.

	Formulas for	
	Series-Parallel	
	Series	**Parallel**
Resistance	$R_T = R_1 + R_2 + R_3 \ldots + R_n$	$1/R_T = 1/R_1 + 1/R_2 + 1/R_3 \ldots + 1/R_n$

© 2014 Cengage Learning

7. $2\,\Omega$

8. a. Find the total resistance (R_T) for the parallel resistors R_1, R_2, R_3, and R_4.

$$\frac{1}{R_T} = \frac{1}{R_1} + \frac{1}{R_2} + \frac{1}{R_3} + \frac{1}{R_4}$$

$$\frac{1}{R_T} = \frac{1}{8} + \frac{1}{8} + \frac{1}{8} + \frac{1}{8}$$

$$R_T = 2\,\Omega$$

9. $R_T = 1636.36\,\Omega$

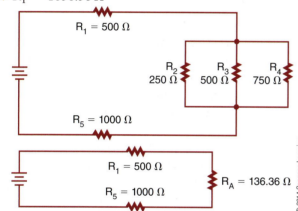

10. The current flows from the negative side of the voltage source through the series components, dividing among the branches of the parallel components, recombining to flow through any more series or parallel components, and then returns to the positive side of the voltage source.

Chapter 14 Ohm's Law

1. Given

I = ?
E = 9 V
R = 4500 Ω

Solution

$$I = \frac{E}{R}$$

$$I = \frac{9}{4500}$$

$$I = 0.002\,\text{A, or 2 mA}$$

2. Given

I = 250 mA = 0.250 A
E = ?
R = 470 Ω

Solution

$$I = \frac{E}{R}$$

$$0.250 = \frac{E}{470}$$

$$\frac{0.250}{1} \bowtie \frac{E}{470}$$

$$(1)(E) = (0.250)(470)$$

$$E = 1175.\text{V}$$

3. Given

I = 10 A
E = 240 V
R = ?

Solution

$$I = \frac{E}{R}$$

$$10 = \frac{240}{R}$$

$$\frac{10}{1} \bowtie \frac{240}{R}$$

$$(1)(240) = (10)(R)$$

$$\frac{240}{10} = \frac{10\,R}{10}$$

$$\frac{240}{10} = 1\,R$$

$$24\,\Omega = R$$

4. a.

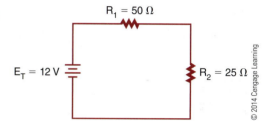

First, find the total resistance of the circuit (series).

$R_T = R_1 + R2$
$R_T = 50 + 25$
$R_T = 75\,\Omega$

Second, redraw the circuit, using the total equivalent resistance.

Third, find the total current of the circuit.

Given

$I_T = ?$
$E_T = 12\,\text{V}$
$R_T = 75\,\Omega$

Solution

$$I_T = \frac{E_T}{R_T}$$

$$I_T = \frac{12}{75}$$

$$I_T = 0.16\,\text{A, or 160 mA}$$

Now, find the voltage drop across R_1 and R_2.

$$I_T = I_1 = I_2$$
$$I_1 = \frac{E_1}{R_1}$$
$$0.16 = \frac{E_1}{50}$$
$$(0.16)(50) = E_1$$
$$8V = E_1$$

$$I_2 = \frac{E_2}{R_2}$$
$$0.16 = \frac{E_2}{25}$$
$$(0.16)(25) = E2$$
$$4V = E_2$$

b.

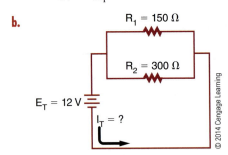

First, find the total resistance of the circuit (parallel).

$$\frac{1}{R_T} = \frac{1}{R_1} + \frac{1}{R_2}$$
$$\frac{1}{R_T} = \frac{1}{150} + \frac{1}{300}$$
$$\frac{1}{R_T} = \frac{2}{300} + \frac{1}{300}$$
$$\frac{1}{R_T} = \frac{3}{300}$$
$$(3)(R_T) = (1)(300)$$
$$\frac{(3)(R_T)}{3} = \frac{300}{3}$$
$$1\,R_T = \frac{300}{3}$$
$$R_T = 100\,\Omega$$

Second, redraw the circuit with equivalent resistance.

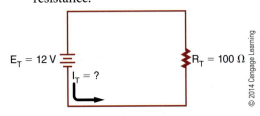

Third, find the total resistance of the circuit.

Given

$I_T = ?$
$E_T = 12\,V$
$R_T = 100\,\Omega$

Solution

$$I_T = \frac{E_T}{R_T}$$
$$I_T = \frac{12}{100}$$
$$I_T = 0.12\,A,\ or\ 120\,mA$$

Now, find the branch currents for R_1 and R_2.

$$E_T = E_1 = E_2$$
$$I_1 = \frac{E_1}{R_1}$$
$$I_1 = \frac{12}{150}$$
$$I_1 = 0.08\,A\ or\ 80\,mA$$

$$I_2 = \frac{E_2}{R_2}$$
$$I_2 = \frac{12}{300}$$
$$I_2 = 0.04\,A,\ or\ 40\,mA$$

c.

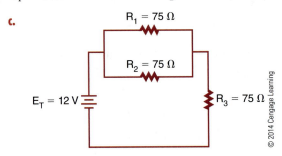

First, find the equivalent resistance for the parallel portion of the circuit.

$$\frac{1}{R_A} = \frac{1}{R_1} + \frac{1}{R_2}$$
$$\frac{1}{R_A} = \frac{1}{75} + \frac{1}{75}$$
$$\frac{1}{R_A} = \frac{2}{75}$$
$$(2)(R_A) = (1)(75)$$
$$\frac{(2)(R_A)}{2} = \frac{(1)(75)}{2}$$
$$1R_A = \frac{75}{2}$$
$$R_A = 37.5\,\Omega$$

Second, redraw the circuit with equivalent resistance.

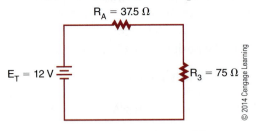

Third, find the total resistance of the circuit.

$$R_T = R_A + R_3$$
$$R_T = 37.5 + 75$$
$$R_T = 112.5\,\Omega$$

Fourth, find the total current of the circuit.

Given

$I_T = ?$
$E_T = 12\,V$
$R_T = 112.5\,\Omega$

Solution

$$I_T = \frac{E_T}{R_T}$$
$$I_T = \frac{12}{112.5}$$
$$I_T = 0.107\,A,\ or\ 107\,mA$$

Fifth, find voltage drops for R_3 and R_4.

$I_T = I_A = I_3$

$I_A = 0.107\,A$

$I_3 = 0.107\,A$

$$I_3 = \frac{E_3}{R_3} \qquad I_A = \frac{E_A}{R_A} \qquad E_A = E_1 = E_2$$

$$0.107 = \frac{E_3}{75} \qquad 0.107 = \frac{E_A}{37.5} \qquad E_1 = 4\,V$$

$$8\,V = E_3 \qquad 4\,V = E_A \qquad E_2 = 4\,V$$

Now, find the branch current for R_1 and R_2.

$$I_1 = \frac{E_1}{R_1} \qquad\qquad I_2 = \frac{E_2}{R_2}$$

$$I_1 = \frac{4}{75} \qquad\qquad I_2 = \frac{4}{75}$$

$I_1 = 0.053\,A$, or $53\,mA$ $I_2 = 0.053\,A$, or $53\,mA$

5. a.

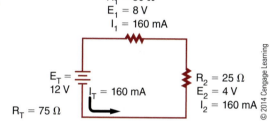

R₁ = 50 Ω
E₁ = 8 V
I₁ = 160 mA

E_T = 12 V
I_T = 160 mA
R_T = 75 Ω

R₂ = 25 Ω
E₂ = 4 V
I₂ = 160 mA

© 2014 Cengage Learning

$I_T = I_1 = I_2$ $\qquad E_T = E_1 + E_2$

$160\,mA = 160\,mA = 160\,mA$ $12\,V = 8\,V + 4\,V$

Note: Rounding may cause a difference in answers.

b.

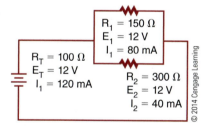

R₁ = 150 Ω
E₁ = 12 V
I₁ = 80 mA

R_T = 100 Ω
E_T = 12 V
I₁ = 120 mA

R₂ = 300 Ω
E₂ = 12 V
I₂ = 40 mA

© 2014 Cengage Learning

$I_T = I_1 + I_2$ $\qquad E_T = E_1 = E_2$

$120\,mA = 80\,mA + 40\,mA$ $12\,V = 12\,V = 12\,V$

5.

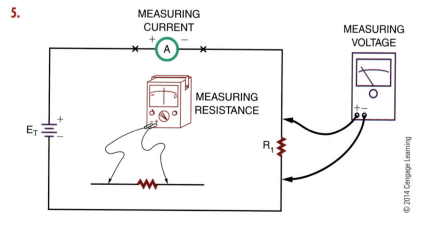

MEASURING CURRENT

MEASURING RESISTANCE

MEASURING VOLTAGE

E_T

R₁

© 2014 Cengage Learning

c.

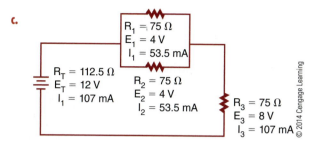

R₁ = 75 Ω
E₁ = 4 V
I₁ = 53.5 mA

R_T = 112.5 Ω
E_T = 12 V
I₁ = 107 mA

R₂ = 75 Ω
E₂ = 4 V
I₂ = 53.5 mA

R₃ = 75 Ω
E₃ = 8 V
I₃ = 107 mA

© 2014 Cengage Learning

$I_T = (I_1 + I_2) + I_3$ $\qquad E_T = (E_1 = E_2) = E_3$

$0.107 = (0.0535 + 0.0535)$ $12\,V = (4 = 4) + 8$

$\qquad = 0.107$

$0.107A = 0.107A = 0.107A$ $12V = 4V + 8V$

Chapter 15 Electrical Measurements—Meters

1. Digital

2. Analog

3. a. 23 volts

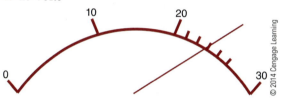

10   20

0   30

b. 220 milliamperes

100   200

0   300

c. 2700 ohms

2000   1000

∞   0

4. One meter can be used to measure voltage, current, and resistance.

6. VOM is an analog multimeter and stands for "volt-ohm-milliammeter." A DMM is a digital multimeter. The DMM is easier to read, with autoranging, autozeroing, and autolock capabilities, but it does require batteries and the meter could be damaged if voltage limitations are exceeded. The VOM simplifies the measuring of instantaneous signal changes, but the range must be manually changed, manually zeroed, and does not offer autolock capabilities.

7. Always set the meter to the highest range being measured.

8. Remove power and reverse the meter leads if the meter deflects to the left side of the meter; if the meter deflects to the right side of the meter, then check to make sure the right range is selected.

9. An ohmmeter measures resistance by placing a voltage across the resistance and measuring the amount of current flowing through the resistor. A high current reading indicates a low resistance, and a low current indicates a high resistance.

10. No, it is autozeroing.

Chapter 16 **Power**

1. **Given**

$P = ?$
$I = 40\,mA = 0.04\,A$
$E = 30\,V$

Solution

$P = IE$
$P = (0.04)(30)$
$P = 1.2\,W$

2. **Given**

$P = 1\,W$
$I = 10\,mA\_0.01\,A$
$E = ?$

Solution

$P = IE$
$1 = (0.01)(E)$
$\dfrac{1}{0.01} = \dfrac{(0.01)(E)}{(0.01)}$
$\dfrac{1}{0.01} = 1E$
$100\,V = E$

3. **Given**

$P = 12.3\,W$
$I = ?$
$E = 30\,V$

Solution

$P = IE$
$12.3 = (I)(30)$
$\dfrac{12.3}{30} = \dfrac{(I)(30)}{30}$
$\dfrac{12.3}{30} = 11$
$0.41\,A = 1$
$I = 0.41\,A,\ or\ 410\,mA$

4. a.

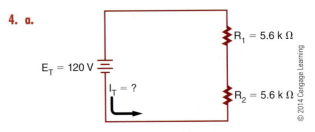

First, find the total resistance of the circuit (series).

$R_T = R1 + R2$
$R_T = 5600 + 5600$
$R_T = 11,200\,\Omega$

Second, redraw the circuit using total resistance.

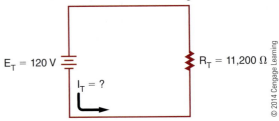

Third, find the total circuit current.

Given

$I_T = ?$
$E_T = 120\,V$
$R_T = 11,200\,\Omega$

Solution

$I_T = \dfrac{E_T}{R_T}$
$I_T = \dfrac{120}{11,200}$
$I_T = 0.0107\,A,\ or\ 10.7\,mA$

Now, find the total circuit power.

$P_T = I_T E_T$
$P_T = (0.0107)(120)$
$P_T = 1.28\,W$

b.

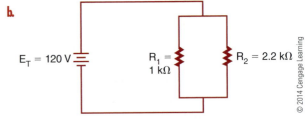

First, find the total resistance of the circuit (parallel).

$\dfrac{1}{R_T} = \dfrac{1}{R_1} + \dfrac{1}{R_2}$
$\dfrac{1}{R_T} = \dfrac{1}{1000} + \dfrac{1}{2200}$
$\dfrac{1}{R_T} = 0.001 + 0.000455$
$\dfrac{1}{R_T} = 0.001455$

$$R_T = \frac{0.001455}{1}$$

$$(0.001455)(R_T) = (1)(1)$$

$$R_T = \frac{1}{0.001455}$$

$$R_T = 687.29\,\Omega$$

Second, redraw the circuit using total resistance.

Third, find the total circuit current.

Given

$I_T = ?$
$E_T = 120\,V$
$R_T = 687.29\,\Omega$

Solution

$$I_T = \frac{E_T}{R_T}$$

$$I_T = \frac{120}{687.29}$$

$$I_T = 0.175\,A,\ or\ 175\,mA$$

Now, find total circuit power.

Given

$P_T = ?$
$I_T = 0.175\,A$
$E_T = 120\,V$

Solution

$$P_T = I_T E_T$$
$$P_T = (0.175)(120)$$
$$P_T = 21\,W$$

c.

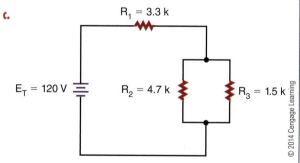

First, find the equivalent resistance for the parallel portion of the circuit.

$$\frac{1}{R_T} = \frac{1}{1500} + \frac{1}{4700}$$

$$\frac{1}{R_A} = 0.000667 + 0.000213$$

$$\frac{1}{R_A} = 0.000880$$

$$(0.000880)\frac{1}{R_A} = \frac{0.000880}{1}$$

$$R_A = \frac{1}{0.000880}$$

$$R_A = 1,136.36\,\Omega$$

Second, redraw the circuit using the equivalent resistance.

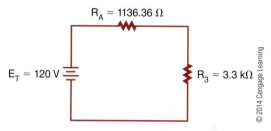

Third, find the total resistance of the circuit.

$$R_T = R_A + R_3$$
$$R_T = 1136.36 + 3300$$
$$R_T = 4436.36\,\Omega$$

Fourth, find the total current for the circuit.

Given

$I_T = ?$
$E_T = 120\,V$
$R_T = 4436.36\,\Omega$

Solution

$$I_T = \frac{E_T}{R_T}$$

$$I_T = \frac{120}{4436.36}$$

$$I_T = 0.027\,A,\ or\ 27\,mA$$

Fifth, find the total power for the circuit.

Given

$P_T = ?$

$I_T = 0.027\,A$

$E_T = 120\,V$

Solution

$$P_T = I_T E_T$$

$$P_T = (0.027)(120)$$

$$P_T = 3.24\,W\ (with\ rounding)$$

$$3.25\,W\ (without\ rounding)$$

Remember that rounding will give a slightly different answer.

5. a. Given

$I_T = 0.010714\,A$
$E_T = 120\,V$
$R_T = 11{,}200\,\Omega$
$P_T = 1.28\,W$

Solution

$$I_1 = \frac{E_1}{R_1}$$
$$0.010714 = \frac{E_1}{5600}$$
$$(0.010714)(5600) = E_1$$
$$60\,V = E_1$$

$$I_2 = \frac{E_2}{R_2}$$
$$0.010714 = \frac{E_2}{5600}$$
$$(0.010714)(5600) = E_2$$
$$60\,V = E_2$$

$$P_1 = I_1 E_1$$
$$P_1 = (0.10714)(60)$$
$$P_1 = 0.64\,W$$

$$P_2 = I_2 E_2$$
$$P_2 = (0.10714)(60)$$
$$P_2 = 0.64\,W$$

b. Given

$I_T = 0.175\,A$
$E_T = 120\,V$
$R_T = 687.29\,\Omega$

Solution

$$I_1 = \frac{E_1}{R_1}$$
$$I_1 = \frac{120}{1000}$$
$$I_1 = 0.12\,A$$

$$I_2 = \frac{E_2}{R_2}$$
$$I_2 = \frac{120}{2200}$$
$$I_2 = 0.055$$

$$P_1 = I_1 E_1$$
$$P_1 = (0.12)(120)$$
$$P_1 = 14.4\,W$$

$$P_2 = I_2 E_2$$
$$P_2 = (0.055)(120)$$
$$P_2 = 6.6\,W$$

c. Given

$I_T = IA = 0.027\,A$
$R_A = 1136.36\,\Omega$

Solution

$$I_A = \frac{E_A}{R_A}$$
$$0.027 = \frac{E_A}{1136.36}$$
$$(0.027)(1136.36) = E_A$$
$$30.672\,V = E_A$$

$$I_1 = \frac{E_1}{R_1}$$
$$I_1 = \frac{30.672}{1500}$$
$$I_1 = 0.0205\,A$$

$$I_2 = \frac{E_2}{R_2}$$
$$I_2 = \frac{30.672}{4700}$$
$$I_2 = 0.00653\,A$$

$I_T = I_3 = 0.027\,A$

$E_T = 120\,V$

$R_T = 4436.36\,\Omega$
$P_T = 3.24\,W$

$$P_1 = I_1 E_1$$
$$P_1 = (0.0205)(30.672)$$
$$P_1 = 0.629\,W$$

$$I_3 = \frac{E_3}{R_3}$$
$$0.27 = \frac{E3}{3300}$$
$$(0.027)(3300) = E_3$$
$$89.1\,V = E_3$$

$$P_2 = I_2 E_2$$
$$P_2 = (0.00653)(30.672)$$
$$P_2 = 0.200\,W$$

$$P_3 = I_3 E_3$$
$$P_3 = (0.0.27)(89.1)$$
$$P_3 = 2.4057\,W$$

Chapter 17 DC Circuits

1. a.

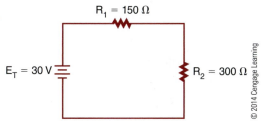

Find the total circuit resistance.

$$R_T = R_1 + R_2$$
$$R_T = 150 + 300$$
$$R_T = 450\,\Omega$$

Redraw the equivalent circuit.

Find the total circuit current.

$$I_T = \frac{E_T}{R_T}$$
$$I_T = \frac{30}{450}$$
$$I_T = 0.0667\,\text{A or } 66.7\,\text{mA}$$

Find the voltage drop across each resistor.
$I_T = I_1 + I_2$ (The current flow in a series circuit is the same throughout the circuit.)

$$I_{R_1} = \frac{E_{R_1}}{R_1}$$
$$0.0667 = \frac{E_{R_1}}{150}$$
$$\frac{0.0667}{1} = \frac{E_{R_1}}{150}$$
$$(1)(E_{R_1}) = (0.0667)(150)$$
$$E_{R_1} = (0.0667)(150)$$
$$E_{R_1} = 10\,\text{V}$$

$$I_{R_2} = \frac{E_{R_2}}{R_2}$$
$$0.0667 = \frac{E_{R_2}}{300}$$
$$\frac{0.0667}{1} = \frac{E_{R_2}}{300}$$
$$(1)(E_{R_2}) = (0.0667)(300)$$
$$E_{R_2} = (0.0667)(300)$$
$$E_{R_2} = 20\,\text{V}$$

Find the power for each resistor.

$$P_{R_1} = I_{R_1}E_{R_1} \qquad\qquad P_{R_2} = I_{R_2}E_{R_2}$$
$$P_{R_1} = (0.0667)(10) \qquad P_{R_2} = (0.0667)(20)$$
$$P_{R_1} = 0.667\,\text{W} \qquad\qquad P_{R_2} = 1.334\,\text{W}$$

Find the total power of the circuit.

$$P_T = I_T E_T \qquad\qquad P_T = P_{R_1} + P_{R_2}$$
$$P_T = (0.0667)(30) \quad\text{or}\quad P_T = 0.667 + 1.334$$
$$P_T = 2.001\,\text{W} \qquad\qquad P_T = 2.001\,\text{W}$$

b.

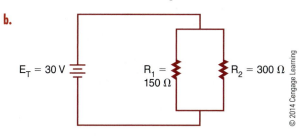

Find the total circuit resistance.

$$\frac{1}{R_T} = \frac{1}{R_1} + \frac{1}{R_2}$$
$$\frac{1}{R_T} = \frac{1}{150} + \frac{1}{300}$$
$$\frac{1}{R_T} = \frac{2}{300} + \frac{1}{300}$$
$$\frac{1}{R_T} = \frac{3}{300}$$
$$(3)(R_T) = (1)(300)$$
$$\frac{(3)(R_T)}{3} = \frac{(1)(300)}{3}$$
$$R_T = \frac{300}{3}$$
$$R_T = 100\,\Omega$$

Redraw the equivalent circuit.

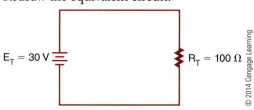

Find the total circuit current.

$$I_T = \frac{E_T}{R_T}$$
$$I_T = \frac{30}{100}$$
$$I_T = 0.3\,\text{A or } 300\,\text{mA}$$

Find the current through each branch of the parallel circuit. The voltage is the same across each branch of the parallel circuit.

$$E_T = E_1 = E_2$$

$$I_{R_1} = \frac{E_{R_1}}{R_1} \qquad\qquad I_{R_2} = \frac{E_{R_2}}{R_2}$$

$$I_{R_1} = \frac{30}{150} \qquad\qquad I_{R_2} = \frac{30}{300}$$

$$I_{R_1} = 0.2\,A \qquad\qquad I_{R_2} = 0.1\,A$$

Find the power for each resistor.

$$P_{R_1} = I_{R_1}E_{R_1} \qquad\qquad P_{R_2} = I_{R_2}E_{R_2}$$
$$P_{R_1} = (0.2)(30) \qquad\qquad P_{R_2} = (0.1)(30)$$
$$P_{R_1} = 6\,W \qquad\qquad P_{R_2} = 3\,W$$

Find the total power of the circuit.

$$P_T = I_T E_T$$
$$P_T = (0.3)(30)$$
$$P_T = 9\,W$$

c.

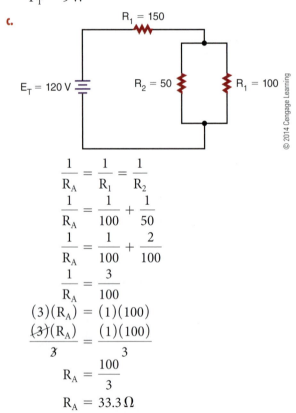

$$\frac{1}{R_A} = \frac{1}{R_1} = \frac{1}{R_2}$$

$$\frac{1}{R_A} = \frac{1}{100} + \frac{1}{50}$$

$$\frac{1}{R_A} = \frac{1}{100} + \frac{2}{100}$$

$$\frac{1}{R_A} = \frac{3}{100}$$

$$(3)(R_A) = (1)(100)$$

$$\frac{\cancel{(3)}(R_A)}{\cancel{3}} = \frac{(1)(100)}{3}$$

$$R_A = \frac{100}{3}$$

$$R_A = 33.3\,\Omega$$

Redraw the circuit.

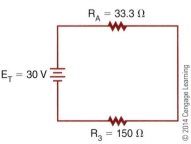

Now find the total circuit resistance.

$$R_T = R_A + R_3$$
$$R_T = 33.3 + 150$$
$$R_T = 183.3\,\Omega$$

Find the current flow (IT) for the equivalent circuit.

$$I_T = \frac{E_T}{R_T}$$

$$I_T = \frac{30}{183.3}$$

$$I_T = 0.164\,A \text{ or } 164\,mA$$

Find the voltage drop across resistors the equivalent circuit. (Current is the same throughout a series circuit.)

$$I_T = IR_A = IR_3$$

$$I_{R_A} = \frac{E_{R_A}}{R_A}$$

$$0.164 = \frac{E_{R_A}}{33.3}$$

$$(1)(E_{R_A}) = (0.164)(33.3)$$

$$E_{R_A} = 5.46\,V$$

$$I_{R_3} = \frac{E_{R_3}}{R_3}$$

$$0.164 = \frac{E_{R_3}}{150}$$

$$(1)(E_{R_3}) = (0.164)(150)$$

$$E_{R_3} = 24.6\,V$$

Find the current across each of the resistors in the parallel portion of the circuit.

$$I_{R_1} = \frac{E_{R_1}}{R_1} \qquad\qquad I_{R_2} = \frac{E_{R_2}}{R_2}$$

$$I_{R_1} = \frac{5.46}{100} \qquad\qquad I_{R_2} = \frac{5.46}{50}$$

$$I_{R_1} = 0.056\,A \qquad\qquad I_{R_2} = 0.109\,A$$

Find the power across each component, and then find the total power.

$$P_T = I_T E_T$$
$$P_T = (0.164)(30)$$
$$P_T = 4.92\,W$$

$$P_{R_1} = I_{R_1}E_{R_1}$$
$$P_{R_1} = (0.056)(5.46)$$
$$P_{R_1} = 0.298\,W$$

$$P_{R_2} = I_{R_2}E_{R_2}$$
$$P_{R_2} = (0.109)(5.46)$$
$$P_{R_2} = 0.595\,W$$

$$P_{R_3} = I_{R_3}E_{R_3}$$
$$P_{R_1} = (0.164)(24.6)$$
$$P_{R_3} = 4.034\,W$$

2. Assume I_4 is 10 mA.

Given   **Solution**

$I_4 = 10\,mA$	$I_4 = E_4/R_4$	$E_3 = 6 - 1.5$	$E_2 = 9 - 7.5$	$E_1 = 12 - 9$
$E_4 = 1.5\,V$	$0.010 = 1.5/R_4$	$E_3 = 4.5\,V$	$E_2 = 1.5\,V$	$E_1 = 3\,V$
	$R_4 = 1.5/0.010$	$13 = 0.010 + 0.150$	$12 = 0.260 + 0.750$	$11 = 1.010 + 0.500$
	$R_4 = 150\,\Omega$	$13 = 0.260\,A$	$12 = 1.01\,A$	$11 = 1.51\,A$

$13 = E_3/R_3$	$12 = E_2/R_2$	$11 = E_1/R_1$
$0.260 = 4.5/R_3$	$1.01 = 1.5/R_2$	$1.51 = 3/R_1$
$R_3 = 4.5/0.260$	$R_2 = 1.5/1.01$	$R_1 = 3/1.51$
$R_3 = 17.31\,\Omega$	$R_2 = 1.5\,\Omega$	$R_1 = 2\,\Omega$

3.

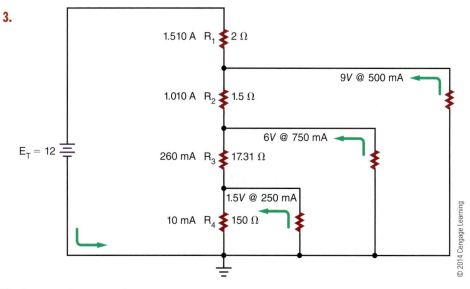

$E_T = 12$

1.510 A R_1 2 Ω

1.010 A R_2 1.5 Ω

9V @ 500 mA

260 mA R_3 17.31 Ω

6V @ 750 mA

1.5V @ 250 mA

10 mA R_4 150 Ω

4. Assume I_2 is 10 mA.

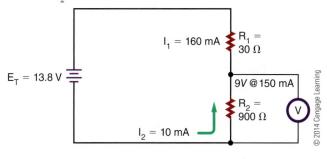

$E_T = 13.8\,V$

$I_1 = 160\,mA$ $R_1 = 30\,\Omega$

9V @ 150 mA

$R_2 = 900\,\Omega$ V

$I_2 = 10\,mA$

Given   **Solution**

$I_2 = 10\,mA$	$12 = E2/R_2$	$E_1 = 13 - 8 - 9$
$E_2 = 9\,V$	$0.010 = 9/R_2$	$E_1 = 4.8\,V$
$R_2 = ?$	$R_2 = 900\,\Omega$	$11 = 0.010 + 0.150$
		$11 = 0.160\,A$

$$11 = E_1/R_1$$
$$0.160 = 4.8/R_1$$
$$R_1 = 4.8/0.160$$
$$R_1 = 30\,\Omega$$

5.

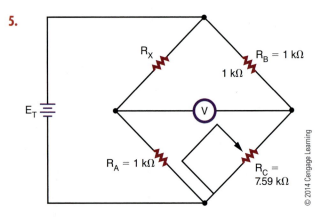

© 2014 Cengage Learning

Given

$R_A = 1\,k\Omega$
$R_B = 1\,k\Omega$
$R_C = 7.59\,k\Omega$
$R_X = ?$

Solution

$R_X/R_A = R_B/R_C$
$(R_X)(R_C) = (R_B)(R_A)$
$R_X = (R_B)(R_A)/R_C$

$R_X = (R_B)(R_A)/R_C$
$R_X = (1000)(1000)/7590$
$R_X = 131.75\,\Omega$

Chapter 18 Magnetism

1. The domain theory of magnetism can be verified by jarring the domains into a random arrangement by heating or hitting with a hammer. The magnet will eventually lose its magnetism.

2. The strength of an electromagnet can be increased by increasing the number of turns of wire, by increasing the current flow, or by inserting a ferromagnetic core in the center of the coil.

3. The left-hand rule for conductors: Grasp the wire in the left hand with the thumb pointing in the direction of the current flow; the fingers will point in the direction of the flux lines.

4. In Figure 18–15 (page 189), when the loop is rotated from position A to position B, a voltage is induced when the motion is at right angles to the magnetic field. As the loop is rotated to position C, the induced voltage decreases to 0 volts. As the loop continues to position D, a voltage is again induced, but the commutator reverses the output polarity so it is the same as was first output by the DC generator. The output pulsates in one direction, varying twice during each revolution between 0 and maximum.

5.

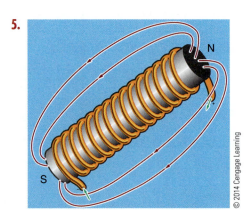

© 2014 Cengage Learning

6. Polarity of an electromagnet can be determined by grasping the coil with the left hand, with the fingers pointing in the direction of current flow; the thumb will point in the direction of the north pole.

7. Left-hand rule for generator: Holding the left hand with the thumb, index, and middle fingers extended at right angles to each other (Figure 18-13, page 188). With the thumb pointing in the direction of movement, the index finger in the direction of flux lines, the middle finger points in the direction of current flow.

8.

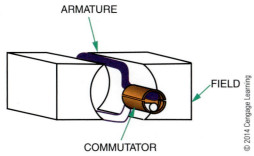

ARMATURE

FIELD

COMMUTATOR

© 2014 Cengage Learning

9. DC motor operation is dependent on the principle that a current-carrying conductor placed at right angles to a magnetic field will move at right angles to the direction of the field. A commutator reverses the direction of the current flow at the top or zero torque position, resulting in the DC motor armature rotating.

10. The basic meter movement relies on the same principles of the DC motor. A pointer is attached to a rotating coil and moves according to current flow. The pointer moves across a graduated scale and indicates the amount of current flow.

Chapter 19 **Inductance**

1. Lenz's law: An induced emf in any circuit is always in a direction to oppose the effect that produced it. When current stops or changes direction in a circuit, an emf (electromotive force) is induced back into the conductor through the collapsing magnetic field. This opposition to current flow is referred to as counter emf. The faster the rate of change the greater the counter emf.

2. All conductors have some inductance depending on the conductor and the shape of it. When a signal is removed, a counter emf is induced back into the conductor.

3. Inductors can be fixed or variable, made with an air core or ferrite or powdered iron core. Toroid cores are round and offer high inductance for a small size and contain the magnetic field within the core.

4. The magnetic field around an inductor can be increased by using an iron core.

5.

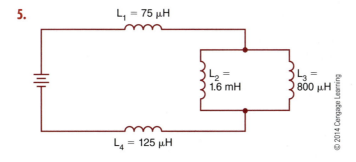

$L_1 = 75\ \mu\text{H}$

$L_2 = 1.6\ \text{mH}$

$L_3 = 800\ \mu\text{H}$

$L_4 = 125\ \mu\text{H}$

© 2014 Cengage Learning

Given

$L_1 = 75\,\mu\text{H}$
$L_2 = 1.6\,\text{mH} = 1600\,\mu\text{H}$
$L_3 = 800\,\mu\text{H}$
$L_4 = 125\,\mu\text{H}$

Solution

$$\frac{1}{L_P} = \frac{1}{L_2} + \frac{1}{L_3}$$

$$\frac{1}{L_P} = \frac{1}{0.0016} + \frac{1}{0.0008}$$

$$\frac{1}{L_P} = 1875$$

$$L_P = 533.33\,\mu\text{H}$$

$$L_T = L_1 + L_P + L_4$$

$$L_T = 75\,\mu\text{H} + 533.33\,\mu\text{H} + 125\,\mu\text{H}$$

$$L_T = 733.33\,\mu\text{H}$$

6. First, draw the circuit:

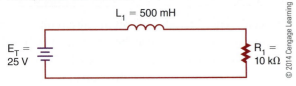

$L_1 = 500 \text{ mH}$

$E_T = 25 \text{ V}$

$R_1 = 10 \text{ k}\Omega$

© 2014 Cengage Learning

Given

$E_T = 25 \text{ V}$
$L_1 = 500 \text{ mH} = 0.5 \text{ H}$
$R_1 = 10 \text{ k}\Omega = 10,000 \Omega$

Solution

$$t = \frac{L}{R}$$

$$t = \frac{0.5}{10,000}$$

$$t = 0.00005$$

$$t = 50 \,\mu\text{sec}$$

$100 \,\mu\text{sec} = 2$ time constants, energized 86.5%

$25 \times 86.5\% = 21.63 \text{ V}$

This voltage represents E_R on rise.

$$E_L = E_T - E_R$$
$$E_L = 25 - 21.63$$
$$E_L = 3.37 \text{ V}$$

7. $t = \dfrac{L}{R}$

a. $t = \dfrac{1}{100} = 0.01 \text{ s}$

b. $t = \dfrac{0.1}{10000} = 0.00001 \text{ s}$

c. $t = \dfrac{0.010}{1000} = 0.00001 \text{ s}$

d. $t = \dfrac{10}{10} = 1 \text{ s}$

e. $t = \dfrac{1}{1000} = 0.001 \text{ s}$

Growth

		Time Constants				
		1	2	3	4	5
		63.2%	86.5%	95%	98.2%	99.3%
a.	0.01 s	0.01	0.02	0.03	0.04	0.05
b.	0.00001 s	0.00001	0.00002	0.00003	0.00004	0.00005
c.	0.00001 s	0.00001	0.00002	0.00003	0.00004	0.00005
d.	1 s	1	2	3	4	5
e.	0.001 s	0.001	0.002	0.003	0.004	0.005

Decay

		Time Constants				
		1	2	3	4	5
		36.8%	13.5%	5%	1.8%	0.7%
a.	0.01 s	0.01	0.02	0.03	0.04	0.05
b.	0.00001 s	0.00001	0.00002	0.00003	0.00004	0.00005
c.	0.00001 s	0.00001	0.00002	0.00003	0.00004	0.00005
d.	1 s	1	2	3	4	5
e.	0.001 s	0.001	0.002	0.003	0.004	0.005

© 2014 Cengage Learning

Chapter 20 Capacitance

1. The charge is stored on the plates of the capacitor.

2. A capacitor is two conductor plates separated by an insulator. When a DC voltage is applied to the capacitor, a current flows until the plates are charged and then current flow stops.

3. When a capacitor is charged and removed from a circuit, it holds the last charge indefinitely.

4. To discharge a charged capacitor, short both leads together.

5. The capacitance of a capacitor is directly proportional to the area of the plate. For example, increasing the plate area increases the capacitance. Capacitance is also inversely proportional to the distance between the plates. If the plates are moved further apart, the strength of the electric field between the plates decreases.

6. Types of capacitors include polarized electrolytic capacitors, paper and plastic capacitors, ceramic disk capacitors, and variable capacitors. Styles of capacitors include radial leads and axial leads.

7. First, draw the circuit:

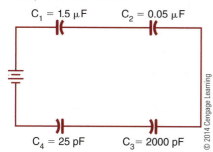

Given

$C_1 = 1.5\,\mu F$
$C_2 = 0.05\,\mu F$
$C_3 = 2000\,pF = 0.002\,\mu F$
$C_4 = 25\,pF = 0.000025\,\mu F$

Solution

$$\frac{1}{C_T} = \frac{1}{C_1} + \frac{1}{C_2} + \frac{1}{C_3} + \frac{1}{C_4}$$

$$\frac{1}{C_T} = \frac{1}{1.5} + \frac{1}{0.05} + \frac{1}{0.002} + \frac{1}{0.000025}$$

$$\frac{1}{C_T} = 0.667 + 20 + 500 + 40000$$

$$\frac{1}{C_T} = 40,520.667$$

$$\frac{1}{C_T} = \frac{40,520.667}{1}$$

$$(40,520.667)(C_T) = (1)(1)$$

$$C_T = \frac{1}{40,520.667}$$

$$C_T = 0.000024678$$

$$C_T = 24.678\,pF$$

8. First, draw the circuit.

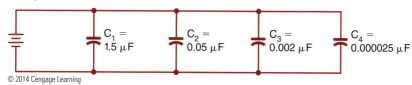

© 2014 Cengage Learning

Given

$C_1 = 1.5\,\mu F$
$C_2 = 0.05\,\mu F$
$C_3 = 2000\,pF = 0.002\,\mu F$
$C_4 = 25\,pF = 0.000025\,\mu F$

Solution

$$C_T = C_1 + C_2 + C_3 + C_4$$
$$C_T = 1.5 + 0.05 + 0.002 + 0.000025$$
$$C_T = 1.552025\,\mu F,\ \text{or}\ 1.55\,\mu F$$

9.

Charge

	Time Constants				
	1	2	3	4	5
	63.2%	86.5%	95%	98.2%	99.3%
100 V	63.2 V	86.5 V	95 V	98.2 V	99.3 V

Discharge

	Time Constants				
	1	2	3	4	5
	36.8%	13.5%	5%	1.8%	0.7%
100 V	36.8 V	13.5 V	5 V	1.8 V	0.7 V

© 2014 Cengage Learning

10.

Charge

	Time Constants				
	1	2	3	4	5
	63.2%	86.5%	95%	98.2%	99.3%
t = 1 sec	1 s	2 s	3 s	4 s	5 s

Discharge

	Time Constants				
	1	2	3	4	5
	36.8%	13.5%	5%	1.8%	0.7%
t = 1 sec	1 s	2 s	3 s	4 s	5 s

© 2014 Cengage Learning

Chapter 21 Alternating Current

1. A conductor must be placed in a magnetic field in order for magnetic induction to occur.

2. To apply the left-hand rule, the thumb is pointed in the direction of the conductor movement, the index finger (extended at right angles to the thumb) indicates the direction of the magnetic lines of flux from north to south, and the middle finger (extended at a right angle to the index finger) indicates the direction of current flow in the conductor. The left-hand rule is used to determine the direction of current flow in a conductor that is being passed through a magnetic field.

3. The peak-to-peak value is the vertical distance between the two peaks of a waveform.

4. The effective value of alternating current is the amount that will produce the same degree of heat in a given resistance as an equal amount of direct current.

5. $E_{RMS} = 0.707 \times E_P$
$E_{RMS} = (0.707)(169)$
$E_{RMS} = 119.48\,V$

6. $E_{RMS} = 0.707 \times E_P$
$85 = 0.707 \times E_P$
$\dfrac{85}{0.707} = E_P$
$120\,V = E_P$

7. $f = \dfrac{1}{t}$
$f = \dfrac{1}{0.02}$
$f = 50\,Hz$

8. $f = \dfrac{1}{t}$
$400 = 10$
$t = \dfrac{1}{400}$
$t = 0.0025\,s$

9. **a.** Square wave:

b. Triangular wave:

c. Sawtooth wave:

© 2014 Cengage Learning

10. Nonsinusoidal waveforms can be considered as being constructed by algebraic addition of sine waves having different frequencies (harmonics), amplitudes, and phases.

Chapter 22 AC Measurements

1. Using rectifiers to convert the AC signal to a DC current allows using a DC meter movement to measure AC.

2. The clamp-on ammeter uses a split-core transformer. The core can be opened and placed around the conductor. A voltage is induced into the core, which is also cut by a coil. The induced voltage creates a current flow that is rectified and sent to a meter movement.

3. An oscilloscope can provide the following information about an electronic circuit: the frequency of a signal, the duration of a signal, the phase relationship between signal waveforms, the shape of a signal's waveform, and the amplitude of a signal.

4. Initially set the oscilloscope controls as follows: intensity, focus, astigmatism, and position controls (set to the center of their range).

Triggering: INT +
Level: Auto
Time/CM: 1 msec
Volts/CM: 0.02
Power: On

Connect the oscilloscope probe to the test jack of the voltage calibrator. Adjust the controls for a sharp, stable image of a square wave; check the values shown on the scope display against those set on the voltage calibrator.

5. Connect the oscilloscope probe to the test jack of the voltage calibrator in order to make sure the probe is not faulty and is calibrated properly.

6. Before hooking the oscilloscope probe to an input signal, set the volt/cm switch to its highest setting.

7. A frequency counter consists of a time base, an input-signal conditioner, a gate control circuit, a main gate, a decade counter, and a display.
   - *Time base*—compensates for the different frequencies being measured.
   - *Signal conditioner*—converts the input signal to a wave shape and amplitude compatible with the circuitry in the counter.
   - *Gate-control circuitry*—acts as the synchronization center of the counter. It opens and closes the main gate and provides a signal to latch the count at the end of the counting period and resets the circuitry for the next count.
   - *Main gate*—passes the conditioned input signal to the counter circuit.
   - *Decade counter*—keeps a running tally of all the pulses that pass through the main gate.
   - *Display*—provides a visual readout of the frequency being measured.

8. The integrated circuit has been the primary force for moving the frequency counter from the laboratory to the workbench. It has reduced the physical size of the counter.

9. A frequency counter measures frequency in a repair shop, an engineering department, or a ham radio shack, on an industrial production line, or anywhere a frequency measurement is necessary.

10. The Bode plotter allows the user to produce a graph of a circuit's frequency-amplitude and phase-angle analysis.

Chapter 23 **Resistive AC Circuits**

1. In a pure resistive AC circuit, the current and voltage waveforms are in phase.

2. **Given**

   $I_T = 25\,mA = 0.025\,A$
   $E_T = ?$
   $R_T = 4.7\,k\Omega = 4700\,\Omega$

   Solution

   $$I_T = \frac{E_T}{R_T}$$

   $$0.025 = \frac{E_T}{4700}$$

   $$\frac{0.025}{1} \diagup\!\!\!\!\diagdown \frac{E_T}{4700}$$

   $$(1)(E_T) = (0.025)(4700)$$

   $$E_T = 117.5\,V$$

3.

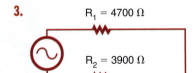

© 2014 Cengage Learning

Given

$I_T = ?$
$E_T = 12\,V$
$R_T = ?$
$R_1 = 4.7\,k\Omega = 4700\,\Omega$
$R_2 = 3.9\,k\Omega = 3900\,\Omega$
$E_1 = ?$
$E_2 = ?$

Solution

$R_T = R_1 + R_2$
$R_T = 4700 + 3900$
$R_T = 8600\,\Omega$

$$I_T = \frac{E_T}{R_T}$$

$$I_T = \frac{12}{8600}$$

$I_T = 0.0014\,A,$
or $1.4\,mA$

$$I_T = I_1 = I_2$$

$$I_1 = \frac{E_1}{R_1} \quad \frac{0.0014}{1} \bowtie \frac{E_1}{4700} \qquad\qquad I_2 = \frac{E_2}{R_2}$$

$$0.0014 = \frac{E_1}{4700} \qquad\qquad 0.0014 = \frac{E_2}{3900}$$

$$(1)(E_1) = (0.0014)(4700) \qquad \frac{0.0014}{1} \bowtie \frac{E_2}{3900}$$

$$E_1 = 6.58 \, V \qquad\qquad (1)(E_2) = (0.0014)(3900)$$

$$E_2 = 5.46 \, V$$

4. Given

$E_T = 120 \, V$

$R_1 = 2.2 \, k\Omega = 2200 \, \Omega$

$R_2 = 5.6 \, k\Omega = 5600 \, \Omega$

$I_1 = ?$

$I_2 = ?$

Solution

$E_T = E_1 = E_2$

$I_1 = \frac{E_1}{R_1}$

$I_1 = \frac{120}{2200}$

$I_1 = 0.055 \, A$, or 55 mA

$I_2 = \frac{E_2}{R_2}$

$I_2 = \frac{120}{5600}$

$I_2 = 0.021 \, A$, or 21 mA

5. In a series circuit, current flow is the same throughout the circuit. In a parallel circuit, voltage is the same throughout the circuit.

6. The rate at which energy is delivered to a circuit or energy (heat) is dissipated determines the power consumption in an AC circuit, just as in a DC circuit.

7. Given

$I_T = ?$

$E_T = 120 \, V$

$R_T = 1200 \, \Omega$

$P_T = ?$

Solution

$I_T = \frac{E_T}{R_T}$

$I_T = \frac{120}{1200}$

$I_T = 0.1 \, A$, or 100 mA

$P_T = I_T E_T$

$P_T = (0.1)(120)$

$P_T = 12 \, W$

8. Given

$I_T = ?$

$E_T = 12 \, V$

$R_T = ?$

Solution

$R_T = R_1 + R_2 + R_3$

$R_T = 47 + 100 + 150$

$R_T = 297 \, \Omega$

$I_T = E_T/R_T$

$I_T = 12/297$

$I_T = 0.040 \, A$

$P_T = I_T E_T$

$P_T = (0.04)(12)$

$P_T = 0.48 \, W$

9. Given

$I_T = ?$

$E_T = 12 \, V$

$R_T = ?$

Solution

$I_1 = E_1/R_1$

$I_1 = 12/47$

$I_1 = 0.255 \, A$

$P_1 = I_1 E_1$

$P_1 = (0.255)(12)$

$P_1 = 3.06 \, W$

10. Given

$I_T = ?$

$E_T = 12 \, V$

$R_T = ?$

$I_T = E_T/R_T$

$I_T = 12/107$

$I_T = 0.112$

Solution

$1/R_A = 1/R_2 + 1/R_3$

$1/R_A = 1/100 + 1/150$

$R_A = 60 \, \Omega$

$I_A = E_A/R_A$

$0.112 = E_A/60$

$(0.112)(60) = E_A$

$6.72 \, V = E_A$

$R_T = R_1 + RA$

$R_T = 47 + 60$

$R_T = 107 \, \Omega$

$I_3 = E_3/R_3$

$I_3 = 6.72/150$

$I_3 = 0.0448 \, A$

$P_3 = I_3 E_3$

$P_3 = (0.0448)(6.72)$

$P_3 = 0.301 \, W$

Chapter 24 Capacitive AC Circuits

1. In a capacitive AC circuit, the current leads the applied voltage.
2. Capacitive reactance is a function of the frequency of the applied AC voltage and the capacitance of the circuit.
3. **Given**
$$X_C = ?$$
$$\pi = 3.14$$
$$f = 60\,Hz$$
$$C = 1000\,\mu F = 0.001\,F$$

   Solution
$$X_C = \frac{1}{2\pi fC}$$
$$X_C = \frac{1}{(2)(3.14)(60)(0.001)}$$
$$X_C = \frac{1}{0.3768}$$
$$X_C = 2.65\,\Omega$$

4. **Given**
$$I_T = ?$$
$$E_T = 12$$
$$X_C = 2.65\,\Omega$$

   Solution
$$I_T = \frac{E_T}{X_C}$$
$$I_T = \frac{12}{2.65}$$
$$I_T = 4.53\,A$$

5. Capacitive AC circuits can be used for filtering, coupling, decoupling, and phase shifting.
6. A low-pass filter consists of a capacitor and resistor in series with the output taken across the capacitor. At low frequencies, the capacitive reactance is higher than the resistance, so most of the voltage is dropped across the capacitor where the output is taken.
7. AC can be removed from a DC source using an RC low-pass filter referred to as a decoupling network.
8. Capacitive coupling circuits allow AC components of a signal to pass through a coupling network while blocking the DC components of the signal.
9. In a phase-shift network, the input is applied across the capacitor and resistor, and the output is taken across the resistor. Current leads the voltage in a capacitive circuit. The voltage across the resistor is in-phase, resulting in the output voltage leading the input voltage.
10. Phase-shift networks are valid for only one frequency because the capacitive reactance varies with changes in frequencies. Changing the reactance results in a different phase shift.

Chapter 25 Inductive AC Circuits

1. In an inductive circuit, the current lags the applied voltage.
2. The inductive reactance of an inductive circuit is affected by the inductance of the inductor and the frequency of the applied voltage.
3. **Given**
$$X_L = ?$$
$$\pi = 3.14$$
$$f = 60\,Hz$$
$$L = 100\,mH = 0.1\,H$$

   Solution
$$X_L = 2\pi fL$$
$$X_L = (2)(3.14)(60)(0.1)$$
$$X_L = 37.68\,\Omega$$

4. **Given**
$$I_T = ?$$
$$E_T = 24\,V$$
$$X_L = 37.68\,\Omega$$

   Solution
$$I_T = \frac{E_T}{X_L}$$
$$I_T = \frac{24}{37.68}$$
$$I_T = 0.64\,A \text{ or } 640\,mA$$

5. Applications for inductors in circuits include filtering and phase shifting.
6. **Given**
$$I_L = 0.086\,A$$
$$L = 0.100\,H$$
$$f = 50\,Hz$$
$$E_L = ?$$

   Solution
$$X_L = 2\pi fL$$
$$X_L = (2)(3.14)(50)(0.1)$$
$$X_L = 31.4\,\Omega$$

$$I_L = E_L/X_L$$
$$0.086 = E_L/31.4$$
$$(0.086)(31.4) = E_L$$
$$2.7\,V = E_L$$

7.

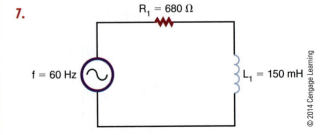

Given

$L = 0.150\,\mathrm{H}$
$R = 680\,\Omega$
$f = 60\,\mathrm{Hz}$
$Z = ?$

Solution

$X_L = 2\pi fL$
$X_L = (2)(3.14)(60)(0.15)$
$X_L = 56.52$

$Z = \sqrt{R^2 + X_L^2}$
$Z = \sqrt{(680)^2 + (56.52)^2}$
$Z = 682.34\,\Omega$

8.

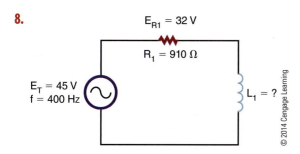

$E_{R1} = 32\,\mathrm{V}$
$R_1 = 910\,\Omega$
$E_T = 45\,\mathrm{V}$
$f = 400\,\mathrm{Hz}$
$L_1 = ?$

© 2014 Cengage Learning

Given

$E_T = 45\,\mathrm{V}$
$R_1 = 910\,\Omega$
$E_{R_1} = 32\,\mathrm{V}$
$f = 400\,\mathrm{Hz}$
$L_1 = ?$

Solution

$I_{R_1} = E_{R_1}/R_1$
$I_{R_1} = 32/910$
$I_{R_1} = 0.0352\,\mathrm{A}$

$E_{X_L} = E_T - E_{R_1}$
$E_{X_L} = 45 - 32$
$E_{X_L} = 13\,\mathrm{V}$

$I_{X_L} = E_{X_L}/X_L$
$0.0352 = 13/X_L$
$X_L = 13/0.0352$
$X_L = 369.32\,\Omega$

$X_L = 2\pi fL$
$369.32 = (2)(3.14)(400)(L)$
$369.32/(2)(3.14)(400) = L$
$0.147\,\mathrm{H},\ \text{or } 147\,\mathrm{mH} = L$

9. In a low-pass filter, the input is fed across the inductor and resistor, and the output is taken from across the resistor. In a high-pass filter, the input is fed across the resistor and inductor, and the output is taken from across the inductor.

10. The frequency above or below the frequencies passed or attenuated in an inductive circuit is called the cutoff frequency.

Chapter 26 Reactance and Resonance Circuits

1.

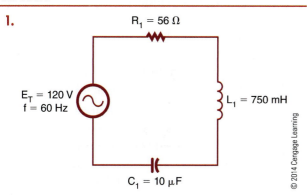

$R_1 = 56\,\Omega$
$E_T = 120\,\mathrm{V}$
$f = 60\,\mathrm{Hz}$
$L_1 = 750\,\mathrm{mH}$
$C_1 = 10\,\mu\mathrm{F}$

© 2014 Cengage Learning

Find the capacitive reactance.

$$X_C = \frac{1}{2\pi fC}$$

$$X_C = \frac{1}{(6.28)(60)(0.000010)}$$

$$X_C = 265.39\,\Omega$$

Find the inductive reactance.

$X_L = 2\pi fL$
$X_L = (6.28)(60)(0.750)$
$X_L = 282.60\,\Omega$

Now, solve for X.

$X = X_L - X_C$
$X = 282.6 - 265.39$
$X = 17.2\,\Omega\ (\text{inductive})$

Using X, solve for Z.

$Z^2 = X^2 + R^2$
$Z^2 = (17.21)^2 + (56)^2$
$Z^2 = 296.18 + 3136$
$Z = \sqrt{3432.18}$
$Z = 58.58\,\Omega$

Solve for total current.

$$I_T = \frac{E_T}{Z}$$

$$I_T = \frac{120}{58.58}$$

$$I_T = 2.05\,\mathrm{A}$$

2.

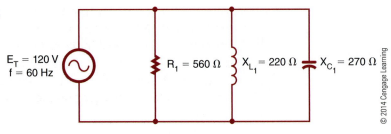

© 2014 Cengage Learning

Find the individual branch current.

$$I_R = \frac{E_R}{R}$$

$$I_R = \frac{120}{560}$$

$$I_T = 0.214\,A$$

$$I_{X_L} = \frac{E_{X_L}}{X_L}$$

$$I_{X_L} = \frac{120}{220}$$

$$I_{X_L} = 0.545\,A$$

$$I_{X_C} = \frac{E_{X_C}}{X_C}$$

$$I_{X_C} = \frac{120}{270}$$

$$I_{X_C} = 0.444\,A$$

Find I_X and I_Z using I_R, I_{X_L}, and I_{X_C}.

$$I_X = I_{X_L} - I_{X_C}$$

$$I_X = 0.545 - 0.444$$

$$I_X = 0.101\,A\ (\text{inductive})$$

$$I_Z^2 = I_R^2 + I_X^2$$

$$I_Z = \sqrt{(0.214)^2 + (0.101)^2}$$

$$I_Z = 0.237\,A$$

$$E_T = 120\,V$$

$$I_T = 2.05\,A$$

$$X = 17.2\,\Omega$$

$$R = 56\,\Omega$$

$$I_X = E_X/X$$

$$2.05 = E_X/17.2$$

$$(2.05)(17.2) = E_X$$

$$35.26\,V = E_X$$

$$P_X = I_X E_X$$

$$P_X = (2.05)(35.26)$$

$$P_X = 72.28\,VA$$

$$I_{R_1} = E_{R_1}/R_1$$

$$2.05 = E_{R_1}/56$$

$$(2.05)(56) = E_{R_1}$$

$$114.8\,V = E_{R_1}$$

$$P_R = IRE_{R_1}$$

$$P_R = (2.05)(114.8)$$

$$P_R = 235.34\,W$$

$$P_A^2 = P_X^2 + P_R^2$$

$$P_A^2 = (72.28)^2 + (235.34)^2$$

$$P_A = 246.19\,VA$$

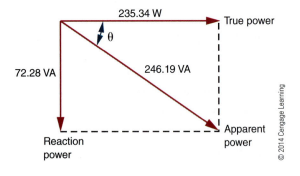

© 2014 Cengage Learning

Chapter 27 Transformers

1. When two electrically isolated coils are placed next to each other and an AC voltage is applied across one coil, the changing magnetic field induces a voltage into the second coil.

2. Transformers are rated in volt-amperes rather than in watts because of the different types of loads that can be placed on the secondary winding. A pure capacitive load causes an excessive current to flow, so a power rating would have little meaning.

3. If a transformer is connected without a load, there is no secondary current flow. The primary windings act like an inductor in an AC circuit. When a load is connected across the secondary winding, a current is induced into the secondary. The current in the secondary establishes its own magnetic field, which cuts the primary, inducing a voltage back into the primary. This induced field expands in the same direction as the current in the primary, aiding it and causing it to increase.

4. The direction in which the primary and secondary windings are wound determines the polarity of the induced voltage in the secondary winding. An induced out-of-phase shift would be 180° out of phase with the primary windings.

5. If a DC voltage is applied to a transformer, nothing occurs in the secondary winding once the magnetic field is established. A changing current in the primary winding is needed to induce a voltage in the secondary winding.

6. Given

$N_P = 400 \text{ tu}$
$N_S = ?$
$E_P = 120 \text{ V}$
$E_S = 12 \text{ V}$

Solution

$$\frac{E_S}{E_P} = \frac{N_S}{N_P}$$

$$\frac{12}{20} \times \frac{N_S}{400}$$

$$1N_S = \frac{(12)(400)}{120}$$

$$N_S = 40 \text{ turns}$$

$$\text{turns ratio} = \frac{N_S}{N_P}$$

$$= \frac{40}{400}$$

$$= \frac{1}{10} \text{ or } 10{:}1$$

7. Given

$N_P = ?$
$N_S = ?$
$Z_P = 16$
$Z_S = 4$

Solution

$$\frac{Z_P}{Z_S} = \left(\frac{N_P}{N_S}\right)^2$$

$$\frac{16}{4} = \left(\frac{N_P}{N_S}\right)^2$$

$$\sqrt{4} = \frac{N_P}{N_S}$$

$$\frac{2}{1} = \frac{N_P}{N_S}$$

The turns ratio is 2:1.

8. Transformers are important for transmitting electrical power because of power loss. The amount of power loss is related to the amount of resistance of the power lines and amount of current. The easiest way to reduce power losses is to keep the current low by stepping up the voltage with transformers.

9. An isolation transformer prevents connecting to ground on either side of the power line for equipment being worked on.

10. An autotransformer is used where the applied voltage needs to be stepped up or stepped down.

Chapter 28 Semiconductor Fundamentals

1. Silicon has more resistance to heat than germanium, making it preferable.

2. Negative temperature coefficient is identified in a material as an increase in temperature that results in a decrease in resistance.

3. Covalent bonding is the process of atoms sharing electrons. When semiconductor atoms share electrons, their valence shell becomes full with eight electrons, thereby obtaining stability.

4. In pure semiconductor materials, the valence electrons are held tightly to the parent atom at low temperatures and do not support current flow. As the temperature increases, the valence electrons become agitated and break the covalent bond, allowing the electrons to drift randomly from one atom to the next. As the temperature continues to increase, the material begins to behave like a conductor. Only at extremely high temperatures will silicon conduct current as ordinary conductors do.

5. Germanium's resistance is cut in half for every 10° Celsius of temperature increase. In some applications, heat-sensitive devices are necessary, and germanium's temperature coefficient can be an advantage.

6. Current flow can be supported in semiconductor material by applying a voltage source to the material. Free electronics are attracted to the positive terminal of the source, and holes will flow to the negative terminal. Current flow is based on the number of free electrons and increases with temperature increase of the material.

7. To convert a block of pure silicon to N-type material, the silicon is doped with atoms having five valence electrons, called pentavalent materials, such as arsenic and antimony.

8. N-type material is identified when the number of free electrons exceeds the electron-hole pair.

9. When a voltage is applied to N-type material, the free electrons contributed by the donor atoms flow toward the positive terminal. Additional electrons break away from their covalent bonds and also flow toward the positive terminal.

10. N- and P-type semiconductor materials have a much higher conductivity than pure semiconductor materials.

11. The conductivity of semiconductor materials can be increased by the addition of impurities.

Chapter 29 P–N Junction Diodes

1. A P–N junction diode allows current to flow in only one direction.

2. A diode conducts when it is forward biased—that is, when the positive terminal of the voltage source is connected to the P-type material, and the negative terminal of the voltage source is connected to the N-type material.

3.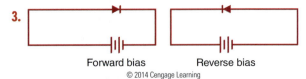

Forward bias   Reverse bias

© 2014 Cengage Learning

4. Joining N- and P-type materials together forms a diode.

5. There are no majority carriers in the depletion region.

6. The barrier voltage can be represented as an external voltage.

7. An external resistor is added when connecting a diode to a voltage source for limiting current flow to a safe value.

8. A silicon diode requires 0.7 volt before it starts to conduct.

9.

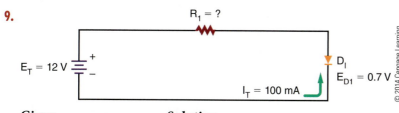

Given

$I_T = 100\,mA$

$E_T = 12\,V$

$E_{D1} = 0.7\,V$

$R_1 = ?$

Solution

$E_{R1} = E_T - E_{D1}$

$E_{R1} = 12 - 0.7$

$E_{R1} = 11.3\,V$

$I_1 = E_{R1}/R_1$

$0.1 = 11.3/R_1$

$R_1 = 11.3/0.1$

$R_1 = 113\,\Omega$

10. No, a diode only conducts in the forward direction and does not conduct until a forward bias is applied that exceeds the barrier voltage.

11. Leakage current increases as the temperature of the diode increases.

12. Diodes can be tested by checking the forward-to-reverse ratio with an ohmmeter. A diode will show a low forward resistance and a high reverse resistance. An ohmmeter can determine the cathode and anode. When the reading is low, the positive lead is connected to the anode and the negative lead is connected to the cathode.

Chapter 30 Zener Diodes

1. When the breakdown voltage is exceeded, a high reversal current (I_Z) flows.

2. The breakdown voltage (E_Z) of a zener diode is determined by the resistivity of the diode.

3. The ability of a zener diode to dissipate power decreases as the temperature increases.

4. Zener diode ratings include the following: Maximum current (I_{ZM}); reverse current (I_R); reverse voltage (E_Z); zener diodes greater than 5 volts have a positive zener voltage temperature coefficient; zener diodes less than 4 volts have a negative temperature coefficient; zener diodes between 4 and 5 volts have either a negative or a positive temperature coefficient.

5.

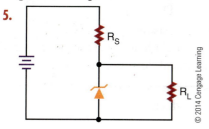

6. In a zener diode voltage regulator, the zener diode is connected in series with a resistor, with the output taken across the zener diode. The zener diode opposes an increase in input voltage because when the current increases, the resistance drops. The change in input voltage appears across the series resistor.

7. In a zener diode voltage regulator, the zener diode is connected in series with a current limiting resistor that is based on the load current and the zener current. The resistor allows enough current to flow in the zener breakdown region of the zener diode. When the load resistance increases, the load current decreases, which results in an increase in voltage across the load resistance. The zener diode opposes any change and conducts more current. The sum of zener current and load current through the zener resistor remains constant, allowing the circuit to regulate for changes in output current and input voltage.

8.

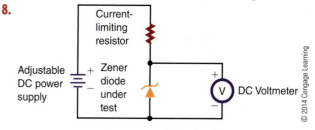

9. A power supply, a current-limiting resistor, an ammeter, and a voltmeter are required for testing a zener diode. The output of the power supply is connected across the limiting resistor in series with the zener diode and ammeter. The voltmeter is connected across the zener diode. The input voltage is slowly increased until the specified current is flowing through the zener diode. The voltmeter reads the zener diode voltage value.

10. Using the test setup in question 8, the voltage is slowly increased until the specified current is flowing through the zener diode. The current is then varied on either side of the specified zener current. If the voltage remains constant, the zener diode is operating properly.

Chapter 31 Bipolar Transistors

1. In NPN transistors, P-type material is sandwiched between two N-type materials, and N-type material is sandwiched between two layers of P-type material to form a PNP transistor.

2. Most transistors are identified by a number, which begins with a 2 and the letter N and has up to four more digits (2Nxxxx).

3. The package of a transistor serves as protection and provides a means of making electrical connections to the emitter, base, and collector. The package also serves as a heat sink.

4. Transistor leads can best be identified by referring to the manufacturer's specification sheet.

5. A diode is a rectifier that allows current to flow in one direction only, and a transistor is an amplifier that provides current amplification of a signal.

6. The emitter-based junction is forward biased, and the collector-based junction is reverse biased.

7. When testing a transistor with an ohmmeter, a good transistor shows a low resistance when forward biased and a high resistance when reverse biased across each junction.

8. A voltmeter, not an ohmmeter, is used to determine whether a transistor is silicon or germanium by measuring the voltage drop across the junction. If an ohmmeter were used, the leads would be difficult to determine because it would be hard to say which end was the emitter or collector. However, the base would be determined as low resistance when forward biased to either emitter or collector and high resistance when reverse biased. A PNP or NPN transistor could be determined.

9. The collector voltage determines whether the device is an NPN or PNP transistor. If the wrong type of transistor is substituted, a failure in the device results.

10. Testing a transistor with a transistor tester, especially an in-circuit tester reveals more information about a transistor than when testing with an ohmmeter.

Chapter 32 Field Effect Transistors (FETs)

1.

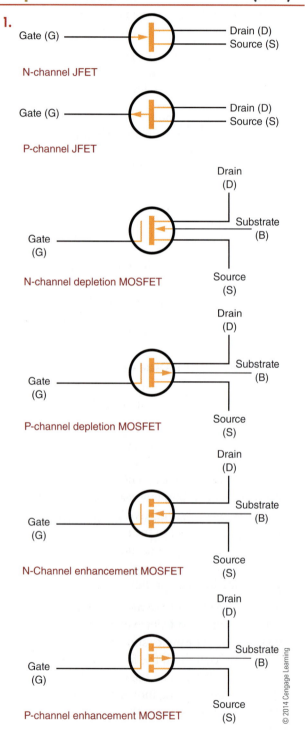

N-channel JFET

P-channel JFET

N-channel depletion MOSFET

P-channel depletion MOSFET

N-Channel enhancement MOSFET

P-channel enhancement MOSFET

© 2014 Cengage Learning

2. The two external bias voltages for a JFET are E_{DS} (the bias voltage between the source and drain) and E_{GS} (the bias voltage between the gate and the source).

3. The pinch-off voltage is the voltage required to pinch off the drain current for a JFET.

4. The pinch-off voltage is given by the manufacturer for a gate-source voltage of 0.

5. Depletion-mode MOSFETs conduct when zero bias is applied to the gate. Depletion-mode MOSFETs are considered to be normally on devices.

6. Depletion MOSFET conducts when zero bias is applied to the gate. An enhancement MOSFET conducts only when a bias voltage is applied to the gate.

7. Enhancement-mode MOSFETs are normally off and only conduct when a suitable bias voltage is applied to the gate.

8. Safety precautions for a MOSFET include the following:
- Keep the leads shorted together before installation.
- Use a metallic wristband to ground the hand being used.
- Use a grounded-tip soldering iron.
- Always ensure the power is off before installation and removal of a MOSFET.

9. N-channel JFETs: Connect the positive lead to the gate and the negative lead to the source or drain. Because a channel connects the source and drain, only one side has to be tested. The forward resistance should be a low reading. P-channel JFETs: Connect the negative lead to the gate and the positive lead to the source or drain. To determine the reverse resistance, reverse the leads. The JFET should indicate an infinite resistance. A low reading indicates a short or leakage.

10. The forward and reverse resistance can be checked with a low-voltage ohmmeter set to its highest scale. The meter should register an infinite resistance in both the forward- and reverse-resistance tests between the gate and source or drain. A lower reading indicates a breakdown of the insulation between the gate and source or drain.

Chapter 33 Thyristors

1. The P–N junction diode has one junction and two leads (anode and cathode); an SCR has three junctions and three leads (anode, cathode, and gate).

2. The anode supply voltage keeps the SCR turned on even after the gate voltage is removed. This allows a current to flow continuously from cathode to anode.

3.

Clipped off
© 2014 Cengage Learning

4. The load resistor is in series with the SCR to limit the cathode-to-anode current.

5. A TRIAC conducts both alternations of an AC input waveform, and an SCR conducts in only one alternation.

6. Disadvantages of TRIACs over SCRs are as follows: TRIACs cannot carry as much current as an SCR; TRIACs have lower voltage rating than SCRs; and TRIACs have lower frequency-handling capabilities than SCRs.

7. A DIAC is used as a triggering device for TRIACs. It prevents the TRIAC from turning on until a certain gate voltage is reached.

8. An SCR can be tested with an ohmmeter or a commercial transistor tester. To test the SCR with an ohmmeter, connect the positive lead to the cathode and the negative lead to the anode. A high-resistance reading in excess of 1 megohm should be read. Reverse the leads so the positive lead is on the anode and the negative lead is on the cathode. Again, a high-resistance reading in excess of 1 megohm should be read. Short the gate to the anode, and the resistance reading should drop to less than 1000 ohms. Remove the short, and the low-resistance reading should remain. Remove the leads and repeat the test.

9. An ohmmeter cannot test for marginal or voltage-sensitive devices in thyristors.

10.

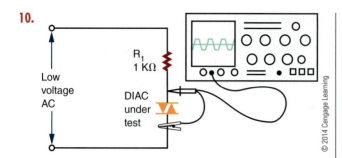

© 2014 Cengage Learning

Chapter 34 Integrated Circuits

1. An integrated circuit consists of diodes, transistors, resistors, and capacitors.

2. Integrated circuits cannot be repaired because the internal components cannot be separated; therefore, problems are identified by individual circuit instead of by individual component.

3. Integrated circuit construction techniques include monolithic, thin film, and thick film.

4. A chip is the semiconductor material that comprises the integrated circuit and is about one-eighth of an inch square.

5. A hybrid integrated circuit contains monolithic, thin-film, and discrete components.

6. If only a few circuits are to be made, it is cheaper to use hybrid integrated circuits.

7. DIPs are manufactured to accommodate the various sizes of integrated circuits: small-scale integration (SSI) with up to 100 electrical components per chip, medium-scale integration (MSI) with 100 to 3,000 electrical components per chip, large-scale integration (LSI) with 3,000 to 100,000 electrical components per chip, very large-scale integration (VLSI) with 100,000 to 1,000,000 electrical components per chip, and ULSI, which stands for ultra large-scale integration.

8. Wafer-scale integration (WSI), which contains entire computer systems including memory, was developed unsuccessfully in the 1980s. This was followed by system on-chip (SOC) design. This process takes components that were typically manufactured on separate chips and designs them to occupy a single chip.

9. Ceramic devices are recommended for severe environmental applications such as in the military and aerospace.

10. A flat-pack is smaller and thinner than a DIP and is used where space is limited; it is made from metal or ceramic.

Chapter 35 Optoelectric Devices

1.

Lightwave	Frequency
Infrared	< 400 THz
Visible	400–750 THz
Ultraviolet	> 750 THz

© 2014 Cengage Learning

2. Because the waveforms are smaller at higher frequency, they have more energy.

3. The photodiode has the fastest response time to light changes of any photosensitive device.

4. A photocell is slow to respond to light change and is not suitable for quick-response applications.

5. The solar cell is a P–N junction device made from semiconductor materials. The P and N layers form a P–N junction. The light striking the surface of the solar cell imparts much of its energy to the atoms in the semiconductor material. The light energy knocks valence electrons from their orbits, creating free electrons. The electrons near the depletion region are drawn to the N-type material, producing a small voltage across the P–N junction. The voltage increases with an increase in light intensity.

6. Solar cells have a low voltage output of approximately 0.45 volt at 50 milliamperes and must be connected in a series and parallel to obtain the desired voltage and current output.

7. The phototransistor lends itself to a wider range of applications because it is able to produce a higher gain. It does not, however, respond to light changes as fast as the photodiode.

8. The light from an LED is produced when the free electrons combine with holes, and the extra energy is released in the form of light.

9. The more current that flows through an LED, the brighter the light emitted. However, a series resistor must be used with LEDs to limit the current flow, or else damage to the LED will result.

10. An optical coupler allows one circuit to transfer a signal to another circuit while being electrically isolated from each other.

Chapter 36 Power Supplies

1. Concerns when selecting a transformer for a power supply include primary power rating, frequency of operation, secondary voltage and current rating, and power-handling capabilities.

2. Transformers are used to isolate the power supply from the AC voltage source. They can also be used to step up or step down voltage.

3. The rectifier in a power supply converts the incoming AC voltage to a pulsating DC voltage.

4. A disadvantage of the full-wave rectifier is that it requires a center-tapped transformer. An advantage is that it requires only two diodes. An advantage of the bridge rectifier is that it does not require a transformer; however, it does require four diodes. Both rectifier circuits are more efficient and easier to filter than the half-wave rectifier.

5. A filter capacitor charges when current is flowing and then discharges when current stops flowing, keeping a constant current flow on the output.

6. Capacitors are selected for filtering to give a long RC time constant. The slower discharge gives a higher output voltage.

7. A series regulator compensates for higher voltages on the input by increasing the series resistance, thereby dropping more voltage across the series resistance so that the output voltage remains the same. It also senses lower voltage on the input, decreasing the series resistance and dropping less voltage, resulting in the output remaining the same. It works in a similar fashion on changes in the load.

8. The voltage and load current requirements must be known when selecting an IC voltage regulator.

9. Voltage multipliers allow the voltage of a circuit to be stepped up without the use of a step-up transformer.

10. The full-wave voltage doubler is easier to filter than the half-wave voltage doubler. In addition, the capacitors in the full-wave voltage doubler are subjected to only the peak value of the input signal.

11. An over-voltage protection circuit, called a crowbar, is used to protect the load from failure of the power supply.

12. Over-current protection devices include fuses and circuit breakers.

Chapter 37 Amplifier Basics

1. A transistor provides amplification by using an input signal to control current flow in the transistor in order to control the voltage through a load.

2. A common emitter circuit provides both voltage and current gain and a high power gain. Neither of the other two circuit configurations provides this combination.

3. A transistor circuit that cannot compensate for changes in the bias current, with no signal applied, becomes unstable. A change in temperature causes the transistor's internal resistance to vary, causing the bias currents to change, resulting in the transistor's operating point shifting, thus reducing the gain of the transistor. This is referred to as thermal instability. It is possible to compensate for temperature changes in a transistor amplifier circuit if a portion

of the unwanted output signal is fed back to the circuit input; the signal opposes the change and is referred to as degenerative or negative feedback.

4. Class B amplifiers are used as output stages of stereo systems. A class B amplifier is biased so that the output current flows for only half of the input cycle.

5. The coupling capacitor blocks the DC voltage and passes the AC signal.

6. Temperature changes affect the gain of a transistor. Degenerative or negative feedback compensates for this condition.

7. Class A amplifiers are biased so that the output flows throughout the entire cycle. Class B amplifiers are biased so that the output flows for only half of the input cycle. Class AB amplifiers are biased so that the output flows for more than half but less than the full input cycle. Class C amplifiers are biased so the output flows for less than half of the input cycle.

8. When connecting two transistor amplifiers together, the bias voltage from one amplifier must be prevented from affecting the operation of the second amplifier.

9. If capacitors or inductors are used for coupling, the reactance of the device will be affected by the frequency being transmitted.

10. A drawback of direct-coupled amplifiers is stability. Any change in the output current of the first stage will be amplified in the second stage because the second stage is biased by the first stage. To improve the stability requires the use of expensive precision components.

Chapter 38 Amplifier Applications

1. DC or direct-coupled amplifiers are used to amplify frequencies from DC (0 hertz) to many thousands of hertz.

2. Temperature stability with DC amplifiers is achieved by using a differential amplifier.

3. Audio voltage amplifiers provide a high voltage gain, whereas audio power amplifiers provide high power gains to a load.

4. The complementary push–pull amplifier requires matched NPN and PNP transistors. The

quasi-complementary amplifier does not require matched transistors.

5. A video amplifier has a wider frequency range than an audio amplifier.

6. A factor that limits the output of a video amplifier is the shunt capacitance of the circuit.

7. An RF amplifier amplifies frequencies from 10,000 hertz to 30,000 megahertz.

8. An IF amplifier is a single-frequency amplifier used to increase a signal to a usable level.

9. An op-amp consists of an input stage (differential amplifier), high-gain voltage amplifier, and an output amplifier. It is a high-gain DC amplifier, capable of output gains of 20,000 to 1,000,000 times the input signal.

10. Op-amps are amplifiers used for comparing, inverting, and noninverting a signal and for summing. They are also used as an active filter and a difference amplifier.

Chapter 39 Oscillators

1. The parts of an oscillator include the frequency-determining circuit called the tank circuit, an amplifier to increase the output signal from the tank circuit, and a feedback circuit to deliver part of the output signal back to the tank circuit to maintain oscillation.

2. A feedback circuit in an oscillator delivers the proper amount of energy to the tank circuit to sustain oscillation by replacing the energy that is dissipated.

3. The difference between the series-fed and the shunt-fed Hartley oscillators is that DC does not flow through the tank circuit of the shunt-fed Hartley oscillator.

4. The crystal provides stability. If the frequency of a tank circuit drifts from the crystal frequency, the impedance of the crystal increases, reducing feedback to the tank circuit, thus allowing the tank circuit to return to the crystal frequency.

5. The tank circuit can sustain oscillation by feeding back part of the output signal in the proper phase to replace the energy losses caused by the resistance of the components in the tank circuit.

6. The major types of sinusoidal oscillators are the Hartley oscillator, the Colpitts oscillator, and the Clapp oscillator.

7. Crystals have a natural frequency of vibration and are ideal for oscillator circuits. The crystal frequency is used to control the tank circuit frequency.

8. Nonsinusoidal oscillators do not produce sine-wave outputs. Typically, all nonsinusoidal oscillators are some form of a relaxation oscillator.

9. Blocking oscillators, multivibrators, RC networks, and integrated circuits are all used in nonsinusoidal oscillators.

10.
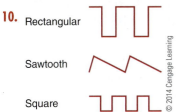

Rectangular

Sawtooth

Square

© 2014 Cengage Learning

Chapter 40 **Waveshaping Circuits**

1. The frequency domain concept states that all periodic waveforms are made up of sine waves. A periodic waveform can be made by superimposing a number of sine waves having different amplitudes, phases, and frequencies.

2. The fundamental frequency is also referred to as the first harmonic.

3. Duty cycle = pulse width/period
Duty cycle = 50,000/100,000
Duty cycle = 0.5 = 50%

4. Overshoot, undershoot, and ringing occur in waveshaping because of imperfect circuits that cannot react in zero time to a rapidly changing input.

5. A differentiator is used to produce a pip or peaked waveform for timing and synchronizing circuits. An integrator is used to produce a raising or falling triangular waveform for waveshaping.

6. A clipping or limiter circuit can change a sine wave into a rectangular waveform.

7. The DC reference level of a signal can be changed by using a clamping circuit to clamp the waveform to a DC voltage.

8. A monostable circuit has only one stable state, and it produces one output pulse for each input pulse. A bistable circuit has two stable stages, and it requires two input pulses to complete a cycle.

9. A flip-flop can produce a square or rectangular waveform for gating and timing signals or for switching applications.

10. A Schmitt trigger is used to convert a sine-wave, sawtooth, or other irregularly shaped waveform to a square or rectangular wave.

Chapter 41 Binary Number System

1.

Digital	Binary	Digital	Binary
0	00000	14	01110
1	00001	15	01111
2	00010	16	10000
3	00011	17	10001
4	00100	18	10010
5	00101	19	10011
6	00110	20	10100
7	00111	21	10101
8	01000	22	10110
9	01001	23	10111
10	01010	24	11000
11	01011	25	11001
12	01100	26	11010
13	01101	27	11011

© 2014 Cengage Learning

2. Seven binary bits are required to represent the decimal number 100 (1100100).

3. To convert a decimal number to a binary number, progressively divide the decimal number by 2, writing down the remainder after each division. The remainders, taken in reverse order, form the binary number.

4. **a.** $100101.001011_2 = 37.171875_{10}$
   b. $111101110.11101110_2 = 494.9296875_{10}$
   c. $1000001.00000101_2 = 65.01953125_{10}$

5. **a.** $011\ 001\ 001\ 011\ 100\ 101\ 010\ 011\ 100\ 001\ 101_2 = 31134523415_8$
   b. $000\ 110\ 001\ 101\ 111\ 100\ 100\ 111\ 100\ 110\ 100_2 = 06157447464_8$
   c. $010\ 011\ 010\ 110\ 001\ 001\ 100\ 111\ 101\ 000\ 110_2 = 23261147506_8$

6. **a.** $653172_8 = 110101011001111010_2$
   b. $773012_8 = 111111011000001010_2$
   c. $033257_8 = 0000110110010101111_2$

7. **a.** $317204_8 = 106116_{10}$
   b. $701253_8 = 230059_{10}$
   c. $035716_8 = 15310_{10}$

8. **a.** $687_{10} = 1257_8$
   b. $9762_{10} = 23042_8$
   c. $18673_{10} = 44361_8$

9. **a.** $1100\ 1001\ 0111\ 0010\ 1010\ 0111\ 0000\ 1101_2 = C972A70D_{16}$
   b. $0011\ 0001\ 1011\ 1110\ 0100\ 1111\ 0011\ 0100_2 = 31BE4F34_{16}$
   c. $1001\ 1010\ 1100\ 0100\ 1100\ 1111\ 0100\ 0110_2 = 9AC4CF46_{16}$

10. **a.** $7B23C67F_{16} = 0111\ 1011\ 0010\ 0011\ 1100\ 0110\ 0111\ 1111_2$
    b. $D46F17C9_{16} = 1101\ 0100\ 0110\ 1111\ 0001\ 0111\ 1100\ 1001_2$
    c. $78F3E69D_{16} = 0111\ 1000\ 1111\ 0011\ 1110\ 0110\ 10011101_2$

11. **a.** $3C67F_{16} = 247423_{10}$
    b. $6F17C9_{16} = 7280585_{10}$
    c. $78F3E69D_{16} = 2029250205_{10}$

12. **a.** $687_{10} = 2AF_{16}$
    b. $9762_{10} = 2622_{16}$
    c. $18673_{10} = 45F1_{16}$

13. Convert each decimal digit to a binary digit $(0 - 9)$ using a 4-bit BCD binary code for decimal digit.

14. **a.** $0100\ 0001\ 0000\ 0110_{BCD} = 4106_{10}$
    b. $1001\ 0010\ 0100\ 0011_{BCD} = 9243_{10}$
    c. $0101\ 0110\ 0111\ 1000_{BCD} = 5678_{10}$

Chapter 42 Basic Logic Gates

1.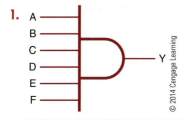

2.

D	C	B	A	Y
0	0	0	0	0
0	0	0	1	0
0	0	1	0	0
0	0	1	1	0
0	1	0	0	0
0	1	0	1	0
0	1	1	0	0
0	1	1	1	0
1	0	0	0	0
1	0	0	1	0
1	0	1	0	0
1	0	1	1	0
1	1	0	0	0
1	1	0	1	0
1	1	1	0	0
1	1	1	1	1

© 2014 Cengage Learning

3.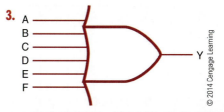

4.

D	C	B	A	Y
0	0	0	0	0
0	0	0	1	1
0	0	1	0	1
0	0	1	1	1
0	1	0	0	1
0	1	0	1	1
0	1	1	0	1
0	1	1	1	1
1	0	0	0	1
1	0	0	1	1
1	0	1	0	1
1	0	1	1	1
1	1	0	0	1
1	1	0	1	1
1	1	1	0	1
1	1	1	1	1

© 2014 Cengage Learning

5. The NOT circuit is used to perform inversion or complementation.

6. The circle or bubble is placed at the input for inversion of the input signal, and it is placed at the output for output inversion.

7.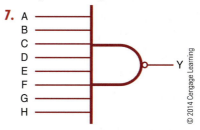

8.

D	C	B	A	Y
0	0	0	0	1
0	0	0	1	1
0	0	1	0	1
0	0	1	1	1
0	1	0	0	1
0	1	0	1	1
0	1	1	0	1
0	1	1	1	1
1	0	0	0	1
1	0	0	1	1
1	0	1	0	1
1	0	1	1	1
1	1	0	0	1
1	1	0	1	1
1	1	1	0	1
1	1	1	1	0

© 2014 Cengage Learning

9.

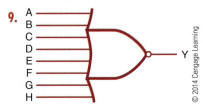

© 2014 Cengage Learning

10.

D	C	B	A	Y
0	0	0	0	1
0	0	0	1	0
0	0	1	0	0
0	0	1	1	0
0	1	0	0	0
0	1	0	1	0
0	1	1	0	0
0	1	1	1	0
1	0	0	0	0
1	0	0	1	0
1	0	1	0	0
1	0	1	1	0
1	1	0	0	0
1	1	0	1	0
1	1	1	0	0
1	1	1	1	0

© 2014 Cengage Learning

11. An XOR gate generates an output only when the inputs are different. If the inputs are both 0s or 1s, the output is a 0.

12. An XNOR gate has a maximum of two inputs.

13.

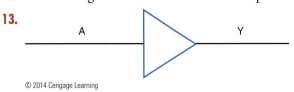

© 2014 Cengage Learning

14. It is a special logic gate that isolates or provides a high current.

15.

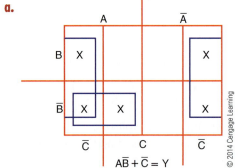

© 2014 Cengage Learning

Chapter 43 Simplifying Logic Circuits

1. Use the Veitch diagram as follows:
   a. Draw the diagram based on the number of variables.
   b. Plot the logic functions by placing an X in each square representing a term.
   c. Obtain the simplified logic function by looping adjacent groups of X's in groups of eight, four, two, or one. Continue to loop until all X's are included in a loop.
   d. "OR" the loops with one term per loop.
   e. Write the simplified expression.

2. a.

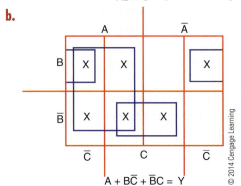

© 2014 Cengage Learning

c.

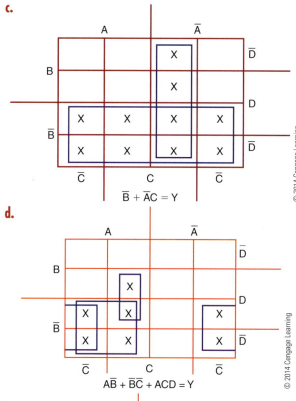

$\overline{B} + \overline{A}C = Y$

d.

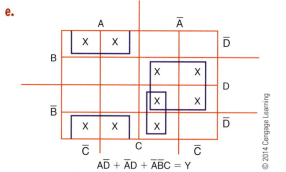

4. a.

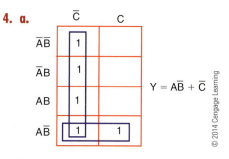

$Y = A\overline{B} + \overline{C}$

b.

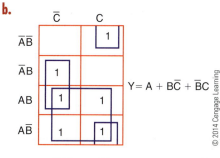

$Y = A + B\overline{C} + \overline{B}C$

c.

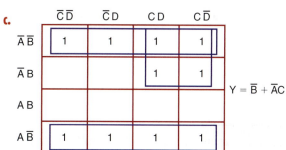

$Y = \overline{B} + \overline{A}C$

d.

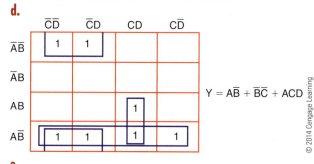

$Y = A\overline{B} + \overline{B}\overline{C} + ACD$

e.

$Y = A\overline{D} + \overline{A}D + \overline{A}\overline{B}C$

3. Use a Karnaugh map as follows:
   a. Draw the diagram based on the number of variables.
   b. Plot the logic functions by placing a 1 in each square representing a term.
   c. Obtain the simplified logic function by looping adjacent groups of 1s in groups of eight, four, or two.
   d. "OR" the loops with one term per loop.
   e. Write the simplified expression.

© 2014 Cengage Learning

Chapter 44 Sequential Logic Circuits

1. To change the output of an RS flip-flop requires a high, or 1, to be placed on the R input. This changes the state of the flip-flop to a 0 on the Q output and a 1 on the $\bar{Q}$ output.

2. The major difference between the D flip-flop and the clocked RS flip-flop is that the D flip-flop has a single data input and a clock input.

3. A counter is constructed of flip-flops connected in either an asynchronous or synchronous count mode. In the asynchronous mode, the Q or $\bar{Q}$ output of the first stage is connected to the clock input of the next stage depending on whether the counter is counting up (Q or down ($\bar{Q}$). In the synchronous mode, all the clock inputs of each of the stages are connected in parallel.

4.

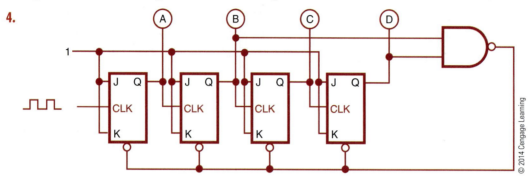

© 2014 Cengage Learning

5. A shift register is designed to store data temporarily and/or change its format. Data can be loaded into the shift register either serially or in parallel, and it can be unloaded either serially or in parallel.

6. Shift registers can be used to store data, for serial-to-parallel and parallel-to-serial data conversion, and to perform such arithmetic functions as division and multiplication.

7. Bit—1 **b**inary dig**it**;
Nibble—4 bits;
Byte—8 bits;
Word—16, 32, 64, and 128 bits

8. Complementary data available at the $\bar{Q}$ output is the opposite or complement of the data available at the Q output.

9. MOS memory can be packed at a higher density, offering more memory locations in the same space as bipolar memory.

10. Static RAM loses data when the power is removed. Dynamic RAM uses capacitors to store data but must be recharged or refreshed to maintain stored data.

11. RAM is random access memory and is used for the temporary storage of programs. ROM is read-only memory and allows data to be permanently stored and not changed.

12. EEPROMs are electrically erasable PROMs and can be reprogrammed.

Chapter 45 Combinational Logic Circuits

1. Encoders allow encoding of keyboard inputs into binary outputs.

2. A decimal-to-binary priority encoder is required for keyboard inputs.

3. Decoders allow the processing of complex binary codes into a recognizable digit or character.

4. Types of decoders include 1-of-10 decoders, 1-of-8 decoders, 1-of-16 decoders, and BCD-to-seven-segment decoders.

5. Multiplexers are used to select and route one of several input signals to a single output.

6. Multiplexers can be used for data line selection and parallel-to-serial conversions.

7.

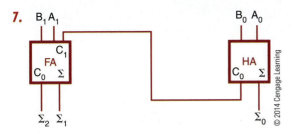

8. The half adder accepts the two binary digits to be added and generates a sum and a carry. The carry is fed to the next stage and added to the two binary digits, generating a sum and a carry. The answer is the result of the carry and the two sum outputs.

9.

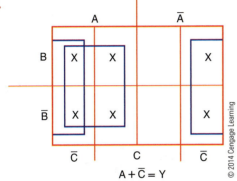

$$A + \overline{C} = Y$$

10. The PAL is easier to program than a PLA. The increased flexibility of the PLA adds extra gates resulting in an array harder to program.

Chapter 46 Microcomputer Basics

1. A computer consists of a control unit, an arithmetic logic unit (ALU), memory, and an input/output unit (I/O) The control unit decodes the instructions and generates the necessary pulses to carry out the specified function. The arithmetic logic unit performs all the math logic and decision-making operations. Memory is where the programs and data are stored. The input/output unit allows data to be entered into and removed from the computer (see Figure 46.1).

2. An interrupt signal from an external device lets the computer know that it would like data or wants to send data.

3. A microprocessor is part of a microcomputer. It consists of four basic parts: registers, arithmetic logic unit, timing and control circuitry, and decoding circuitry.

4. The microprocessor performs the control functions and handles the math logic and decision making for a microcomputer.

5. A microcontroller includes a microprocessor, memory, and input/output interfacing. It is a stand-alone system.

6. *Data movement instructions*—moves data from one location to another within the microprocessor and memory.
Arithmetic instructions—allows the microprocessor to compute and manipulate data.
Logic instructions—perform AND, OR, exclusive OR functions, and complement.
Input/output instructions—control I/O devices.

7. Microcontrollers are single-chip computers that are ideal for monitoring and controlling devices that do not require interface.

8. Home devices and appliances use microcontroller to remove human interaction include: Television, DVD player, Washing Machine, Dish washer, Coffee maker, and Microwave.

9. Microcontrollers are used in automobiles in any situation that requires sensory data to control events in the automobile such as engine fuel and emissions. Also, devices such as radio control, seat control, climate control, and antitheft are further applications.

10. RISC—reduced instruction set computer is based on the strategy that fewer basic instructions allow for higher performance.

Self-Test Answers

Chapter 47 **Project Design**

1.

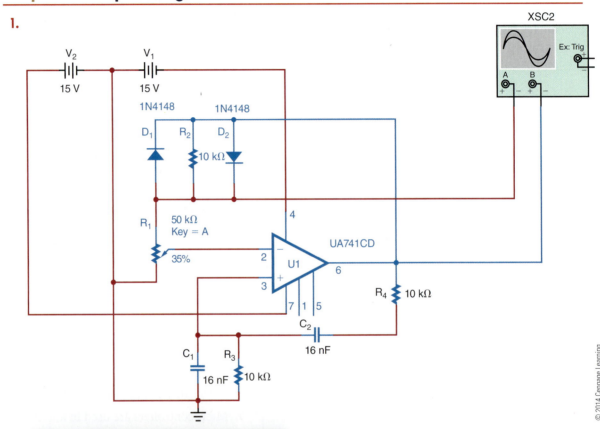

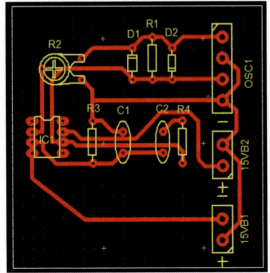

2. The schematic diagram is not copyrighted. The PCB layout is copyrightable providing it is original artwork. However, the copyright is weak unless the whole circuit is developed in a single proprietary chip or multiple chips.

3. Yes, providing it is the original artwork.

4.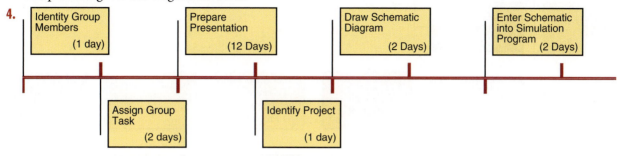

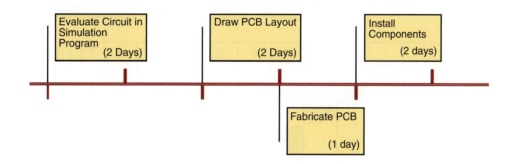

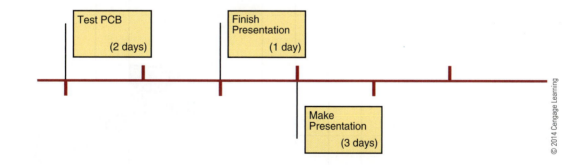

© 2014 Cengage Learning

5. 20 days

6.

Wein Bridge Oscillator—Budget			
Item	**Description**	**Reference Designator**	**Approximate Cost**
Resistor	10 kΩ 1/4 Watt 2%	R2, R3, R4	$0.25
Resistor, variable	50 kΩ Trim Pot	R1	$2.50
Capacitor	16 nF or 0.016 μF	C1, C2	$1.25
Diode	1N4148	D1, D2	$0.25
Integrated circuit	UA741	U1	$1.25
		Total	$5.50

© 2014 Cengage Learning

7. The project is cost-effective.

8.

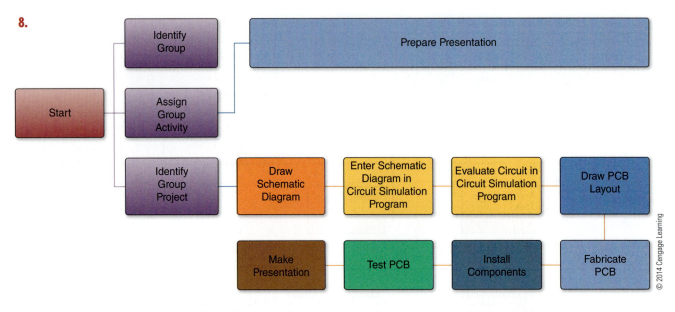

© 2014 Cengage Learning

9.

	Activity/Day	1	2	3	4	5	6	7	8	9	10	11	12	13	14
1	Identify group	█													
2	Assign group task	█	█												
3	Identify project			█											
4	Draw schematic diagram				█	█									
5	Enter schematic diagram in circuit simulation program					█	█								
6	Evaluate circuit in circuit simulation program						█	█							
7	Draw PCB layout (manual or with PCB program)							█	█						
8	Fabricate PCB									█	█				
9	Install components										█	█			
10	Test PCB											█	█		
11	Prepare presentation		█	█	█	█	█	█	█	█	█	█	█	█	
12	Make presentation														█

© 2014 Cengage Learning

10.

Wein Bridge Oscillator—Actual Construction Expenses				
Item	**Description**	**Reference Designator**	**Approximate Cost**	**Vendor**
Resistor	10 kΩ 1/4 Watt 2%	R2, R3, R4	$0.09	Mouser
Resistor, variable	50 kΩ Trim Pot	R1	$0.45	Jameco
Capacitor	16 nF or 0.016 μF	C1, C2	$0.68	Mouser
Diode	1N4148	D1, D2	$0.08	Mouser
Integrated circuit	UA741	U1	$0.29	Jameco/Digikey
		Total	$1.59	

This cost does not include the PCB.
© 2014 Cengage Learning

Chapter 48 Printed Circuit Board Fabrication

1. A printed circuit board supports the copper traces on an insulating base material.

2.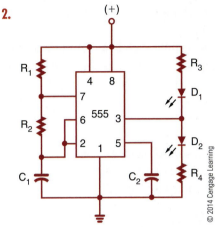

3.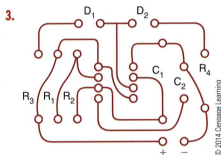

4.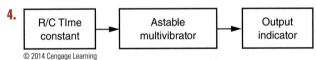

5. Basic rules for laying out a printed circuit board include the following:
   - Avoid running parallel traces closer than necessary.
   - Segregate analog and digital areas.
   - Put ground traces next to sensitive analog lines to act as a shield to reduce electromagnetic interference (EMI) and crosstalk.
   - Pay close attention to parts placement to simplify routing.

6. Gerber files include the following:
   1. Top trace layout
   2. Bottom trace layout
   3. Silkscreen (if used)
   4. Solder mask (if used)
   5. Aperture file
   6. Drill file

7. Hand-draw design transfer includes the following steps:
   1. Transfer the printed circuit board artwork to the back of graph paper, and write "BOTTOM."
   2. Wrap the artwork around a properly sized copper-clad board that has been cleaned with steel wool or with powder cleanser and water. The copper side of the board should be toward the artwork with the side identified as bottom facing outward.
   3. Using a scribe, locate the center of each pad where an X was placed on the bottom side. Before removing the artwork, run a finger over each pad to ensure it was marked with the scribe.
   4. Remove the artwork, being careful not to get fingerprints on the copper.
   5. With a resist pen or other fine-tip permanent marker, such as a black Sharpie®, draw the pads at each of the scribe marks. Using the bottom-view artwork, connect the pads with traces. Go over each of the traces and pads several times to build up a thick layer of resist.

8. To use the tray method, the copper-clad board with applied resist is etched in a glass tray by dropping the board onto the surface of the acid. The copper side of the board should face toward the acid for approximately 15 to 20 minutes.

9. Printed circuit board trace defect repairs include the following methods:
   - Shorted traces are separated using a knife.
   - Open traces are bridged with wire and solder.

10. Storage: Keep away from sources of ignition. Store in a cool, dry, well-ventilated area, away from incompatible substances. Heating: Do not expose the container to heat or flame or temperatures over 104°F (40°C).

Chapter 49 Printed Circuit Board Assembly and Repair

1. ICs can be removed from a printed circuit board with a small, flat-blade screwdriver or an IC chip remover.

2. A printed circuit board is used to mechanically support and electrically connect components for a circuit.

3. Oscilloscope

4. A logic pulser injects a logic level into a circuit, which can be picked up by a logic probe.

5. Soldering flux helps the solder alloy to flow around the connections by cleaning the component leads and pads of oxide and film allowing the solder to adhere.

6. Solder guns are designed to supply heat to large electrical connections and are used for joining large wires, connectors, etc.

7. Temperature-controlled soldering iron

8. Chisel tip

9. A soldering iron that is resting in a stand to remove excess heat

10. Security, to protect the copyright of a circuit

Chapter 50 Basic Troubleshooting

1. Isolation transformer

2. Troubleshooting technique include those listed here:
   a. Check the circuit over thoroughly before applying power.
   b. Apply power and determine whether the circuit is operating properly.
   c. Use senses to troubleshoot defective circuit.
   d. Expose circuit to the operating environment to determine proper operation.

3. Test equipment for effective troubleshooting includes the following:

   a. Ohmmeter—used to determine open or short circuits and measure resistance values.
   b. Voltmeter—used to measure voltage drops across components while they are in the circuit.

   c. Ammeter—used to measure current flow in a circuit or across individual components.
   d. Oscilloscope—used to measure and/or align amplitude and/or frequency of a signal and compare waveforms.

4. Electronics technicians extend their senses in the following ways:
   - Looking for trouble such as a component that is receiving excess current, is split open, or is charred
   - Smelling for burnt insulations from components that receive excess current such as transformers, coils, or components with insulated leads
   - Listening for sounds such as a snap, crackle, or pop that indicate loose wires, bridged connections, or frying components
   - Touching components to determine whether they are running hot or cold

5. A short circuit has 0 ohms resistance, and an open circuit has an infinite (∞) resistance.

6. Visually check the printed circuit board over thoroughly before power is applied.

7. An open trace acts like a resistor that is open.

8. Electronics technicians should chart and record any measurements taken for future troubleshooting references.

9. If a problem is difficult to solve, the electronics technician should step away from it for a while and think about the problem.

10. A schematic diagram can show voltage measurements, waveforms in a circuit, and how the components relate to each other.

INDEX

E